CHEMICAL ZOOLOGY

Volume VII

MOLLUSCA

Contributors to This Volume

S. BRICTEUX-GRÉGOIRE

ROBERT ENDEAN

MARCEL FLORKIN

F. GHIRETTI

A. GHIRETTI-MAGALDI

R. GILLES

T. W. GOODWIN

ESTHER M. GOUDSMIT

C. GRÉGOIRE

MICHELINE MARTOJA

E. H. MERCER

C. P. RAVEN

E. SCHOFFENIELS

CHARLES R. STASEK

P. A. VOOGT

KARL M. WILBUR

CHEMICAL ZOOLOGY

Edited by MARÇEL FLORKIN
DEPARTMENT OF BIOCHEMISTRY
UNIVERSITY OF LIÈGE
LIÈGE, BELGIUM

and

BRADLEY T. SCHEER
DEPARTMENT OF BIOLOGY
UNIVERSITY OF OREGON
EUGENE, OREGON

Volume VII

MOLLUSCA

ACADEMIC PRESS New York and London 1972

ACADEMIC PRESS, INC.
111 Fifth Avenue, New York, New York 10003

United Kingdom Edition published by
ACADEMIC PRESS, INC. (LONDON) LTD.
24/28 Oval Road, London NW1

LIBRARY OF CONGRESS CATALOG CARD NUMBER: 67-23158

PRINTED IN THE UNITED STATES OF AMERICA

Contents

List of Contributors

Numbers in parentheses indicate the pages on which the authors' contributions begin.

S. Bricteux-Grégoire (301), Department of Biochemistry, University of Liège, Liège, Belgium

Robert Endean (421), Zoology Department, The University of Queensland, Brisbane, Queensland, Australia

Marcel Florkin (301), Department of Biochemistry, University of Liège, Liège, Belgium

F. Ghiretti (201), Istituto de Biologia Animale e Centro per lo Studio della Fisiologia e Biochimica delle Emocianine, Università di Padova, Padua, Italy

A. Ghiretti-Magaldi (201), Istituto di Biologia Animale e Centro per lo Studio della Fisiologia e Biochimica delle Emocianine, Università di Padova, Padua, Italy

R. Gilles (393, 467), Department of Biochemistry, University of Liège, Liège, Belgium

T. W. Goodwin (187), Department of Biochemistry, Johnson Laboratories, University of Liverpool, Liverpool, England

Esther M. Goudsmit (219), Marine Biological Laboratory, Woods Hole, Massachusetts

C. Grégoire (45), Department of Biochemistry, University of Liège, Liège, Belgium

Micheline Martoja (349), Institut Océanographique, Paris, France

E. H. Mercer* (147), Honorary Member of the Scientific Staff of the Chester Beatty Institute, Institute of Cancer Research, Royal Cancer Hospital, London, England

C. P. Raven (155), Zoological Laboratory, University of Utrecht, Utrecht, The Netherlands

* Present address: 2763 Girasol Avenue, Palm Springs, California

E. SCHOFFENIELS (393), Department of Biochemistry, University of Liège, Liège, Belgium

CHARLES R. STASEK (1), Department of Biological Science, The Florida State University, Tallahassee, Florida

P. A. VOOGT (245), Laboratory of Chemical Animal Physiology, University of Utrecht, Utrecht, The Netherlands

KARL M. WILBUR (103), Department of Zoology, Duke University, Durham, North Carolina

Preface

Zoology is currently undergoing a period of transition in which chemical knowledge is progressively integrated with the more classic knowledge of morphology and systematics. Biochemical studies of species, as well as of higher taxa, open new disciplines to the zoologist and offer new viewpoints in considering problems of structure, function, development, evolution, and ecology. The biochemist has considerable opportunities for broadening his sphere of investigation because of the enormous selection of animal species available for study from which a great variety of compounds can be obtained and reactions observed. There are abundant prospects for fruitful collaboration between the biochemist and zoologist in studies in which the characteristics of the animal and the biochemical constituents and processes interact in significant ways.

Very often the initial obstacle in undertaking investigations in new fields is the complexity and scattered character of the literature. This treatise is aimed primarily at making it possible for zoologists and chemists, who have a limited knowledge of the literature in fields other than their own, to gain a valid impression of the present state of knowledge in chemistry and zoology and an introduction to the existing literature. Thus, we have invited research workers who have contributed significantly to problems involving combined chemical and zoological approaches to summarize the knowledge in their specific disciplines of interest and competence. The authors have been encouraged to be critical and synthetic and to include mention of gaps in knowledge as well as the established information.

The treatise is arranged by phyla, an arrangement which seemed most suitable for presenting chemical information of zoological significance and for bringing to the attention of chemists those aspects of biochemical diversity of greatest potential interest. Each section, dealing with a major phylum, is introduced by a discussion of the biology and systematics of the group. This is followed by chapters dealing with various aspects of the biochemistry of the group. In general, the authors of individual chapters have been given full freedom, within the limitations of space, to develop their assigned topic. We thought that in this way the reader would have the advantage of the author's personal experience in and attitude toward his field, and that this would more than compensate for any unevenness in coverage that might result.

We are grateful to Professor K. M. Wilbur for his help in the early planning of this treatise, to the authors for their cooperation and patience, and to the staff of Academic Press for their careful work.

MARCEL FLORKIN
BRADLEY T. SCHEER

Contents of Other Volumes

Volume I: PROTOZOA

Volume II

Section I: PORIFERA

Section II: COELENTERATA, CTENOPHORA

Section III: PLATYHELMINTHES, MESOZOA

Volume III

Section I: ECHINODERMATA

Volume IV: ANNELIDA, ECHIURA, AND SIPUNCULA

Volume V: ARTHROPODA Part A

Volume VI: ARTHROPODA Part B

CHAPTER 1

The Molluscan Framework

Charles R. Stasek

I. Introduction

The Chordata and the Mollusca exhibit much more variation in structure among their higher taxa than other free-living metazoans. But while all major diagnostic features of chordates are found in every species at least at some time during its life cycle, there is no single key character that is present throughout the Mollusca. Instead, a mollusk is recognized as such by the possession of any one of an array of traits, or more usually, of a combination of traits. Correlated with the chief differences among the subgroups of these phyla, diversity within the Chordata is initially sorted out by taxonomists at the level of subphylum, yet the principal variations upon the molluscan framework are usually regarded as being at the relatively low rank of class.

This latter situation seems to reflect the fact that those features thought to be the primitive bases of each of the several molluscan traits are commonly combined in a model, called the archetype or the hypothetical ancestral mollusk. Following the construction of the archetype, each actual molluscan class is usually represented as having sprung forth independently from it by loss or gain of characters, as illustrated by Morton (1967, p. 21), without indication of greater or lesser interclass relationships. Recently, the artificiality of the archetype and the intellectual danger of its use in deriving an evolutionary concept of the phylum have been recognized (Morton, 1967, p. 14; Salvini-Plawen, 1969b). The model may remain a convenient tool in teaching the comparative morphology of this difficult group, but an appellation implying its primary role in phylogeny has been misleading and predisposes one to consider molluscan evolution from a limited point of view.

This chapter presents an interpretation of the functional bases underlying the evolution of the major features of the Mollusca and the derivation of its classes. Primary aims have been to put information into an order that fosters comparative investigations and to reflect in a taxonomy of the higher categories the relationships inferred from an interpretation of phylogeny. Descriptive synopses of the several classes have not been included; they would be redundant, for good ones have already been published by Portmann (1960), Morton and Yonge (1964), and Morton (1967).

II. Phylogenetic Origins

Ideas on the phylogenetic origins of the Mollusca have been reviewed by Clark (1964, p. 249ff), Vagvolgyi (1967), and Salvini-Plawen (1968a, 1969b). Rather than describe and evaluate the disparate views hitherto proposed, as already especially well done by Vagvolgyi, I shall present here only the general problem facing those who may further attempt to resolve the phylogenetic relationships of the Mollusca to other phyla and reach a conclusion believed to best interpret existing information.

The flatworms, mollusks, annelids, and arthropods have long been separate and highly diversified phyla, each characterized by a distinctive groundplan or framework that sets it off from the others to varying degrees. Yet, evidence from comparative anatomy and embryology indicates that these groups are phylogenetically interrelated. The fossil record has been of little help in unraveling the early interrelationships, since the soft-bodied flatworms are extremely poor candidates for preservation and are unknown as fossils, and with the exception of a few well differentiated annelids from the Australian Precambrian, remains of all the remaining groups are unrepresented prior to Cambrian times (Glaessner, 1962).

The essential, if exceedingly distant relationships among these phyla are generally acceded to, and one would expect the Precambrian stem groups to have been much less different from one another in what we regard as basic features than are their descendants today. They would have been less different because, first, intergrading groups existed, and second, the actual magnitude of the differences among the morphological types within each stem group would have been less at the mutual base of the lineages than it is now after more than five hundred million additional years of segregated evolution. It follows that at least some of the very traits now held to be basic in defining the individual phyla were, in their incipient stages, nothing more than variable minor traits of specialization integrated into a single broad framework based upon other and now recondite functional criteria.

To attain a grasp of phylogeny, one must have insight into the following points. Characteristics that are now reasonably stable in a lineage, such as the number of pairs of pedal muscles in most molluscan classes, were more variable among earlier representatives of the groups. Features whose trend has been stabilized are taxonomically significant at high levels, but, going back in time, their systematic value is reduced in relation to increased variability. Even if certain stabilized features are not utilized in formal systematics, they may strongly influence our preconceptions of a group at any hierarchic or phyletic level. The anatomy even of the conceptualized molluscan archetype was based not upon primitive features, but upon the end products of stabilization. Therefore, the archetype automatically fails to provide indications of molluscan evolution prior to the relatively advanced stage it represents.

If the tendency of a structural system has not been toward stabilization, but towards greater diversity, as with radular types in the Gastropoda (Termier and Termier, 1968), the system is taxonomically significant at levels correspondingly lower than those represented by stable traits. But at earlier times, prior to their diversification into discrete lineages, the structures would have held higher systematic importance in organisms variable in other ways.

Variability in some features has persisted in certain lineages (as gill number of chitons and *Neopilina*) but stabilized in others (as gill number in bivalves and higher cephalopods). Furthermore, combinations of correlated traits that now set one higher taxon apart from others were conceivably unsorted and overlapping in the ancestral groups (see Simpson, 1953, p. 346). *Neopilina,* a monoplacophoran, shares certain basic traits with chitons, gastropods, and bivalves, and is a relict from one such group.

Application of these ideas back through time leads one to consider the possibility that all the stem groups that eventually radiated into the antecedents of existing protostome phyla could have been placed in an expanded concept of the turbellarian Platyhelminthes or in a single, now defunct phylum that would have included the extant Nemertinea. That is, at the earliest phases of protostome radiation, features such as incipient segmentation, a complete digestive tract, or development via a trochophore-like larva could have evolved to thoroughly recognizable, if unfamiliar, levels among the ancestral turbellariform populations. But these qualities would not have been held to signify separate taxa at the rank of phylum, or even class, had systematists been alive to classify those early faunas. Recent turbellarians may be considered to be the products of an early carnivorous sidebranch of this radiation. The proto-annelid–arthropodan lineages may be supposed to have arisen, perhaps independently, from a more cohesive fraction of the early populations, their physiognomies then being characterized by a mosaic of traits, some new and distinctive, some retained in modified form from the ancestral types. In a brief abstract, Harry (1968) seemed to arrive at similar conclusions.

Specifically, the phylogenetic origins of the Mollusca probably lay near that of the pre-annelids in small, ciliated, acoelomate vermiform organisms in which serial repetition of various organ systems (pseudometamerism) was variously expressed. Trends toward increase in size later led to the partial abandonment of locomotory support by a solid body wall and the appearance of hydrostatic skeletal elements through cavitation, mainly of the hemocoel in the molluscan line and of the coelom in that of the annelids. The pseudometamerism of the stem groups may be viewed as having become regularized in the annelids, with the repetitive nature of most of the organ systems becoming numerically correlated, thus giving rise to eumetamerism. In the mollusks, correlation among the organ systems was never wholly achieved. Pseudometamerism sometimes became strikingly expressed within the lineage, and even approached a low level of numerical correlation in the Monoplacophora. Its expression gradually declined as fewer pairs of presumably more efficient organs assumed the total functions previously performed by multiple, less efficient ones. Traces of pseudometamerism remained in evidence until independently lost or obscured within the lineage of each of the higher classes.

III. Unfolding of the Framework

Because the initial evolutionary phases of the Mollusca took place during Precambrian times and are unrecorded by fossil evidence, their probable characteristics and sequence of appearance must be pieced together indirectly. This intellectual process employs the rational extension of data obtained from comparative studies of existing conditions and of fossil material, limited though these studies and material may be, and is entirely retrospective in nature. This comparative information is evaluated in such a manner that present conditions are conceived to have unfolded through natural selection in a relatively probable, comprehensible, and functional continuum.

The following hypothetical development of an evolutionary rationale is mostly new, but comes closest to the views expressed by Fretter and Graham (1962). Basically, it appears to me that most of the broad features that distinguish the Mollusca as a phylum could have arisen in direct relation to a single evolutionary accomplishment—the secretion of a cuticle over the dorsal body surface, together with trends toward increase in size.*

Beedham and Trueman (1967, 1968, 1969) were probably correct in concluding that the cuticle was preceded by a protective mucoid coat produced over the dorsal surface of early representatives of the molluscan lineage (Fig. 1) and that it was the precursor of the calcified shell that today characterizes the Mollusca. The initial cuticle was added to peripherally, and perhaps in thickness as growth took place (Fig. 1B, c). The epithelium underlying and secreting the cuticle can be regarded as equivalent to the mantle (pallium) of more advanced forms (Fig. 1B, ma). The early forms were only a few millimeters long and acoelomate. They had a ciliated epidermis, a diverticulated gut, a subterminal dorsal anal pore (a), four longitudinal nerve cords (n), a series of dorsoventral muscles (mu), and a pair of mesodermally derived tubules (tu) somewhere dorsolateral to the gut (gu.) Ciliary locomotion on a mucous track was characteristic.

To accommodate conditions found in the existing classes, the edge of the cuticle of more highly advanced pre-Mollusca is envisaged as having become set into a groove at the margins of the secretory epithelium (Fig. 1C, pg) (Stasek, 1972).

Predisposition to the secretion of calcium may have existed in the pre-Mollusca, with spicules being laid down within the cuticle, as occurs in aplacophorans and chitons, where, apparently, separate cell types

* The following statement by Lang (1896, p. 268) was brought to my attention after this chapter was submitted to the press: "If such a hypothetical [turbellarian] racial form were to secrete a dorsal shell, perhaps at first in the form of a thick cuticle containing calcareous particles, a typical Molluscan organisation would be produced. The development of a shell would deprive the greater part of the surface of the body of its original respiratory function, and would lead to the formation of localised gills. By means of the development of a mantle fold these delicate-skinned organs could be brought under the protection of the shell. . . . A part of the dorsoventral musculature would be changed into the shell muscle."

While the idea that mollusks had turbellarian ancestry has had its proponents (see Vagvolgyi, 1967), Lang's view on the functional bases underlying the evolution of the Mollusca appears to have been entirely ignored over the past three quarters of a century until independently developed in the present chapter.

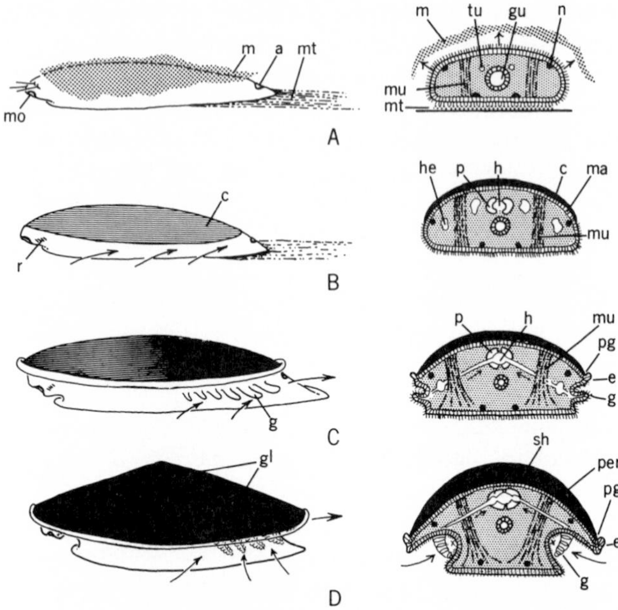

FIG. 1. Four stages in the establishment of the molluscan framework from a turbellariform organism, all shown in lateral and cross-sectional aspects. (A) Proposed ancestral form with complete gut and ability to secrete mucus as a protective measure and as a locomotory track. (B) Transitional turbellariform stage with radula and cuticule. The cuticle led to development of a ciliary respiratory mechanism (arrows) and of a hemocoel consisting of sinuses and dorsal heart and pericardium, the latter by modification and enlargement of the gonoducts. (C) Transitional molluscan stage with gills contained in an incipient mantle cavity underlying the eaves of the mantle. (D) Advanced molluscan stage with calcified cuticle and folded gills. Dorsoventral muscles have become pedal retractors. Key: a = anus, c = cuticle, e = eaves of mantle, g = gills, gl = growth lines, gu = gut, h = heart, he = hemocoelic sinuses, m = mucous covering, ma = mantle, mo = mouth, mt = mucous track, mu = dorsoventral muscles, n = longitudinal nerve cords, p = pericardium, per = periostracum, pg = periostracal groove, r = radula, sh = shell, tu = gonoducal tubules.

secrete the spicules and cuticular substance. Later deposition of calcium carbonate as layers raised the original cuticle above the general body surface to become the periostracum of the shell of the higher Mollusca (Fig. 1D, per). The periostracum has since remained relatively thin because its general undersurface is no longer in contact with the secretory epithelium.

With the appearance of underlying shell deposits, the sites of attachment and secretion of the periostracum became restricted to the groove (periostracal groove) around the periphery of the body (Fig. 1D, pg). Indirect evidence concerning present conditions (Mutvei, 1964) indicates that secretory cells within this groove arose from a generative zone in the depths of the groove and emigrated to the outer surface of the growing mantle. As they passed outward from the groove, the cells secreted first organic periostracal material and then, as their position was displaced by growth, a sequence of calcium carbonate in one or more forms within an organic matrix. The location of the periostracal groove in relation to the folds along its sides is important in determining the homologies of the mantle edge in the extant classes (Stasek, 1972).

Any general theory accounting for the deposition of calcium carbonate throughout the Mollusca is premature, but among aquatic species this process requires the preliminary secretion of the conchiolin matrix of the shell (Wilbur, 1964). The organic component of the cuticle therefore probably appeared prior to the evolution of a shell composed of layered calcium carbonate. This step also occurs ontogenetically (Raven, 1958).

The appearance of a persistent cuticle, especially an inflexible one containing calcified layers, had concurrent or subsequent effects on the elaboration of the molluscan framework as discussed below.

While the cuticle is here envisaged as fitting closely everywhere upon the dorsal surface (Fig. 1B, c), superficial folds of the molluscan body wall typically extend the mantle and shell as eaves beyond the body (Fig. 1C,D, e). This arrangement forms an underlying space that, besides increasing the protective attributes of the shell, promotes greater freedom of movement of the soft parts. In contrast to the prior stage (Fig. 1B), the evolution of eaves would increase the ease of locomotory turning by contractions of the body in the absence of appendages such as those found in arthropods, another heavily cuticularized group.

The space between the body and its eaves is termed the mantle cavity, and the entire double-walled fold, the shell-secreting outer surface, as well as the inner nonsecretory one, is said to comprise the mantle in these instances. The locomotory portion of the body thus separated from the mantle is called the foot.

The edges of the mantle and the periostracal groove became tucked under the edges of the eaves as an advanced condition, but in a more primitive state, retained in modified form by the chitons, the mantle edge was normally exposed around the margins of the shell (Fig. 1,C and D), from which position it was probably retractible.

Two subsequent developments were closely related to the appearance of a cuticle and mantle cavity and finally led to the complete breaking away from the conservative turbellariform framework. Flatworms are limited in size by dependence upon simple diffusion of gases to all parts of the body and lack of a blood circulatory system and gills. Food is carried to all regions of the body by the gut, which in higher orders is pseudometamerously branched to varying degrees. Such branching serves some functions of a circulatory system and has likely enabled the higher Turbellaria to attain a relatively large size.

In molluscan evolution, even if a rudimentary circulatory system originated prior to the cuticle, as seems possible, the secretion of a dorsal covering diminished the efficiency of respiration by simple diffusion to all regions of the body, although sufficient gas exchange could have taken place through the noncuticularized and unfolded epithelium of the body wall in very small organisms (Fig. 1B). One may suppose, therefore, that the trend toward larger size was a primary factor behind the elaboration of hemocoelic spaces (Fig. 1B, he) with cycling of blood brought about by a pumping heart (Fig. 1B, h). Later, in relation to increases in size, vascularized outpouchings of the body wall formed gills (Fig. 1C, g). These latter organs probably would develop in noncuticularized regions other than the head in free-moving organisms, hence from the walls of the mantle cavity, and were perhaps a single pair of complexly folded structures in some protomolluscan populations, and as maintained in some aplacophorans (Salvini-Plawen, 1969b). However, in the populations giving rise to more successful molluscs, the gills were originally several pairs of simpler outpouchings, in keeping with the pseudometamerous nature of the body. The mantle cavity at this stage was too shallow for a single pair of large gills, and any complex formation would likely have been preceded by several less complicated structures. Water currents across the gills were created by the beating of cilia on their surfaces.

In higher Mollusca with deeper mantle cavities, subsequent trends involved reduction of the number of gills to one or two pairs, probably in relation to their increased folding, individual size, and efficiency. In chitons, where a long, rather than a deep mantle cavity exists, there was an increase in gill number with increase in body size (p. 17).

Typically, the gills divide the mantle cavity into inhalant and exhalant chambers (Fig. 1D), and their sides have become progressively more folded, forming a series of flattened filaments (platelets) on either side of a central axis, which contains afferent and efferent blood vessels (Fig. 2). The ciliation on the broad faces of the filaments, the so-called lateral cilia (lc), produce the water current, while that on the edges, the frontal and abfrontal cilia (fc, ac), cleanse the gill of particulate material contained in the respiratory current. Terminal cilia (tc) connect the tips of the platelets to adjoining structures. In the higher classes with one or two pairs of gills, the filaments are characteristically elongated, stand-

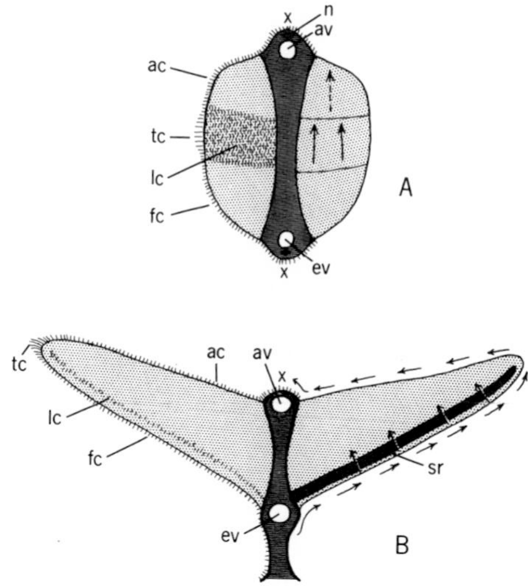

FIG. 2. (A) Gill of a chiton in cross section showing central axis, with blood vessels and nerves, and face view of the short gill platelets. Pattern of ciliation on left; direction of ciliary beat on right. (After Yonge, 1939a.) (B) Generalized molluscan ctenidium in cross section. Elongated platelets are supported by rodlets. (After Yonge, 1947.) Key: ac = abfrontal cilia, av = afferent blood vessel, ev = efferent blood vessel, fc = frontal cilia, lc = lateral cilia, n = nerve, sr = supporting rodlet, tc = terminal cilia, x = site of rejectory tract of cilia.

ing well out from the axis (Fig. 2B). Each filament is then supported by a pair of internal chitinous rodlets (Fig. 2B, sr). A molluscan gill possessing all of the above traits, or a gill deduced to have been derived by loss of one or more traits, is termed a ctenidium.

The ctenidium of the Mollusca, either in its preliminary phases exhibited by the lower groups, or as modifications of the generalized, fully formed structure, came to play a central role in evolutionary diversification, as reflected in those taxonomies employing "branchia" as a suffix for groups ranked as subclasses or orders. Yonge (1947) discussed the basic form and diversity of the ctenidium throughout the molluscan classes.

If one pair of gills was present as an early evolutionary phase, blood draining from them through the efferent vessels passed to a posterodorsal heart, the muscularized ventricle of which sent the blood out to general sinuses in the tissues. From the tissues, the blood percolated back to the gills via the afferent vessels. If more than one gill pair existed in the early Mollusca, the blood from all of them on either side emptied into a common, unlined lateral sinus and then proceeded to the heart.

Regardless of the original number of gills, the mechanism for circulation of blood through them and the body sinuses was attained by expansion of the mesodermal, probably gonoducal tubules present in early members of the lineage (Fig. 1A, tu). This expansion formed a pericardial space for the beating heart, which was itself primitively formed by slightly muscularized invaginations of the pericardial walls (Fig. 1B,C, p,h). Were it not for this nearly incidental development of a coelomic space, the Mollusca would today be regarded as an acoelomate phylum. They have escaped that designation on a technicality.

The gonoducts were finally regionated into three components of which the proximal region retained the original reproductive function. The central and distal components formed the molluscan "kidney" through which the gametes continued to pass. Judged from what little is known of the excretory physiology of extant molluscan classes, the central region, which had formed the pericardium and heart, developed a filtration mechanism that drove body fluids into the pericardial space. The walls of the distal portion of the gonoduct evolved secretory and reabsorptive capacities that modified the filtrate prior to its extrusion as urine to the outside medium. Later developments in several independent lineages led to the separation of reproductive and excretory functions with the evolution of separate openings from the gonads into the mantle cavity. Martin and Harrison (1966) summarized molluscan excretory systems from the above point of view. Salvini-Plawen (1969b), in contrast, figured the separation of excretory and reproductive systems as primitive. The vestigial connections remaining between gonad and pericardium in gastropods and cephalopods, and the origin of the gonads from the pericardial walls in chitons and bivalves, all with gonoducts separate from the kidneys, suggest the older view to be a more correct interpretation.

The evolution of a firm cuticle favored the development of yet another molluscan trait. In modern turbellarians, pseudometamerously arranged dorsoventral muscles attach to the basement membrane of the epithelium (Hyman, 1951, p. 78), and, presuming a similar organization in their predecessors, gave rise to the dorsoventral pedal retractor muscles of the primitive Mollusca (Fig. 1A,C, mu) (Salvini-Plawen, 1969b). The attachment sites of the pedal muscles appear by light microscopy to insert in the basement membrane of the mantle epithelium, or by tonofibrils in the epithelial cells lying between the shell and muscle, but at least in the pulmonate *Acroloxus lacustris,* electron microscopy reveals that the actual insertions lie in a connective tissue layer immediately underlying the epithelium (Hubendick, 1958). The firm connection between soft parts and shell is achieved by a brush border of the outer epithelial surfaces interposed between the muscle and shell. This myoadhesive epithelium has been discussed by Mutvei (1964), but comparative work is required. Functional attachments of the dorsoventral muscles to the cuticle provided a firm base for locomotory control and permitted the organisms to clamp down protectively upon solid substrata, or, as a later reciprocal development on soft substrata, to withdraw the body into the shell cavity.

Initial stages of all the major molluscan traits, save the radula, have thus far been accounted for in a gradually emerging complex; the Mollusca did not spring forth as an archetype, but their framework was gradually derived from forms that could today be retrospectively regarded as extremely specialized turbellarians.

The appearance of the radula in the molluscan framework needs mention. Because this feeding structure is found in nearly all Mollusca, including representatives of the aplacophorans, which appear to be closest to the stem groups, it must have originated somewhere among the earliest forms (Fig. 1B, r). The radula probably appeared prior to all other distinctly molluscan features. Judged from the simpler states in aplacophorans, the early radula consisted of a few rows of cuticularized teeth placed directly upon the epithelium of the floor of an eversible pharynx, perhaps with the intervention of a cuticular layer between the teeth and the epithelium (see Hyman, 1967, p. 32; Salvini-Plawen, 1967). Protrusible pharynges of various kinds are found in the Turbellaria, but hardened teeth upon the pharyngeal wall are not found within that class. The newly erected class or phylum Gnathostomulida, however, is a group of herbivorous "flatworms" characterized by cuticular jaws (Riedl, 1969). The existence of these worms lends some credence to the assertion that cuticular teeth could have arisen within a flatworm framework, although implications of phylogenetic relationships between gnathostomulids and mollusks obviously is not intended.

A major trend among the molluscan prototypes is thought to have been that of increase in size. The organisms conceived to represent the cuticularized stem groups were probably less than 1 cm. long, and probably ranged about half that size or less. The lower limit is admittedly a guess, but the upper limit is indicated by the fact that the first representatives of the Monoplacophora and the Gastropoda, first appearing in the Lower Cambrian, and of the Bivalvia, whose earliest occurrence is in Middle Cambrian rocks, measured, in general, less than 1½ cm. anteroposteriorly, as surveyed from several volumes of the *Treatise on Invertebrate Paleontology*. Bretsky and Bermingham (1970) reported the earliest known occurrence of the Scaphopoda from the Upper Ordovician; the maximum length of 480 specimens was 4.9 mm. (mean, 3.1 mm.). There are a few species of Recent scaphopods of equivalent small size, but most are from two to thirty times the length of the Ordovician representatives. The earliest Cephalopoda, which appeared in the Upper Cambrian, were also comparatively tiny. The conical, chambered shells were less than 25 mm. long, and the body chamber was only about one-third that size. There were general and independent trends toward larger size in many lineages of these classes, suggesting that the first fossil representatives of each of them were themselves intermediate components of these trends.

In contrast, some of the first recorded Polyplacophora from the Upper Cambrian were larger than most Recent species, reaching about 100 mm. long (Bergenhayn, 1960). The chiton fauna of the Upper Cambrian was probably the result of a very early and separate trend from much smaller ancestral forms whose remains have not yet been found or recognized. Subsequently, with the development of gastropods that competed with them, and the later appearance of more advanced lines of generally smaller chitons, the larger primitive Polyplacophora were replaced ecologically and taxonomically.

IV. Derivation of the Classes

The unrecorded Precambrian creatures were antecedent to about ten classes of mollusks. Of these, three are entirely without living representatives. They are the Hyolitha, Mattheva, and Stenothecoida, discussed by Marek and Yochelson (1964) and Yochelson (1966, 1969), respectively. The suggested life forms and habits of these lower Paleozoic classes remain questionable, and the relation of the Hyolitha to the Molluscan framework is dubious in my estimation. The relationships of the classes with extant members to the unfolding of the framework are retrospectively inferred to be as discussed below.

A. Aplacophora

Of the classes with living species, only the aplacophorans lack a fossil record, yet of all the classes the Aplacophora appear to be closest to the stem groups (Salvini-Plawen, 1969b). All other earlier types have become extinct, probably through ecological replacement by more advanced forms.

Extant aplacophorans are vermiform (Fig. 3A,C) and entirely marine, usually living as selective deposit feeders in or upon soft mud or ooze or associated with sessile organisms upon which they may feed. Interstitial species have been reported (Swedmark, 1956).

With reference to the cuticularized pre-Mollusca (Fig. 1B,C), the body form of the aplacophorans can be inferred to have been derived by enrollment of the dorsal surface and concomitant lateral restriction of the mantle cavity and foot to a ventral groove (Fig. 3A,B, f). A more highly evolved condition in the class is one in which the cuticle completely encircles the body (Fig. 3C). Calcareous spicules are laid down within the cuticle by the epithelium, and many protrude through it, often lending a furry appearance to the organism.

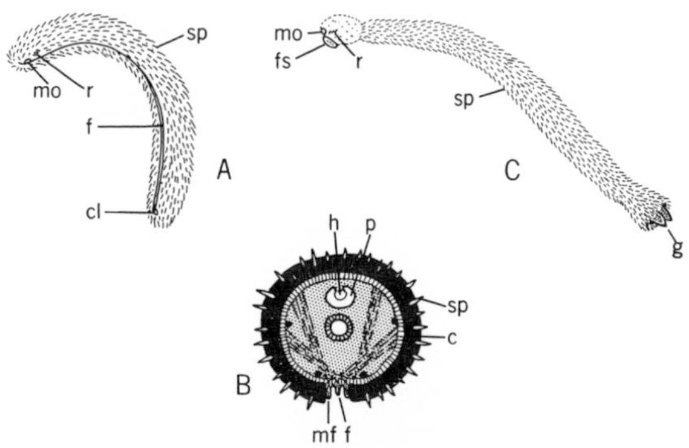

FIG. 3. Aplacophora. (A) Subclass Ventroplicida, generalized representative. (B) Cross section of A, illustrating enrollment of cuticle and restriction of foot to narrow groove. (C) Subclass Caudofoveata, generalized representative. Key: c = cuticle, cl = cloaca, f = foot, fs = foot shield, g = gills, h = heart, mf = middle mantle fold, mo = mouth, p = pericardium, r = position of radula, sp = spicule in cuticle; muscles, nerves, and gut as in Fig. 1.

The presence or absence of a footfold, together with correlative features, has been utilized as the basis for the separation of the aplacophorans either into two orders (Boettger, 1956; Hyman, 1967, p. 58) or two classes (Salvini-Plawen, 1968b, 1969a,b). Those with a footfold, the Ventroplicida, glide about on a mucous track by pedal ciliary waves, and respiratory gas exchange occurs either through the epithelium of the constricted mantle cavity flanking the foot or through pallial folds in a larger portion of the mantle cavity (cloaca) at the posterior end (Fig. 3A, cl) (Salvini-Plawen, 1968d). Those without a ventral footfold, the Caudofoveata, burrow in ooze by peristaltic contractions of the anterior part of the body, and the foot is reduced to the so-called oral shield (or foot shield, fs; Salvini-Plawen, 1969b). In correlation with the complete enfoldment of the body by the cuticle and the burrowing habit, a pair of ciliated gills is located in the posterior mantle cavity (Fig. 3C, g) (Salvini-Plawen, 1968c, 1969b).

These two groups comprise separate lines of evolution (Boettger, 1956; Salvini-Plawen, 1969b), but their points of origin were closely linked, and the variations of body plan do not seem to me to reflect those at the rank of class, as proposed by Salvini-Plawen. An intermediate course has been taken in recognizing them as subclasses, using Boettger's unaltered terminology.

B. POLYPLACOPHORA

The chitons (Fig. 4A) are pseudometameric in the arrangement of the shell plates, musculature, gills, nervous system, and, in some, the

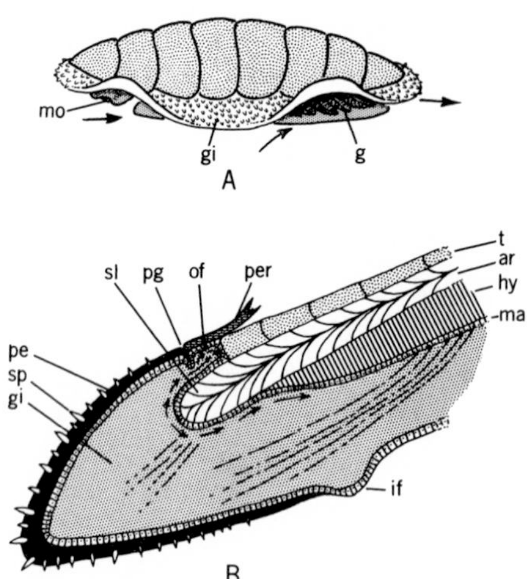

FIG. 4. Polyplacophora. (A) A generalized chiton illustrating primitive elongate form, with head fused to overlying body, and pseudometamerous shell plates and gills. Arrows indicate respiratory current. (B) Schematic section of shell and mantle showing probable locations of generative zones of the mantle (arrows), one in the periostracal groove, another along the margin of sutural lamina. (Modified from Mutvei, 1964.) Key: ar = articulamentum, g = gills, gi = girdle, hy = hypostracum, if = inner mantle fold, ma = mantle, mo = mouth, of = outer mantle fold, pe = pellicle, per = periostracum, pg = periostracal groove, sl = sutural lamina, sp = spicule, t = tegmentum.

heart. The repetitive but uncorrelated nature of these systems is not to be confused with the segmentation of annelids (Hunter and Brown, 1965). The head is not well demarcated, but is completely roofed over by and fused to the mantle and shell, which, together with the elongate, flattened form, reflects the proposed ancestral conditions (Fig. 1B,C).

The Polyplacophora are inhabitants of solid substrata upon which they crawl by slow muscular waves of the broad foot. Mucus aids in maintaining a firm grasp upon the substratum. The tips of the larger radular teeth are hardened with iron deposited as magnetite, its only known occurrence in a biological system (Carefoot, 1965). While most chitons are herbivores, some are omnivorous (Barnawell, 1960), and *Placiphorella* is a strict carnivore (McLean, 1962).

Larval life involves a trochophore larva that metamorphoses directly to a juvenile chiton (Raven, 1958, p. 132); a veliger stage is lacking. Some chitons brood their young in the mantle cavity (Smith, 1966).

Ontogenetically, the shell of the Polyplacophora begins as an undivided sheet of periostracal cuticle under which calcification later takes place in separate transverse regions, thus giving rise to the adult condition in which there are eight overlapping shell plates (Fig. 4A) (Raven, 1958, p. 150), a state that lends anteroposterior flexibility to the elongate body. These plates are surrounded by a tough, yet flexible girdle (Fig. 4A,B, gi) in which calcareous spicules are embedded, recalling the condition in aplacophorans (Fig. 3B). The location of the periostracal groove (Fig. 4B, pg) indicates that the girdle is homologous with the middle mantle lobe of other mollusks (Stasek, 1972) and that the arrangement of girdle and shell is essentially comparable to that envisaged for the pre-Mollusca (Fig. 1C), except that the exposed mantle rim became covered by a cuticular secretion, here termed the pellicle (Fig. 4B, pe), as likewise found in some aplacophorans.

As described by Bergenhayn (1955, p. 28), the most primitive chitons (Paleoloricata) had shell valves consisting of three layers: periostracum, tegmentum, and hypostracum (Fig. 5A). The overlap of a posterior valve by the next anterior one was accomplished in this primitive order by the scale-like "dead" portions (apical area) of the more anterior valve (Fig. 5A,C, aa). The valves were not strongly anchored in the girdle, and the relationship of valves to girdle must have been relatively

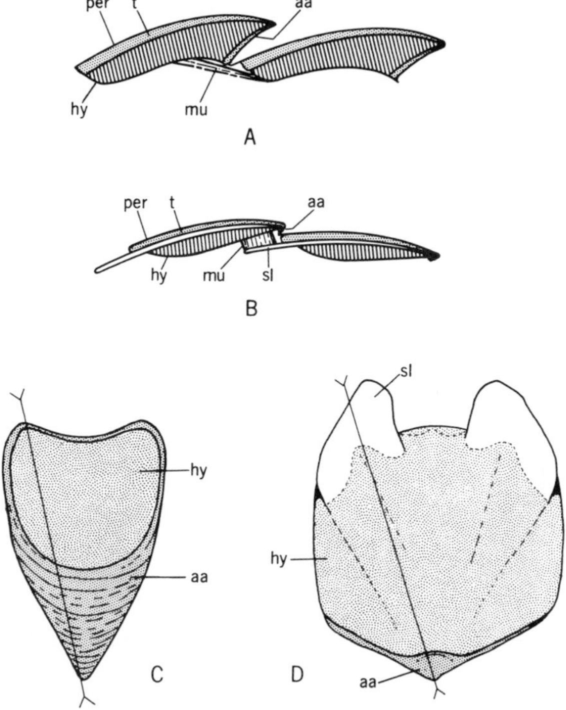

Fig. 5. Intermediate valves of primitive chitons. (A) Section through two valves of a paleoloricate showing probable relationships. Note lack of articulamentum and overlap of "dead" scale-like apical areas. (B) Similar section through two valves of a neoloricate. Note that overlap is formed by "living" articulamentum (C) Ventral aspect of one valve of *Chelodes,* a paleoloricate. (D) Ventral aspect of one valve of *Helminthochiton,* a primitive neoloricate. Straight lines in C and D indicate position of sections shown in A and B. Key: aa = apical area, hy = hypostracum, per = periostracum, mu = muscle, sl = sutural lamina, t = tegmentum.

tenuous. The hypostracum was secreted as a thick callus, apparently both by the general mantle surface and by the epithelium at the sites of muscle insertions.

With the appearance of the Neoloricata in the Ordovician (?) (Smith, 1960), a major advance in polyplacophoran shell structure was achieved; namely, the production in each valve of two extensions (sutural laminae) reaching anteriorly from below the anterior border of the tegmentum to underlie the next anterior valve (Fig. 5B,D, sl). An inner hypostracal layer continued to be laid down as callus on the general inner surfaces of the valves (Fig. 5B,D, hy). In contrast with the Paleoloricata, the apical area (Fig. 5B,D, aa) of modern types is minute, and overlap of the valves involves "living" shell.

Although their phylogenetic source is uncertain, the sutural laminae probably were derived through local organization of the hypostracal layer (Beedham and Trueman, 1967), and this through specialization of the mantle into two pockets at the anterior border of each of the valves but the first. A generative epithelial zone additional to that in the periostracal groove and located at the anterior edge of each sutural lamina (Fig. 4B) is thought to account for the conformities of the laminae (see Mutvei, 1964, p. 237).

Muscles (Fig. 5B, mu) join the overlapping regions of the valves and, probably upon disturbance, impart rigidity to the body. Speculatively, homologous muscles existed in the primitive types (Fig. 5A). If so, they were longitudinally oriented and took part in postural movements related to straightening out the body after flexure.

The muscle fibers of modern chitons terminate as tonofibrils within the epithelial cells (Von Knorre, 1925). This relationship needs to be investigated by electron microscopy.

Principal evolutionary advances in the Neoloricata mainly involved the establishment of a firmer relationship between the valves and girdle, and the hypertrophy of the latter into a more highly muscularized structure. These advances were achieved later in the Paleozoic through differentiation, again apparently of the hypostracum, into additional laminae (insertion plates) anchoring the valves to the girdle and providing sites for muscle attachment (Fig. 6A, ip). Phylogenetically, the development of insertion plates occurred first on the anterior valve, then on the tail valve, and finally on all the intermediate valves. Subdivision of the insertion plates (Fig. 6B) and roughening of their edges, and of those of the tegmentum as well, are further advances arrived at convergently in various lineages. All specialized portions of the hypostracum are termed articulamentum (Fig. 5B, ar), which consists of several sublayers.

In addition to the above trends, the Mopaliidae (Fig. 6C) and the

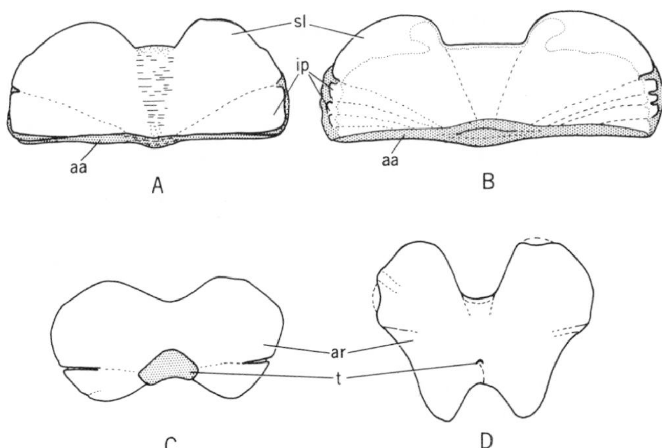

FIG. 6. (A and B) Ventral aspects of valves of *Tonicella lineata* and *Ischnochiton evanida,* respectively, showing one and three pairs of insertion plates. (C and D) Dorsal aspects of valves of *Amicula vestita* and *Cryptochiton stelleri,* respectively, illustrating trend toward dominance of the articulamentum in the mopaliid–acanthochitonid lineage. Key: aa = apical area, ar = articulamentum, ip = insertion plates, sl = sutural laminae, t = tegmentum.

Acanthochitonidae have emphasized development of the articulamentum and coverage of the dorsal body surface by the girdle. These changes were concomitant with decrease in the size of the exposed tegmentum and culminate in *Cryptochiton,* the whole shell of which gradually becomes covered by the girdle during ontogeny. In each mature valve, the tegmentum survives only as a minute vestige at the bottom of a pit or completely buried in the articulamentum (Fig. 6D, t).

The foregoing discussion has been based upon information gleaned from Pilsbry (1892–1894), Bergenhayn (1955, 1960), and Smith (1960), and upon original observations of preserved and fossil material at The California Academy of Sciences.

Although qualitatively somewhat simpler, the gills of chitons bear clear resemblance to the bipectinate ctenidia of primitive gastropods (Fig. 2A,B) (Yonge, 1939a). The leaflets of the gill of the former lack supportive rods, which characterize higher molluscan groups.

By relating the number of gills to body length in six species of chitons, Johnson (1969) found that at any comparable ontogenetic stage, larger species bear a greater number of gills than smaller ones. She further stated that small species have shorter, wider gill cavities than do larger species, and concluded that chitons of different adult sizes have the same effective gill surface area per unit length of the animals. The data underlying the latter conclusion are not convincing, but there is a suggestion that the number and structure of the gills is somehow functionally related to the size of the animal. The common idea that the more primitive chitons (Order Lepidopleurida) possess fewer gills than advanced types needs to be appraised from this point of view.

The Polyplacophora are interpreted as having parted early from the mainstream of the phylum. They reflect a common origin with the Monoplacophora in a pseudometamerous, cuticularized ancestor in which the head was fused to the overlying mantle (Fig. 1C). The gill structure of both groups, but especially that of the chitons, represents an intermediate stage prior to the emergence of more typical ctenidia. Several pairs of gills are inferred to have hung in the posterolateral portions of the mantle cavity. The development of the exposed mantle rim to form the spiculose girdle and the divided nature of the shell point to a history long separate from that of other mollusks, although the former trait is reminiscent of that in aplacophorans.

The development of the articulamentum only after the chitons were a differentiated group indicates that this layer is not homologous with the inner layer of the shell of other mollusks, as it was believed to be by Beedham and Trueman (1968, 1969).

Although the ancestral pre-polyplacophorans may have "experimented" with the deposition of layered calcium carbonate, there is no known fossil evidence that such organisms existed. The relatively delayed appearance in the Late Cambrian of already highly refined members of the class and the lack of more primitive calcified types of multivalved mollusks in the fossil record—except, perhaps, the Mattheva (Yochelson, 1966)—suggest the possibility that shells containing layers of calcium carbonate were independently derived in the lineages leading, on the one hand, to the Polyplacophora and, on the other hand, to the Monoplacophora. Beedham and Trueman (1967, p. 228 and Fig. 6) implied but did not state this conclusion in their preferred evolutionary interpretation of the chiton shell as a structure without univalved ancestry.

Both the Aplacophora and the Polyplacophora bear spiculose cuticles and have been united in the Subphylum Aculifera by Salvini-Plawen (1969b). But the spiculous cuticle of the former (Fig. 3B, c) is homologous with the aspiculose periostracum of the chitons (Fig. 4B, per). On the other hand, the spicule-bearing girdle of the Polyplacophora is equivalent to the occasionally cuticularized but aspiculous mantle fold lateral to the foot of aplacophorans (Fig. 3B, mf). Perhaps both regions of the cuticle were spiculose in some of the common ancestral groups, as indicated by Beedham and Trueman (1967). A discussion of the homologies of the mantle edge is presented elsewhere (Stasek, 1972).

While the phylogenetic relationships of the Polyplacophora appear to be distinct from any other class, their point of departure from the main evolutionary course of events was closer to that of the Monoplacophora than to those of the diphyletic Aplacophora. The reasons for aligning the chitons with the aplacophorans in a single class, the Amphineura, were the spurious report of the transient appearance of seven dorsal shell plates in the developing young of the aplacophoran *Nematomenia,* discussed by Thompson (1960) and Hyman (1967, p. 53), and the mutual presence of variously arranged pseudometameric ladder-type nervous systems. Nervous systems of this kind also occur in the Monoplacophora and in primitive gastropods, so that the rationale behind a Class Amphineura is dissipated.

C. Monoplacophora

The evolution of mollusks with univalve shells consisting of a thin periostracum overlying layers of calcium carbonate that were added to peripherally and in thickness gave rise to and determined the evolutionary mainstream of the phylum. The first recorded shells of this kind are assigned to two classes, the Monoplacophora and the Gastropoda, found in Lower Cambrian rocks (Glaessner, 1962). Members of both classes undoubtedly existed long before those strata were laid down, but it is in Precambrian monoplacophoroids that the Gastropoda and the remaining classes are inferred to have found their origins. In situations where polyphyly is thought to have played a part in constituting a class, as in the Gastropoda (p. 25), the Monoplacophora seem always to have been the parental stock from which all the lineages have sprung.

The Monoplacophora were diverse during the Lower and Middle Paleozoic, but the class is now represented by six or seven known species of a single genus, *Neopilina,* a deep water genus without a fossil record. This genus is no doubt modified in many regards from its Paleozoic relations, but on the basis of comparative studies, many of its general features may reasonably be thought to be primitive, and in the following discussion these qualities have been extrapolated backwards to the earliest monoplacophorans.

Among the Lower Cambrian types, the Palaeacmaeidae were small (about 12–15 mm. long), bilaterally symmetrical, and cap-shaped. Six pairs of muscle scars (Rasetti, 1954; Knight and Yochelson, 1960) provide the clue to the pseudometamerous nature of the organisms. A reconstruction appears in Fig. 7A,B. The respiratory, circulatory, excretory, and nervous systems, in addition to the musculature, probably were repetitive, but in more or less uncorrelated fashion. The muscle scars were located near the edge of the shell, and there were large gaps

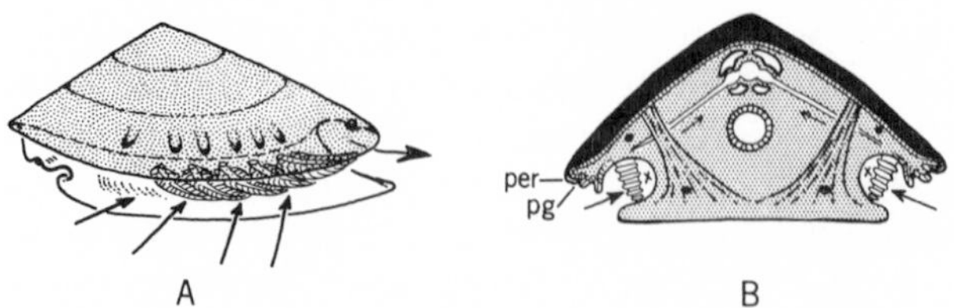

FIG. 7. Lateral view (A) and cross section (B) of a pseudometamerous univalved mollusk; shell type based upon the Lower Cambrian Palaeacmaeidae. Note undertucked nature of the periostracal groove and periostracum. Arrows indicate respiratory currents; per = periostracum, pg = periostracal groove; other aspects of anatomy as in Fig. 1.

anteriorly and posteriorly between the ends of the series of scars. There could not have existed a very capacious lateral mantle cavity in that case, but several pairs of gills probably hung from its roof, creating an enclosed exhalant chamber along the foot. In this regard they would have resembled chitons and Lemche and Wingstrand's (1959) interpretation of *Neopilina*. Alternatively, the posterior gap between the muscle scars may indicate a heightened mantle chamber in that region; if so, one or more pairs of fairly large gills could have been accommodated there. If the gills of the Palaeacmaeidae were ctenidia, they were bipectinate as in chitons and unlike the virtually monopectinate gills of *Neopilina*. The leaflets of the gills probably lacked supportive rodlets as in recent Monoplacophora and chitons.

The central position of the umbo indicates that subequal marginal increments account for the peaked shell with the muscle scars arranged with one pair on the anterolateral slope, two pairs laterally, and three pairs of smaller muscles on the posterolateral slope, an arrangement that became significant later in the Paleozoic. The mantle edge probably was divided into three lobes (Fig. 7B), as in *Neopilina* and most of the higher Mollusca. Phylogenetically, the outer lobe became distinct when the exposed mantle rim of the prototypes became tucked under the margin of the shell. The groove between the outer two lobes thus became the site of the generative zone of the shell-secreting epithelium and of periostracal secretion. The undertucked nature of the periostracal groove completed the process, begun long before, of dorsal coverage of the body by the cuticle and provided for the first time a facade composed entirely of protective shell. The periostracal groove itself came to occupy a less hazardous position during this process. As in modern *Neopilina* and the chitons, the head is presumed to have been fused to the overlying mantle and shell and to bear rudimentary sense organs.

In the later Cambrian, a dichotomy arose in the Monoplacophora.

One group, the Subclass Cyclomya, unlike any molluscan groups so far discussed, emphasized the dorsoventral axis of the body, a character foreshadowed by the Palaeacmaeidae and having enormous significance to subsequent evolution of the phylum. The muscle scars were reduced in number or partly fused to form anterior, posterior, and lateral pairs arranged in a ring, with the coiled apex of the shell projecting above it (Fig. 8A) (Rollins, 1969). The muscle insertions were well within

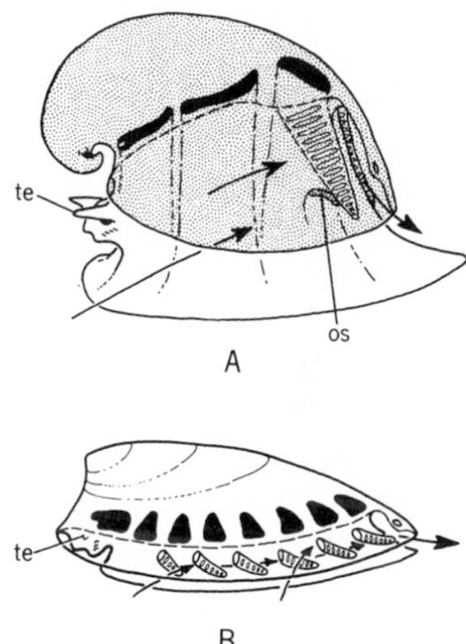

FIG. 8. Generalized Monoplacophora. (A) Cyclomyan with emphasis on dorsoventral axis, few pedal muscles, and proposed to have a well developed head and a deep mantle cavity, hence fewer but larger gills. (B) Tergomyan with primitive elongate body, many pedal muscles, poorly differentiated head, and shallow mantle cavity, hence a larger number of smaller gills. Arrows indicate proposed respiratory currents; os = osphradium, te = sensory tentacle.

the aperture, indicating the presence of an extensive mantle cavity and the ability of the animal to withdraw into it. The heightened mantle cavity also suggests the potential for a reduction in the number of gills from several smaller to fewer larger ctenidia, as possibly inherited from the Palaeacmaeidae. Supporting rodlets in the ctenidial filaments are inferred to have made their appearance as the filaments increased in size.

A development suggested by the steep anterior slope of the cyclomyan shell is that a distinct head existed in these organisms (Fig. 8A). A head that could be moved about freely under the overhanging shell and project somewhat from beneath its rim would have had great potential consequences. It could lead to greater development of the cephalic sense organs, especially eyes, but it could lead especially to greater facility of movement. While the Cyclomya became extinct in the Devonian, their lineage gave rise earlier to the Gastropoda, which amplified these favorable qualities.

The shell of a second subclass of Monoplacophora, the Tergomya, is characterized by a larger number of muscle scars (five to eight) and a greatly diminished anterior slope (Fig. 8B). Probably neither of these qualities is primitive, although Horný (1965) considered the terminal apex of the Tergomya to be ancestral to the centralized one in the Cyclomya. The elongate and relatively low Tergomya range as fossils from Cambrian to Devonian times and gave rise to *Neopilina* but, in our estimation, to no innovative molluscan types. *Neopilina* has survived the general ecological replacement experienced by the Monoplacophora probably because it is a specialized deposit feeder living on ooze in very deep water.

An alternative view of evolution within the Monoplacophora was presented by Starobogatov (1970), who proposed that the initial members of the class were spirally coiled and had two to three shell muscles and several pairs of gills, kidneys, and gonads. These spiralled forms yielded (1) cap-shaped types with many muscle scars and a reduced number of gills, and (2) spiral types with but one pair each of retractor muscles and gills. The latter group gave rise to the gastropods. Starbogatov's scheme was based upon the supposition that the planispiral protoconch figured from one specimen of *Neopilina galatheae* (Lemche and Wingstrand, 1959) indicated that spiralling was the ancestral adult condition of the class. However, Menzies (1968) reported a specimen in which the protoconch was a simple bulbous, uncoiled structure. This undermines the view offered by Starobogatov, whose paper is nevertheless an excellent treatment of the probable morphology of the early monoplacophorans.

Except for the muscle scars preserved on fossil material, that which is known of the soft parts of the Tergomya is based upon the investigation of Lemche and Wingstrand (1959) on ten poorly preserved and damaged specimens of *N. galatheae,* the type species. Of the ten specimens with soft parts, only two were utilized in the descriptions of internal anatomy other than musculature. Subsequently, four other species have been retrieved from deep water in the Eastern Pacific Ocean (Menzies, 1968) and one from the Gulf of Aden (Tebble, 1967). Lemche and Wingstrand (1960) presented a few comparative observations on one of these, *Neopilina ewingi.* Living specimens of *Neopilina* have yet to be observed.

A crystallographic study of the shell of *N. galatheae* was published by Schmidt (1959), and the ultrastructure, histochemistry, and amino acid content has been described by Meenakshi and co-workers (1970).

In *Neopilina,* there are eight pairs of pedal retractor muscles, five or six pairs of gills in the lateral mantle cavity, six pairs of nephridia (the middle two of which also act as ducts for the two pairs of gonads), two pairs of auricles, and a ladder-type nervous system. There is some correspondence of the organs, lending a kind of segmented appearance to this genus.

In view of what has been presented earlier as a molluscan framework functionally evolving from acoelomate stock, a most peculiar feature of *Neopilina* is the presence of a pair of extensive, but flattened, dorsal cavities lying between the gut and mantle. These spaces were thought by Lemche and Wingstrand (1959) to be coelomic and were postulated to be perhaps homologous with the gonadal cavities of chitons. Tenuous connections were described between the dorsal cavities and nephridia, thus lending credence to the coelomic nature of the cavities, but in a footnote (Lemche and Wingstrand, 1959, p. 56), and in subsequent work by Lemche and Wingstrand (1960) and Lemche (1962), they were observed in *N. ewingi* to bear distinct connections with the pharynx. *Neopilina* is not near the base of the Mollusca, as often stated, and the group of Monoplacophora to which it belongs seems to have been evolutionarily sterile. Furthermore, as far as present knowledge extends, there is no functional rationale that has been offered for the existence of large coelomic spaces in organisms that appear to move not by burrowing but by locomotor pedal waves on mucous trails. Our conclusion, in accord with that of Steinbock (1963, p. 394), is that the dorsal spaces in *Neopilina* are not coelomic, but further investigation is needed before a decision can be reached as to their homologies.

Before leaving the Monoplacophora, I should like to point out that it was probably within this class that the veliger larva appeared as a typical feature of the molluscan life cycle. This conclusion rests upon the fact that a veliger stage characterizes all descendant classes but the Cephalopoda. Although usually lost, the minute protoconchs adherent to the shells of two specimens suggest that *Neopilina* probably passes through a veliger stage during its life history.

In the Polyplacophora, the primordia of the major qualities of the molluscan framework appear during later development and metamorhosis of the trochophore larva, or of its equivalent in the Aplacophora (Thompson, 1960), leading directly to the settled, preadult condition. The comparable period of transition in primitive groups of higher Mollusca is not direct, but is altered to form the extended, dispersive veliger stage. That which is a preadult condition of the chitons and aplacophorans is a second larval type in higher classes in which the shell, mantle cavity, and foot seem to develop precociously.

The main feature of the trochophore, the ciliary band of prototroch, is not lost in the veliger, but becomes more or less expanded onto lobes anterior to the mouth. This velum functions as a swimming and food-gathering organ (Fretter and Montgomery, 1968; Mapstone, 1970).

Molluscan life histories are highly varied (Raven, 1958; Morton, 1967), but it may be said that the veliger larva, having once evolved, became the dominant larval phase of the phylum, with the trochophore stage often becoming suppressed or eliminated. Partial or complete suppression of the veliger larva also has occurred either functionally, as in advanced gastropods where it is retained nondispersively in an egg case, or entirely, as in cephalopods with yolky eggs. It may be that cephalopods evolved prior to the appearance of a veliger larva in the monoplacophoran line.

D. Introduction to the Higher Classes

The innovation of a calcified univalve shell that retained all growth stages to form a functional unit (Stasek, 1963a) was of primary importance in altering the course of molluscan evolution. The main axis of growth was primitively the anteroposterior one, as in aplacophorans, chitons, and *Neopilina,* but all advanced classes stress the dorsoventral axis and incorporate to advantage the simple growth parameters that provide a protective chamber contained within the shell. The initial evolutionary phases of each of the higher classes involved heightening of the body through ventral extension of the free eaves of the mantle. This led to high, short organisms rather than to low elongate ones. Topographical or anatomical reversions toward elongation have occurred, as in the squids and opisthobranch gastropods.

Originally, the trends towards increase in height reflect increases in the relative size of the chamber housing the visceral mass and posterior portions of the mantle cavity containing the gills. In addition, such heightening of the body and mantle chamber was correlated with reductions in the number of gills to one or two pairs of larger, true ctenidia.

Reductions in the number of pedal muscles, insertions of which came to be set well back within the aperture (Fig. 8A), also took place. However, the final low number of these muscles in the Bivalvia and Cephalopoda was not arrived at until these classes had been otherwise established as independent lineages, for multiple pedal muscles have been reported in members of the former by McAlester (1966) and Driscoll (1964) and in the latter by Mutvei (1963). It is this character of multiple muscle scars that especially recalls the monoplacophoran ancestry of these classes. The Cyclomya exhibit the above trends within the monoplacophoran lineage itself.

Early monoplacophoroids probably were herbivorous or omnivorous grazers upon solid substrata in shallow water, but various populations soon penetrated other adaptive zones within which the genesis of the remaining classes took place. The facies of each of the resultant classes reflects adaptations to the settings of those initial adaptive zones, in the sense of Simpson (1953, p. 199ff). Each major adaptive zone involved a unique mode of feeding, and except for the primitive gastropods, which replaced the Monoplacophora, new locomotory relationships to the substratum. In all classes but the Cephalopoda, the hemocoelic spaces of the foot became important in hydraulic mechanisms, particularly in burrowing forms such as the bivalves (Trueman, 1966; Trueman *et al.*, 1966) and the scaphopods (Trueman, 1968). Hydraulic locomotor systems probably had their functional sources in the primitive means whereby an animal extended its body after having been retracted within the shell cavity.

The major diagnostic characters of each class thus are viewed as having evolved first as specializations within the monoplacophoran framework and only later, through diversification within each lineage, to have come to characterize taxa that biologists now rank at class level. There is great probability that certain of the classes are polyphyletic (Gastropoda: Rollins and Batten, 1968; Bivalvia: McAlester, 1966).

There follows a structural and functional rationale behind the appearance of each class, but not of their subsequent adaptive radiations, which often involved ecological reversions and convergent evolution.

E. Gastropoda

Concise morphological descriptions of the Gastropoda and its subgroups may be found in Morton and Yonge (1964), Morton (1967), and Purchon (1968). Extensive discussions of anatomy, physiology, ecology, and systematics have been presented by Hyman (1967) and Franc (1968). In this section we shall introduce new perspective into an old problem: the origin of and functional bases behind the peculiar gastropodan body form.

The suggestion was made previously that the cyclomyan Monoplacophora had evolved a head that could be moved independently of the overlying body. The formation of such a head ostensibly took place by the gradual intrusion and expansion of the anterior portion of the mantle cavity over the "neck" of the animal concurrently with the withdrawal of the pedal muscle insertions to positions deeper within the shell (Fig. 8A). A narrower "waist" thus came to separate the head and foot from the upper, shell-covered portion of the body. The Gastropoda probably would not have come into being had there not been previous evolution of a free head, for it is at the resultant "waist" that torsion, the hallmark of the Gastropoda, takes place.

Torsion is an ontogenetic process resulting at first from fairly rapid contractions of the asymmetrical (right) larval retractor muscle that passes from the shell to insert mainly in the velum, in combination with slower effects of differential growth of the larva, or from differential growth alone (Crofts, 1955; Fretter, 1969). During this process, the anus and subsequently developed organs of the mantle cavity come to lie above the head in the prosobranch condition (Fig. 9). The orienta-

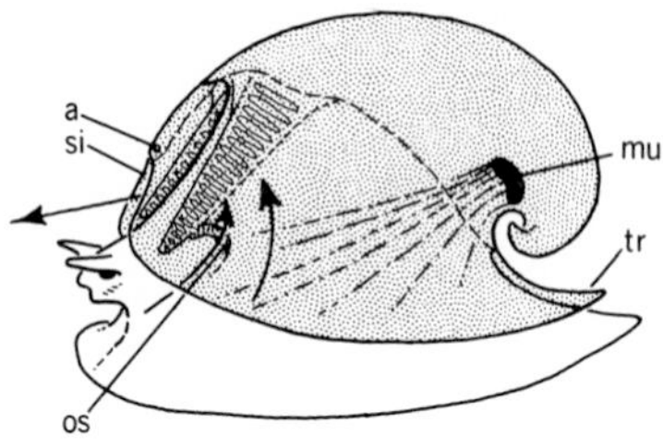

FIG. 9. An archaeogastropod illustrating the torted condition resulting in the prosobranch condition and an extensive mobile head region. Pedal muscles reduced to one pair. Key: a = anus, mu = muscle insertion, os = osphradium, si = marginal excurrent sinus in shell, tr = "train" of shell.

tion of all the internal organs within the shell cavity, and of the shell itself, are reversed, even when certain of them develop after torsion is otherwise completed. For example, following torsion, the nervous commissures passing through the waist of the body (the pleurovisceral connectives) simply grow into a figure of eight (streptoneury).

The definitive pedal retractor muscles, which are paired in some primitive gastropods (e.g., Scissurellidae, Haliotidae) (Fretter and Graham, 1962, p. 140) or horseshoe-shaped in others (e.g., Patelloida), give no indication in the adult that they have undergone torsion, for they pass from their insertions on the shell into the pedal mass without twisting about or changing sides as they traverse the region of the waist (see figures in Crofts, 1955; Hyman, 1967, p. 198). As described later, this arrangement comes about through early emigration of the left and right muscles to the sides opposite that in which they originated.

The extinct Superfamily Bellerophontacea, now generally thought to be part of the mainstream of gastropodan evolution, possessed planispiral shells (Fig. 9), the precise bilaterality of which gives little indication as to their orientation relative to the foot. In this respect, bellerophont shells could as easily be those of cyclomyan monoplacophorans with one pair of retractor muscles, and are, in fact, occasionally considered to be untorted (Termier and Termier, 1968).

Yonge (1947) and Knight (1952) concluded that the bellerophonts bore an anteriorly placed mantle cavity, as evidenced by the presence of an emargination or sinus in the aperture of the shell (Fig. 9, si). This sinus with the anus recessed in its midline was regarded as equivalent to the excurrent slits or holes that direct feces and excretory wastes away from the head of such gastropods as keyhole limpets (Fissurellidae) and abalones (Haliotidae). As further aid in deciding the nature of the bellerophonts, Knight (1952, p. 51) referred to his previous discovery that they had a single pair of pedal retractor muscles on the post-torsional posterior slope of the coiled shell (Fig. 9, mu). The bellerophonts were therefore gastropods deduced to have one symmetrically disposed pair of each of the organs of the pallial complex, which includes the ctenidia, auricles of the heart, and nephridia, in keeping with living members of the lower Archaeogastropoda.

Living groups of prosobranch gastropods are usually considered to have arisen through the development of profound asymmetries in shell form and in the organs of the pallial complex primarily in relation to increased "sanitary" mechanisms originally necessitated by having excrement deposited over the head. These trends have been described in detail by Yonge (1947), whose report should be read in light of the papers by Knight (1952) and Crofts (1955).

The prosobranchs are presently thought to have given rise to the two remaining subclasses of gastropods, the Opisthobranchia and Pulmonata, by trends towards detorsion. But all of these radiations go beyond the purpose of this chapter. A concise description of current views in this regard was presented by Cox (1960, p. 141ff), Morton (1963), and Morton and Yonge (1964).

Difficulty in clearly delimiting the bellerophont Gastropoda from the Monoplacophora arose with the discovery that certain planispiral species with a marginal sinus also possess three symmetrical pairs of muscle scars, as in the Cyclomya (Rollins and Batten, 1968). Prior to the discovery of the multiple scars, these shells were regarded as those of bellerophonts, but following it, Rollins and Batten concluded that the shells were monoplacophoran in origin and that the marginal sinus was positioned posteriorly. Their hypothesis was that ". . . the sinus-slit developed *prior* to torsion and only because it did exist when torsion occurred was viability and a positive selective ['sanitary'] advantage assured." Placement of the main muscle scars on the expanded portion of the body whorl (Fig. 8A), rather than back within the spire as in gastropods (Fig. 9) further indicates the untorted condition of these organisms (Starobogatov, 1970).

If transitions exist directly between the coiled Monoplacophora and the coiled gastropods, class assignment of small, but relatively high cap-shaped and spiraled shells (Helcionellacea) from the Lower Cambrian raises the question of polyphyletic origin of the Gastropoda. In the absence of visible muscle scars, which if multiple would probably indicate that the shells were monoplacophoran, other qualities of the shell have been used by paleontologists to decide whether torsion had taken place.

A flaired "train" on one side of the aperture suggests that location to be the posterior region overlying the foot, as in some bellerophonts (Fig. 9, tr) (Knight *et al.,* 1960). In combination with an apex that bends or spirals towards the "train," the inference is that the apex, which faces forward in the Monoplacophora, now projects posteriorly; therefore, the visceral mass and mantle had rotated relative to the foot, bringing the anus and mantle cavity over the head. These forms are thus to be grouped with the Gastropoda, although Yochelson (1967) indicated that they may comprise a separate class. That torsion could have occurred more than once in evolution seems impossible at face value, and resolution with the probable fact of gastropodan polyphyly demands further discussion.

The cyclomyan Monoplacophora probably resembled early coiled gastropods in every regard except for those reflecting the process of torsion. Following the appearance of torsion, the lineage of the Cyclomya declined after a duration of at least 140 million years, apparently being ecologically replaced by their torted, but otherwise comparable descendants. Torsion was obviously of enormous selective value to the gastropods of early and mid-Paleozoic times.

Consideration of the probable selective advantages conferred by torsion has resulted in two major schools of thought, reviewed by Ghiselin (1966), Fretter (1967), Thompson (1967), and Purchon (1968). One school fosters the belief that torsion had adaptive significance for the swimming veliger larva by bringing the mantle cavity into a position where, upon retraction, the tender velum rather than the tougher foot would first enter it. As a larval adaptation, it created the "unsanitary" anatomical configurations to which gastropods have had to adapt, beginning with the development of the excurrent emargination discussed earlier. The belief that torsion arose as a larval adaptation by which the adult was inconvenienced has been virtually discarded in recent years. The principal reasons for this are that most of the torsional process in primitive archaeogastropods is completed long after the larva has given up its planktonic mode of life and that the foot, which enters the shell cavity last post-torsionally, is as delicate as, not tougher than the velum.

The alternative school of thought holds that an anteriorly placed mantle cavity is of value primarily to the adult in bringing forward the sensory equipment (osphradia) (Figs. 8A and 9, os) contained in the mantle cavity, and in allowing the animal to circulate fresh water rather than that disturbed during locomotion. A flaw in this view is that torsion probably originated in mollusks living on solid substrata upon which there would have been little deposited material. If deposits were present, mucous tracks would have inhibited their disturbance, and slow progression of the mollusks would not have promoted agitation greater than that produced by environmental currents and waves. Furthermore, the respiratory water probably was drawn in not posteriorly, as attributed to the Archetype, but anterolaterally and so would not have pulled in water through which the animal had already passed. The osphradia would have tested the water into which the animal was moving in that case.

A third hypothesis accounting for the evolution and function of torsion is figured and summarized in Fig. 10. In the depicted scheme, partial torsion began in the Monoplacophora as a temporary, reversible swing of the larval and adult shell, the cavity of which had been phylogenetically deepening as the eaves developed (Fig. 10B). The potential for increasing the lateral movement was in correlation with withdrawal of the insertions of the pedal muscles away from the shell margins and with enlargement of the head. A swing to left or right in the adult may have provided a somewhat larger lateral cavity into which the head could be accommodated when the animal was threatened by predators or by desiccation during intertidal exposure. The process is envisaged as beginning in the larval stage through the inception of a slight delay in the development of the left retractor muscles. The delay, even if slight, would have resulted in a temporary condition in which the right muscles were stronger than the left ones and would have assured consistent, fast withdrawal without the matter of "choice" as to which way to turn the shell.

During later stages in the evolution of torsion the difference in the time of development of the left and right muscles became progressively more extended until in the higher archaeogastropods an actual sequence in the appearance of the muscles was manifested (Fig. 10F). The gastropods may be distinguished from the Monoplacophora essentially by permanency and post-settling continuance of the torsional process in the former.

Feliksiak (1959) hypothesized that *Neopilina,* then only recently announced to the world, undergoes a temporary but fixed torsion during late larval life. This suggestion was based upon the asymmetrical shape and position of the larval shell adherent to one of the specimens described by Lemche and Wingstrand (1959, fig. 34). Feliksiak seemed to conclude that bilateral symmetry is restored to *Neopilina* by differential growth during early phases of benthic life, as evidenced by growth lines. That asymmetries in the shell necessarily reflect the torsional process is to be questioned (Crofts, 1955), but the facts in relation to *Neopilina* must be left to future investigations.

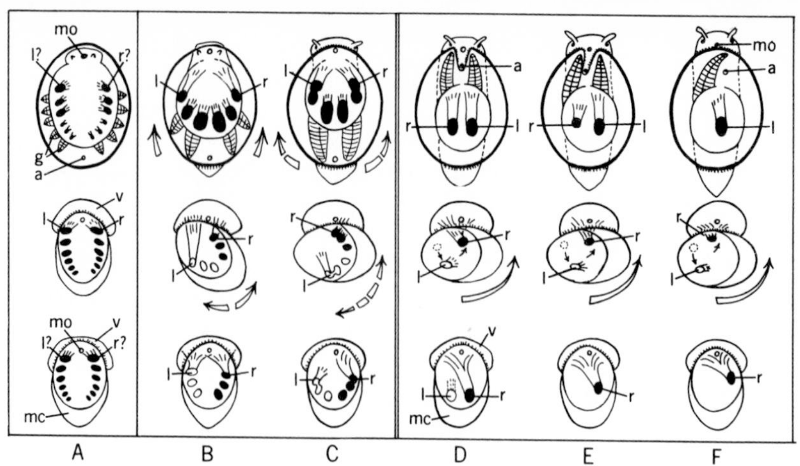

FIG. 10. A schematic proposal for the phylogenetic origin and function of torsion. (Columns A–C) Dorsal views of three possible ontogenetic stages in each of three types of monoplacophorans, adult at top; (D–F) same for three primitive gastropods. Horizontal rows indicate one of several possible phylogenetic sequences of the ontogenetic stages. Key: l, r = insertions of left and right larval retractor muscles and their derivatives in the adult; a = anus; f = foot; g = gills; mc = mantle cavity; mo = mouth; v = velum. (A) PALAEACMAEIDAE. *Lower figures:* Multiple left and right muscles equally developed, retraction symmetrical. Probably all retractors appeared simultaneously in ontogeny. *Upper figure:* Adult with narrow mantle cavity. No, or little, lateral swing of shell possible since head and body are broadly fused to shell. (B) INTERMEDIATE CYCLOMYAN. *Lower figures:* Difference in muscle symbols indicates that right muscles develop somewhat sooner than left ones, leading to asymmetrical, but reversible contraction, which swings shell counterclockwise. *Adult:* Development of left muscles catches up with right ones, giving rise to symmetrically disposed adult. Adult with developing head and sense organs, longer foot, deeper mantle cavity. Any amount of lateral swing of shell to left or right increases size of cavity into which head can be withdrawn upon disturbance. (C) ADVANCED CYCLOMYAN. *Lower figures:* Prolonged delay of development of left muscles. Effects of reversible contraction greater than in B in correlation with fusion of anterior muscles. Adult with well developed head and foot, deep mantle cavity correlated with reduction of gills to one large pair. Potential for lateral swing of shell increased inversely as attachments of head and body to shell decrease in area. (D) PRIMITIVE ARCHAEOGASTROPOD (bellerophont). *Lower figures:* Muscles reduced to one pair; delay in development of left muscle continued; fibers of right muscle cross to left side. Permanent torsion occurs in two phases, but more slowly than in extant archaeogastropods. Contraction of right muscle brings posterior of shell and visceral mass around 90°; right muscle insertion moves to post-torsional left, while primordium of pretorsional left retractor moves to post-torsional right (small arrows). Adult symmetrical, but left and right are reversed. No swing of shell necessary to provide large shell cavity into which the highly developed head can be withdrawn. Head also thus becomes even more free from visceral mass and has correspondingly greater facility of movement, which could have lent additional adaptive significance to complete permanent torsion. (E) INTERMEDIATE ARCHAEOGASTROPOD (based on *Haliotis*, Crofts, 1955). *Lower figures:* Delay of development of left muscle profoundly increased, but it becomes the predominant muscle in adult. (F) ADVANCED ARCHAEOGASTROPOD (based on *Calliostoma*, Crofts, 1955). *Lower figures:* Insertion of right larval retractor more on right side, leading to greater lateral displacement upon contraction. Right muscle disappears during ontogeny. Adult asymmetry marked.

If evidence proves Feliksiak's deductions to be correct, torsion probably first appeared as a partial but fixed orientation in young monoplacophorans. It remained a larval trait in the Tergomya, but the initial function, whatever it was, would have been one potentially benefiting the adult as well as the larva in certain lineages of the Cyclomya and was finally carried to completion in forms we now recognize as archaeogastropods.

The major probable function of the torted state, whether temporary or fixed, is here interpreted as one permitting greater freedom of movement for the head, which evolved simultaneously with torsion, and of providing a protective cavity into which it could be withdrawn (compare Figs. 8A and 9).

Temporary or permanent torsion is held to have begun as adaptive neither solely to the larva nor to the adult, but to the entire ontogeny. The major feature that impressed itself upon the gastropods probably began as a differential developmental process in organisms that as adults were untorted. These ancestral monoplacophorans existed for hundreds of millions of years. I hypothesize that during that extended period of time the temporary ontogenetic process of torsion based upon differential developmental rates of the retractor muscles came to hold adaptive significance to several lineages and was carried through to the adult phases as a permanent quality in more than one of them. Therefore, the proposed phylogenetic scheme outlined in Fig. 10 offers a rationale for the polyphyletic evolution of torted mollusks.

F. Bivalvia

By and large, the monoplacophorans and primitive gastropods of the early Cambrian seem to have been sluggish grazers of surface films or larger benthic algae. It was earlier inferred that within that herbivorous adaptive zone, and while the phylogenetically fertile Monoplacophora were still less than 1 cm. long, some side groups were experiencing anatomical trends toward increased efficiency of individual pairs of their pseudometamerous organ systems. In relation to the heightened form of the body and mantle cavity, some of these monoplacophorans had already successfully reduced the number of ctenidia to one pair. Ciliation upon the surfaces of the gill filaments continued to function in creating a respiratory water current and in removing particulate matter from it and the gills. This unwanted material undoubtedly included detritus and living plankton; that is, it comprised a quantity of material drawn from the same bank of organic energy that, from their earliest history, entire other phyla, especially the sponges and brachiopods, had tapped as a source of food. It is not surprising that one or more of the archaic monoplacophoran populations should have gradually come to exploit the same bank for its food supply, since a collecting mechanism already existed in the ciliary cleansing device of its gills. The source, but not the kind of food would have been changed, for the original filter-feeding types probably retained the essentially herbivorous habits of their ancestors.

Some gastropods, such as *Crepidula,* utilize similar mechanisms for collecting food, but having arisen late, found their potential for radiation somewhat stifled by preexisting and highly diversified filter-feeding mollusks. The Cambrian filter-feeding types, on the other hand, had entered an adaptive zone that had been untried by previous members of the phylum. The Lower Cambrian class Stenothecoida have been interpreted as filter-feeding mollusks (Yochelson, 1969); if so, they were unsuccessful in that they have left no descendant taxa.

Wherever among the invertebrates a ciliary filter-feeding device has evolved, the mechanism for pumping water and collecting food has become hypertrophied with respect to its homologs in relatives that do not filter-feed. Filter-feeding mollusks have not been exceptions to this generality, and in fact, provide some of the best examples (Stasek, 1966). Jørgensen (1966) presented an excellent survey and evaluation of filter-feeding mechanisms among the invertebrates.

The filter-feeding Monoplacophora are envisaged to have undergone trends toward increasing the length of the gill axes and of the filaments upon them in correlation with ventral extension of the eaves of the mantle and shell, which housed the expanding mantle cavity (Fig. 11A). Passage of mucus-bound material anteriorly may originally have been by way of ciliary tracts on the surface of the body, but channeling devices and flaps of the body wall soon funneled the potential food to the mouth. These flaps, the labial palps (lp) later expanded and acquired a sorting mechanism based upon relative sizes of the particles gathered by the ctenidia. Lower organisms often have more than one mode of obtaining nourishment, and it seems probable that the monoplacophorans that came to rely upon filter-feeding also continued for some time as grazers and had a radula. The importance and size of this scraping organ became reduced, and it finally was lost in relation to the degree of dependence upon the filter-feeding habit. The early types of filter-feeding monoplacophorans, therefore, probably lived upon solid substrata. As in other early Monoplacophora, the head was not clearly delimited from the overhanging eaves of the mantle.

Retrospectively, the Bivalvia (Fig. 11B) are the descendants of these hypothetical monoplacophorans. During the transition, lateral sites of calcification evolved on either side of an intermediate, nearly uncalcified

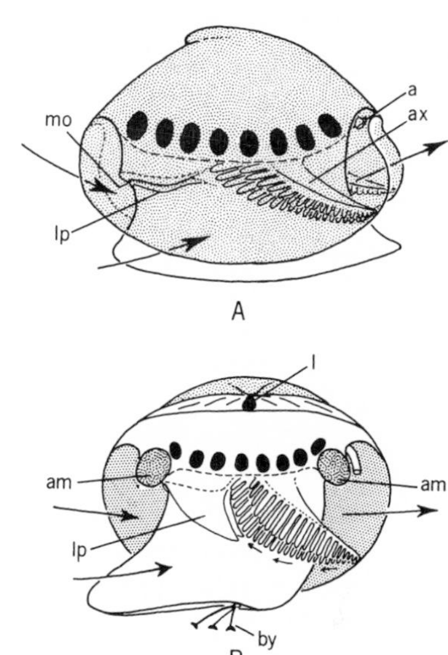

FIG. 11. (A) Hypothetical filter-feeding monoplacophoroid with undivided shell ancestral to the Bivalvia (B) with divided shell and multiple muscle scars, as in the Ordovician *Babinka*. Key: a = anus, am = adductor muscles, ax = gill axis, by = byssus, l = ligament, lp = labial palp, mo = mouth.

elastic band of conchiolin, the ligament, and formed two hinged shell valves. Probably as described by Yonge (1953), the bivalve shell began with lateral compression of the body and the appearance of two embayments in the edge of the mantle—an anterior inhalant notch and a posterior exhalant one, as illustrated in Fig. 11A. Yonge described additional factors that were involved in the derivation of the Bivalvia, such as the attachment of the flaplike mantle eaves to the shell by way of a ring of submarginal pallial muscles and the modification of the anterior and posterior ends of these to form adductor muscles that close the shell valves antagonistically against the opening resiliancy of the ligament (Trueman, 1964, p. 64), with which they concurrently evolved.

The earliest bivalves, such as *Babinka*, retained pseudometameric pedal muscles (McAlester, 1966, p. 233) but probably had lost other imprintings of pseudometamerism. Some modern clams, McAlester noted, still possess multiple pairs of pedal muscles.

In summary, the evolution of the filter-feeding habit in a univalved ancestral group of Monoplacophora allowed subsequent complete enclosure of the body through enlargement and ventral extension of the lateral mantle eaves. While the head had never been more than poorly delimited in the ancestral monoplacophorans, decephalization ensued and the radula and buccal muscle mass were lost.

Stasek (1961) and Allen and Sanders (1969) considered the possibility that filtration was the original method of feeding in the class, but an alternative view of the origin of filter-feeding bivalves was fully developed by Yonge (1939b) and has been long espoused. That view was predicated upon the idea that the gills of those bivalves believed to be most primitive, the Nuculacea, were not involved in feeding and that deposit feeding by way of extensile appendages of the labial palps was the original mode of gathering food in the class. Filter-feeding was thought to have evolved only after all other features of the bivalves had been established. Although subsequent evidence has shown the expressed bases for this view are invalid (Stasek, 1961, 1963b, 1965), the concept is deeply ingrained in the malacological literature (e.g., Purchon, 1968, p. 101).

G. Scaphopoda

The evolutionary ties of a third lineage are puzzling, but possibly there was some alliance with the pre-Bivalvia. This lineage now comprises a small class, the Scaphopoda (Fig. 12), all of which are selective

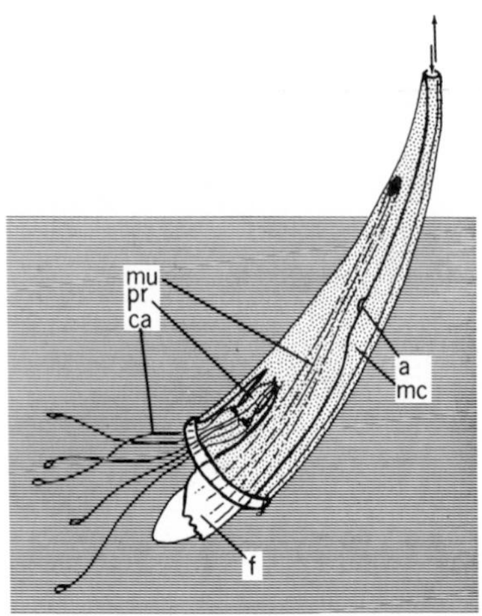

Fig. 12. Scaphopod in feeding position. Key: a = anus, ca = captacula, f = foot, mc = mantle cavity, mu = pedal muscles, pr = probocis.

deposit feeders on microorganisms and detritus. Recent studies are those of Morton (1959), Dinamani (1964), and Davis (1968). Fischer-Piette and Franc (1968) and Purchon (1968) presented admirable synopses of the class, but little comparative work exists. Relationship to the Bivalvia is suggested by similarities in the nervous system and dubiously evidenced by the fact that the mantle of the larva is extended ventrally as two lateral lobes.

Functioning in the selective gathering of food organisms from below the surface of the substratum in which these mollusks lie nearly buried are bunches of extensile, ciliated, filamentous outgrowths, the captacula (ca), originating near the mouth. Particles collected by the swollen ends of the captacular filaments are moved by cilia along the filaments and deposited in a groove on the foot, from which position they are transferred to the mouth, which is located at the end of a proboscis (pr). This structure is muscular, but does not comprise a head. In keeping with the nature of the material with which it deals, the large but simple radula is specialized as a grappling device rather than as a scraping one.

Shell growth and form in the class is unique and has profoundly influenced the nature of the mantle cavity. The posteroventral margins of the heightened mantle eaves fuse in the ventral midline of the postlarval scaphopod, forming a hornlike shell open at both ends. Subsequent growth takes place through addition of tubular increments at the larger end, while an unknown mechanism of resorption or dissolution of shell material enlarges the opening at the smaller, posterior end.

Probably in association with the constriction of the mantle cavity, especially the posterior end, there is no continuous flow of water through the mantle cavity, and water is first moved inwardly through the posterior shell opening and into the mantle cavity by ciliary activity. Rhythmic plunger-like contractions of the foot apparently drive the water, together with faeces, back out through the same aperture.

This group lacks ctenidia, and respiratory gas exchange takes place through the vascularized walls of the mantle cavity. The absence of a pericardium and true heart is interpreted as simplification. The two pairs of pedal muscles (mu) may be the sole remnant of pseudometamerism in the class.

H. Cephalopoda

Yet a fourth line of evolution, that leading to the Cephalopoda (Fig. 13), emphasized an actively carnivorous mode of life. The morphological and behavioral trends within this class reflect a highly active life style, both in relation to feeding and to escape from predators, although there has been some secondary adaptive radiation into microphagous and sessile niches. Morton (1963), Donovan (1964), Morton and Yonge (1964), and Teichert (1967) have summarized the major lines of evolution. This class is traditionally treated last in surveys and taxonomies because of the complex structure and behavior of its members, but the lineage was an early one and may have separated from the monoplacophorids before the other higher classes were clearly delineated.

The food capturing and escape activities of the minute, early precephalopods probably involved rapid, momentary leaps off the sub-

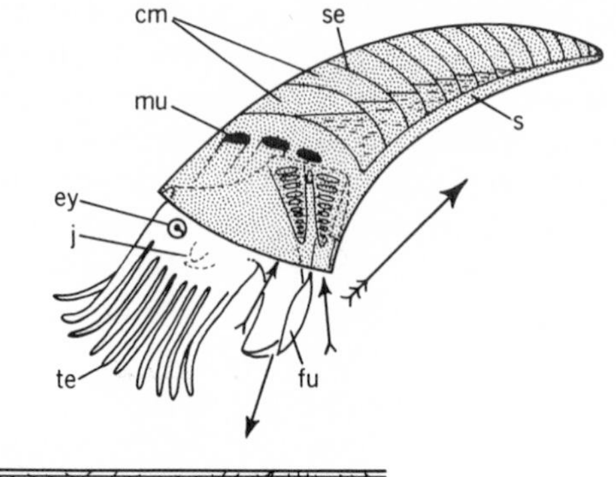

FIG. 13. Reconstruction of a primitive cephalopod (Shell type based upon Upper Cambrian *Plectronoceras*) with multiple muscle scars and two pairs of gills. Smaller arrows indicate water currents; larger arrow indicates direction of swimming movement. Water content within camerae is suggested. Key: cm = camerae, ey = eye, fu = foot (funnel), j = jaws, mu = pedal muscles, s = siphuncle, se = septum, te = tentacle.

stratum. At an early stage, the anterior part of the foot became a grasping organ (Mutvei, 1957) composed of several muscular tentacles (te). Cuticular jaws (j), present in the ancestral group, were exaggerated as a beak, yet a radula was retained. Sense organs, particularly the eyes (ey), were enlarged and became highly developed by comparison with those of other mollusks (Wells, 1966a). The ganglia of the nervous system tended to aggregate as a complex brain in a distinct head region (Wells, 1966b).

Subsequent developments led to sustained periods of freedom from the substratum and to increased directional capacities while in "flight." The earliest recorded cephalopods radiated to become the Subclass Nautiloidea, eventually with 2500 species, and from them derived the Ammonoidea, of about 5000 species. These groups, now extinct but for a few species of *Nautilus*, were extremely successful in their time. Living squids and octopuses comprise a relatively small subclass, the Coleoidea, of about 650 species. These are the end results of a lineage of nautiloids tending, in relation to their strongly natant habits, toward reduction and loss of the shell and increased streamlining, speed, and muscularization of the mantle.

An early innovation was the evolution of a chambered shell as a buoyancy mechanism. The body of a shelled cephalopod periodically shifts its position toward the growing edge of the shell in conjunction with growth. After a shift has occurred, a shelly septum (se) is secreted by the mantle in such a way that a liquid-filled chamber (camera, cm) is formed behind it. When the new septum has been produced, the liquid in the chamber is gradually pulled out by an unknown, but probably osmotic mechanism through the porous calcareous walls of a tube, the siphuncle (s). This tube contains a vascularized extension of the mantle, the epithelium of which must be actively involved in the transport of the liquid from the chamber to the blood of the organism. Alterations in buoyancy are brought about by varying the amount of liquid in the older camerae. The gas that remains in the chambers is primarily nitrogen under pressure of just less than an atmosphere, regardless of the volume of the gas. Functional aspects of buoyancy of the cephalopod shell have been described by Denton and Gilpin-Brown (1963, 1966) and Denton (1964).

The position of the siphuncle in the earliest recorded cephalopods was along the anatomically posterior wall of the shell (Fig. 13). This suggests that the shell was tipped, or perhaps horizontally disposed in life, for in a vertical position the liquid in the partly filled camerae would not have been in contact with the siphuncle. In later cephalopods, the position of the siphuncle varied from posterior to anterior (topographically ventral to dorsal, according to specialists), the septa became folded to varying degrees, and the shell was coiled in a variety of ways. Devices that funnel or channel the fluid in the camerae to ensure contact with the siphuncle at strategic regions of the shell is a topic that has not been treated in the literature. Such devices would be of special importance to swiftly swimming species adapted to move vertically, as well as horizontally, through the sea.

Ventral enrollment and later fusion of the foot to form a funnel (fu) through which water could be shot with propulsive force, first solely by muscular waves of the foot, as in *Nautilus* (Bidder, 1965), and later of the walls of the mantle cavity in forms less incased by their shells, also reflects the tendency to be free from the substratum. In accordance with the muscular mode of jet propulsion that evolved in early cephalopods, and with the increased needs for maximum exchange of respiratory gases in highly motile organisms, the slower ciliary mechanism for creating a water current was lost; the ctenidia lack cilia.

The coelom of the Cephalopoda is of much greater extent than in other Mollusca. The large size of this cavity may be viewed as a flotation mechanism allowing for increase in body size without increasing the quantity of sinkable material in its construction. Flotation in some squids, the Cranchiidae, is aided by a huge coelom in which the ionic constituents of the fluid are composed of relatively light ammonium chloride rather than of heavier ions (Denton, 1964, p. 427). That the quantity and composition of body fluids of cephalopods are regulated by secretions of neurovenous tissues was discussed by Young (1970). These still enigmatic tissues originate adjacent to ganglia and pass to endings in the walls of veins or in the organ of the anterior chamber of the eye. One may conjecture that these secretions may also be involved in the function of the siphuncle and perhaps play a considerable role in flotation or derived activities.

The coelom in cephalopods may perform another function in relation to swimming. As discussed by Trueman and Packard (1968), the high pressures developed in the mantle cavity during jet propulsion are transferred to the hemocoelic spaces, momentarily enlarging them. These pressures cause distortion of the eye in *Octopus* (Boycott and Young, 1956), but more severe effects on the organs of the body are possibly buffered by the liquid in the large coelomic spaces. This is the only molluscan class to which a large coelom has functional significance. Rather than being a primitive quality, the size of these spaces is more reasonably viewed as an adaptation to pelagic existence.

Together with the multiple pairs of pedal muscle scars in several fossil nautiloids (Mutvei, 1963), the dual kidney–auricle–gill complex of *Nautilus* is here regarded as a remnant of pseudometamerism. The presence of two pairs of each structure of the pericardial complex is usually explained as a probable secondary duplication from the archetypal condition (see Morton and Yonge, 1964, p. 56), but the following rationale speaks against this view.

There is but a single pair of gills present in all cephalopods except *Nautilus*. Each member of the pair has at its base a separate pumping structure, the branchial heart, which in these vigorous forms propels blood through the capillary bed of the gill to venous blood vessels. *Nautilus*, a sluggish creature by comparison with higher cephalopods, lacks branchial hearts, and blood returns to the heart by way of sinuses, not by well-formed veins. Rather than envisaging in *Nautilus* the loss of such highly efficient organs as the branchial hearts and the concurrent embryologically difficult task of producing not only a second pair of gills, but additional members of the pericardial complex as well, one can more reasonably expect that there were at least four relatively inefficient gills present in the pseudometameric protocephalopods and that the faster metabolic tempo experienced by more highly organized pelagic forms led to increased efficiency of one pair of gills through the development of branchial pumping structures and increased vascularization. The remaining gills were subsequently lost as the single pair increased its respiratory capacity.

V. The Molluscan Subphyla

An evaluation of the interrelationships of the molluscan classes is shown in Fig. 14. Because the entire Precambrian base of the phylogenetic tree and the stages retrospectively deduced to have been in-

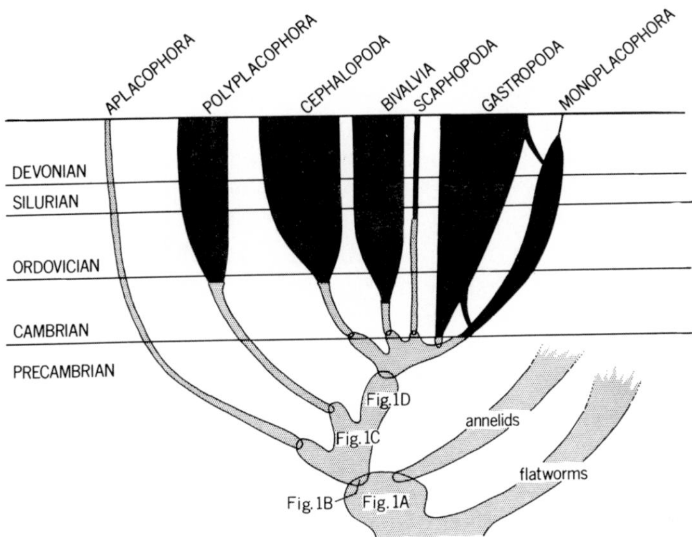

FIG. 14. Phylogeny of the molluscan classes during Precambrian and Paleozoic times. Known fossil record indicated by darker areas, inferred early history by lighter regions. Overlap of areas within the figure suggests that early members of a descendant group could be considered specialized representatives of its ancestral stock. Polyphyly of the Gastropoda is indicated. Relative sizes of the classes is only very roughly suggested by the widths of the lines of descent.

volved in the derivation of the molluscan framework (Fig. 1) are necessarily hypothetical, I anticipate that some malacologists might insist that I have but substituted four "hypothetical ancestral mollusks" for the usual Archetype. That allegation may be technically correct, but it can hardly be regarded as support for the classic view. The scheme presented in this chapter is not held to be the final answer to the question of molluscan phylogeny, but it has the following advantages:

1. It bears a functional and temporal rationale lacking in the Archetypal approach.

2. The stages figured are not simply the result of combining features that have stabilized during advanced phases of molluscan evolution.

3. It does not have a fully formed mollusk at the base of the evolutionary tree.

4. It harmoniously, rather than confusingly and ineffectually, incorporates pseudometamerism into the total picture.

5. It takes into account rather than disregards the fossil record.

6. It fosters rather than inhibits the determination of relationships above the rank of class.

The last point has special importance in that extant Mollusca are inferred to be the progeny of three separate lineages stemming from unrecorded pre-Mollusca. There are no known intermediate forms, fossil or Recent, bridging the enormous gaps between any two of the three lineages; therefore, each is here considered a subphylum.

One lineage, that of the Class Aplacophora, bearing only incipiently developed elements of the molluscan framework, became otherwise specialized along two long-separate lines the ranking of which is open to question. These organisms comprise the Subphylum Aculifera.

A second major lineage more thoroughly realized the potential of the molluscan framework, but did so with emphasis upon pseudometamerism. The nature of the shell and mantle edge especially indicate that the chitons have gone their own way from that of other mollusks. The sole Class Polyplacophora is here recognized as constituting the Subphylum Placophora. This innovation will likely meet resistance until the repetitive nature of polyplacophoran anatomy is admitted to be a manifestation of pseudometamerism and not merely secondary duplication from imagined archetypal simplicity.

The third lineage, the Subphylum Conchifera, is composed of the Monoplacophora and its four descendant classes. The evolution of a cap-shaped univalve shell gave impetus to the full realization of the molluscan framework as we now know it. This development was followed in each conchiferan class by diminution of pseudometameric structure until the anatomy of the phylum can be said to be characteristically, but not basically, unrepetitive in nature.

The following list summarizes the relationships indicated in Fig. 14:

- Phylum Mollusca Cuvier, 1795
  - Subphylum Aculifera Hatscheck, 1891
    - Class Aplacophora von Jhering, 1876
  - Subphylum Placophora von Jhering, 1876
    - Class Polyplacophora de Blainville, 1816
  - Subphylum Conchifera Gegenbaur, 1878
    - Class Monoplacophora Wenz, 1940
    - Class Gastropoda Cuvier, 1795
    - Class Bivalvia Linné, 1758
    - Class Scaphopoda Bronn, 1862
    - Class Cephalopoda Cuvier, 1795

Milburn (1960) gave succinct historical perspective to categories above class rank and agreed with Naef's recognition of two subphyla, the Amphineura (containing the Aplacophora and Polyplacophora) and the Conchifera. Milburn's reconstruction of a phylogenetic tree and that proposed here bear general similarity, although derived independently and from different points of view. Furthermore, he parenthetically suggested that separate subphyla might be required for aplacophorans and chitons.

An impermeable dorsal cuticle has been proposed as the initial factor leading to the evolution of the molluscan body plan and the eventual establishment of the group as a phylum. That the major variations assumed by the body covering during early radiation were the primary qualities upon which the higher subgroups were to be founded is therefore not surprising.

REFERENCES

Allen, J. A., and Sanders, H. L. (1969). *Malacologia* **7**, 381–396.

Barnawell, E. B. (1960). *Veliger* **2**, 85–88.

Beedham, G. E., and Trueman, E. R. (1967). *J. Zool.* **151**, 215–231.

Beedham, G. E., and Truman, E. R. (1968). *J. Zool.* **154**, 443–451.

Beedham, G. E., and Trueman, E. R. (1969). *Proc. Malacol. Soc. London* **38**, 550–551.

Bergenhayn, J. R. M. (1955). *Skr. Mineral. Paleontol.-Geol. Inst.* **22**, 1–42.

Bergenhayn, J. R. M. (1960). *J. Paleontol.* **34**, 168–178.

Bidder, A. M. (1965). *Proc. Eur. Malacol. Congr., 1st, 1962,* p. 111.

Boettger, C. R. (1956). *Zool. Anz., Suppl.* **19**, 223–256.

Boycott, B. B., and Young, J. Z. (1956). *In* "Bertil Hanström: Zoological Papers in Honour of his Sixty-fifth Birthday" (K. G. Wingstrand, ed.), pp. 76–105. Zool. Inst., Lund.

Bretsky, P. W., and Bermingham, J. J. (1970). *J. Paleontol.* **44**, 908–924.

Carefoot, T. H. (1965). *Proc. Malacol. Soc. London* **36**, 203–212.

Clark, R. B. (1964). "Dynamics in Metazoan Evolution." Oxford Univ. Press (Clarendon), London and New York.

Cox, L. R. (1960). *In* "Treatise on Invertebrate Paleontology," (I) Mollusca 1, (R. C. Moore, ed.), pp. 84–169. Univ. of Kansas Press, Lawrence.

Crofts, D. R. (1955). *Proc. Zool. Soc. London* **125**, 711–750.

Davis, J. D. (1968). *Proc. Malacol. Soc. London* **38**, 135–138.

Denton, E. J. (1964). *In* "Physiology of Mollusca" (K. M. Wilbur and C. M. Yonge, eds.), Vol. 1, pp. 425–434. Academic Press, New York.

Denton, E. J., and Gilpin-Brown, J. B. (1963). *J. Physiol. (London)* **168**, 44–45.

Denton, E. J., and Gilpin-Brown, J. B. (1966). *J. Mar. Biol. Ass. U.K.* **46**, 723–759.

Dinamani, P. (1964). *Proc. Malacol. Soc. London* **36**, 1–5.

Donovan, D. T. (1964). *Biol. Rev.* **39**, 259–287.

Driscoll, E. G. (1964). *J. Paleontol.* **38**, 61–66.

Feliksiak, S. (1959). *Bull. Acad. Pol. Sci., Ser. Sci. Biol.* **7**, 427–431.

Fischer-Piette, E. and Franc, A. (1960). *In* "Traité de Zoologie" (P.-P. Grassé, ed.), Vol. 5, Part 2, pp. 1654–1700. Masson, Paris.

Fischer-Piette, E., and Franc, A. (1968). *In* "Traité de Zoologie" (P.-P. Grassé, ed.), Vol. 5, Part 3, pp. 987–1017. Masson, Paris.

Franc, A. (1968). *In* "Traité de Zoologie" (P.-P. Grassé, ed.), Vol. 5, Part 3, pp. 1–893. Masson, Paris.

Fretter, V. (1967). *Proc. Malacol. Soc. London* **37**, 357–366.

Fretter, V. (1969). *Proc. Malacol. Soc. London* **38**, 375–386.

Fretter, V., and Graham, A. (1962). "British Prosobranch Molluscs." Ray Society, London.
Fretter, V., and Montgomery, M. C. (1968). *J. Mar. Biol. Ass. U.K.* **48**, 498–520.
Ghiselin, M. T. (1966). *Evolution* **20**, 337–348.
Glaessner, M. F. (1962). *Biol. Rev.* **37**, 467–494.
Harry, H. W. (1968). *Mar. Biol. Ass. India. Symp. Mollusca* **1**, 42–43.
Horný, R. (1965). *Sb. Narodni Muz. Praze* **21**, 57–70.
Hubendick, B. (1958). *Ark. Zool.* [2] **11**, 31–36.
Hunter, W. R., and Brown, S. C. (1965). *Biol. Bull.* **128**, 508–521.
Hyman, L. H. (1951). "The Invertebrates: Platyhelminthes and Rhynchocoela. The Acoelomate Bilateria." McGraw-Hill, New York.
Hyman, L. H. (1967). "The Invertebrates: Mollusca I." McGraw-Hill, New York.
Johnson, K. M. (1969). *Veliger* **11**, 272–276.
Jørgensen, C. B. (1966). "Biology of Suspension Feeding." Pergamon, Oxford.
Knight, J. B. (1952). *Smithson. Misc. Coll.* **117**, 1–56.
Knight, J. B., and Yochelson, E. L. (1960). *In* "Treatise on Invertebrate Paleontology," (R. C. Moore, ed.), (I) Mollusca 1, pp. 77–84. Univ. of Kansas Press, Lawrence.
Knight, J. B., Cox, L. R., Keen, A. M., Batten, R. L., Yochelson, E. L., and Robertson, R. (1960). *In* "Treatise on Invertebrate Paleontology" (R. C. Moore, ed.), Vol. 1, pp. 169–310. Univ. of Kansas Press, Lawrence.
Lang, A. (1896). "Text-book of Comparative Anatomy." Part 2. Macmillan, London.
Lemche, H. (1962). *Scientia (Milan)* [6] Jan., 1–8.
Lemche, H., and Wingstrand, K. G. (1959). *Galathea Rep.* 3, 9–71.
Lemche, H., and Wingstrand, K. G. (1960). *In* "Traité de Zoologie" (P.-P. Grassé, ed.), Vol. 5, Part 2, pp. 1787–1821. Masson, Paris.
McAlester, A. L. (1966). *Malacologia* **3**, 433–439.
McLean, J. H. (1962). *Proc. Malacol. Soc. London* **35**, 23–26.
Mapstone, G. M. (1970). *Helgolaender Wiss. Meeresunters.* **20**, 565–575.
Marek, L., and Yochelson, E. L. (1964). *Science* **146**, 1674–1675.
Martin, A. W., and Harrison, F. M. (1966). *In* "Physiology of Mollusca" (K. M. Wilbur and C. M. Yonge, eds.), Vol. 2, pp. 353–386. Academic Press, New York.
Meenakshi, V. R., Hare, P. E., Watabe, N., Wilbur, K. M., and Menzies, R. J. (1970). *Anton Bruun Rep.* No. 2, pp. 3–12.
Menzies, R.-J. (1968). *Mar. Biol. Ass. India Symp. Mollusca* **1**, 1–9.
Milburn, W. P. (1960). *Veliger* **3**, 43–47.
Morton, J. E. (1959). *J. Mar. Biol. Ass. U.K.* **38**, 225–238.
Morton, J. E. (1963). *Proc. Linn. Soc. London* **174**, 53–72.
Morton, J. E. (1967). "Molluscs." Hutchinson University Library, London.
Morton, J. E., and Yonge, C. M. (1964). *In* "Physiology of Mollusca" (K. M. Wilbur and C. M. Yonge, eds.), Vol. 1, pp. 1–58. Academic Press, New York.
Mutvei, H. (1957). *Ark. Mineral. Geol.* **2**, 219–254.
Mutvei, H. (1963). *Palaeontol. Z.* **37**, H. 1–2, 16.
Mutvei, H. (1964). *Ark. Zool.* [2] **16**, 221–278.
Pilsbry, H. A. (1892–1894). *Manual Conchol.* **14**, 1–350; **15**, 1–133.
Portmann, A. (1960). *In* "Traité de Zoologie" (P.-P. Grassé, ed.), Vol. 5, Part 2, pp. 1625–1654. Masson, Paris.
Purchon, R. D. (1968). "The Biology of the Mollusca." Pergamon, Oxford.
Rasetti, F. (1954). *J. Paleontol.* **28**, 59–66.

Raven, C. P. (1958). "Morphogenesis: The Analysis of Molluscan Development." Pergamon, Oxford.
Riedl, R. J. (1969). *Science* **163**, 445–452.
Rollins, H. B. (1969). *J. Paleontol.* **43**, 136–140.
Rollins, H. B., and Batten, R. L. (1968). *Paleontology* **11**, 132–140.
Salvini-Plawen, L. (1967). *Z. Morphol. Oekol. Tiere* **59**, 318–340.
Salvini-Plawen, L. (1968a). *Syst. Zool.* **17**, 192–208.
Salvini-Plawen, L. (1968b). *Mar. Biol. Ass. India. Symp. Mollusca* 3, 248–256.
Salvini-Plawen, L. (1968c). *Sarsia* **31**, 105–126.
Salvini-Plawen, L. (1968d). *Sarsia* **31**, 131–142.
Salvini-Plawen, L. (1969a). *Mitt. Zool. Ges. Braunau* **1**, 85–98.
Salvini-Plawen, L. (1969b). *Malacologia* **9**, 191–216.
Schmidt, W. J. (1959). *Galathea Rep.* **3**, 73–77.
Simpson, G. G. (1953). "The Major Features of Evolution." Columbia Univ. Press, New York.
Smith, A. G. (1960). *In* "Treatise on Invertebrate Paleontology," (I) Mollusca 1 (R. C. Moore, ed.), pp. 41–76. Univ. of Kansas Press, Lawrence.
Smith, A. G. (1966). *Proc. Calif. Acad. Sci.* **32**, 433–446.
Starobogatov, Y. I. (1970). *Paleontol. J.* **4**, 293–302.
Stasek, C. R. (1961). *Proc. Zool. Soc. London* **137**, 511–538.
Stasek, C. R. (1963a). *J. Morphol.* **112**, 215–231.
Stasek, C. R. (1963b). *Veliger* **6**, 91–97.
Stasek, C. R. (1965). *Malacologia* **2**, 349–366.
Stasek, C. R. (1966). *Malacologia* **5**, 67–68.
Stasek, C. R. (1972). *Veliger* In press.
Steinbock, O. (1963). *Verh. Deut. Zool. Ges., Zool. Anz., Suppl.* **26**, 385–403.
Swedmark, B. (1956). *Not. and Rev.* **2**, 70–97.
Tebble, N. (1967). *Nature (London)* **215**, 663–664.
Teichert, C. (1967). *In* "Essays in Paleontology and Stratigraphy" (C. Teichert and E. L. Yochelson, eds.), Spec. Publ. 2, pp. 162–210. Univ. of Kansas Press, Lawrence.
Termier, G., and Termier, H. (1968). *In* "Traité de Zoologie" (P.-P. Grassé, ed.), Vol. 5, Part 3, pp. 894–925. Masson, Paris.
Thompson, T. E. (1960). *Proc. Roy. Soc., Ser. B* **153**, 263–278.
Thompson, T. E. (1967). *Malacologia* **5**, 423–430.
Trueman, E. R. (1964). *In* "Approaches to Paleoecology" (J. Imbrie and N. Newell, eds.), pp. 45–74. Wiley, New York.
Trueman, E. R. (1966). *Science* **152**, 523–525.
Trueman, E. R. (1968). *J. Zool.* **154**, 19–27.
Trueman, E. R., and Packard, A. (1968). *J. Exp. Biol.* **49**, 495–507.
Trueman, E. R., Branch, A. R., and Davis, P. (1966). *J. Exp. Biol.* **44**, 469–492.
Vagvolgyi, J. (1967). *Syst. Zool.* **16**, 153–168.
von Knorre, H. (1925). *Jena. Z. Naturwiss.* **61**, (new ser. 54), 469–632.
Wells, M. J. (1966a). *In* "Physiology of Mollusca" (K. M. Wilbur and C. M. Yonge, eds.), Vol. 2, pp. 523–545. Academic Press, New York.
Wells, M. J. (1966b). *In* "Physiology of Mollusca" (K. M. Wilbur and C. M. Yonge, eds.), Vol. 2, pp. 547–590. Academic Press, New York.
Wilbur, K. M. (1964). *In* "Physiology of Mollusca" (K. M. Wilbur and C. M. Yonge, eds.), Vol. 1, pp. 243–282. Academic Press, New York.
Yochelson, E. L. (1966). *U.S., Geol. Surv., Prof. Pap.* **523-B**, 1–9.

Yochelson, E. L. (1967). *In* "Essays in Paleontology and Stratigraphy" (C. Teichert and E. L. Yochelson, eds.), Spec. Publ. 2, pp. 141–161. Univ. of Kansas Press, Lawrence.

Yochelson, E. L. (1969). *Lethaia* **2**, 49–62.

Yonge, C. M. (1939a). *Quart. J. Microsc. Sci.* **81**, 367–390.

Yonge, C. M. (1939b). *Phil. Trans. Roy. Soc. London, Ser. B* **230**, 79–147.

Yonge, C. M. (1947). *Phil. Trans. Roy. Soc. London, Ser. B* **232**, 443–518.

Yonge, C. M. (1953). *Trans. Roy. Soc. Edinburgh* **62**, 443–478.

Young, J. Z. (1970). *Phil. Trans. Roy. Soc. London, Ser. B* **257**, 309–321.

CHAPTER 2

Structure of the Molluscan Shell

C. Grégoire

I. Introduction

Excellent monographs in the past concerned the shell and pearl structures in individual families and species—*Nautilus, Spirula,* and *Sepia* (Owen, 1832; Appellöf, 1892–1893); various gastropods and bivalves, including Mytilidea (Tullberg, 1882; Ehrenbaum, 1885); Moynier de Villepoix, 1892; List, 1902); *Anodonta* (Moynier de Villepoix, 1892; Rassbach, 1912); *Margaritana,* and *Meleagrina* (Römer, 1903; Rubbel, 1911); the Ceylon pearl oyster *Margaritifera vulgaris,* shell and pearls (Jameson, 1912); and the shells of the Viviparidae (Kessel, 1933). The monumental review of Fritz Haas (1929–1935) in Winterstein's Handbuch furnishes a wealth of information on the literature prior to 1935. The considerable work of W. J. Schmidt (1921a,b, 1923, 1924, 1928, 1931, 1932a,b) contributed greatly to the understanding of many structures so far incorrectly interpreted. Bøggild (1930) by his classification of the shell structures types, clarified the extreme diversity and complexity of the architecture of the molluscan shell.

Within the last 25 years, and especially recently, the molluscan shell has been selected as an ideal tool for the study of the processes of biomineralization, predominantly focussed for many years on bones and teeth. The results collected during that period have been recently summarized in several reviews and papers on the physiological, biochemical, molecular, and crystallographic aspects of calcification, shell formation, and regeneration (e.g., Watabe and Wada, 1956; Wilbur, 1960, 1964; Abolinš-Krogis, 1963; Wilbur and Simkiss, 1968; Bevelander and Nakahara (1969a), on the ultrastructure of the organic matrices (Grégoire *et al.*, 1955; Grégoire, 1967b) and on the mineral architecture in various groups (McClintock, 1967; Taylor *et al.*, 1969; Kennedy *et al.*, 1969; Wise, 1970).

Recent monographs concern the shells of *Nautilus* (Grégoire, 1962; Mutvei, 1964, 1969, 1970), bivalves (Grégoire, 1960; Oberling, 1964; Taylor *et al.*, 1969; Kennedy *et al.*, 1970 Wise, 1971), the American oyster (Galtsoff, 1964), Muricidae (Petitjean, 1965), and patelloid and bellerophontoid gastropods (MacClintock, 1967). Wilbur and Simkiss (1968) have reviewed the background information on the presence of polysaccharides and mucopolysaccharides in the shell.

The results so far obtained with the scanning electron microscope promise a rapid and considerable development of the researches in the field of molluscan shell architecture. However, owing to the relatively poor resolution of this instrument at the present time, it remains necessary to use the transmission electron microscope when details of ultrastructure needing high resolution must be recorded.

II. Types of Mineral Architecture and Organic Ultrastructure in Molluscan Shells

In a systematic study with the polarization microscope of the different modes of crystalline aggregation in molluscan shells, Bøggild (1930) recognized eight types of aragonitic or calcitic structures. Bøggild's classification has been used with various modifications in recent works (MacClintock, 1967; Oberling, 1964; Taylor *et al.*, 1969; Kennedy *et al.*, 1969).

The shells of several families of mollusks are characterized by the presence in the same shell of more than one of these types of structure.

They consist, for instance, of an upper prismatic layer, and a lower aragonitic layer. An uncalcified structure, the periostracum, covers the prismatic layer. This association of different layers in the same shell had already been described by Moynier de Villepoix (1892) in the shell of *Anodonta.* The diagram of this architecture, established by this author, has been reproduced, more or less modified, in several handbooks as representing the typical architecture of a molluscan shell. However, this arrangement in two calcified layers covered with an organic periostracum characterizes only the shells from a few families of bivalves—Pinnidae, Aviculidae, Pernidae, Mytilidae, Vulsellidae, Nayadidae, Margaritanidae, and Trigoniidae.

A previously undescribed calcareous layer (mosaicostracum) has been recently detected by Hamilton (1969) between the periostracum and the outermost shell layer of bivalves and of some gastropods. This layer displays distinct surface patterns, which appear to be taxonomically consistent at the species level. According to Hamilton, the ultrastructure of the mosaicostracum is a potential tool for specific identification and for the study of evolution in living and fossil bivalves.

A. The Homogeneous Structure

According to Bøggild (1930), this kind of aggregation, examined in the conventional light microscope, is amorphous in its typical form. This structure is especially developed in the upper layer of the Limidae and in some layers of the Mytilidae. It includes portions of shells which contain traces of other structural types—crossed lamellar, prismatic, foliated, and grained (see below).

In the descriptions of Taylor *et al.* (1969) and Kennedy *et al.* (1969), the homogeneous structure is always aragonitic. It consists of minute carbonate granules, all with a similar crystallographic orientation. In the electron microscope, the granules appear as elongate, lenticular, or irregular blocks of carbonate.

B. The Prismatic Structure

In several mollusks, especially in bivalves, the prismatic structure forms the middle layer of the shell wall, between the outer fibrolamellar periostracum and the inner nacreous layer. The prisms are elongated mineral columns, (Fig. 1), single or branched, polygonal or showing indented boundaries in transverse section. The prisms are wrapped in sheaths of organic substance. They are disposed parallel, side by side, at right angles, or obliquely to the nacreous lamellae. The orientation of their axes and that of their optic axes in relation to the shell varies

Figs. 1–4. *Atrina* (*Pinna*) *nigra* Durhenn (Bivalvia, Pinnidae).

greatly. Most commonly, their axes are normal to the surface of the shell. However, in many instances they have an oblique position.

For more than a century, the structure of the prisms and the nature of their mineral components have been extensively studied with polarization and conventional microscopes (Carpenter, 1844, 1848; Rose, 1858; von Nathusius-Königsborn, 1877; Moynier de Villepoix, 1892; Tullberg, 1882; List, 1902; Biedermann, 1902a,b, 1914; Römer, 1903; Bütschli, 1908; Rubbel, 1911; Rassbach, 1912; Karny, 1913; W. J. Schmidt, 1921a,b, 1924, 1932a,b), by X-ray diffraction analysis (Tsutsumi, 1928, 1929), and more recently with transmission and scanning electron microscopes (Watabe and Wada, 1956; Tsujii *et al.*, 1958; Wada, 1961a,b; Grégoire, 1961a,b; Watabe, 1965; Travis *et al.*, 1967; Travis, 1968a,b; Travis and Gonsalves, 1969; Wise, 1970).

In his crystallographic studies, Bøggild (1930) divided the prismatic structures into three groups: the normal (simple prismatic structure, Taylor *et al.*, 1969), the complex, and the composite structures (composite prisms, Taylor *et al.*, 1969). In the normal, or simple prismatic structure, the prisms are calcitic (e.g., in many gastropods and marine bivalves) or aragonitic (e.g., in Unionacea and Trigonacea). The calcitic prisms have a transverse striation, the aragonitic prisms a transverse and a longitudinal diverging striations, which give them a feathery appearance (W. J. Schmidt, 1921a,b, 1924, 1931; Grégoire, 1961b; Taylor *et al.*, 1969). On the polygonal surfaces of the prisms, the electron microscope reveals a mosaic of microcrystals variously oriented in the different species of mollusks (Watabe and Wada, 1956; Tsujii *et al.*, 1958; Grégoire, 1961b, Figs. 25, 36–38, and 41; Watabe, 1965).

In the prisms of *Pinctada martensii*, Watabe and Wada (1956) described superimposed carbonate lamellae separated by conchiolin sheets. Observation of the successive stages of a decalcification process in *Atrina nigra* has shown that the prisms are composed of mineral disks or lenticular segments piled upon each other and associated with organic material (Grégoire, 1961b, Pl. 4, Figs. 5 and 6) (Figs. 2 and 3). In *Margaritana margaritana*, Taylor *et al.* (1969) observed division of the

---

FIG. 1. Three fragments of pink prisms after boiling for 15 minutes in glycerin–potassium hydroxide. The residues of the destroyed organic sheaths appear as an amorphous substance (arrow) (× 300). (Grégoire, 1961d.)

FIGS. 2 and 3. Prisms in the process of decalcification (titriplex). Successive steps of fragmentation of the mineral prisms—transverse circular notches, lenticular segments composed of stacked disks. Phase contrast (× 300). (Grégoire, 1961b, Figs. 15 and 16.)

FIG. 4. Broken, flattened fragments of the organic sheaths from decalcified pink prisms. The irregular, transverse striation is shown on many fragments. Phase contrast (× 140). (Grégoire, 1961.)

prisms into longitudinal blocks, intersecting the conchiolin prism wall obliquely. This type of structure had also been observed in *Atrina nigra* by Grégoire (1961b, Figs. 39 and 40).

According to Bøggild (1930) the complex or composite prismatic structure, always aragonitic, consists (e.g., in Nuculacea) of prisms of the first order, each of them composed of prisms of the second order arranged in a feathery manner and which radiate outward from the axis of the first order prisms. This type of prism has been recently restudied by Taylor *et al.* (1969) and Kennedy *et al.* (1969) in the bivalve *Codakia.*

In the shell of *Mytilus edulis,* the prisms are thin calcitic needles. The size of these needles is about a hundred times smaller than that of the prisms of other species, such as those of the Pinnidae (von Nathusius-Königsborn, 1877; Tullberg, 1882; Ehrenbaum, 1885; List, 1902; W. J. Schmidt, 1921a; Grégoire, 1961a,b).

As shown on thin sections by Travis (1968a,b) in *Mytilus edulis,* by Travis and Gonsalves (1969) in *Mytilus, Crassostrea,* and *Mercenaria,* the mineral prisms are composed of well ordered parallel rows of microcrystals rectangular in shape. The three-dimensional morphological orientation of these microcrystals coincides with a well defined three-dimensional crystallographic orientation. The prisms are wrapped in organic sheaths (von Nathusius-Königsborn, 1877; Moynier de Villepoix, 1892). After decalcification of the prisms, these sheaths appear in the form of soft transparent or rigid tubes, polygonal in section (Grégoire, 1961b, Figs. 2–17), and showing a transverse striation (Fig. 4). Protracted ultrasonic disintegration of these sheaths in *Turbo* leaves fibrils mixed with an unidentified amorphous substance (Grégoire, 1961b). Some of these fibrils seem to be composed of filaments in which spherical granules are alined.

In the marginal area of prism growth—*Pinctada martensii* (Watabe and Wada, 1956; Wada, 1957b, 1961a), *Crassostrea virginica* (Tsujii *et al.,* 1958), *Turbo* sp. and twenty-one species of bivalves (Grégoire, 1961b), and in transverse sections (replicas of polished and etched surfaces) (Grégoire, 1961b)—the organic sheaths appear in the form of flat or bulging cords surrounding the mineral polygons. In certain complex prisms (Karny, 1913; W. J. Schmidt, 1921a; Watabe and Wada, 1956), these polygons are subdivided into chambers.

The thin organic sheaths which surround the *Mytilus* prisms (Moynier de Villepoix, 1892) appear in the electron microscope in the form of perforated granular membranes containing fibrils (Grégoire, 1961a,b). These perforated membranes resemble closely the lacelike reticulated sheets which characterize the nacreous conchiolin of the Mytilidae

(Grégoire *et al.*, 1955; Grégoire, 1957a, 1960, 1961a,b). Travis *et al.* (1967) confirmed the identity of structure between nacreous and prism conchiolins in the shell of *Mytilus*. As shown on thin sections by Travis (1968a,b) and Travis and Gonsalves (1969), the organic intraprismatic matrix which surrounds the microcrystals (see above) is a striking fingerprint of the mineralized matrix. It is organized into closely packed sheet-like compartments and subcompartments in which the crystals are deposited. This intraprismatic organic matrix is chiefly composed of filaments which average 28 Å in width.

On the other hand, according to Uozumi and Iwata (1969) in *Mytilus coruscus*, the structure of the prism sheaths is amorphous; therefore entirely different from that of nacreous conchiolin. The latter is identical to that described by Grégoire (1961a,b) in *Mytilus edulis*.

The transverse striation, or growth lines, which are visible on the lateral facets of the mollusk prisms and of those of their sheaths results from discontinuous secretion of conchiolin, which alternates with deposition of crystalline material (Moynier de Villepoix, 1892; Biedermann, 1902a,b; Römer, 1903; Bütschli, 1908; Rassbach, 1912; W. J. Schmidt, 1921a,b, 1924). The conceptions concerning the nature of the growth lines (Römer, 1903; Biedermann, 1902a; Bütschli, 1908; Rubbel, 1911; Rassbach, 1912) have been reviewed elsewhere (Grégoire, 1967b). On the basis of observations with the electron microscope (Grégoire, 1961b), the growth lines consist of ring-shaped transverse thickenings of the prism sheaths, which agrees with former suggestions of Rubbel (1911) and of Rassbach (1912). More recent observations (Grégoire, 1967b, Pl. 4, Fig. 3) have shown that these thickenings consist of conchiolin shreds disposed in ring-shaped clusters or strands anchored at irregular intervals on the inner surface of the sheaths. These clusters are separated from each other by smooth or fibrillar surfaces. This organic material represents the marginal remains of the organic layers (intraprismatic matrices) deposited periodically in greater amounts inside the prisms, a process which is accompanied by a temporary decrease or interruption in the deposition of mineral substance. In *Mytilus* (Travis *et al.*, 1967), the intraprismatic matrices would consist of fibrils.

C. The Foliated Structure

In the calcitic shells of Ostreidae, Pectinidae, Anomiidae, and other marine mollusks, a substance called pseudonacre, subnacreous layer, or calcitostracum (W. J. Schmidt, 1924, 1932b) forms the inner layer of the shell valves. Dull or bright, this substance does not present the iridescence of mother-of-pearl. According to W. J. Schmidt (1924, 1932b) and Bøggild (1930) the calcitostracum is built up of more or less regular parallel leaves. In Spondylidae, Pectinidae, Ostreidae, and Anomiidae, the orientation of the leaves is very irregular, sometimes horizontal, sometimes oblique, or quite vertical. The leaves are composed of laths, tablets, blades, reglets, or needles, forming aggregates (W. J. Schmidt, 1932b). In each aggregate, the crystals are parallel, and in the plane of each lamella orientation of the aggregates varies greatly. The crystallographic orientation of the calcite elements in the foliated structure has been studied by Wada (1964b), Tsujii *et al.* (1958), and Taylor *et al.* (1969). In the electron microscope (Watabe *et al.*, 1958; Tsujii *et al.*, 1958; Grégoire, 1958a; Watabe and Wilbur, 1961; Taylor *et al.* 1969) and in the scanning electron microscope (Taylor *et al.*, 1969), the tabular calcite crystals described formerly with the conventional microscope form semi-horizontal layers imbricated like the tiles of a roof (Pl. 3, Fig. 4: Grégoire, 1967b) (Fig. 5). The imbrication of calcite crystals on the surface of the layer results from dendritic growth and the creeping of crystals over adjacent crystals (Watabe *et al.*, 1958; Wilbur, 1960). The calcite crystals consist of aggregates of mosaics of crystalline subunits (Wada, 1963a). Wada (1961a,b, 1963b,c, 1964a,b) studied the orientation of the crystal structures on the inner shell surface in the growth zones of calcitostracum and observed (in *Anomia* and *Ostrea,* but not in *Chlamys,* Wada, 1964a), a spiral arrangement of the calcite crystals forming stepped pyramidal growth hills. These growth hills consist of a spherical aggregation of foliated calcite crystals.

In contrast with the dense and compact mother-of-pearl (see below), lacunes in the mineral substance, filled with fluid in the living animal, are scattered in the calcitostracum (Tullberg, 1882). In *Ostrea,* the calcitostracum also contains irregular areas of a white porous substance, composed of unoriented calcite crystals, which give the inner surface of the valves a chalky appearance. The nature and the function of these chalky deposits have been analyzed in a substantial literature (Tullberg, 1882; Orton and Amirthalingam, 1926–1927; Ranson, 1939, 1940, 1941; Korringa, 1951, 1952; Galtsoff, 1964; see Figs. 4 and 5, Pl. 28, Taylor *et al.*, 1969).

The organic components of the calcitostracum are scarcer than in mother-of-pearl (W. J. Schmidt, 1924) (see page 58). They amount in weight to 0.5–0.6% of the substance (*Ostrea edulis,* Korringa, 1951; *Crassostrea virginica,* Galtsoff, 1964). In contrast with mother-of-pearl, the topography of these organic constituents in the architecture of the calcitostracum is discontinuous (W. J. Schmidt, 1924; Amirthalingam, 1927; Orton and Amirthalingam, 1926–1927).

Orton and Amirthalingam (1926–1927) observed in the shell of *Ostrea edulis* the deposition of thin brown lamels on the inner surface of the valves in autumn at the end of the growth season of the shell. These lamels are later embedded in the mineral substance (Ölflecken: W. J. Schmidt, 1932b). Finely granular organic substances have been also observed by Tsujii *et al.* (1958) on replicas of the inner surfaces of the shell of *Crassostrea virginica.*

The scarce organic residues left by decalcification of the calcitostracum are composed of thin transparent glassy leaflets. These organic shreds are more resistant than nacreous conchiolin to mechanical disintegration by ultrasonic vibrations (Grégoire, 1958a, 1967).

In the electron microscope, these leaflets differ in their appearance from the nacreous conchiolin matrices (see below), and consist of transparent, amorphous or granular veils and membranes, which disintegrate in part into fibrils—*Ostrea edulis* Linné (Grégoire *et al.*, 1955); *Ostrea tulipa* Lamarck, *Anomia ephippium, Ostrea iridescens* (*margaritacea*) Lamarck, *Placuna orbicularis* Retzius; chalky chambers of *Ostrea edulis* (Grégoire, 1958a, 1967b, Pl. 3, Figs. 5, 6 and 7) (Figs. 6 and 7). A fenestration, less characteristic than that described in the tight lacelike nacreous conchiolin matrices of bivalves (see below) has been occasionally observed in some membranes resulting from decalcification of the shells of *Ostrea edulis, O. tulipa, O. iridescens,* and *Anomia ephippium* (Grégoire, 1958d). A conchiolin membrane showing an oriented network with spacings of 60–80 Å has been observed by Watabe and Wilbur (1961) in *Crassostrea virginica.*

All these organic residues of the calcitostracum appear to be essentially composed of fibrillar networks in which the fibrils, 28 Å in average diameter, are frequently concealed and embedded in an unidentified amorphous substance (Grégoire, 1958a, 1967b, Pl. 3, Fig. 5). In ultrathin sections of the shell of *Crassostrea virginica,* Watabe (1965) reported the absence of interlamellar matrices and the presence of pericrystalline matrices in the form of thin membranes. The latter structures probably represent the veils and membranes detected previously between the calcite crystals on replicas of surfaces of fracture of the calcitic shells and of *Placuna* pearls (Grégoire, 1958a) (Fig. 8). According to Watabe (1965), intracrystalline membranes separate the subunits in the tabular calcite crystals of the calcitostracum.

D. The Nacreous Structure

Mother-of-pearl forms the innermost layer of the shell in several families of mollusks, and is never found outside the mollusks (Bøggild, 1930). In bivalves, the nacreous substance characterizes the Mytilidae, Pteriidae, Pinnidae, Pernidae, Vulsellidae, Unionidae, Margaritanidae, Trigoniidae, Mutelidae, Aetheriidae, and Anatinidae. In gastropods, it

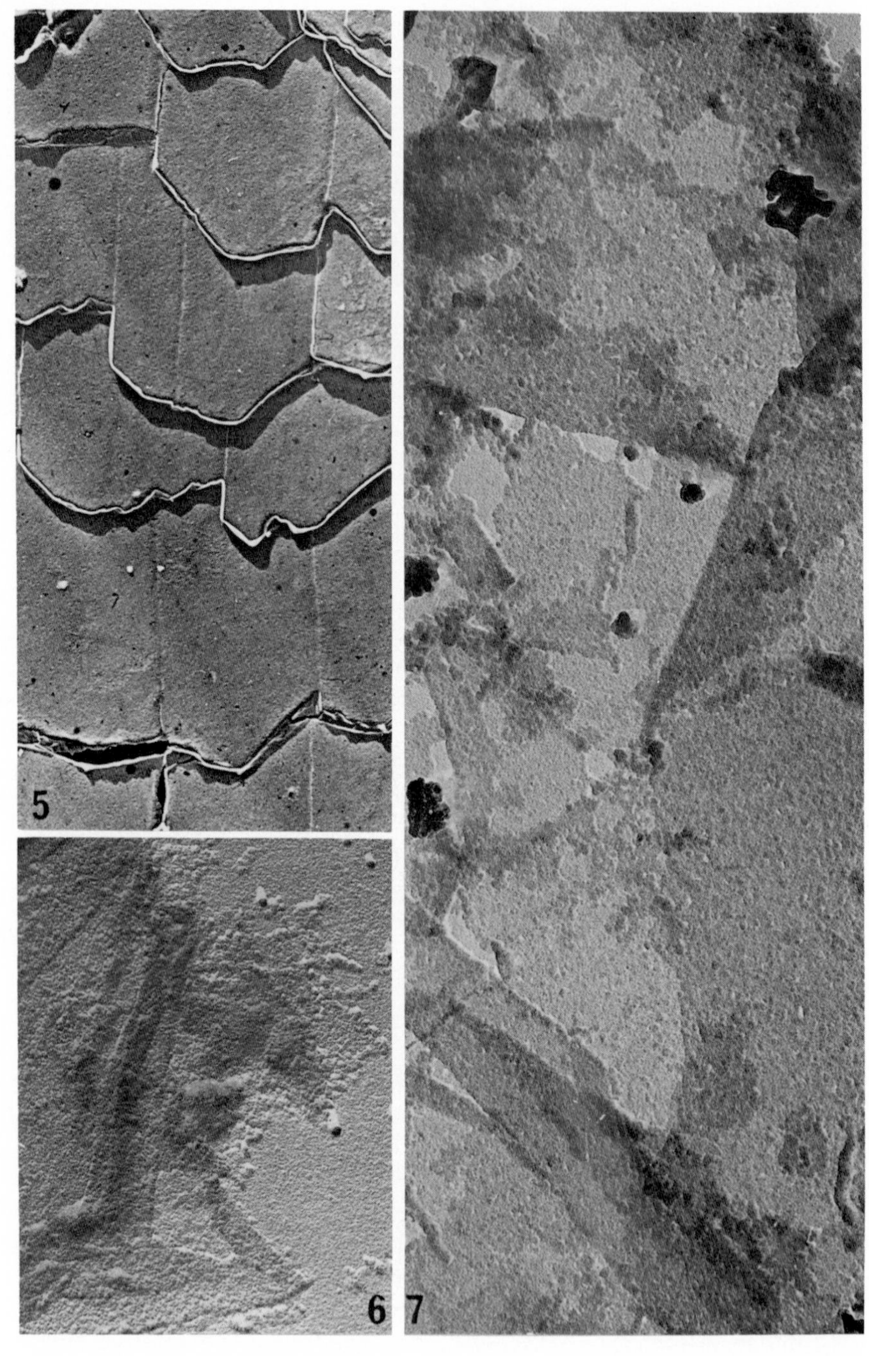
5
6
7

is found in the Haliotidae, Stomatiidae, Turbinidae, Trochidae, Umboniidae, and Angariidae. In cephalopods, mother-of-pearl is found in the inner layer of the shell wall and in the septa of *Nautilus* and of the ammonoids, and only in the septa of *Spirula.*

Mother-of-pearl (W. J. Schmidt, 1923, 1924, 1928), always aragonitic, consists of numerous mineral leaves or lamellae, parallel to the inner surface of the shell, and superimposed horizontally. The lamellae are composed of tabular, euhedral, or rounded crystals of aragonite disposed in a single layer. The weakly developed *c*-axis of the crystals is oriented at right angles to the inner depositional surface. In tangential view, a lamella appears as a flagging of polygonal slabs. The lamellae alternate with sheets of organic matter (interlamellar conchiolin, see below). The still incompletely known mechanism of alternation of lamellae and of organic matrices has been discussed by Wilbur and Simkiss (1968) and by Towe and Hamilton (1968a,b).

The architecture of mother-of-pearl has been compared to brickwork (Ehrenbaum, 1885) in which the crystals correspond to the bricks and the conchiolin matrices to the concrete (Fig. 9). The number of leaves varies with the species, between 450 and 5000 per millimeter of thickness of mother-of-pearl (Pfund, 1917; Neumann, 1927; W. J. Schmidt, 1923, 1928; Ahrberg, 1935; Wada, 1961a,b; Watabe, 1963, 1965). As in brickwork, the limits of the crystals do not coincide in adjacent lamellae (sheet nacre, Taylor *et al.*, 1969), except in gastropods (e.g., see Wise and Hay, 1968; Wise, 1970), in *Nautilus* (von Nathusius-Königsborn, 1877; Biedermann, 1902a; W. J. Schmidt, 1923, 1924; Ahrberg, 1935; Grégoire, 1962; Erben *et al.*, 1968; Mutvei, 1969, 1970; Wise, 1970) and in the middle shell layer of Nuculacea, Unionacea, Trigonacea, and Pandoracea (Taylor *et al.*, 1969). In these shells, the crystals are piled upon each other like coins. As revealed with the scanning electron microscope, in a growth surface of a shell characterized by this columnar arrangement (Wise and Hay, 1968, Figs. 12 and 13) the stacks of crystals appear much taller than would be expected from data obtained with

FIG. 5. *Ostrea iridescens* (Bivalvia, Ostreidae). Positive carbon replica, palladium shadowed, of the polished and etched inner surface of a valve, showing the overlapping sheets of calcite crystals (× 14,000). (Grégoire, 1958a, 1967b.)

FIG. 6. *Ostrea edulis* Linné (Bivalvia, Ostreidae). Organic matrix from the calcitostracum consisting of thin granular membranes or veils. Pores absent or not visible. The edge of the fragment appears dissociated into nodular particles. Platinum shadowed (× 48,000). (Grégoire, 1958a, 1967b.)

FIG. 7. *Placenta* (*Placuna*) *orbicularis* Retzius (Bivalvia, Anomiidae). Calcitostracum from the inner portion of a valve. Organic matrix, stained with osmium tetroxide. Fragments of granular membranes. Pores absent of obliterated (above). Platinum shadowed (× 42,000). (Grégoire, 1958a, 1967b.)

Fig. 8. Pearls of *Placenta* (*Placuna*) *orbicularis* Retzius ("pearls of Trincomalee," Ceylon). Positive carbon replica, palladium shadowed, of a polished and etched section of a pearl, showing about forty superimposed mineral sheets built up of large laths. Note the large irregularities in the thickness of the crystals in the successive sheets. The organic substance is scattered between the laths in the form of folded shreds (× 16,400). (Grégoire, 1957b.)

FIG. 9. *Unio rectus* Lamarck (Bivalvia, Unionidae). Direct, carbon–platinum replica of a surface of transverse fracture of mother-of-pearl. Traces of the crystalline subunits and of the interlamellar conchiolin matrices appear on the 001 planes of the tabular aragonite crystals (× 21,000). (Grégoire, 1967b.)

the aid of the transmission electron microscope (See Grégoire, 1962, Figs. 8, 9, and 45).

The arrangement of the aragonite crystals in the growth regions of the inner shell surface of several bivalves has been studied by W. J. Schmidt (1924) and by Wada (1957a,b, 1958, 1960a,b, 1961a,b, 1966a). As described by Wada (1966a, 1968b,c), three different crystallographic types characterize the formation of a new lamella in mother-of-pearl of *Pinctada fucata* (spiral, close, and loop patterns). These patterns, observed frequently on the (001) face of the aragonite crystals suggest that the single crystals have grown by the spiral mechanism around screw dislocation. In other instances, single crystals seem to grow by epitaxial settlement of many minute crystallites on the basal face of a large crystal (see Fig. 3 in *Nautilus*, Grégoire, 1962) or by coalescence in parallel orientation on interlamellar matrix. The aragonite crystals are composed of parallel subunits (Grégoire, 1966b; Grégoire *et al.*, 1969) (Fig. 10) or aragonitic laths oriented parallel to the crystallographic *a*-axes (Mutvei, 1970).

The organic components of mother-of-pearl (nacreous conchiolin) were traditionally described with the conventional microscope as consisting of a linear network of extremely thin interlamellar sheets held together by transverse bridges (intercrystalline sheets) (Tullberg, 1882; Ehrenbaum, 1885; Moynier de Villepoix, 1892; List, 1902; Römer, 1903; Rubbel, 1911; Rassbach, 1912; Jameson, 1912; W. J. Schmidt, 1923, 1924). The electron microscope confirmed the interlamellar and intercrystalline topography of the conchiolin components in the nacre architecture (Grégoire, 1957a; Wada, 1961a,b; Towe and Hamilton, 1968a,b). The presence or absence of an intracrystalline conchiolin has been a matter of controversy in the recent years (see Wada and Sakai, 1963; Watabe, 1965; Towe and Hamilton, 1968a). There is now a general agreement that an intracrystalline matrix does exist in the aragonite crystals. According to Mutvei (1970), this intracrystalline conchiolin would wrap the crystallites of small size from which the large tabular crystals are composed. In the electron microscope, the organic sheets freed by decalcification of the nacreous layer appear in the form of lacelike networks of trabeculae, delimiting a fenestration or pore patterns (Grégoire *et al.*, 1949, 1950, 1955; Grégoire, 1957a, 1958b,c, 1960, 1962, 1967b). The interlamellar conchiolin matrices form mosaics of polygonal fields delimited by cords of intercrystalline conchiolin. These structures reveal the outlines of the original polygonal crystals between which the matrices were sandwiched before decalcification and were called "crystal imprints" (Grégoire, 1959a,b, 1962, 1966b; Grégoire and Teichert, 1965) or "crystal scars" (Mutvei, 1969).

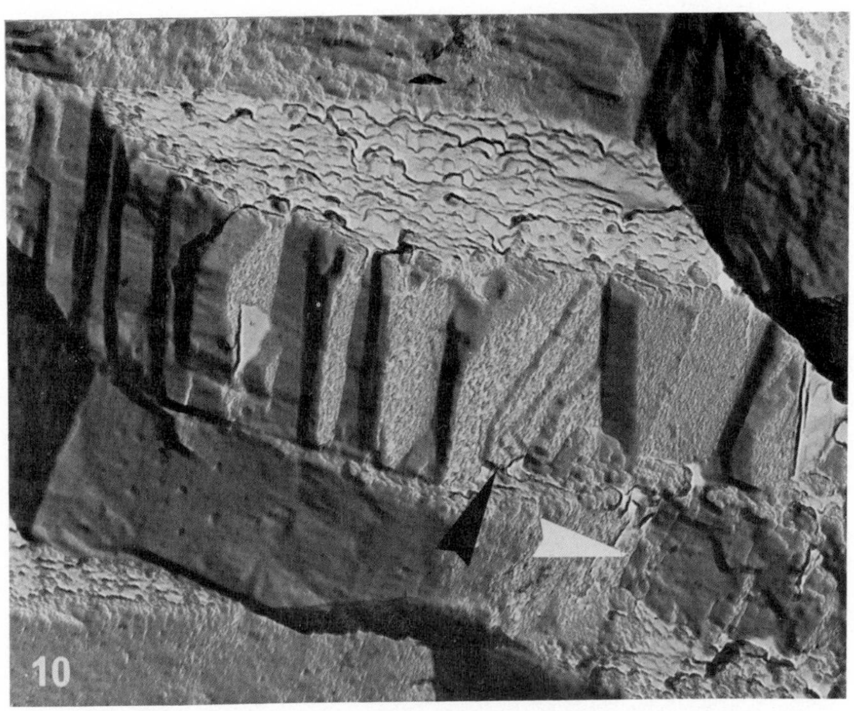

FIG. 10. *Unio rectus* Lamarck (Bivalvia, Unionidae). Direct carbon platinum replica of a surface of transverse fracture of mother-of-pearl, showing traces of the interlamellar matrix undulating over crystalline subunits on the 001 plane of a tabular crystal of aragonite. On the surface of transverse fracture of the same crystal, parallel oriented cliffs reveal its inner architecture (black arrow). Note, in the lamella below, the different orientation of the cleavage planes in an adjacent crystal (white arrow) (× 30,000).

The structure of these nacreous conchiolin matrices differs in taxonomic groups of mollusks. Three patterns of structure (nautiloid, gastropod, and pelecypod) based on differences in the size, form of the trabeculae, average diameter, frequency of distribution, and relative surfaces of the openings or pores in the fenestration have been recognized in nacreous conchiolin. Measurements of these details of structure showed that the differences recorded in the patterns were statistically characteristic at the class level of taxonomy. Examination of the reticulated sheets of conchiolin in thirty specimens from nine families of bivalves revealed significant differences between the patterns on a subclass scale (family) (Grégoire, 1960: cf. Figs. 13, 14, and 15).

In *Nautilus* (two species), the reticulated sheets of conchiolin are characterized by a rather loosely knit texture, sturdy, irregularly cylindri-

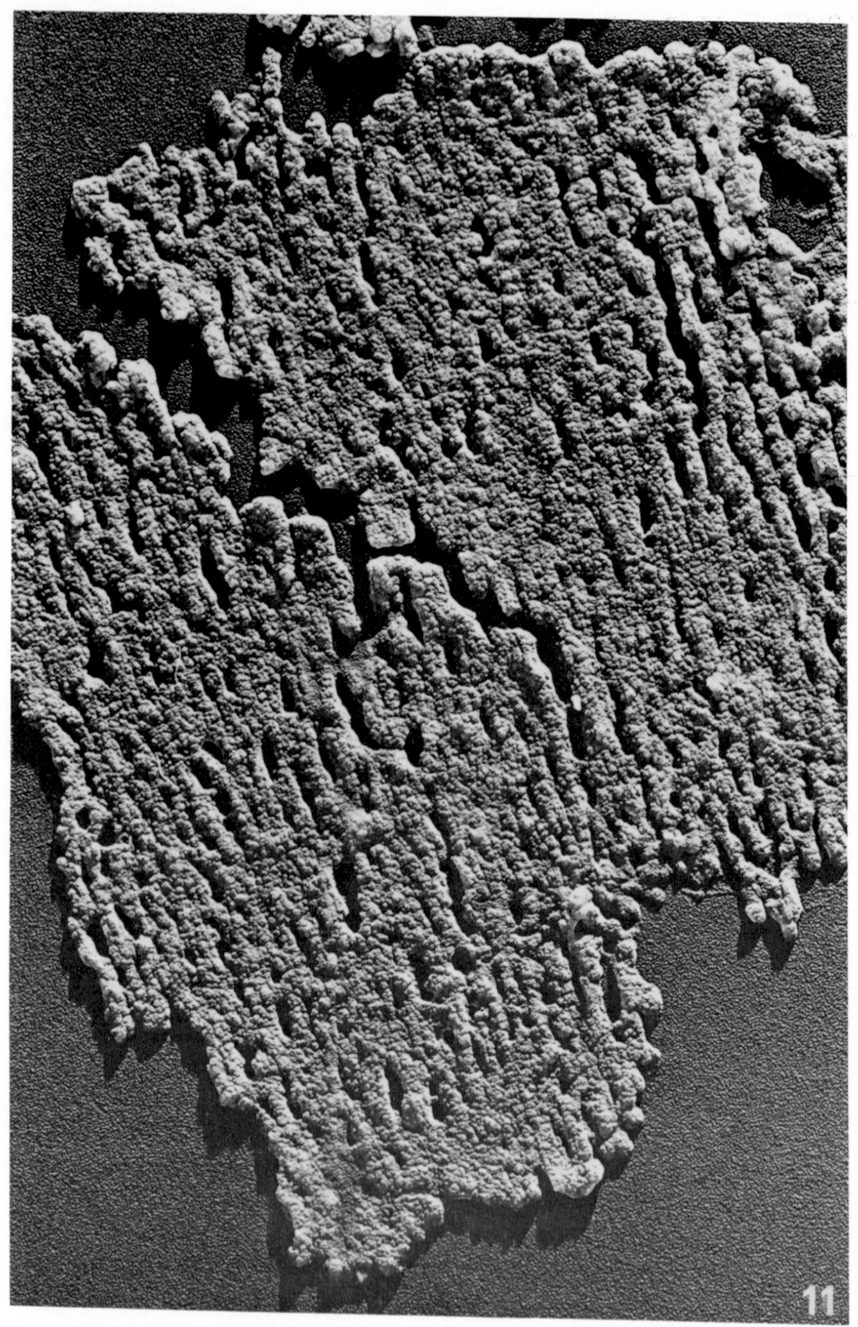

FIG. 11. *Nautilus pompilius* Linné (Cephalopoda, Nautiloidea). Decalcified mother-of-pearl. Fragments of interlamellar matrix (nautiloid pattern) stained with osmium tetroxide and palladium shadowed (× 43,000). (Grégoire, 1962, Fig. 39).

cal trabeculae sprinkled with hemispheral protuberances or tuberosities and by a broad, frequently elongated fenestration (Fig. 11). In several families of gastropods (Angariidae, Stomatiidae, Turbinidae, Umboniidae, Trochidae, Haliotidae), the trabeculae are less robust, and appear frequently in the form of flattened ribbons. The openings of the fenestration are rounded and more regularly disposed than those in *Nautilus* (Figs. 12 and 13). In the pelecypod pattern, the reticulum is more compact than in the two other patterns. The trabeculae are more slender and twisted. The openings of the fenestration, of irregular outlines and size and generally much smaller than the holes in the two other patterns, are unequally dispersed in the sheets (Figs. 14–16). In Pteriidae and in Mutelidae (bivalves), the conchiolin sheets show some resemblance to those of the gastropods studied. However, measurements indicate differences in structure between the patterns in these bivalves and that of the gastropods that are statistically significant with the criteria used (see details in Grégoire, 1960).

The organic residues left by decalcification of the aragonitic pearls of *Pinctada martensii* Dunker are more resistant to mechanical disintegration than the normal shell nacreous conchiolin. The pearl conchiolin showed the characteristic pattern of the conchiolin from the nacreous layer of the same species (Grégoire *et al.*, 1955, Pl. 21).

Fragments of shells, collected in graves from archaeological sites (Tikal, Maya I) (Grégoire, 1960) and too small to be recognizable by standard zoological criteria, could be identified on the family scale by the patterns characterizing this family in the conchiolin residues of mother-of-pearl. These observations suggest that the electron microscope could bring useful information into this field of investigation.

The structure characteristics of the pelecypod pattern have been confirmed by Tanaka *et al.* (1963) in *Pinctada,* and by Travis *et al.* (1967) in *Mytilus.* On the other hand, according to Mutvei (1969), there is no fenestration in the nacreous conchiolin matrices. Thin membranes bridge the intertrabecular areas. Instead of three patterns, Mutvei describes two main structure types of conchiolin in mollusks: (1) a nautiloid–gastropod type in which the trabecular pattern varies considerably in the different parts of the nacreous layer and within the limits of the matrix corresponding to single crystals of aragonite and (2) a more uniform pelecypod type. The divergences between the results of Grégoire *et al.* (1949, 1950, 1955) and Grégoire (1957a, 1960, 1967b) and those of Mutvei concerning the nautiloid and gastropod patterns (two different patterns or a single pattern) are probably due to the fact that Mutvei based his conclusions on analysis of two species of gastropods only whereas the other authors studied 17 species.

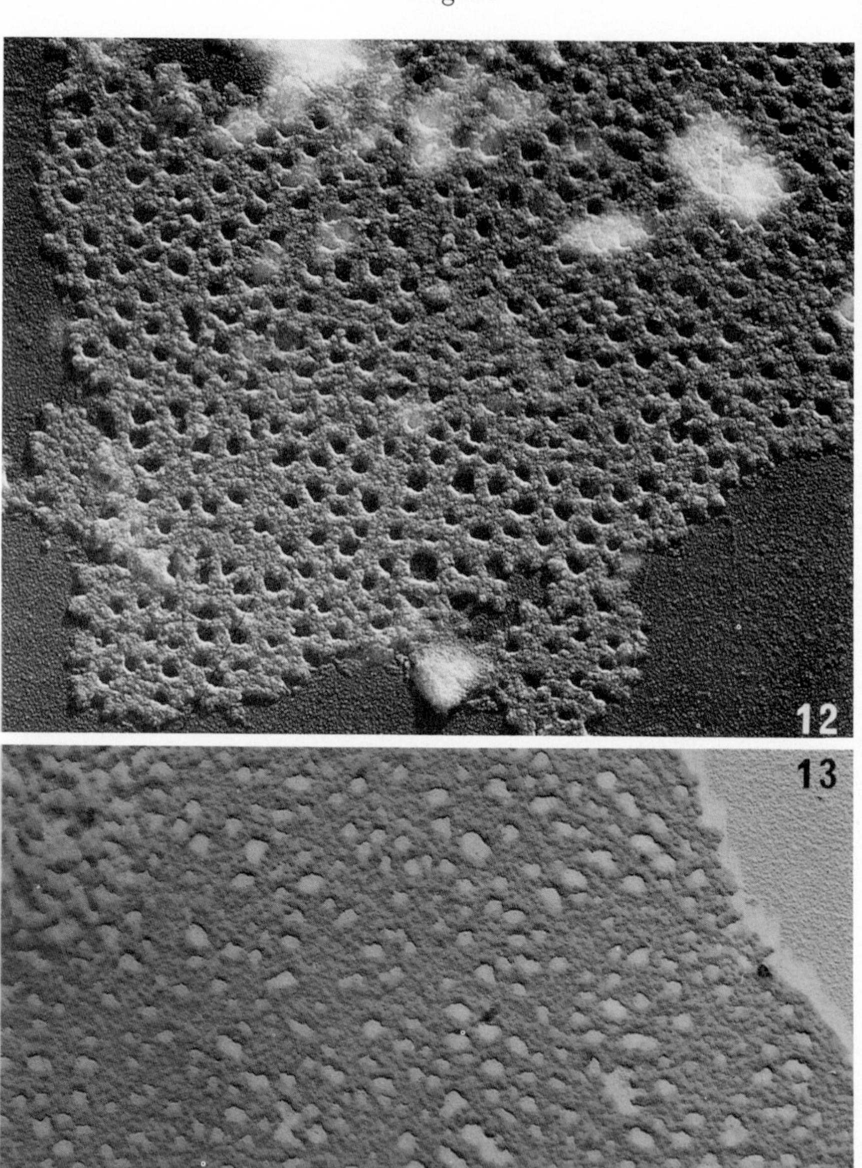
12
13

The inner layer of the shell in *Neopilina galatheae* Lemche is built up of mother-of-pearl (W. J. Schmidt, 1959) in which the aragonite crystals are made up of subunits encased in organic matrix (Meenakshi *et al.*, 1970). Decalcification of this layer leaves tough rigid shreds. In the electron microscope (Grégoire, 1962, 1967b, Pl. 1, Figs. 7 and 8) these shreds appear in the form of thick membranes with scattered scarce perforations. These membranes seem to be composed of dense fibrillar networks embedded in a granular or nodular substance. Erben *et al.* (1968) observed similar structures in the same material by means of the scanning electron microscope.

In the septa of *Spirula* (Grégoire, 1961c, 1967b, confirmed by Mutvei, 1969, 1970), the structure of mother-of-pearl differs from that of all the mollusks so far studied (Figs. 17 and 18). In contrast with the soft conchiolin membranes freed by decalcification of the nacreous layer in many mollusks, the organic components of a decalcified dome-shaped septum of *Spirula* is a semirigid, elastic material, which preserves the form of the original mineralized septum. This material is cleavable into thinner rigid lamellae. The lamellae are composed of planes of sturdy parallel fibers, which are themselves aggregates of microfibrils, about 40 Å in diameter (Pl. 2, Figs. 1–3 in Grégoire, 1967b). The successive planes of parallel fibrils are differently oriented. Adjacent planes form angles varying from 19° to 90°. In replicas of polished and etched transverse sections of the septa, the mineral lamellae seem to be composed of curved, parallel, imbricated pads, rounded flakes, or lenticular segments. These structures might represent crystalline subunits. In portions of the septa, two or three planes of fibers seem to alternate with one or more crystalline layers. Some fibers seem to penetrate between the mineral flakes inside the lamellae. According to Mutvei (1970), the horizontal mineral lamellae do not consist of aragonite crystals but of parallel aragonite rods with different orientation in successive lamellae.

While the topic is outside the scope of this chapter, it may be briefly mentioned that the three patterns of structure of nacreous conchiolin, variously altered, have been detected in mother-of-pearl of more than 250 species of fossil mollusks (Grégoire, 1958c, 1959a,b, 1966b; Grandjean *et al.*, 1964; Grégoire and Teichert, 1965; Grégoire and Voss-Foucart, 1970) from various ages (15,000 to 450 million years old).

FIG. 12. *Umbonium giganteum* Lesson (Gastropoda, Umboniidae). Decalcified mother-of-pearl. Fragment of interlamellar matrix (gastropod pattern) stained with osmium tetroxide and platinum shadowed (× 48,000). (Grégoire, 1960.)

FIG. 13. *Angaria delphinus* Linné (*laciniata* Lamarck) (Gastropoda, Angariidae). Decalcified mother-of-pearl. Fragment of interlamellar matrix (gastropod pattern). Platinum shadowed (× 42,000). (Grégoire, 1960.)

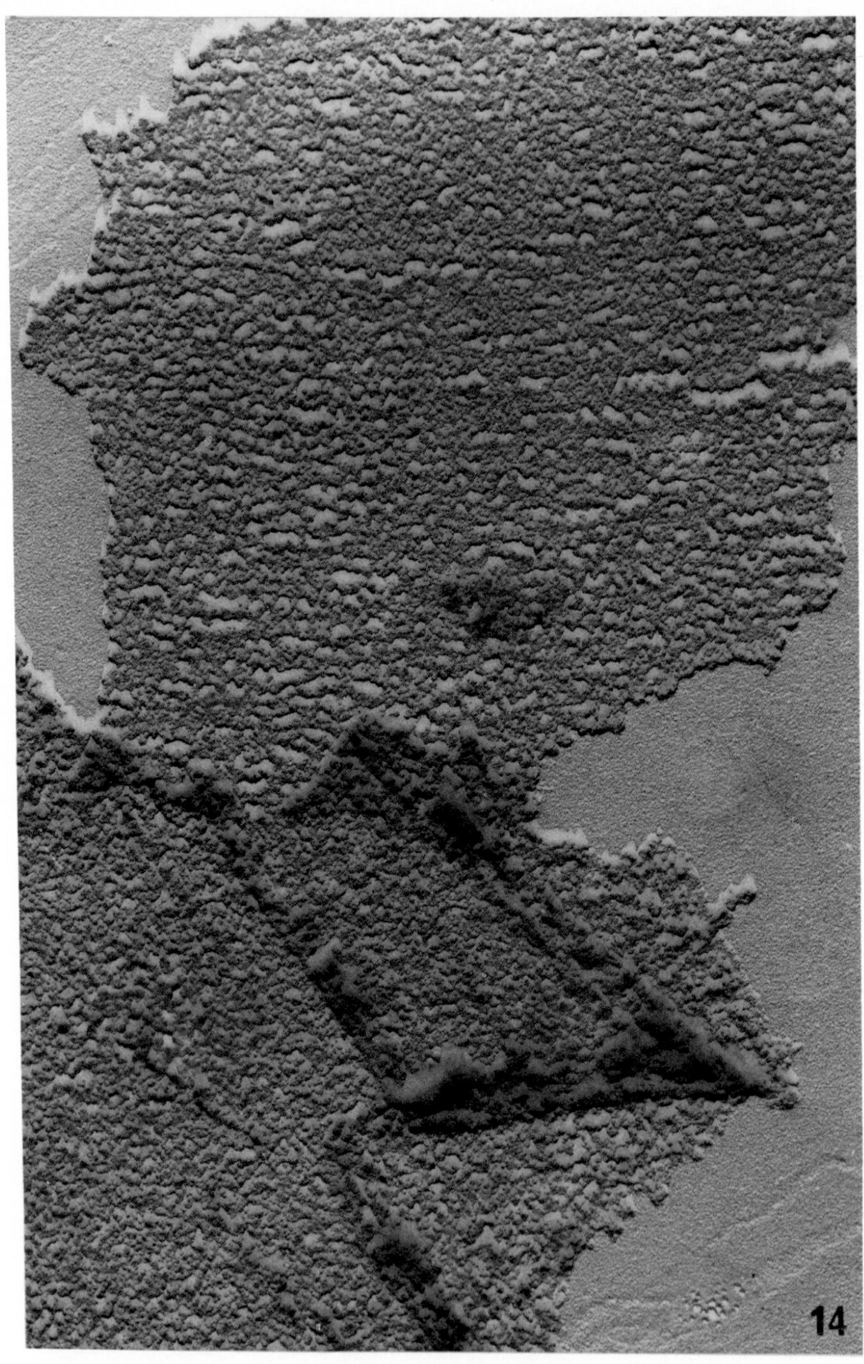

FIG. 14. *Aspatharia* (*Spathopsis*) *wismannii* von Martens (Bivalvia, Mutelidae). Decalcified mother-of-pearl. Fragment of interlamellar matrix (pelecypod pattern), platinum shadowed ($\times$ 42,000). (Grégoire, 1960.)

In these shells, the original aragonite had been preserved or had been transformed into calcite. These conchiolin remains were biuret positive, which indicates preservation of peptidic bonds, in agreement with detection of portions of the original proteins by Florkin *et al.* (1961) in Oligocene and Eocene specimens of this material in which the electron microscope had revealed structurally recognizable conchiolin matrices (Grégoire, 1958c). The alterations recorded in fossil conchiolin of Paleozoic and Mesozoic nautiloids and Ammonoids were reproduced experimentally on modern conchiolin material in exposing modern *Nautilus* shells to diagenetic and metamorphic factors involved in fossilization, such as heat and pressure (Grégoire, 1964, 1966a,b, 1968; Grégoire and Voss-Foucart, 1970; Voss-Foucart and Grégoire, 1971).

E. The Myostracum

It has long been observed, especially in mother-of-pearl, that the innermost shell layer underlying the muscles has another architecture than the adjacent portions of this layer—*durchsichtige Substanz* (Tullberg, 1882; Ehrenbaum, 1885; List, 1902); *Stäbchenschicht* (Müller, 1885); *Hypostrakum* (Thiele, 1893; Jameson, 1912; Brück, 1913; Lowenstam, 1964); *helle Schicht* (Rubbel, 1911; Rassbach, 1912; W. J. Schmidt, 1923, 1924); *myostrakum* (Oberling, 1955, 1964; Taylor *et al.*, 1969; Taylor and Kennedy, 1969a).

As shown by W. J. Schmidt (1923, 1924) in his extensive study of the *helle Schicht* in mother-of-pearl, this layer is a modified form of the nacreous structure in which the *c*-axis of the crystals of aragonite is considerably developed and the crystals assume a prismatic structure. These prisms, which vary greatly in size and shape, are oriented normally to the surface of the layer (Figs. 19 and 20) (Grégoire, 1958d). More recently, the area of insertion of the muscles has been studied by Grégoire (1962) in the *Nautilus* shell, in the patelloid shells by MacClintock (1967) in various bivalves by Taylor *et al.* (1969), Taylor and Kennedy (1969), and Kennedy *et al.* (1969). According to the last authors, the pallial myostracum, a highly distinct myostracal structure, separates shell layers of differing architecture.

In the horseshoe shaped region of insertion of the shell muscles and in the annulus of the *Nautilus* shell, an organic substance is frequently preserved in the dried shells in the form of a rigid, semitransparent membraneous disk. In the living animal, this membrane is sandwiched between the epithelium layer of the mantle and the inner shell layer. The literature on this disk and on its connections with the muscle fibrils, the fibrils of the specialized pallial epithelium, and the components of the myostracum has been reviewed elsewhere in detail (see Grégoire,

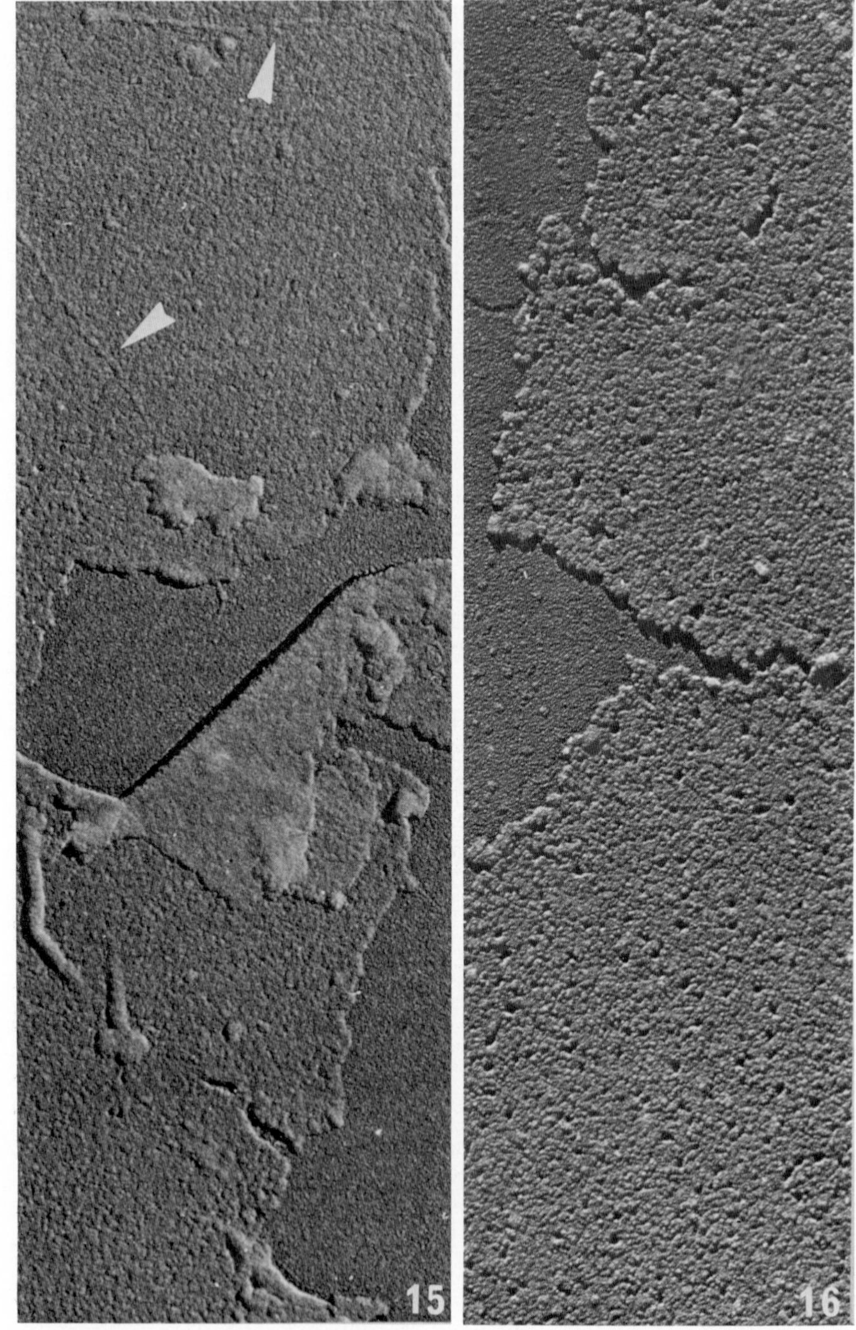
15
16

1962, pp. 6 and 37). In the electron microscope, the disk is found to be composed of several layers of extremely small microfibrils (Grégoire, 1962). This material is thoroughly distinct in its structure from the nacreous conchiolin. Waagen (1867–1870) had already recognized this difference. On the other hand, this disk has been described as conchiolin by other authors (Crick, 1898; Kessler, 1923; Lange, 1941; M. Schmidt, 1925; Mutvei, 1957).

In *Nautilus*, the structure of the inner shell surface differs greatly on either side of the bulging ridge marking the demarcation between the edges of the muscle insertions and the free inner surface of the shell (Grégoire, 1962). The latter shows parallel growth, concentric overgrowth, and aggregation of tabular crystals of aragonite. In the area of muscle attachment, the surfaces underlying the membranous disk are covered in many areas with swarming small lenticular bodies resembling seed crystallites (Grégoire, 1962, Figs. 10, 12, and 13). Two to 126 million of these corpuscles were counted per square centimeter in these areas of insertion of the muscles. In other fields, bundles of columnar crystals of aragonite, resembling conellen developed on the keels of ammonites, protrude over the background in various orientations (Grégoire, 1962, Fig. 11). All these formations reflect the influence of the environment (traction) on the mode of growth of the aragonite crystals.

As pointed out elsewhere (Grégoire, 1962), these microcrystals might be the earliest stages of development of a hundred times or more larger corpuscles described with the conventional microscope in the areas of muscle attachment by Herdman and Hornell (1903), Rubbel (1911), and Jameson (1912) (hypostracum muscle pearls).

The intense proliferation of seed crystallites in the areas of muscle insertion might represent in the posterior, adapical region, the steps of fast deposition of new nacreous substance on the adapical areas of resorption left by the most recent forward motion of the muscles during the growth of the shell. Simultaneously, along the anterior adoral ridge of the muscle scars, the clusters of seeds might reflect an especially

---

FIG. 15. *Pandora trilineata* Say (Bivalvia, Anatinidae). Decalcified mother-of-pearl. Fragment of interlamellar matrix (pelecypod pattern). Thinly granular membranes with traces of the crystals (crystal imprints: white arrows). Palladium shadowed (× 40,000). (Grégoire, 1960.)

FIG. 16. *Pinna nobilis* (Bivalvia, Pinnidae). Decalcified mother-of-pearl. Fragment of interlamellar matrix (pelecypod pattern). Dense meshworks of fibrillar trabeculae. Scattered pores of small size. Palladium shadowed (× 40,000). (Grégoire, 1960, Fig. 5).

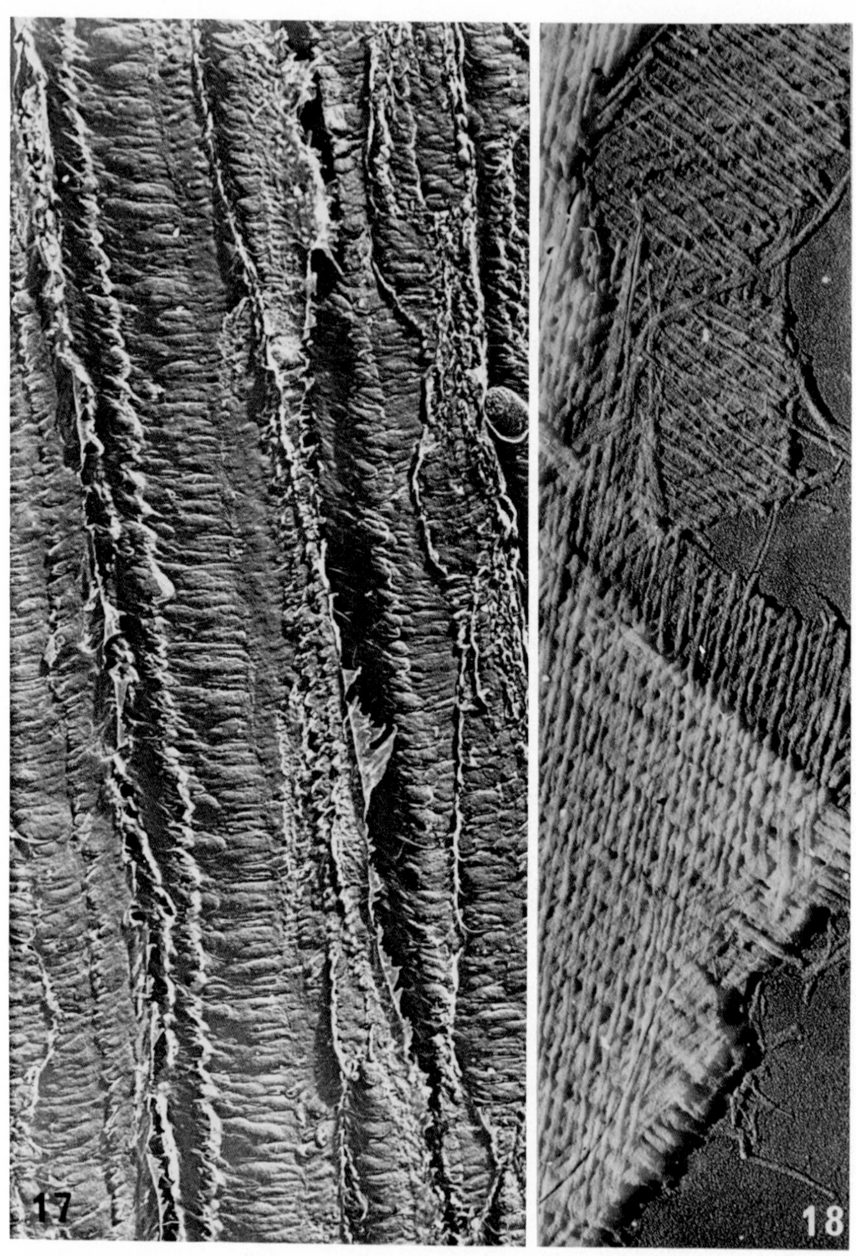

FIG. 17. *Spirula spirula* Linné (Cephalopoda, Spirulidae). Positive carbon replica, palladium shadowed, of the polished and etched transverse section of one of the last adoral septa. (× 14,000.) (Grégoire, 1961c, 1967b.)

FIG. 18. *Spirula spirula* Linné (Cephalopoda, Spirulidae). Decalcified septal mother-of-pearl. Organic matrix, palladium shadowed (× 33,000). (Grégoire, 1961c, 1967b.)

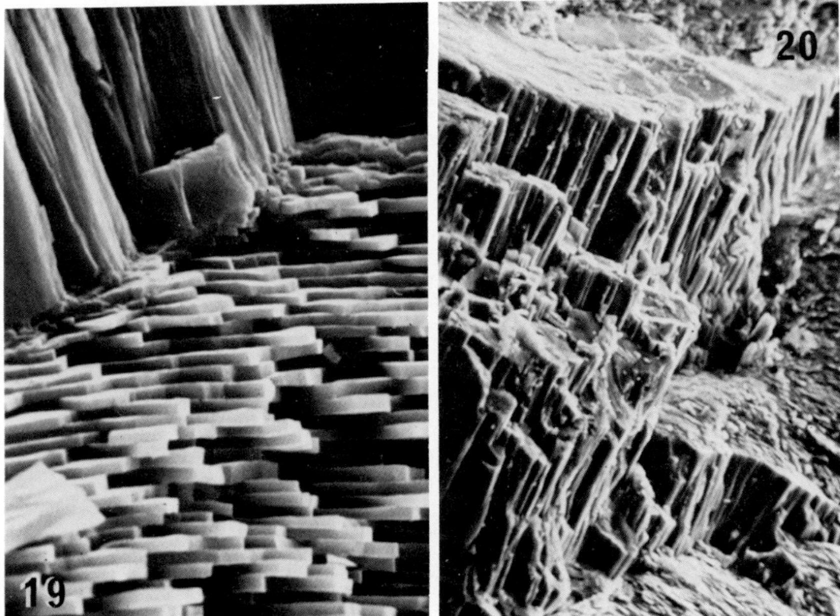

FIGS. 19 and 20. *Aetheria elliptica forma plumbea* (Bivalvia, Aetheriidae). Surfaces of fracture of the shell in the area of muscle attachment. In these fields, the *helle Schicht* or myostracum appears to be composed of parallel columnar crystals of aragonite or myostracal prisms. Fig. 19 shows the boundary between the helle Schicht and mother-of-pearl, which is built up like a brickwork. Stereoscan (Cambridge Scientific Instruments Limited) Fig. 19, × 2900; Fig. 20, × 280 (Grégoire, unpublished observations).

active formation of new shell substance induced by the particular environmental conditions existing in the close vicinity of the shell muscles (Grégoire, 1962).

F. THE GRAINED STRUCTURE

According to Bøggild (1930), this structure is uncommon in mollusk shells. It consists of more or less irregularly formed grains. Finely grained layers are found in parts of the shells of *Sepia* (aragonite) and of *Argonauta* (calcite). Unoriented grains may penetrate layers characterized by another structure (e.g., homogeneous structure), especially in the upper layers of some bivalves.

G. THE CROSSED LAMELLAR STRUCTURE

This structure is the most common of the structures of mollusk shells. It is found in many marine bivalves and in freshwater and marine gastropods. The common aragonitic crossed lamellar structure has long been known (de Bournon, 1808; Carpenter, 1844, 1848; Rose, 1858; von Nathusius-Königsborn, 1877; von Gümbel, 1884; Ehrenbaum, 1885; Biedermann, 1902a,b, 1914; W. J. Schmidt, 1924; Bøggild, 1930; Haas, 1935; Kessel, 1935–1936, 1950; Reyne, 1951; Barker, 1964; I. Kobayashi, 1964, 1966; Petitjean, 1965; MacClintock, 1967; Taylor *et al.*, 1969; Taylor and Kennedy, 1969).

According to Bøggild (1930) the most perfect crossed lamellar structure is made up of aragonite. The corresponding calcitic structure occurs in Aviculidae, Ostreidae, and Patellidae (crossed foliated structure) (MacClintock, 1967). As described by Bøggild (1930), this structure is built up of large lamels (first order lamels). Each lamel has a nearly rectangular form with the longer axis placed in the horizontal direction, parallel to the surface of the shell, whereas the shorter axis, in most instances, has a vertical position. The lamels of first order are made up of numerous smaller inclined lamels (second order lamels). The latter consist of tiny elongate crystals (third order lamels) which appear in the form of thin parallel laths, rods, or fibrils (I. Kobayashi, 1964, 1966). The second order lamels are oriented normally to the surface of the first order lamels. In two adjacent first order lamels, the second order lamels are inclined in opposite direction. This produces a characteristic crossing of these lamels. The type of architecture realized in the crossed lamellar structure, and especially the arrangement in crossed position of adjacent lamels, gives the whole shell a great solidity. The organic matter in the crossed lamellar structure is scarce (about 2%) (Bøggild, 1930).

H. The Complex and the Complex Crossed Lamellar Structures

According to Bøggild (1930) and Taylor *et al.* (1969), the complex structure always consists of aragonitic material and is composed in most instances of sublayers of two kinds, one of which is finely prismatic, in which the prisms present a radiating disposition, and the other one is a complex crossed lamellar structure. This description of the complex structure has been recently amended by MacClintock (1967) and by Taylor *et al.* (1969). According to Kessel (1950), the spheritic disposition of the prisms is predominant in the gastropods. The complex structure can involve cone-in-cone arrangement of crystal aggregates (MacClintock, 1967) in which each cone is built of radiating needles. According to Kessel (1950), the radiating spheritic disposition of the aragonite crystals is predominant in gastropods.

In Bøggild's classification of the types of crystallization in the mollusk shell, the porcelainous substance of the earlier authors corresponds at least in part to the crossed lamellar structure or to the complex crossed lamellar structure. In the shell of *Nautilus*, a part of the outer layer of the shell wall (W. J. Schmidt, 1924; Grégoire, 1962; Mutvei, 1964), the umbilical callus (Grégoire, 1962), and a part of the substances interposed between the shell wall and the mural portions of the septa (sutural substances) (Grégoire, 1962) consist of porcelainous substance.

The structure of the organic components in the crossed structure and porcelainous substance is little known. The scarce residues of decalcification of the outer layer in the shell wall and of the umbilical callus in *Nautilus* appear in the form of semirigid pellicles (Grégoire, 1962). In the electron microscope, fibrillar veils were detected in these pellicles (Fig. 21). The fibrils, 4–6 mμ in diameter, seem to be composed of thinner filaments, 15–30 Å in diameter. These veils might correspond to the conchiolin accumulations described by Mutvei (1964) in the spherulitic prismatic layer of the outer portion of the shell wall.

Fig. 21. *Nautilus pompilius* Linné. Decalcified porcelainous (outer) layer of the shell wall. The residues, semirigid, lusterless membranes, consist of fibrillar material. Palladium shadowed (× 37,500).

III. The Cuttle Bone (*Sepia officinalis* L.)

In the lamellar ventral zone of the cuttle bone, which is aragonitic (Kelly, 1901; Bütschli, 1908), parallel lamellae or septa are held apart by organic–calcareous pillars delimiting cavities or chambers. Membranes parallel to the septa and disposed at right angles to the interseptal pillars are anchored in these chambers (Appellöf, 1892–1893; W. J. Schmidt, 1924; Denton, 1961; Denton and Gilpin-Brown, 1961, 1966). The organic portion of these structures contains chitin (Appellöf, 1892–1893; Rudall, 1955: chitin β; Hackman, 1960; Stegemann, 1961b), mostly (84%) in a free form (Jeuniaux, 1963).

Decalcification of the septa and of the pillars leaves organic "ghosts" in the form of elongated conical tubes with distinct growth lines, described by Appellöf (1892–1893, p. 22) and by W. J. Schmidt (1924) (see Grégoire, 1967b, Pl. 5, Fig. 1). In the electron microscope, this material appears to be composed of substantial networks of microfibrils, partly embedded in a nodular or amorphous material (Grégoire, 1967), destroyed by sodium hydroxide at 100°C. Fibrillar networks characterize the structure of the decalcification remnants of the dorsal shield. The structure of the cuttle bone has been recently studied with the scanning electron microscope (Reumuth, 1968).

IV. Mineralogy and Crystallography

The literature on mineralogy and crystallography of mollusk shells has been reviewed by W. J. Schmidt (1921a,b, 1923, 1924, 1928, 1931, 1932a,b), Clarke and Wheeler (1917, 1922), Prenant (1927), Mayer (1930), Böggild (1930), Haas (1935), Petitjean (1965), Vinogradov (1953), Hall and Kennedy (1967), Taylor *et al.* (1969), and Kennedy *et al.* (1969). Several early works were performed with the aid of the polarisation microscope and on etched preparations (e.g., Kelly, 1901; Karny, 1913). A great advance in this field was the introduction by Tsutsumi (1928, 1929) of the more precise technique of X-ray diffraction.

Aragonite and calcite are the chief minerals involved in the structure of mollusk shells. Bøggild studied the distribution of these minerals in the different taxonomic groups of mollusks. Aragonite and calcite are found in different layers of the same shell in several superfamilies. Aragonitic regions (muscle pads, resilium) have been detected in oyster shells long considered mineralogically to be composed entirely of calcite (Stenzel, 1962, 1963; Taylor *et al.*, 1969).

The factors which influence the aragonite–calcite ratio and the development of aragonitic and calcitic layers in shells (organic matrix, temperature, and salinity) have been reviewed elsewhere (Wilbur and Watabe, 1963; Wilbur and Simkiss, 1968; Grégoire, 1967b; Taylor *et al.*, 1969).

The influence of the organic shell matrices on the regulation of the crystal type has long been recognized. In 1932, W. J. Schmidt already had suggested that crystal growth in the prisms was periodically inhibited by deposition of organic material (1932a). Differences in the amino acid patterns of conchiolin also play a role (Roche *et al.*, 1951; Tanaka *et al.*, 1960a,b; Hare, 1963; Kitano and Hood, 1965). According to Taylor and Kennedy (1969b), the extension of the periostracum as the interprismatic wall in the outer shell layer of bivalves strongly suggests that it exercises some control on the presence and formation of the prisms.

As pointed out by Watabe (1965) the regulation of structures as alternating layers of crystals and interlamellar matrix in nacre may mean a periodic deposition of the interlamellar matrix layer. The presence of this layer over the crystals inhibit their growth, and if the matrix deposition is periodic, then the observed ordered structure of crystal layers of similar thickness might well be formed. Crystal growth of the calcitostracum with its reduced amount of matrix will be less regulated, with the result that multilayers of calcite crystals can be formed at one time (Watabe and Wilbur, 1961; Watabe, 1965; Wada, 1968b,c).

Watabe and Wilbur (1960, 1961) and Wilbur and Watabe (1963) investigated experimentally the influence of the organic matrix on the crystal type. Using mantle–shell preparations (W. K. Brooks, 1905; de Waele, 1930; Bevelander and Benzer, 1948; Hirata, 1953), they inserted decalcified matrix from aragonitic shells at the site of shell formation in the oyster *Crassostrea virginica,* which normally deposits calcite. Aragonite crystals formed along with calcite crystals in four out of seventeen cases. Only calcite was formed on other substrata, such as calcitic matrix, glass, and plastics. Similarly, when calcium carbonate was precipitated from solutions *in vitro,* aragonite crystals were formed on aragonitic matrix from two species of mollusks but not on other materials. Thus, the organic matrix appears to be one of the factors in the determination of crystal type.

As shown by Lowenstam (1954a,b, 1964), Dodd (1963, 1964, 1966), and Malone and Dodd (1967) in a number of bivalves which have an outer calcitic and an inner aragonitic layers, the aragonite–calcite ratio of the shells increases with elevation in environmental temperature. Similarly, salinity affects the mineralogy of some shells (Lowenstam, 1954b; Dodd, 1963, 1966). As shown by Dodd (1963), increasing salinity is connected with a decrease in aragonite shell content. The results of Dodd have been questioned by Eisma (1966).

It has long been suggested that orientation of the growth of the crystal seeds on the inner shell surfaces is controlled by factors such as direction of growth of the outer palleal epithelium (Rose, 1858; Kelly, 1901; Karny, 1913; W. J. Schmidt, 1923, 1924). In modern investigations, one becomes more and more aware of the biological importance of the processes of epitaxy—*inotrope Verkalkung* (Royer, 1928; W. J. Schmidt, 1955) *Matrizenprinzip* (Seifert, 1961a,b; Wilbur and Simkiss, 1968)—in which the nature of the substrate is a major factor determining the nucleation, size, and orientation of the growing crystals. In mother-of-pearl, for instance, the orientation of the crystallographic *b*-axis of the aragonite crystals in the growth regions of the inner surface of the shell depends on the direction of elongation and of growth of the mantle (W. J. Schmidt, 1924; Wada, 1960a, 1961a,b, 1963c, 1968b), on the direction of the currents in the extrapallial fluid (Wada, 1960a), and on the orientation of the trabeculae of the conchiolin matrix underlying the crystals (Grégoire, 1962). However, as pointed out by Wilbur and Simkiss (1968), more data must be gathered before safe conclusions may be drawn in this field.

Traces elements, especially strontium, magnesium, and manganese, contained in the mollusk shells have been extensively studied in recent years (Lowenstam, 1954a,b, 1963, 1964; Chave, 1954; Krinsley, 1959; Turekian and Armstrong, 1960; Leutwein and Waskowiak, 1962; Pilkey and Goodell, 1963; Dodd, 1965; Harris, 1965; Hallam and Price, 1966, 1968a,b; Price and Hallam, 1967). The data obtained, reviewed by Dodd (1967) and Taylor *et al.* (1969) are important for phylogeny, evolution, and paleoecology.

V. Uncalcified or Partly Calcified Organic Structures of Mollusk Shells

A. The Periostracum

The uncalcified organic structure called periostracum covers the outermost calcified layer of the shell (prismatic layer in bivalves). It is frequently worn away (e.g., *Nautilus*). The literature on the structure and the chemical nature of the periostracum will be found in Haas (1935), Kessel (1940), Trueman (1949), Brown (1952), Beedham (1958), Dunachie (1963), Wada (1966e), Beedham and Owen (1965), Wilbur and Simkiss (1968), Meenakshi *et al.* (1969), and Taylor and Kennedy (1969). The formation of the periostracum has been recently studied with the electron microscope by Bevelander and Nakahara (1967) and by Wada (1968a).

The periostracum is composed of various numbers of layers (Manigault, 1939; Kawaguti and Ikemoto, 1962; Dunachie, 1963; Beedham and Owen, 1965). Freshwater specimens of gastropods have more layers than marine species (Meenakshi *et al.*, 1969). As pointed out by Brown (1952) the periostracum originates as a single homogeneous layer and soon becomes differentiated into two or more layers (Meenakshi *et al.*, 1969). The middle layer becomes differentiated into compact inner and outer zones separated by a vacuolated zone.

The periostracum differs in its chemical constitution from that of the organic matrices of the mineralized layers. (See Table I, Part 2). It consists of a quinone-tanned protein (Brown, 1952, confirmed by Beedham, 1958; Degens *et al.*, 1967). The periostracum contains fifteen to seventeen amino acids (Wilbur and Simkiss, 1968; Meenakshi *et al.*, 1969). Lipids and neutral polysaccharides are also present. Acid mucopolysaccharides, commonly associated with calcification, are absent in periostracum.

As in the calcified portions of the shell, the chemical composition of the periostracum varies greatly in relation to different degrees of protein tanning (Wilbur and Simkiss, 1968) and is influenced by taxonomical and environmental factors (see Table I, Parts 2 and 3) (Degens *et al.*, 1967; Meenakshi *et al.*, 1969).

B. The Ligament

The literature on the structure and chemical composition of the ligament has been reviewed by Haas (1935), Mitchell (1935), Trueman (1942, 1949, 1950, 1951, 1953a,b, 1964), and Galtsoff (1964). The principal function of the ligament in the bivalves is to open the shell when the adductor muscles relax (Dall, 1889) and to hold the dorsal margins of the valves together (see Trueman, 1964). The ligament is secreted by the mantle tissues in the same manner as the valves. Its formation has been recently studied with the electron microscope by Bevelander and Nakahara (1969b) in *Mytilus edulis* and *Pinctada radiata*.

The ligament consists of two main layers with two subsidiary layers. The outer and inner layers of the ligament represent local modifications of the same two layers of the shell (Mitchell, 1935; Owen *et al.*, 1953; Beedham, 1958; see Trueman, 1964). Mineralization of the ligament, which occurs in many bivalves (e.g., Pteriomorphia) (Kennedy *et al.*, 1969), is invariably in the form of aragonite.

As reported by Trueman (1964), the outer layer of the ligament (lamellar layer) (Newell, 1937) is composed of nearly parallel lamellae of a fibrous protein. It has undergone hardening by aromatic tanning,

TABLE I
RESULTS OF STUDIES ON CONCHIOLIN COMPOSITION

Investigators and material	Abstracts of results
1. *Biochemical Composition of the Conchiolins and Extrapallial Fluid* Early literature: Frémy, 1855; Schlossberger, 1856 (scleroprotidic nature of conchiolin); Voit, 1860; Krukenberg 1881–1882, 1885; Abderhalden (1908); see references and reviews in Haas (1935) and Galtsoff (1964).	
Wetzel (1900) *Mytilus, Pinna* (bivalves)	Gly, Leu, Tyr, Try, Arg were detected in conchiolin hydrolyzates. The composition of prism conchiolin differs from that of nacre conchiolin.
Meyer (1913). *Sepia, Loligo* (cephalopods)	Detected chitin in mollusk shells.
Strauss (1921)	Detected reductive substances in conchiolin.
Stary and Andratschke (1925). *Mytilus edulis* (bivalve)	Scleroprotidic nature of conchiolin (Millon, xanthoprotein, and ninhydrin reactions positive).
Friza (1932) *Mytilus, Anodonta, Pinna* (bivalves)	Total nitrogen percentage in shell: *Mytilus,* 16.66%; *Anodonta* nacre, 15.29%. Total sulfur content: *Mytilus*; 0.61%; *Anodonta,* 0·75%. Tyr: *Mytilus* conchiolin, 3.75%; *Anodonta,* 1.46%. Try: *Pinna,* 2.62%; *Mytilus,* 2.62%. Arg: *Mytilus,* 5.28%; *Anodonta,* 5.31%. During shell decalcification, a conchiolin fraction is soluble in a 2% solution of hydrochloric acid.
Turek (1933) Bivalves, gastropods, cephalopods (*Nautilus, Spirula, Argonauta*)	The content in organic substances differs significantly in gastropods (0.17–2.20%) and in cephalopods (3.07–6.99%). The values in bivalves are in between. On the basis of analyses of several substances (N, Ca, Mg, Fe, P, SiO_2), the author recorded differences between taxonomic groups inside a single class, and discusses the phylogenetic implications of the findings.
Bevelander and Benzer (1948) *Anodonta, Codakia, Venus* (bivalves) and other mollusks	Histochemical detection of mucopolysaccharides in mollusk shells.
Comfort (1951a,b)	Reviewed the distribution of pigments in mollusk shells. The pigments of shells are divisible into acid-soluble and acid-insoluble groups. The acid-soluble pigments, found in Archaeogastropods, Opisthobranchs, and lower bivalves, are dissolved from the organic matrix during decalcification. These pigments include porphyrins, melanins, pyrroles, and other unidentified pigments of small molecular size. In

TABLE I (*Continued*)

Investigators and material	Abstracts of results
	most higher mollusks and all pulmonates, the acid–insoluble pigments of shell conchiolin are firmly attached to the protein components. These pigments appear to be incorporated during shell secretion, possibly by a process of quinone tanning.
Roche *et al.* (1951); Ranson (1952) *Meleagrina and Pinna* (bivalves). Prismatic (calcitic) and nacreous (aragonitic) layers. *Aragonitic shells: Turbo* (Gastropoda); *Calcitic shells: Gryphaea, Ostrea, Pecten* (bivalves)	Calcitic prismatic layer: higher content in Tyr and Gly, and lower content in Arg, Ser, Glu, than in aragonitic nacreous layer. The conchiolins from calcitic and aragonitic shells differ in their amino acid patterns. However, a general relation between crystalline form of calcium and amino acid pattern of conchiolin does not distinctly appear.
Tanaka and Hatano (1953, 1955) *Pinctada martensii* (bivalve)	Lower concentration in Ala and Asp in prismatic than in nacreous layer. In the latter, detection of Hydroxypro, and Cys. Detection of mucopolysaccharides in conchiolin.
Beedham (1954) *Anodonta* (bivalve)	The shell conchiolin consists mainly of protein, although lipoid material and possibly some polysaccharide may be present. The conchiolin from the inner shell layers differs from that of the outer shell layers in its composition.
Grégoire *et al.* (1955) *Nautilus macromphalus* (cephalopod); *Pinctada margaritifera* (bivalve); *Trochus* sp. (Gastropod)	The nacre conchiolin consists of three biochemical components: (1) A protein fraction soluble in water or in sodium borate—nacrin (soluble nacrin: Florkin, 1966) containing 15 amino acids in which nitrogen represents 23.64% of the total nitrogen conchiolin. (2) A fraction soluble in sodium hydroxide: nacrosclerotin (insoluble nacrin: Florkin, 1966) containing 15 amino acids in which nitrogen represents 70.57% of the total nitrogen conchiolin. (3) A fibrous residual "polypeptide" fraction—nacroin, in which nitrogen represents 5.58% of the total nitrogen conchiolin. In nacroin, the total nitrogen of Ala and Gly amounts to 83.6% of the total amino nitrogen in *Pinctada margaritifera;* the total nitrogen amino acid represents 70% of the total nacroin nitrogen. A nitrogen fraction remains unidentified (see Goffinet, 1965). Nacroin hydrolyzates give positive orcin and Fehling reactions, which suggests the presence of glucidic radicals.

TABLE I (*Continued*)

Investigators and material	Abstracts of results
Tsujii (1955,1960) Marine and freshwater bivalves	Histochemical detection of mucopolysaccharides in the mantle epithelium and in conchiolin from decalcified pearls.
Abelson (1956, 1959, 1963) *Mercenaria mercenaria* (bivalve)	Identifies 15 amino acids in the organic matrix of the shell.
Abolinš–Krogis (1958) *Helix pomatia* (gastropod)	Reports the detection of a polysaccharide material in the organic matrix.
Tanaka *et al.* (1960b) *Pinctada martensii* (pearl and shell) (bivalve)	In prism conchiolin, more Phe and Pro and less Ala than in nacre conchiolin.
Tanaka *et al.* (1960a) *Pinctada martensii* (pearl and shell)	Separation of several fractions from nacreous conchiolin with 70% acetic acid, 2 *N* ammonia, copper ethylenediamine, and Co(py)4 Cl2. The nitrogen terminal amino acids of these fractions were Gly, Ser, and Asp. The carbon terminal amino acids of conchiolin treated by 70% acetic acid and 2/*N* ammonia were Gly, Ser, and Thr.
Piez (1961) *Australorbis glabratus*	The proteins from snail shell are related to collagen and also contained hydroxylysine, which may have an important role in calcification.
Piez (1963) *Australorbis glabratus* (gastropod)	In the shell protein, high content of Gly, low content of hydroxyproline. Lys amounts as in invertebrate collagen.
Stegemann (1961, 1963) *Sepia* (cuttle bone), *Loligo, Octopus, Alloteuthis* (cephalopods)	The protein fraction (conchagen) can be converted to soluble protein (gelatin) with steam and or with 0.01 *N* sodium hydroxide. An insoluble chitin fraction consists of 2% protein which contains three unknown phenolic amino acids. The chitin also contains a smaller percentage of amino sugars, not identical with glucosamine.
Hare (1962) *Mytilus californianus; Mytilus edulis diegensis* (bivalves)	The proteins from various shell layers in *Mytilus* contain very large amounts of Gly and Ala (50%) of the total residues. The relative amounts of 16 amino acids do not sensibly differ in the outer and inner prismatic and in the nacreous layers (Fig. 11). This result is in agreement with previous records of similar ultrastructures in prism sheaths and nacreous matrices in *Mytilus edulis* (Grégoire, 1960).
Hare (1963) *Mytilus californianus* (bivalve)	The amino acid composition of the organic matrix in mineralized layers differs markedly in quantity but not in quality from that of the matrix in nonmineralized layers. Slightly

TABLE I (*Continued*)

Investigators and material	Abstracts of results
	higher proportion of acidic amino acids in calcitic layers than in aragonitic layers (not confirmed by Wada, 1963a,b,c). Hydroxylysine and hydroxyprolin absent in calcified proteins of *Mytilus*.
Tanaka *et al.* (1963) *Pinctada martensii* (bivalve)	The amino acid composition of the nacreous layer differs from that of the prismatic layer.
Jeuniaux (1963) Amphineura (chiton); various cephalopods; *Helix pomatia* (gastropod); *Ostrea edulis* (bivalve)	Using a purified chitinase confirms the presence of chitin in shells of cephalopods (*Sepia, Loligo*) (Meyer, 1913; Lotmar and Picken, 1950; Rudall, 1955; Hackman, 1960; Stegemann, 1961). Detects chitin in dorsal plates of chitons, in shell wall, and in septa of *Nautilus,* in septa of *Spirula,* in the shells of *Helix pomatia* and of *Ostrea edulis.* The organic components of the calcified layers contain free chitin in higher proportions than bound chitin.
Meenakshi (1963) *Pila globosa* (gastropod)	The two calcareous layers (ostracum, hypostracum) differ from the periostracum in their chemical nature. These two layers consist mostly of an acid mucopolysaccharide with traces of a protein, soluble in dilute acids and with a very low Tyr content. These layers do not contain quinone-tanned proteins.
S. Kobayashi (1964a) 11 calcitic shells; 5 aragonitic shells	The extrapallial fluid of species forming calcitic shells contains a single protein fraction. In the species in which the shell calcium carbonate occurs as aragonite or as both aragonite and calcite, three or more protein fractions are present in the extrapallial fluid.
S. Kobayashi (1964b) 15 molluscan species	The extrapallial fluid of the species with aragonite shell and aragonite and calcite shells contains a much more complex system of proteins, acid mucopolysaccharides, and PAS-positive materials than that of the species with calcite shells. In all the species examined, hyaluronate-like substances and PAS-positive materials, possibly neutral heteropolysaccharides, were commonly found in the extrapallial fluid.
Goffinet (1965) *Nautilus macromphalus* (Cephalopod); 16 species of bivalves	Confirms the results of Grégoire *et al.* (1955) regarding soluble and insoluble nacrins. Identifies in nacroin, in addition to the polypeptidic fraction detected by the formers authors, a substantial polyaminosaccharidic

TABLE I (*Continued*)

Investigators and material	Absracts of results
	fraction. The prosthetic group of this fraction (35% of the nacroin) consists of chitin. In nacroin, the polyaminosaccharidic fraction represents the stable component. The polypeptidic fraction is easily destroyed. Detects chitin, glucosamine, galactosamine in nacre, calcitostracum, porcelainous layer and in certain prismatic layers. Chitin is a permanent component of nacreous conchiolin.
Simkiss (1965) *Crassostrea* (bivalve)	The organic matrix consists of two fractions: (1) a protein with high amounts of Gly and of decarboxylic amino acids, (2) a sulfate-polysaccharide (8% by weight).
Rucker (1965) Various marine mollusks	The amino acids patterns in conchiolins of marine shells are strikingly different even for closely related taxa.
Wada (1966c) Various cephalopods, gastropods, and bivalves	The aragonitic conchiolin is characterized by relatively high concentrations of Asp, Ser, Gly, and Ala. The calcitic conchiolin (calcitostracum) is characterized by relatively high concentrations of Asp, Ser, and Gly, and by lower concentration of Ala.
Hare (1965–1966) *Mercenaria, Crassostrea virginica* (bivalves)	The relative proportions of most of the amino acids in the soluble proteins in the extrapallial fluid and in the largely insoluble shell matrix are similar. These proportions differ regarding His (higher in extrapallial fluid) Pro and Cys (lower in extrapallial fluid).
Mitterer (1966) *Mercenaria* (bivalve)	Asp, Pro, Gly, Ser, Glu, and Ala are the most abundant amino acids in the insoluble fraction of conchiolin.
Wada (1966) Cephalopods, gastropods, bivalves	The conchiolin of the mineralized layers are a chondroitin sulfate–protein complex or acid mucoprotein with sulfate groups. The conchiolin from aragonitic layers or from calcitostracum have in common a high concentration in Asp, Ser, and Gly. In aragonitic layers, the proportions of Ala, Pro, and Leu in conchiolin are higher than in calcitostracum. Ser is more abundant in calcitic conchiolin.
Akiyama (1966e) Various bivalves	Studied the particularities of the amino acid patterns in the conchiolins associated with the various shell architectures found in bivalves (Bøggild, 1930; I. Kobayashi, 1968)—prismatic layer, predominance of Gly; nacreous

TABLE I (*Continued*)

Investigators and material	Abstracts of results
	layer, predominance of Ala (except in *Pinctada martensii* where there is predominance of Gly); foliated layer, predominance of Gly; crossed lamellar, composite prismatic, and homogeneous layers, Gly ranges from 10 to 17%; complex layers, Leu in larger amounts (29%), Gly in the range of 15%.
Travis *et al.* (1967) *Mytilus* (bivalve)	The amino acid composition of the structural proteins of adjacent nacreous and prismatic layers are remarkably similar regardless of whether calcium carbonate is deposited as calcite (prisms) or as aragonite (nacre). Gly, Ala, Asp and Ser account for 80–85% of the total amino acid residues.
Wada (1967a) *Pinctada martensii* (marine bivalve); *Hyriopsis schlegelii* (freshwater bivalve)	Ala and Gly, the main amino acids of nacrosclerotin and nacroin (Grégoire *et al.*, 1955) are five times more abundant in nacreous conchiolin than in the extrapallial fluid. This suggests that the protein fraction in the extrapallial fluid, which corresponds to nacrosclerotin and nacroin, may be incorporated at a very high rate in the organic matrix of the nacre.
Wada (1967b) *Crassostrea gigas; Chlamys nobilis* (bivalves)	The amino acid pattern of the extrapallial fluid in contact with the calcitostracum differs from that of the conchiolin associated with this layer.
Wada (1967c) *Pinctada martensii; Chlamys nobilis* (bivalves)	The amino acid pattern of the inner mucus does not differ from that of the extrapallial fluid (*Pinctada martensii*). Asp, Ser, and Gly are in similar amounts in the mucus of the two species.
Degens *et al.* (1967) Bivalves (8 sp.); gastropods (4 sp.); cephalopods (1 sp.)	The composition in amino acids differs in taxonomic groups, including species.
Meenakshi *et al.* (1970) *Neopilina ewingi, Neopilina* sp.	The conchiolin consists of tanned protein, lipids, a neutral mucopolysaccharide, and acid mucopolysaccharide, but no chitin. The distribution of these substances varies in the different shell layers. This matrix contains 17 amino acids, in which Gly is predominant (42–45%), with large amounts of Leu and Val. This amino acid composition of the organic matrix differs from that of the other mollusks and resembles that of the periostracum of the modern mollusks (see Wilbur and Simkiss, 1968).

TABLE I (*Continued*)

Investigators and material	Abstracts of results
Voss-Foucart (1968) *Nautilus pompilius*	The conchiolin of mother-of-pearl in *Nautilus* consists of a glycoprotidic fraction (nacrosclerotin + nacroin) (see Grégoire *et al.*, 1955; Goffinet, 1965), associated with a soluble fraction (nacrin). The protidic component of the glycoprotidic fraction is characterized by a high content of Gly and Ala.
Florkin (1971)	Proposes (p. 11) to divide the constituents of nacre conchiolin into water-soluble nacrin and insoluble nacroin, a scleroprotein.
Bricteux-Grégoire *et al.* (1968) *Pinna nigra, Pinna nobilis* (bivalves)	In global nacre conchiolin, Gly, Ala, Ser, and Asp are predominant. In global prism conchiolin, Gly and Asp are predominant.
Kawahara and Yoshiaki (1968) *Patinopecten* and *Spisula* (bivalves); *Haliotis* (gastropod)	All three species have in common a large number of Asp residues in their conchiolin. In the calcitostracum conchiolin, Ser is higher, Tyr and Phe are lower than in aragonitic conchiolin. A characteristic amino acid pattern was found in the aragonitic resiliums of the two bivalves. *Haliotis* has a higher percentage of amino sugars (80% glycosamine) than the two bivalves.
Voss-Foucart *et al.* (1969) *Aetheriidae, Unionidae, Mutelidae, Margaritiferidae* (bivalves)	The insoluble residue of decalcification of the nacre is only a portion of the total organic substance of this layer. A fraction soluble in hydrochloric acid has a composition which differs from that of the insoluble residue. In the shell of the *Aetheriidae*, the soluble fractions of the total nacreous substance, the fibrillar membranes which coat the vesicles or intranacreous cavities, and the periostracum are nearly identical. The papyraceous membranes embedded in the nacre of the *Aetheridae* are probably expansions of the periostracum and have the same amino acid composition.
Goffinet (1969) *Nautilus macromphalus*	The fibrillar structure of the nacroin is destroyed by purified chitinases. The mechanically resistant intact nacroin becomes fragile after treatment with purified chitinases.
Goffinet and Jeuniaux (1969) *Nautilus pompilius*	Nacroin is a mucoprotein (or glycoprotein). The polysaccharidic fraction composing the prosthetic group of this mucoprotein consists at least in part of chitin, identified by means of purified chitinases. In addition to the polyacetylglucosaminic fraction, an important amount of glucosamine has been identified in the nacroin residue.

TABLE I (*Continued*)

Investigators and material	Abstracts of results
Saleuddin and Hare (1970) *Helix* (gastropod)	The amino acid composition of the regenerated shell is more related to that of the nacreous layer than to that of the outer layers of the normal shell.
Crenshaw (1970) *Mercenaria mercenaria* (bivalve)	A water soluble glycoprotein amounting to 18% by weight of the organic matrix was isolated by decalcification with EDTA. 30% of the total amino acid residues are 30% Asp, 16% Gly, and 10% Ser. This soluble matrix is the portion of the total organic matrix most intimately associated with the mineral phase.
2. Taxonomic and Phylogenetic Variations in Amount and Composition of Conchiolins	
Wada (1966d) 3 shells of gastropods; 5 shells of bivalves; 2 shells of cephalopods	Conchiolin has a specific amino acid pattern which varies in mineralized and nonmineralized layers of the same shell. The amino acid patterns distinctly differ in the three groups of mollusks.
Jeuniaux (1963)	In the phylogenetic evolution of the mollusks, the proportion of chitin decreases and the proportion of calcium carbonate increases.
Goffinet (1965) 16 species of bivalves (Taxodonta, Anisomyaria, Eulamellibranchia); 3 gastropods (*Haliotis tuberculata, Monodonta lineata, Mühlfeldtia sanguinea*)	Large variations in the relative amounts of chitin in the organic components of different shell layers of a single species and in the same layer in different species do not permit to evaluate systematically the participation of chitin to the constitution of the different portions of the shell. Chitin is in low concentration in the shells of Anisomyaria, in high concentration in those of Adapedonta. In the shells of the latter group, chitin forms a complete cuticular envelope.
Hare and Abelson (1963–1964, 1964–1965) 200 species of mollusks	Proteins from closely related species or from layers having the same architecture have a similar composition. Species not closely related or differing in their architecture differ in their protein composition. More recently evolved groups (*Conus, Parapholas*) have a smaller percentage of organic matrices than do the more primitive groups (*Astraea, Mytilus, Nautilus*) (see Turek, 1933). In these primitive groups, large amounts of Gly and Ala (45–50% of the total amino acids). Glucosamine is in much larger amounts in primitive

TABLE I (*Continued*)

Investigators and material	Abstracts of results
	shells than in recently evolved shells. With the phylogenetic evolution of the mollusks, there is a decrease of the proportion of chitin and an increase of the proportion of calcium carbonate.
Akiyama (1966) Various bivalves	Shells of similar structure but from different families may have a similar pattern of amino acid concentrations.
Degens and Spencer (1966) Bivalves (global analyses)	Thr, Glu, and Asp are predominant in the Recent groups, Gly in the primitive groups.
Degens *et al.* (1967) *Cryptochiton, Nautilus* (cephalopod) (global analyses); *Haliotis, Achatinella* (gastropod)	Throughout the molluscan phylum, the urea–hydroxylamine fraction is consistently enriched in Asp, Lys, and amino sugars. It is suggested that the peptide fraction contains the active sites for the deposition of the mineral in calcification. The shells of Archaeogastropods and of *Nautilus* contain much larger amounts of organic components than the shells of neogastropods.
3. *Variations in Biochemical Composition of Conchiolin in Relation to Environmental Factors (Temperature, Salinity)*	
Degens and Parker (1965); Degens *et al.* (1967) Gastropods, pelecypods, cephalopods (global analyses)	The same environmental factors control to a certain degree the chemical composition of the shell organic matrix. The amino acid composition in calcified and uncalcified tissues of mollusks can be described by a limited number of independent factors which are related to both phylogeny and environment.
Hare (1962) *Mytilus californianus, Mytilus edulis diegensis* (bivalves)	Elevation of temperature and decrease in salinity are accompanied with an increase of the acidic amino acid residues in the aragonitic layers. Thus, the differences in matrix composition between the nacreous aragonitic layer and the outer prismatic calcitic layer become progressively smaller. The calcitic organic matrix composition does not seem to vary significantly with temperature.
Degens and Spencer (1966) bivalves (global analyses)	Individuals of the same species taken from the same environment have almost identical amino acid composition in their shells. Conchiolin composition in amino acids differs in shells of single species in different environments.

TABLE I (*Continued*)

Investigators and material	Abstracts of results
Ghiselin *et al.* (1967) Amphineura, cephalopods, Monoplacophora, gastropods, bivalves	In conchiolin, a peptide nucleus composed chiefly of neutral amino acids is not influenced by environment, whereas basic amino acids, namely the relative proportions of Gly, Ala, Ser, and Tyr, are affected.
Mitterer (1966) *Mercenaria* (bivalve)	There is little variability in the amino acid composition of the organic matrix due to environmental effects.
Meenakshi *et al.* (1969) *Potamopyrgus* (gastropod)	Increase in salinity induces variations in proportion of certain amino acids namely a decrease in Gly and an increase in Glu of the periostracum.
Voss-Foucart *et al.* (1969) Aetheriidae (bivalves)	In agreement with electron microscopic results (Grégoire, 1960), the genus *Aetheria* constitutes a polymorphic species in which the various forms are ecophenotypes with an identical biochemical composition.
4. *Biochemical Composition of Uncalcified or Partly Calcified Organic Structures of Mollusk Shells (Periostracum, Ligament, Byssus)*	
Friza (1932) *Anodonta cygnaea* (bivalve)	In the periostracum, Tyr amount differs from that in nacre conchiolin: 3.3% against 1.40% in nacre. Try amount differs from that in nacre conchiolin: 2.62% against 2.78%. Arg amount differs from that in nacre conchiolin: 5.53% against 5.31%
Brown (1952) *Mytilus edulis* (bivalve)	The byssus threads, the periostracum and the hinge consist of a quinone-tanned protein. Precursor of the quinone is probably an aromatic acid or a protein.
Beedham and Trueman (1958) *Anodonta cygnaea, Sphaerium corneum, Mytilus edulis* (bivalves)	The iodine associated with the shell occurs principally in the periostracum (utilization of ^{131}I).
Beedham (1958) *Anodonta cygnaea, Mytilus, Ostrea* (bivalves)	Confirms Brown (1952)—the periostracum consists of a quinone-tanned protein. It differs in its composition from the conchiolin of the inner shell layers which contain more Asp and Glu.
Hillman (1961)	The epithelium secreting the periostracum synthesises a phenolic substrate. Oxidation of this substrate produces tanning quinones.
Hare (1962, 1963) *Mytilus californianus* (bivalve)	In small shells, there are profound changes in the periostracum composition as the animal grows during the first year. The periostracum has 400–500 Gly and 150 Tyr residues per

TABLE I (*Continued*)

Investigators and material	Abstracts of results
	thousand. Nonmineralization of certain periostracums could be ascribed to their low amounts in dicarboxylic amino acids. The uncalcified byssus contains over 300 Gly residues per thousand. It also contains hydroxyprolin. The ligament is composed of $\frac{2}{3}$ aragonite and $\frac{1}{3}$ organic matrix. This matrix has extremely high amounts of Met (327‰), Prol (71‰), Lys (92‰), and very low amounts of Arg (7.8‰) and Tyr (not detected).
Meenakshi (1963) *Pila globosa* (apple snail) (gastropod)	The periostracum differs in its chemical nature from the organic components of the inner calcareous layers.
Jeuniaux (1963) *Ostrea edulis* (bivalve)	Chitin constitutes 0.58% of the organic components of the periostracum, mostly in a free form. (81%).
Goffinet (1965) 16 species of bivalves (*Taxodonta, Anisomyaria* Eulamellibranchia); 3 species of gastropods.	Chitin has been consistently detected in periostracum. Large variations in the amounts of this substance in periostracum of bivalves cast doubt as regards the homologies between the periostracums in taxonomic groups.
Mitterer (1966) *Dosinia discus* (bivalve)	In periostracum, the amino acid composition is quite different from that of the shell matrix. Gly and Tyr account for 70% of the total residues. Gly alone comprises 50% of periostracum.
Wada (1966e) *Sepia, Haliotis, Hyriopsis*	In conchiolin of nonmineralized layers, including the periostracum, lower Ser, higher Tyr and Prol than in mineralized layers (see above). The sum of acidic amino acids is consistently smaller in nonmineralized conchiolin in single species. The nonmineralized layers have a pattern of amino acids characteristic of quinone-tanned proteins.
Degens *et al.* (1967) *Tagelus divisus*	The composition of the periostracum differs in specimens of single species collected from different latitudes.
Meenakshi *et al.* (1969). 27 species of freshwater and marine bivalves and of freshwater and terrestrial gastropods.	15 amino acids are present in the periostracum of each of the 27 species. Gly represents 40% or more of the amino acids of periostracum of marine and freshwater bivalves and of freshwater and terrestrial gastropods. In marine gastropods there are lower concentration of Gly and higher ranges of Lys, Arg, Asp, and Ser than in freshwater species. Acidic am-

TABLE I (*Continued*)

Investigators and material	Abstracts of results
	ino acids residues are in high concentration in marine gastropods, but in lower concentration in freshwater gastropods and in marine bivalves.
Voss-Foucart *et al.* (1969) Aetheriidae (bivalves)	In the periostracum and in its expansions in the nacre (papyraceous membranes), Gly amounts to 66%, Tyr to 12%, and Ala to less than 1% of the total amino acids.

giving molecular cross linkage of adjacent polypeptide chains (Trueman, 1950; Beedham, 1958), and contains no calcium carbonate. The inner layer—(cartilage or resilium) (Dall, 1889; Jackson, 1890), (fibrous ligament) (Newell, 1937)—is fibrous in structure and typically consists of relatively little tanned protein and calcium carbonate.

The function of the ligament in various bivalves has been extensively studied by Trueman (1942, 1944, 1949, 1950, 1951, 1953a,b) and Owen *et al.* (1953). According to Trueman (1964), the molecular arrangement of the scleroprotein of the outer layer, in which chemical bands unite adjacent polypeptide chains, appears to be well suited to its mechanical role of withstanding tensile stress.

C. Organic Materials Involved in the Hydrostatic Mechanism of *Nautilus*

In the *Nautilus* shell, the surfaces of the chambers and especially those of the convex posterior adapical sides of the septa are covered with dark brown and soft membranes, which also wrap the outer surface of the siphuncle in its intracameral portion (Owen, 1832; Barrande, 1857; Appellöf, 1892–1893; Grégoire, 1962; Mutvei, 1964). Chitin constitutes about 30% of the dry weight of these membranes (Jeuniaux, 1963). On the siphuncle, these structures play a role in regulation of permeability (Denton and Gilpin-Brown, 1966).

In the electron microscope (Grégoire, 1962), these membranes appear to be built up of dense networks of unoriented fibrils, aggregates of microfilaments (40–50 Å in diameter in shadowed preparations) (Fig. 22). Certain microfilaments appear to be composed of chains of corpuscles. Nodular patches of an apparently amorphous material, probably protein in nature, are scattered among the fibrils and partly embed them. Sodium hydroxide at 100°C destroys this nodular substance and does not alter the microfibrillar networks, which are probably chitin.

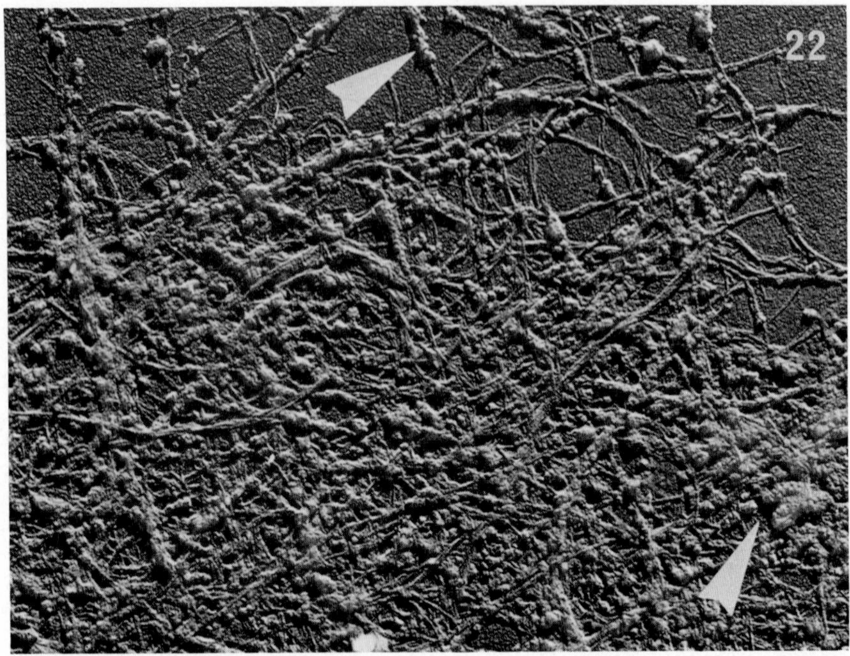

Fig. 22. *Nautilus pompilius* Linné. Portion of the intracameral membrane coating the adapical convex side of a septum in the anterior part of the phragmocone. Networks of fibrillar material, mixed with nodular substance (white arrows) which embedd in part the fibrils. Staining with osmium tetroxide and palladium shadowing (× 52,000). (Grégoire, 1962.)

The intracameral portion of the siphuncle is composed of two layers. The outer spicular layer, between two septal necks, coated by the brown membrane described above, is a porous sheath (H. Brooks, 1888; Appellöf, 1892–1893; Mutvei, 1964; Denton and Gilpin-Brown, 1966). This sheath consists of spindle-shaped calcareous crystals, gathered in small groups in which they are frequently radiating from a center. These crystals are embedded in an organic substance. The inner layer is a mahogany brown tube which is known by various names (horny tube: Buckland, 1837; inner conchiolin tube: H. Brooks, 1888; *unverkalkte innere Chitinmembran:* Appellöf, 1892–1893; inner conchiolin layer: Mutvei, 1964; porous horny siphuncular tube: Denton and Gilpin-Brown, 1966). It is composed of several concentric organic membranes. According to Mutvei (1964), this tube is an uncalcified nacreous layer. However, as shown in the electron microscope (Grégoire, 1958d, 1967b, Pl. 5, Figs. 3, 4, 5a,b), these membranes have not the lacelike reticulated structure characteristic of the septal conchiolin (see Grégoire, 1962, Fig. 59). They appear in the form of dense networks of fibrils clustered in bundles of fibers (Fig. 23). The microfibrils consist, as in the pericameral brown membranes, of microfilaments (40–80 Å in diameter). These networks of fibrils are similarly embedded in a nodular or amorphous material which is also dissolved in hot sodium hydroxide (Grégoire, 1967b) (Fig. 24).

VI. Colors of Shells

The colors of shells are due to physical factors and to pigments. As shown by Pfund (1917) in *Obovaria*, the colors of mother-of-pearl (iridescence) have two causes: (1) diffraction of light due to a grating-like structure of terminal edges of successive lamellae; (2) interference of light due to reflection from numerous parallel lamellae of sensibly equal thickness (0.4–0.6 μ). The literature on the pigments in mollusk shells has been reviewed by Comfort (1951a,b) (see Table I).

In *Pinctada fucata*, formation of yellow nacre and pearls seems to occur by the secretion and synthesis of the yellow pigment in the epithelial cells of the mantle and of the pearl sac (Wada, 1969); yellow pearls are produced by the pearl sac epithelium derived from the mantle tissue of the yellow nacre oyster which is transplanted into the gonad of pearl oysters. This form never occurs with the pearl sac epithelium derived from the mantle tissue of the white nacre oyster.

(See Goodwin, this volume, chapter 6, for a detailed review of this topic.)

VII. Biochemical Composition of the Nacreous Conchiolin in Relation to Its Ultrastructure*

W. J. Schmidt (1924), using physical analytical methods, found conchiolin to be anisotropic and birefringent. Wada (1961a) and Tanaka *et al.* (1963), using X-ray diffraction, described the nacreous conchiolin of *Pinctada martensii* as an amorphous substance. According to Wilbur and Watabe (1963), conchiolin of the normal nacreous matrix of *Elliptio complanatus* has the structure of β-keratin, and the structure of α- and β-keratin are found in the regenerating matrix of the same species.

On the basis of X-ray data (lack of wide angle diffraction patterns) (Tanaka *et al.*, 1963), electron microscopy (absence of crossed-striated fibrils) (Grégoire *et al.*, 1955), and biochemical data (absence of hydroxyprolin) (Hare, 1963; Hare and Abelson, 1963–1964) (absence of

* See Table I.

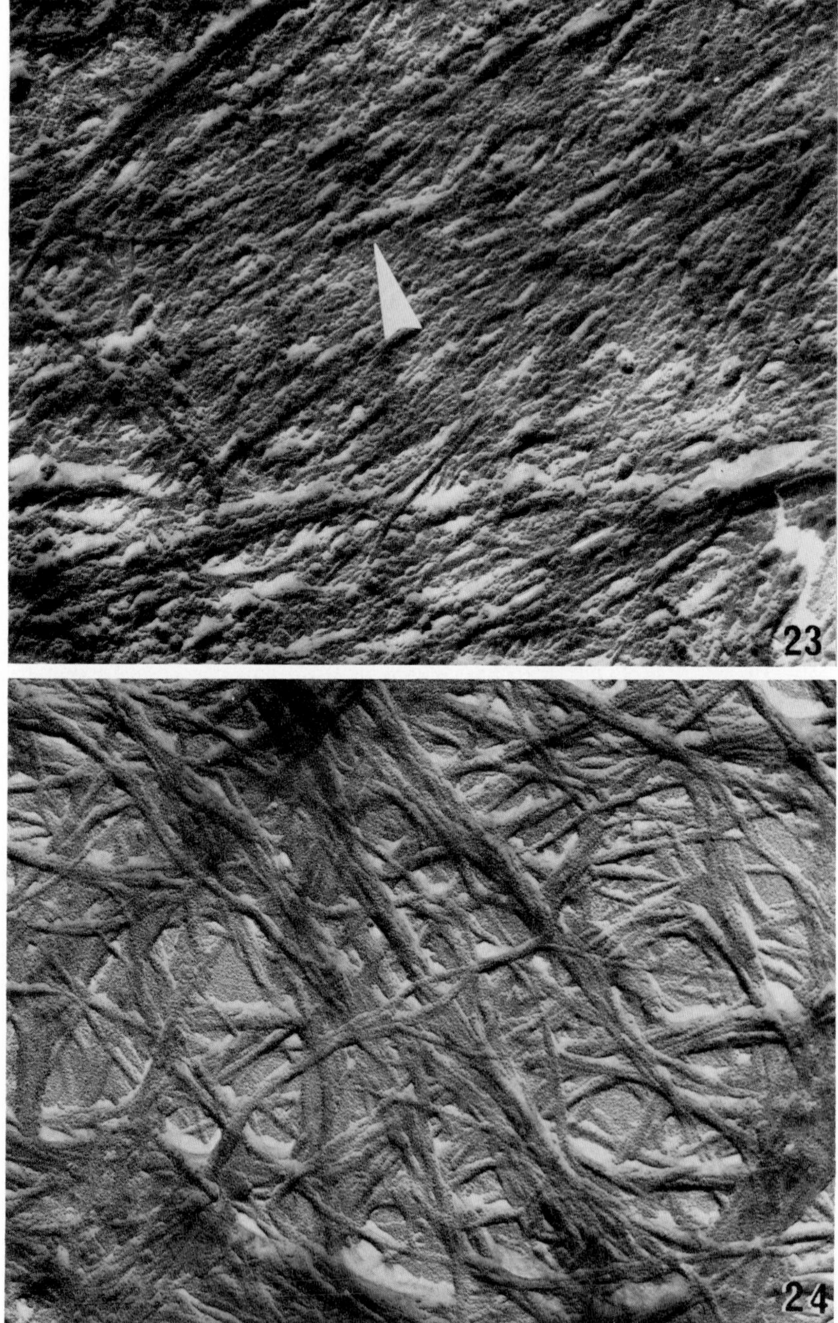
23
24

hydroxylysine except for *Helix pomatia* (Piez, 1961) and of cystine (Roche *et al.*, 1951; Stegemann, 1961a,b), conchiolin differs from collagen and resembles the keratin-myosin-epidermin-fibrin group of proteins (Wilbur and Watabe, 1963; Degens *et al.*, 1967).

Frémy's conchiolin (1855), long considered as a scleroprotein (Schlossberger, 1856). is a complex structure. Successive extractions (Grégoire *et al.*, 1955) (See Table I) decompose the conchiolin of the decalcified mother-of-pearl into three fractions—a water-soluble protein (soluble nacrin), a scleroprotein fraction (nacrosclerotin or insoluble nacrin), and a residual fraction (nacroin). As shown by Goffinet (1965, 1969) and Goffinet and Jeuniaux (1969) using a specific enzymatic method, nacroin contains not only the polypeptides characterized by large amounts of alanine and glycine and detected by Grégoire *et al.* (1955), but constitutes a complex, at least in the *Nautilus* conchiolin, in which the prosthetic group consists of polyaminosaccharides. In the prosthetic group, chitin, formerly detected by Jeuniaux (1963) in the wall and the septa of the *Nautilus* shell, represents about 35% of the nacroin fraction. These conceptions of the conchiolin fraction have been modified by the results of more recent analyses by Voss-Foucart (1970). The basic glycoprotidic complex of the conchiolin appears to be composed of the combined fractions nacrosclerotin (or insoluble nacrin) and nacroin identified by Grégoire *et al.* (1955), and by Goffinet (1965, 1969) (see Table II).

The fractions of conchiolin have been examined with the electron microscope. The original lacelike reticulated structure of the normal conchiolin matrix, variously altered, is still recognizable after elimination of the soluble nacrin. In nacroin, the reticulated structure is disintegrated and many trabeculae are replaced by clusters of fibers, fibrils, and aggregates of microfilaments. Certain microfilaments seem to be composed of chains of beads 20–22 Å in diameter. On the basis of these observations, the glycoprotein complex of mother-of-pearl conchiolin might include the fibrillar elements. Early observations of conchiolin matrices with the electron microscope (Grégoire *et al.*, 1955; Grégoire, 1960)

FIG. 23. *Nautilus macromphalus* Sowerby (Cephalopoda, Nautiloidea). Delaminated sheet from the portion of the inner uncalcified and stratified connecting ring adjacent to the outer spicular layer. These sheets consist of meshworks of fibrils embedded in part in an amorphous or nodular material (white arrow). Platinum shadowed (× 48,000).

FIG. 24. Same material as in Fig. 23, heated to 100°C for 60 hours in a normal solution of sodium hydroxide. The fibrillar meshworks appear cleaned from the nodular or amorphous material. Compare with Fig. 23. Platinum shadowed (× 54,000).

TABLE II
AMINO ACID CONTENT OF NACRE CONCHIOLINS AFTER NACRINE EXTRACTION[a]

Amino acids	Microbiological determination[b] (amino acid nitrogen indicated in percent of total nitrogen)			Chromatographic determination[c] (amino acid residues in percent total amino acid residues)
	Pinctada margaritifera	*Nautilus macromphalus*	*Trochus* sp.	*Nautilus pompilius*
Aspartic acid	6.5	3.9	5.9	6.0
Threonine	0.5	1.0	1.1	1.2
Serine	Could not be recorded with the method used			8.8
Glutamic acid	0.6	3.7	2.9	4.0
Proline	—	0.05	0.3	0.3
Glycine	30.6	26.2	10.6	36.8
Alanine	24.1	25.1	20.5	26.8
Valine	1.6	1.1	2.2	1.3
Methionine	1.8	0.02	0.2	0.3
Isoleucine	1.0	2.2	1.2	1.2
Leucine	5.8	1.8	1.5	2.2
Tyrosine	0.4	0.0	0.0	0.6
Phenylalanine	1.0	3.6	1.8	5.0
Lysine	0.5	0.2	1.1	0.2
Histidine	0.06	0.3	—	0.2
Arginine	12.9	16.0	12.3	5.0

[a] Glycoprotein complex: Voss-Foucart (1970); nacroin: Florkin (1971).
[b] Grégoire *et al.* (1955), p. 14, Table IV.
[c] Voss-Foucart (1970).

already suggested the participation of fibrils to the constitution of the trabeculae of nacreous conchiolin. Fibrils were also observed in fossil remains of nacreous conchiolin (Grandjean *et al.*, 1964; Grégoire, 1966b, 1967b, Pl. 6, p. 661). The fibrillar structure of the nacreous conchiolin has been confirmed in *Pinctada* and *Elliptio* by Watabe (1965), in *Mytilus* by Travis *et al.* (1967), and by Mutvei (1969) in *Nautilus*.

The inner structure of the trabeculae of the normal modern (Grégoire *et al.*, 1955; Wada, 1961a; Tanaka *et al.*, 1963; Grégoire, 1967b) and fossil conchiolins (Grégoire, 1966b) appears to be amorphous, granular, or marbled, with slightly denser rounded areas. Staining in negative contrast reveals ill-defined networks of unoriented microfibrils or bundles of parallel short microfilaments scattered at random (Travis *et al.*, 1967; Grégoire, 1967b). In the *Nautilus* conchiolin, Mutvei (1969) reported a parallel orientation of the microfibrils in the trabeculae.

As reported above, the structure of the organic matrix differs in the various types of mineral architecture of the mollusk shells: perforated lacelike reticulated sheets (in mother-of-pearl and in the sheaths of *Mytilus* prisms), veils, granular and fibrillar membranes (in calcitostracum), networks of microfibrils (in sheaths of calcitic and aragonitic prisms, in porcelainous substance, in interlamellar pillars of the cuttle bone), or sturdy networks of fibers, bundles of microfibrils (in septal nacre of *Spirulidae*). Fibrils seem also to compose uncalcified shell materials, as the brown membrane which coats the chambers in *Nautilus* and the inner concentric membranes of the intracameral portion of the *Nautilus* siphuncle.

At the present time, there is no major objection to the former suggestion (Grégoire, 1960) that a microfibrillar structure made up of filaments 20–40 Å in diameter and frequently resolved in chains of beads might constitute the basic ultramicroscopic framework common to the organic matrices associated or not with the two main polymorphs of calcium carbonate in the mollusk shells. These microfibrils might be concealed and embedded in different organic substances in the lacelike matrices of the nacre, in the veils and membranes of calcitostracum and porcelainous substance, in the sheaths and scattered pellicles of the prisms, and in the membranes of the cuttle bone.

As an example, in the nacre, the taxonomic, generic, or specific differences in the described patterns of structure of the conchiolin matrices, the form of the trabeculae, and the size and distribution of the openings in the fenestration might be determined by variations in the spatial combination of glycoprotein filaments forming the basic framework (nacroin + nacrosclerotin) and the substances (nacrin) embedding these filaments. The tuberosities consistently observed to sprinkle the trabeculae of the nacreous matrices might be one of the morphological aspects of these additional substances. The exact topography and arrangement of the microfibrillar networks within the trabeculae of the nacreous matrices must be accurately determined before one would be able to detect in this type of shell architecture preferential areas of nucleation of the mineral phase on specific biochemical structures (e.g., the ends of lateral chains of amino acids) (Hare, 1963).

VIII. Biochemical Data on Organic Components of Mollusk Shells

Abstracts of the literature are presented in chronological order in Table I. The present situation in this field may be tentatively summarized as follows:

1. The amounts of organic matter in mollusk shells varies greatly, from 0.01% in certain neogastropod shells to 90% in certain cephalopods (Clarke and Wheeler, 1917, 1922; Turek, 1933; Vinogradov, 1953; Akiyama, 1966; Hare and Abelson, 1964–1965).

2. The biochemical composition of the conchiolin generally differs in the different layers and architectures characterized by one of the two types of mineralogical forms of calcium carbonate, aragonite and calcite. When these two polymorphs of calcium carbonate coexist in the same shell (as in bivalves—calcitic prismatic and aragonitic nacreous layers), it has been shown that the relative amounts of these polymorphs are influenced by environmental factors such as temperature and salinity. The diversity in the biochemical composition of the different layers precludes using total shells instead of isolated single layers for biochemical studies on taxonomic and phylogenetic problems.

3. Several reports indicate that there are taxonomic differences at different scales of the hierarchy of the mollusk phylum in the amino acid composition of the shells.

4. The amino acid composition of the extrapallial fluid and that of the periostracum and other nonmineralized organic portions of the shells differ from that of the conchiolin matrices of the calcified portions of the shells. Environmental changes also affect the biochemical composition of the periostracum.

5. In the phylogenetic evolution of the molluscs, the proportion of chitin decreases, that of calcium carbonate increases.

IX. Conclusion

A survey of the literature of the last years shows that among the problems which will be studied in the near future the following topics will probably be preferentially selected:

1. Origin and nature of the precursors of the conchiolin matrices, in the body, in the mantle cells and in the extrapallial fluid, in the different taxonomic groups; influence of changes in environmental conditions on these precursors; and study of the physicochemical processes of polymerization of these precursors leading to the structuration of the conchiolin matrices.

2. Biochemical identification of the conchiolin fractions. The old conception that the insoluble conchiolin material left by decalcification of the shell layers is representative of the organic matrices must be revised;

Friza (1932), Grégoire *et al.* (1955), Meenakshi *et al.* (1969), and Voss-Foucart *et al.* (1969) obtained fractions soluble in water, in borate buffer (pH 9.2), and in decalcifiers. Quite recently, Crenshaw (1970) isolated by decalcification with EDTA a water-soluble glycoprotein amounting to 18% by weight of the organic matrix from the *Mercenaria* shell. In future studies on the conchiolin composition, these soluble fractions should be now carefully considered in order to obtain reliable information on the whole organic components of the shells.

3. Systematic phylogenetic and taxonomic studies in the phylum Mollusca of the variations in the biochemical and ultrastructural composition of the organic matrices associated with the mineralogical forms of calcium carbonate.

4. Studies on the problems of nucleation, crystallographic growth, and orientation of the mineral components of the shells under the influence of the organic matrices (epitaxy).

5. Studies on the factors involved in the mechanism of alternance of the mineral lamellae and the organic matrices in mother-of-pearl.

ACKNOWLEDGMENTS

I thank the following persons for gifts of shell specimens used in these studies: Prof. W. Adam, Prof. P. L. J. Benoit, Dr. R. Catala-Stucki, Dr. S. P. Dance, the late Dr. E. Dartevelde, Dr. Paul S. Galtsoff, and Dr. C. Monty. I also thank Prof. A. Capart, editor of the *Bulletin de l'Institut Royal des Sciences Naturelles de Belgique,* for authorizing reproduction of three figures of former papers.

REFERENCES

Abelson, P. H. (1956). *Sci. Amer.* **195,** 83.

Abelson, P. H. (1959). *In* "Researches in Geochemistry" (P. H. Abelson, ed.), pp. 79–103. New York.

Abelson, P. H. (1963). *In* "Organic Geochemistry" (I. A. Breger, ed.), Monogr. No. 16, p. 451. Earth Sci. Ser., Symp. Publ. Div., New York.

Abolinš-Krogis, A. (1963). *Acta Univ. Upsal.* No. 20, p. 1.

Abolinš-Krogis, A. (1958). *Acta Zool. (Stockholm)* **39,** 19.

Ahrberg, P. (1935). *Arch. Molluskenk.* **67,** 1.

Akiyama, M. (1966). *Proc. Jap. Acad.* **42,** 800.

Amirthalingam, C. (1927). *Nature (London)* **119,** 854.

Appellöf, A. (1892–1893). *Kgl. Sv. Vetenskaps akad., Handl.* **25,** 1.

Barker, R. M. (1964). *Malacologia* **2,** 69.

Barrande, J. (1857) *Neues Jahrb. Mineral. Geog. Geol. Petrefak.,* 679–688.

Beedham, G. E. (1954). *Nature (London)* **174,** 750.

Beedham, G. E. (1958). *Quart. J. Microsc. Sci.* **99,** 341.

Beedham, G. E., and Owen, G. (1965). *Proc. Zool. Soc. London* **145,** 405.

Beedham, G. E., and Trueman, E. R. (1958). *Quart. J. Microsc. Sci.* **99,** 199.

Bevelander, G., and Benzer, P. (1948). *Biol. Bull.* **94,** 176.

Bevelander, G., and Nakahara, H. (1967). *Calcif. Tissue Res.* **1,** 55.

Bevelander, G., and Nakahara, H. (1969a). *Calcif. Tissue Res.* **3**, 84.
Bevelander, G., and Nakahara, H. (1969b). *Calcif. Tissue Res.* **4**, 101.
Biedermann, W. (1902a). *Jena. Z. Naturwiss.* **36**, 1.
Biedermann, W. (1902b). *Z. Allg. Physiol.* **1**, 154.
Biedermann, W. (1914). *In* "Handbuch der vergleichenden Physiologie" (H. Winterstein, ed.), Vol. III, Part I, pp. 319–1188. Jena.
Bøggild, O. B. (1930). *Kgl. Dan. Vidensk. Selsk., Skr.* [9] **2**, 233.
Bricteux-Grégoire, S., Florkin, M., and Grégoire, C. (1968). *Comp. Biochem. Physiol.* **24**, 567.
Brooks, H. (1888). *Proc. Boston Soc. Natur. Hist.* **23**, 380.
Brooks, W. K. (1905). "The Oyster." Baltimore, Maryland.
Brown, C. H. (1952). *Quart. J. Microsc. Sci.* **93**, 487.
Brück, A. (1913). *Zool. Anz.* **42**, 7.
Buckland, W. (1837). "Geology and Mineralogy Considered with Reference to Natural History," Vol. 1. London (quoted by Denton and Gilpin-Brown, 1966).
Bütschli, O. (1908). *Abh. K. Ges. Wiss. Göttingen, Math.-Phys. Kl.* [2] **6**, 1.
Carpenter, W. B. (1844). *Rep. Brit. Ass. Advan. Sci.* p. 24.
Carpenter, W. B. (1848). *Rep. Brit. Ass. Advan. Sci.* p. 93.
Chave, K. E. (1954). *J. Geol.* **62**, 266.
Clarke, F. W., and Wheeler, W. C. (1917). *U.S., Geol. Surv., Prof. Pap.* **102**, 1.
Clarke, F. W., and Wheeler, W. C. (1922). *U.S., Geol. Surv., Prof. Pap.* **124**, 1.
Comfort, A. (1951a). *Biol. Rev.* **26**, 285.
Comfort, A. (1951b) *Proc. Malacol Soc. London* **28**, 79.
Crenshaw, M. A. (1970). *Akad. Wiss. Lit. Mainz, Abh. Math.-Naturwiss. Kl.* July, 1970, (Abstract).
Crick, G. C. (1898). *Trans. Linn. Soc. London* (*Zool.*) [2] **7**, 71.
Dall, W. H. (1889). *Amer. J. Sci.* **38**, 445.
de Bournon, E. (1808). "Traité complet de la chaux carbonatée et de l'aragonite," 3 vols. Londres (quoted by Biedermann, 1914).
Degens, E. T., and Parker, R. H. (1965). *Geol. Soc. Amer., Bull.* **76**, 43 (abstr.).
Degens, E. T., and Spencer, D. W. (1966). *Tech. Rep. Woodshole Oceanogr. Inst.* Ref. No. 66.
Degens, E. T., Spencer, D. W., and Parker, R. H. (1967). *Comp. Biochem. Physiol.* **20**, 553.
Denton, E. J. (1961). *Progr. Biophys. Biophys. Chem.* **11**, 177.
Denton, E. J., and Gilpin-Brown, J. B. (1961). *J. Mar. Biol. Ass. U.K.* **41**, 319.
Denton, E. J., and Gilpin-Brown, J. B. (1966). *J. Mar. Biol. Ass. U.K.* **46**, 723.
de Waele, A. (1930). *Mem. Acad. Roy. Belg. Cl. Sci.* **10**, 1.
Dodd, J. R. (1963). *J. Geol.* **71**, 1.
Dodd, J. R. (1964). *J. Paleontol.* **38**, 1065.
Dodd, J. R. (1965). *Geochim. Cosmochim. Acta* **29**, 385.
Dodd, J. R. (1966). *J. Geol.* **74**, 85.
Dodd, J. R. (1967). *J. Paleontol.* **41**, 1313.
Dunachie, J. F. (1963). *Trans. Roy. Soc. Edinburgh* **65**, 383.
Ehrenbaum, E. (1885). *Z. Wiss. Zool.* **41**, I.
Eisma, D. (1966). *J. Geol.* **74**, 89.
Erben, H. K., Flajs, G., and Siehl, A. (1968). *Abh. Math.-Naturwiss. Kl. Akad. Wiss. Mainz* p. 1.

Florkin, M. (1966). *"Aspects moléculaires de l'adaptation et de la phylogénie."* Masson, Paris.
Florkin, M. (1971). *In* "Chemical Evolution and the Origin of Life" (R. Buvet and C. Ponnamperuma, eds.), p. 10. North-Holland Publ., Amsterdam.
Florkin, M., Grégoire, C., Bricteux-Grégoire, S., and Schoffeniels, E. (1961). *C.R. Acad. Sci.* **252**, 440.
Frémy, E. (1855). *Ann. Chim. Phys.* [3] **43**, 93.
Friza, F. (1932). *Biochem. Z.* **246**, 29.
Galtsoff, P. S. (1964). *U.S., Fish Wildl. Serv., Fish. Bull.* **64**, 1.
Ghiselin, M. T., Degens, E. T., Spencer, D. W., and Parker, R. H. (1967). *Breviora* **262**, 1.
Goffinet, G. (1965). "Conchioline, nacroïne et chitine dans la coquille des mollusques." Mém. Lic. Sci. Zool. Fac. Sci., Univ. Liège. (unpublished).
Goffinet, G. (1969). *Comp. Biochem. Physiol.* **29**, 835.
Goffinet, G., and Jeuniaux, C. (1969). *Comp. Biochem. Physiol.* **29**, 277.
Grandjean, J., Grégoire, C., and Lutts, A. (1964). *Bull. Cl. Sci., Acad. Roy. Belg.* [5] **50**, 562.
Grégoire, C. (1957a). *J. Biophys. Biochem. Cytol.* **3**, 797.
Grégoire, C. (1957b). Unpublished data.
Grégoire, C. (1958a). *Arch. Int. Physiol. Biochim.* **66**, 658.
Grégoire, C. (1958b). *Arch. Int. Physiol. Biochim.* **66**, 667.
Grégoire, C. (1958c). *Arch. Int. Physiol. Biochim.* **66**, 674.
Grégoire, C. (1958d). Unpublished observations.
Grégoire, C. (1959a). *Bull. Inst. Roy. Sci. Natur. Belg.* **35**, 1.
Grégoire, C. (1959b). *Nature (London)* **184**, 1157.
Grégoire, C. (1960). *Bull. Inst. Roy. Sci. Natur. Belg.* **36**, 23, 1.
Grégoire, C. (1961a). *J. Biophys. Biochem. Cytol.* **9**, 395.
Grégoire, C. (1961b). *Bull. Inst. Roy. Sci. Natur. Belg.* **37**, 1.
Grégoire, C. (1961c). *Arch. Int. Physiol. Biochim.* **69**, 374.
Grégoire, C. (1961d). Unpublished observations.
Grégoire, C. (1962). *Bull. Inst. Roy. Sci. Natur. Belg.* **38**, 1.
Grégoire, C. (1962). Unpublished results.
Grégoire, C. (1964). *Nature (London)* **203**, 868.
Grégoire, C. 1966a). In "Advances in Organic Geochemistry" (G. D. Hobson and G. C. Speers, eds.), p. 429. Pergamon, Oxford.
Grégoire, C. (1966b). *Bull. Inst. Roy. Sci. Natur. Belg.* **42**, 1.
Grégoire, C. (1967a). Unpublished data.
Grégoire, C. (1967b). *Biol. Rev. Cambridge Phil. Soc.*, **42**, 653.
Grégoire, C. (1968). *Bull. Inst. Roy. Sci. Natur. Belg.* **44**, 1.
Grégoire, C., and Teichert, C. (1965). *Okla. Geol. Notes* **25**, 175.
Grégoire, C., and Voss-Foucart, M. F. (1970). *Arch. Int. Biochim. Physiol.* **78**, 191.
Grégoire, C., Duchâteau, G., and Florkin, M. (1949). *Arch. Int. Physiol.* **57**, 121.
Grégoire, C., Duchâteau, G., and Florkin M. (1950). *Arch. Int. Physiol.* **58**, 117.
Grégoire, C., Duchâteau, G., and Florkin, M. (1955). *Ann. Inst. Océanogr. (Paris)* **31**, 1.
Grégoire, C., Gisbourne, C. M., and Hardy, A. (1969). *Beitr. Elektronenmikrosk. Direktabbild. Oberflächen* **2**, 223.
Haas, F. (1935). *In* "Klassen und Ordnungen des Tierreichs" (H. G. Bronn, ed.), Vol. III, No. 3, Sect. 3, Part. I, pp. 188–409. Akad. Verlag, Leipzig.
Hackman, R. H. (1960). *Aust. J. Biol. Sci.* **13**, 568.

Hall, A., and Kennedy, W. J. (1967). *Proc. Roy. Soc., Ser. B* **168**, 377.
Hallam, A., and Price, N. B. (1966). *Nature (London)* **212**, 25.
Hallam, A., and Price, N. B. (1968a). *Geol. Mag.* **105**, 52.
Hallam, A., and Price, N. B. (1968b). *Geochim. Cosmochim. Acta* **32**, 319.
Hamilton, G. H. (1969). *Veliger* **11**, 185.
Hare, P. E. (1962). Ph.D. Thesis, California Institute of Technology (quoted by Hare, 1963).
Hare, P. E. (1963). *Science* **139**, 216.
Hare, P. E. (1965–1966). *Carnegie Inst. Washington Yearb.*, p. 364.
Hare, P. E., and Abelson, P. H. (1963–1964). *Carnegie Inst. Wash., Yearb.* p. 267.
Hare, P. E., and Abelson, P. H. (1964–1965). *Carnegie Inst. Wash., Yearb.* p. 223.
Harris, R. C. (1965). *Bull. Mar. Sci.* **15**, 265.
Herdman, W. A., and Hornell, J. (1903). *Brit. Ass. Rep., Southport* p. 695.
Hillman, R. E. (1961). *Science* **134**, 1754.
Hirata, A. (1953). *Biol. Bull.* **104**, 394.
Jackson, R. T. (1890). *Mem. Boston Soc. Natur. Hist.* **4**, 277.
Jameson, L. H. (1912). *Proc. Zool. Soc. London* 260.
Jeuniaux, C. (1963). "Chitine et Chitinolyse." Masson, Paris.
Karny, H. (1913). *Sitzungsber. Akad. Wiss. Wien. Math-Naturwiss. Kl., Abt. 3* **122**, 207.
Kawaguti, S., and Ikemoto, N. (1962). *Biol. J. Okayama Univ.* **8**, 21 and 31.
Kawahara, H., and Yoshiaki, M. (1968). *Bull. Fac. Fish. Hokkaido Univ.* **19**, 52.
Kelly, A. (1901). *Jena. Z. Naturwiss.* **35**, 429.
Kennedy, W. J., Morris, N. J., and Taylor, J. D. (1970). *J. Palaeontol.* **13**, 379.
Kennedy, W. J., Taylor, J. D., and Hall, A. (1969). *Biol. Rev.* **44**, 499.
Kessel, E. (1933). *Z. Morphol. Oekol. Tiere* **27**, 129.
Kessel, E. (1935–1936). *Z. Morphol. Oekol. Tiere* **30**, 774.
Kessel, E. (1940). *Z. Morphol. Oekol. Tiere* **36**, 581.
Kessel, E. (1950). *Zool Anz.* **145**, Ergsbd, 373.
Kessler, P. (1923). *N. Jahrb. Min. Centralbl.* pp. 499 and 689.
Kitano, Y., and Hood, D. W. (1965). *Geochim. Cosmochim. Acta* **29**, 29.
Kobayashi, I. (1964). *Sci. Rep. Tokyo Kyoiku Daigaku, Sect. C* **8**, 295.
Kobayashi, I. (1966). *Sci. Rep. Tokyo Kyoiku Daigaku, Sect. C.* **9**, 189.
Kobayashi, I. (1968). *Venus* **27**, 111.
Kobayashi, S. (1964a). *Biol. Bull.* **126**, 414.
Kobayashi, S. (1964b). *Bull. Jap. Soc. Sci. Fish.* **30**, 893.
Korringa, P. (1951). *Proc. Calif. Acad. Sci.* [4] **27**, 133.
Korringa, P. (1952). *Quart. Rev. Biol.* **27**, 266.
Krinsley, D. (1959). *J. Paleontol.* **34**, 744.
Lange, W. (1941). *Palaeontographica* **93**, Part A, 1.
Leutwein, F., and Waskowiak, R. (1962). *N. Jahrb. Mineral., Abh.* **99**, 45.
List, T. (1902). *Pubbl. Staz. Zool. Napoli, Monogr.* **27**, 1.
Lotmar, W., and Picken, L. E. R. (1950). *Experientia* **6**, 58.
Lowenstam, H. A. (1954a). *Proc. Nat. Acad. Sci. U.S.* **40**, 39.
Lowenstam, H. A. (1954b). *J. Geol.* **62**, 284.
Lowenstam, H. A. (1963). *In* "The Earth Sciences" (T. W. Donnelly, ed.), pp. 137–195. Univ. of Chicago Press, Chicago.
Lowenstam, H. A. (1964). *In* "Recent Researches in the Fields of Hydrosphere,

Atmosphere and Nuclear Geochemistry" (Committee of Sugawara Festival Volume, ed.), p. 373. Maruzen, Tokyo.
MacClintock, C. (1967). *Peabody Mus. Natur. Hist. Bull.* **22**, 140.
Malone, P. G., and Dodd, J. R. (1967). *Limnol. Oceanogr.* **12**, 432.
Manigault, P. (1939). *Ann. Inst. Oceanogr. Monaco* **18**, 331.
Mayer, F. K. (1930). *Chem. Erde* **6**, 2.
Meenakshi, V. R. (1963). *Proc. Int. Congr. Zool., 16th, 1963* No. 2, p. 79.
Meenakshi, V. R., Hare, P. E., Watabe, N., and Wilbur, K. M. (1969). *Comp. Biochem. Physiol.* **29**, 611.
Meenakshi, V. R., Hare, P. E., Watabe, N., Wilbur, K. M., and Menzies, R. J. (1970). *Sci. Results Southeast Pac. Exped.*
Meyer, W. T. (1913). "Tintenfische" (Monogr. Einheimischer Tiere, Leipzig, 1913).
Mitchell, H. D. (1935). *J. Morphol.* **58**, 211.
Mitterer, R. M. (1966). Ph.D. Thesis, Florida State University.
Moynier de Villepoix, R. (1892). *J. Anat. Physiol, Norm. Pathol. Homme Anim.* **28**, 461.
Müller, F. (1885). *Zool. Beitr., Berlin* **1**, 206.
Mutvei, H. (1957). *Ark. Mineral. Zool.* **2**, 219.
Mutvei, H. (1964). *Ark. Zool.* **16**, 221.
Mutvei, H. (1969). *Stockholm Contrib. Geol.* **20**, 1.
Mutvei, H. (1970). *Biomineralization* **2**, 48.
Neumann, P. (1927). "Ueber die optischen Erscheinungen der Perlmutter." Marburger, PhD. Dissert.
Newell, N. D. (1937). *Kans. State Geol. Surv. Bull.* **10**, 1.
Oberling, J. J. (1955). *J. Wash. Acad. Sci.* **45**, 128.
Oberling, J. J. (1964). *Mitt. Naturforsch. Ges. Bern* **20**, 1.
Orton, J. H., and Amirthalingam, C. (1926–1927). *J. Mar. Biol. Ass. U.K.* **14**, 935–53.
Owen, G., Trueman, E. R., and Yonge, C. M. (1953). *Nature (London)* **171**, 73.
Owen, R. (1832). "Memoir on the Pearly Nautilus (Nautilus pompilius Linné) with Illustrations of its External Form and Internal Structure," pp. 1–68. London.
Petitjean, M. (1965). Thesis, Fac. Sci., Univ. Paris.
Pfund, A. H. (1917). *J. Franklin Inst.* **183**, 453.
Piez, K. A. (1961). *Science* **134**, 841.
Piez, K. A. (1963). *Ann. N.Y. Acad. Sci.* **109**, 256.
Pilkey, O. H., and Goodell, H. S. (1963). *Limnol. Oceanogr.* **8**, 137.
Prenant, M. (1927). *Biol. Rev.* **2**, 365.
Price, N. B., and Hallam, A. (1967). *Nature (London)* **215**, 1272.
Ranson, G. (1939). *Bull. Mus. Hist. Natur., Paris* **11**, 467.
Ranson, G. (1940). *Bull. Mus. Hist. Natur., Paris* **12**, 426.
Ranson, G. (1941). *Bull. Mus. Hist. Natur., Paris* **13**, 49.
Ranson, G. (1952). *C.R. Acad. Sci.* **234**, 1485.
Rassbach, R. (1912). *Z. Wiss. Zool.* **103**, 363.
Reumuth, H. (1968). *Mikrokosmos* **57**, 1.
Reyne, A. (1951). *Arch. Neer. Zool.* **8**, 206.
Roche, J., Ranson, G., and Eysseric-Lafon, M. (1951). *C.R. Soc. Biol.* **145**, 1474.
Römer, O. (1903). *Z. Wiss. Zool.* **75**, 437.
Rose, G. (1858). *Abh. Akad. Wiss. Berlin* p. 63.
Royer, M. L. (1928). *Bull. Soc. Fr. Mineral. Cristallogr.* **51**, 7.
Rubbel, A. (1911). *Zool. Jahrb., Abt. Anat. Ontog. Tiere* **32**, 287.

Rucker, J. B. (1965). *Can. J. Zool.* **43**, 351.
Rudall, K. M. (1955). Symp. Soc. Exper. Biol., **9**, 49.
Saleuddin, A. S. M., and Hare, P. E. (1970). *Can. J. Zool.* **48**, 886.
Schlossberger, J. E. (1856). "Die Chemie der Gewebe des gesamten Thierreiches," Winter Verlag, Leipzig, Heidelberg.
Schmidt, M. (1925). *Fortschr. Geol.* **0**, 272.
Schmidt, W. J. (1921a). *Biol. Zentralbl.* **41**, 135.
Schmidt, W. J. (1921b). *Z. allg. Physiol.* **19**, 191.
Schmidt, W. J. (1923). *Zool. Jahrb., Abt. Anat. Ontog. Tiere* **45**, 1.
Schmidt, W. J. (1924). "Die Bausteine des Tierkörpers in polarisiertem Lichte." Bonn.
Schmidt, W. J. (1928). *In* "Die Rohstoffe des Tierreichs" (F. Pax and W. Arndt, eds.), Vol. II, pp. 122–160. Berlin.
Schmidt, W. J. (1931). *Z. Morphol. Oekol. Tiere* **21**, 789.
Schmidt, W. J. (1932a). *Z. Morphol. Oekol. Tiere, Abt. A* **25**, 235.
Schmidt, W. J. (1932b). *Jena Z. Naturwiss.* **67**, 1.
Schmidt, W. J. (1955). *Nova Acta Leopold.* [N.S.] **17**, 497.
Schmidt, W. J. (1959). *Galathea Rep.* **3**, 73.
Schmidt, W. J. (1962). *Z. Zellforsch. Mikrosk. Anat.* **57**, 848.
Seifert, H. (1961a). *Naturwiss. Rundsch.* **14**, 7.
Seifert, H. (1961b). *Chem.-Ing.-Tech.* **33**, 210.
Simkiss, K. (1965). *Comp. Biochem. Physiol.* **16**, 427.
Stary, Z., and Andratschke (1925). *Z. Physiol. Chem.* **148**, 83.
Stegemann, H. (1961a). *Naturwissenschaften* **48**, 501.
Stegemann, H. (1961b). *Z. Phys. Chem. (Leipzig)* **331**, 269.
Stenzel, H. B. (1962). *Science* **136**, 1121.
Stenzel, H. B. (1963). *Science* **142**, 232.
Tanaka, S., and Hatano, H. (1953). *J. Chem. Soc. Jap., Pure Chem. Sect.* **74**, 193.
Tanaka, S., and Hatano, H. (1955). *J. Chem. Soc. Jap., Pure Chem. Sect.* **76**, 602.
Tanaka, S., Hatano, H., and Suzue, G. (1960). *J. Biochem. (Tokyo)* **47**, 117.
Tanaka, S., Hatano, H., and Itasaka, O. (1960b). *Bull. Chem. Soc. Jap.* **33**, 543.
Tanaka, S., Hatano, H., and Ganno, S. (1963). *Rep. Nippon Inst. Sci. Res. Pearls* No. 74, p. 1.
Taylor, J. D., and Kennedy, W. J. (1969a). *Veliger* **11**, 391.
Taylor, J. D., and Kennedy, W. J. (1969b). *Calcif. Tissue Res.* **3**, 274.
Taylor, J. D., Kennedy, W. J., and Hall, A. (1969). *Bull. Brit. Mus. Natur. Hist., Zool., Suppl.* **3**, 1.
Thiele, J. (1893). *Z. Wiss. Zool.* **55**, 220.
Towe, K. M., and Hamilton, G. H. (1968a). *Calcif. Tissue Res.* **1**, 306.
Towe, K. M., and Hamilton, G. H. (1968b). *J. Ultrastruct. Res.* **22**, 274.
Travis, D. F. (1968a). *J. Ultrastruct. Res.* **23**, 183.
Travis, D. F. (1968b). *In* "Biology of the Mouth," Publ. No. 89, p. 237. Amer. Ass. Advan. Sci., Washington, D.C.
Travis, D. F., and Gonsalves, M. (1969). *Amer. Zool.* **9**, 635.
Travis, D. F., François, C. J., Bonar, L. C., and Glimcher, M. J. (1967). *J. Ultrastruct. Res.* **18**, 519–50
Trueman, E. R. (1942). *J. Roy. Microsc. Soc.* [3] **63**, 69.
Trueman, E. R. (1944). *Nature (London)* **153**, 142.
Trueman, E. R. (1949). *Proc. Zool. Soc. London* **119**, 717.

Trueman, E. R. (1950). *Quart. J. Microsc. Sci.* **91**, 225.
Trueman, E. R. (1951). *Quart. J. Microsc. Sci.* **92**, 129.
Trueman, E. R. (1953a). *Quart. J. Microsc. Sci.* **94**, 193.
Trueman, E. R. (1953b). *J. Exp. Biol.* **30**, 453.
Trueman, E. R. (1964). *In* "Approaches To Paleoecology" (J. Imbrie and N. Newell, eds.), pp. 45–74. New York.
Tsujii, T. (1955). *Bull. Biogeogr. Soc. Jap.* **16–19**, 88.
Tsujii, T. (1960). *J. Fac. Fish., Prefect. Univ. Mie* **5**, 1.
Tsujii, T., Sharp, D. G., and Wilbur, K. M. (1958). *J. Biophys. Biochem. Cytol.* **4**, 275.
Tsutsumi, J. (1928). *Mem. Coll. Sci., Kyoto Imp. Univ.* **11**, 217 and 401.
Tsutsumi, J. (1929). *Mem. Coll. Sci., Kyoto Imp. Univ.* **12**, 199.
Tullberg, T. (1882). *Kgl. Sv. Vetenskapsakad., Handl.* **19**, 1.
Turek, R. (1933). *Arch. Naturgesch.* [N.S.] **2**, 291.
Turekian, K. K., and Armstrong, R. L. (1960). *J. Mar. Res.* **18**, 133.
Uozumi, S., and Iwata, K. (1969). Ultrastructure of the conchiolin in *Mytilus coruscus* Gould. Unidentified Japanese periodical **23**, 1.
Vinogradov, A. P. (1953). "The Elementary Chemical Composition of Marine Organisms," Sears Found. Marine Res., Mem. 2. Yale University, New Haven, Connecticut.
von Gümbel, J. (1884). *Z. Deut. Geol. Ges.* **36**, 386.
von Nathusius-Königsborn, W. (1877). "Untersuchungen über nichtcelluläre Organismen, namentlich Crustaceen-panzer Molluskenschalen und Eihüllen." Berlin [quoted by Biedermann (1902ab, 1914) and by W. J. Schmidt (1923)].
Voss-Foucart, M. F. (1968). *Comp. Biochem. Physiol.* **26**, 877.
Voss-Foucart, M. F. (1970). Doctoral Thesis, University of Liége.
Voss-Foucart, M. F., Laurent, C., and Grégoire, C. (1969). *Arch. Int. Physiol. Biochim.* **77**, 901.
Waagen, W. (1867–1870). *Palaeontographica* **17**, 185.
Wada, K. (1957a). *Bull. Nat. Pearl Res. Lab.* **2**, 74 and 86.
Wada, K. (1957b). *Bull. Jap. Soc. Sci. Fish.* **23**, 302.
Wada, K. (1958). *Bull. Jap. Soc. Sci. Fish.* **24**, 422.
Wada, K. (1960a). *Bull. Jap. Soc. Sci. Fish.* **26**, 549.
Wada, K. (1960b). *J. Electronmicrosc.* **9**, 21.
Wada, K. (1961a). *Bull. Nat. Pearl Res. Lab.* **7**, 703.
Wada, K. (1961b). *Venus* **21**, 204.
Wada, K. (1963a). *J. Electronmicrosc.* **12**, 224.
Wada, K. (1963b). *Bull. Jap. Soc. Sci. Fish.* **29**, 320 and 447.
Wada, K. (1963c). *Venus* **22**, 281.
Wada, K. (1964a). *Bull. Jap. Soc. Sci. Fish.* **30**, 127.
Wada, K. (1964b). *Bull. Jap. Soc. Sci. Fish.* **30**, 132.
Wada, K. (1964c). *Bull. Jap. Soc. Sci. Fish.* **30**, 393.
Wada, K. (1966a). *Electron Microsc., Proc. Int. Congr., 6th, 1966* p. 559.
Wada, K. (1966b). *Nature(London)* **211**, 427.
Wada, K. (1966c). *Bull. Jap. Soc. Sci. Fish.* **32**, 295.
Wada, K. (1966d). *Bull. Jap. Soc. Sci. Fish.* **32**, 253.
Wada, K. (1966e). *Bull. Jap. Soc. Sci. Fish.* **32**, 304.
Wada, K. (1967a). *Bull. Jap. Soc. Sci. Fish.* **33**, 613.
Wada, K. (1967b). *Bull. Jap. Soc. Sci. Fish.* **33**, 1002.
Wada, K. (1967c). *Bull. Jap. Soc. Sci. Fish.* **33**, 1007.

Wada, K. (1968a). *Bull. Nat. Pearl Res. Lab.* **13**, 1540.
Wada, K. (1968b). *Bull. Nat. Pearl Res. Lab.* **13**, 1561.
Wada, K. (1968c). *Nature (London)* **219**, 62.
Wada, K. (1969). *Bull. Nat. Pearl Res. Lab.* **14**, 1765.
Wada, K, and Sakai, T. (1963). *Bull. Jap. Soc. Sci. Fish.*, **29**, 658.
Watabe, N. (1963). *J. Cell Biol.* **18**, 701.
Watabe, N. (1965). *J. Ultrastruct. Res.* **12**, 351.
Watabe, N., and Wada, K. (1956). *Rep. Fac. Fish. Prefect. Univ. Mie* **2**, 227.
Watabe, N., and Wilbur, K. M. (1960). *Nature (London)* **188**, 334.
Watabe, N., and Wilbur, K. M. (1961). *J. Biophys. Biochem. Cytol.* **9**, 761.
Watabe, N., Sharp, D. G., and Wilbur, K. M. (1958). *J. Biophys. Biochem. Cytol.* **4**, 281.
Wetzel, G. (1900). *Z. Phys. Chem.* **29**, 386.
Wilbur, K. M. (1960). *In* "Calcification in Biological Systems," Publ. No. 64, pp. 15–40. Amer. Ass. Advan. Sci., Washington, D. C.
Wilbur, K. M. (1964). *In* "Physiology of Mollusca (K. M. Wilbur and C. M. Yonge, eds.), Vol. 1, chapter 8, pp. 242–282. Academic Press, New York.
Wilbur, K. M., and Simkiss, K. (1968). *Comp. Biochem.* **26a**, 229–295.
Wilbur, K. M., and Watabe, N. (1963). *Ann. N.Y. Acad. Sci.* **109**, 82.
Wise, S. W., and Hay, W. W. (1968). *Trans. Amer. Microsc. Soc.* **87**, 419.
Wise, S. W. (1970). *Eclogae geol. helv.*, **63**, 775.
Wise, S. W. (1971). *Eclogae geol. helv.*, **64**, 1.

Note added in proof:

Meenakshi, V. R. and Scheer, B. T. (*Comp. Biochem. Physiol.*, 1970, **34**, 953) performed chemical studies on the organic matrix of the internal shell of *Ariolimax*. Meenakshi, V. R., Hare, P. E. and Wilbur, K. M. (*Comp. Biochem. Physiol.*, 1971, **40**, 1037) established the amino acid pattern of the organic matrix of six neogastropod shells. S. Kobayashi (*Int. Rev. Cytol.*, 1971, **30**, 327) reviewed the role of acid mucopolysaccharides in calcified tissues, especially in mollusk shells. Sh. W. Wise Jr. and J. de Villiers (*Trans. Amer. Microsc. Soc.*, 1971, **90**, 376) showed with the scanning electron microscope that in the pearl oyster *Pinctada radiata* screw dislocation is an important mechanism in the development of the nacreous layer.

CHAPTER 3

Shell Formation in Mollusks

Karl M. Wilbur

I. Introduction

The mollusk in forming its shell carries out a series of integrated processes—physiological, biochemical, and crystallographic—which result in a highly organized structure of calcium carbonate crystals in an organic matrix. A remarkable feature of the formation of shell is that its complex construction is accomplished extracellularly. The organ immediately responsible is the mantle that lines the inner surface of the shell. Our discussion in this chapter will give primary attention to the mineralogical aspects of shell formation. We shall be concerned with the transfer of materials through the mantle to the growing shell surface and with processes of crystal growth and the formation of crystalline layers as the shell increases in area and thickness. In addition, the replacement of shell following injury and the incorporation of minor and trace elements into the shell will be considered.

The past century has provided a rich literature on the mineralogical aspects of molluscan shell structure. Reviews of earlier studies will be found in Schmidt (1921), Bøggild (1930), and Haas and Bronn (1935). More recent studies are summarized in Mutvei (1964, 1970), Petitjean (1965), MacClintock (1967) Wilbur (1964), Wilbur and Simkiss (1968), Taylor *et al.* (1969), and Grégoire, this volume, Chapter 2.

II. The System of Shell Formation—Compartments

In discussing the processes by which the mineral components of shell are deposited, it is convenient to consider the shell-forming system as consisting of four compartments in linear arrangement (Figs. 1 and 2). The first compartment represents the medium—seawater, freshwater, or soil. Adjoining this, in order, are the tissue compartment, which is the mantle, the extrapallial fluid compartment, and the shell. The compartments are interconnected, permitting ions to pass inward from the medium toward the shell and from the shell to the other compartments. A number of studies have demonstrated these intercompartmental interchanges in freshwater and marine mollusks (Wilbur, 1964; van der Borght and van Puymbroek, 1967; Greenaway, 1971a, b). The freshwater snail *Lymnaea stagnalis* can move calcium from very low concentrations

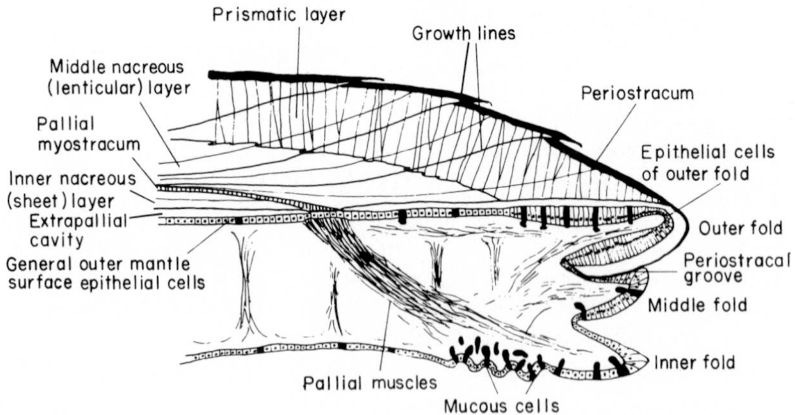

FIG. 1. Diagram of relations between shell and mantle in the peripheral region of the freshwater clam *Anodonta cygnea,* seen in radial section (Taylor *et al.*, 1969, based on Beedham, 1958).

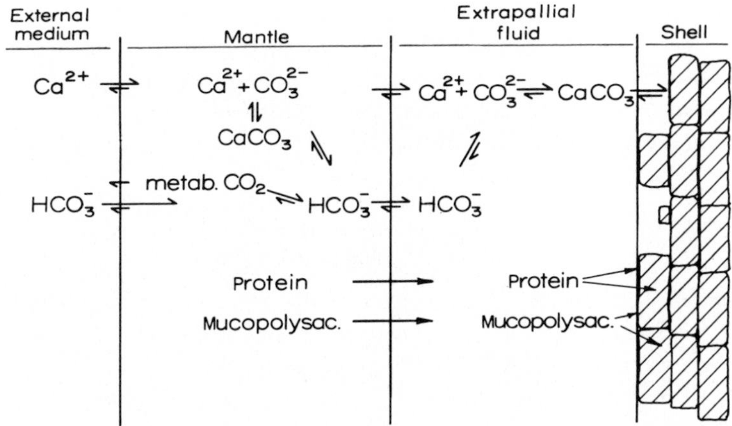

FIG. 2. Diagram of compartments in mollusks showing ions and organic compounds concerned in shell formation (After Wada, 1970a and Greenaway, 1971b).

of the medium into the tissues, blood, and shell against an electrochemical gradient (van der Borght and van Puymbroek, 1967; Greenaway, 1971a). Above 0.5 m*M* Ca/1 there is no adverse gradient (Greenaway 1971a). Movement from the tissues to the medium takes place in this species when the external medium is sufficiently low (van der Borght, 1963; Greenaway, 1971a) or the temperature is markedly reduced (van der Borght and van Puymbroek, 1967). Interestingly, the loss of calcium to a Ca-free medium did not result in low tissue or blood calcium, indicating replacement of blood and tissue calcium by calcium from the shell (Greenaway, 1971a).

Figure 2 illustrates some of the relationships of the materials of individual compartments to the shell. The calcium of shell which normally comes from the medium may also be supplied from solubilized solid calcium carbonate present in the mantle and other tissues. The shell carbonate derives from three sources of bicarbonate: the medium, metabolic carbon dioxide, and tissue carbonate. The organic matrix of shell, consisting mainly of protein and mucopolysaccharide, is secreted by the mantle cells into the extrapallial fluid where it becomes a part of the growing inner surface of the shell. Clearly, the incorporation of ions into the crystalline structure of the shell is directly dependent upon their concentration in the extrapallial fluid compartment at the site of crystallization. The extrapallial fluid concentration, in turn, depends upon concentrations in the blood and tissues which are governed by the rate of intake from the medium, sediment, and food as compared with the rate of movement and excretion to the medium (Odum, 1957). Two additional sources of ions which may participate in shell formation should be mentioned. One is the secreted organic material to which ions may be bound. The other is the external medium which may be in direct contact with the shell at its edge. In some species in which the mantle is not fixed to the shell at the pallial line, there may be access to the inner shell surface generally.

III. The Mantle and Calcium

The form, structure, and location of the mantle are favorable for the transfer of shell materials to the site of deposition. The organ is thin and bounded by a single layer of epithelium on both the side facing the shell and that facing the mantle cavity. Calcium and other materials are carried to the mantle in the blood which passes through an interstitial region between the two epithelial layers (Fig. 3). In addition, ions can pass into the mantle across the inner epithelium facing the mantle cavity and probably at the mantle periphery, where there is contact with the external medium. In freshwater snails calcium is present in the mantle as ions, in bound form, and as calcium spherites (Fig. 4). All

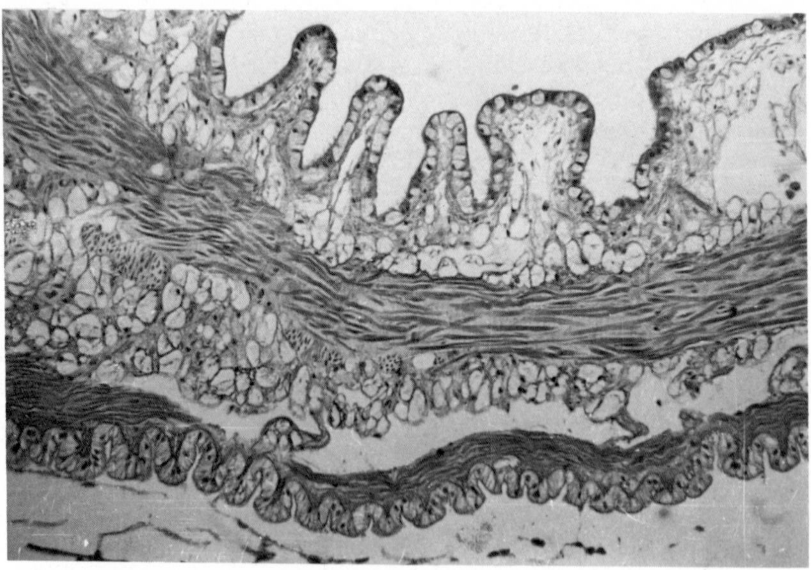

FIG. 3. Cross section through the central portion of the mantle showing the upper epithelium, cavity side (see mucous and ciliary cells), and lower epithelium, shell side. Hemolymph circulates between these two unicellular layers (×300) (Istin and Kirschner, 1968).

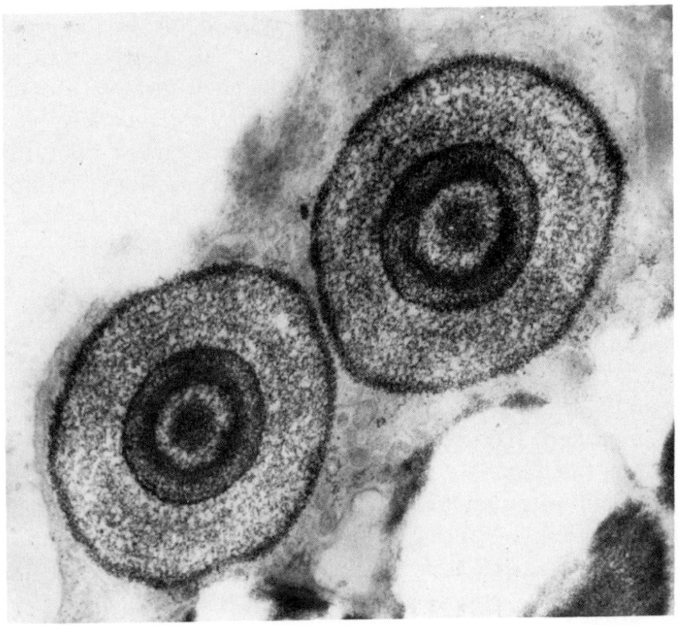

FIG. 4. Calcified granules from the mantle of the freshwater clam *Anodonta cygnea* (× 36,600) (With permission of Dr. Istin, Dr. Masoni, and Dr. Fossat).

three forms probably play a role in shell formation. We shall discuss each in turn, beginning with ionic calcium.

Calcium can move across the epithelium of both sides of the mantle, but movement toward the shell surface is favored by greater permeability on that side. This was shown by using ^{45}Ca and ^{47}Ca to measure the simultaneous movement of calcium in both directions through the isolated mantle of freshwater bivalves (Istin and Maetz, 1964). Movement appears to occur by diffusion. The potentials across the mantle can likewise be interpreted as resulting from a more rapid diffusion toward the shell side. When both sides of the mantle are bathed in Ringer solution of low calcium content, the potential difference may be 25–50 mV, the shell side being positive. Potential measurements on the epithelium of each side demonstrated that most of the potential difference of the whole mantle was due to the external membrane of the cells next to the shell, presumably because of their greater permeability to calcium ions. The change in potential with calcium concentration was slightly less than that predicted by the Nernst equation.

Mucoproteins known to bind calcium (Tanaka and Hatano, 1955; Crenshaw and Young, 1971) provide a method in addition to diffusion for movement of calcium from mantle cells. Some of these substances form the organic shell matrix, and the bound calcium that they transport, may contribute to crystal nucleation on the shell surface. Kinetic studies of calcium in the mantle of marine and freshwater bivalves have shown that the diffusible fraction does not amount to more than 13% of the total and is frequently much less (Istin and Maetz, 1964; Wilbur, 1964). The major portion of the mantle calcium in the freshwater bivalves *Unio* and *Anodonta* can be sedimented readily at low centrifugal acceleration (Istin and Kirschner, 1968; Istin and Girard, 1971a) and presumably consists largely of spherites. The calcareous spherites (Fig. 4) occurring in the mantle are also found in the hepatopancreas and in the connective tissue of the foot of the freshwater gastropod *Helisoma* (Kapur and Gibson, 1968) and visceral sac of the snail *Lymnaea stagnalis* (see Timmermans, 1969, for other references). Ionic calcium entering the tissues from the blood or medium will be deposited on the spherites when the equilibrium favors the precipitation of calcium carbonate. They exhibit a concentric multilayered arrangement which includes fibrous material. Histochemical tests indicate the presence of protein and acid mucopolysaccharide, and perhaps lipid and RNA (Abolins-Krogis, 1960, 1965; Kapur and Gibson, 1968. See also pp. 130 and 131.).

Bound calcium of the mantle is readily converted to ionic calcium, as Istin and Girard (1971a) demonstrated by acidifying a mantle homogenate with 5% carbon dioxide. It appears likely that the calcium spherites in the mantle and other tissues can function as reserves of calcium carbonate, which on ionization can diffuse to the growing shell surface and be deposited as calcium carbonate. In addition to providing reserve calcium carbonate, the calcium spherites may be expected to have a buffer action like that of the calcium carbonate of the shell (see p. 127).

The spherites, consisting of both mineral and organic components, (Fig. 4) merit further attention as calcification systems. Also, the relation of mantle mitochondria to molluscan calcification should be examined in view of the uptake of calcium by mammalian mitochondria (Lehninger *et al.*, 1967; Lehninger, 1970) and the apparent participation of mitochondria of the hen's oviduct in the deposition of calcium carbonate in the eggshell (Schraer and Schraer, 1970).

IV. Environment of Shell Formation

The components of the shell, both crystalline and organic, are provided by the mantle. These materials diffuse from the epithelial cells covering the mantle surface into the extrapallial fluid lying between the mantle and the shell, where they are deposited on the shell surface. Thus, the extrapallial fluid, bounded on one side by the mantle and on the other by the inner shell surface, is the compartment and the medium in which shell formation takes place. Since only a thin layer of extrapallial fluid may exist between the mantle and the shell, the distance involved in transfer of materials from the epithelium to the surface of deposition is small or may be essentially direct. Figure 1 (Taylor *et al.*, 1969) illustrates the relationships between mantle, extrapallial fluid, and shell.

The organic and inorganic constituents of the extrapallial fluid include proteins, mucopolysaccharides, glycoprotein, organic acids, and several inorganic ions. The proteins and mucopolysaccharides in the fluid of many species of marine and freshwater mollusks have been studied by paper and cellulose acetate electrophoresis (Kobayashi, 1964a,b). In mollusks with a calcitic shell, one protein and one or two mucopolysaccharide bands were present, whereas in species with shells of aragonite or both aragonite and calcite, three or more proteins and three or more mucopolysaccharides were evident. The results suggest, but do not demonstrate, that crystal type may be influenced by the organic constituents of the extrapallial fluid. Binding between proteins and mucopolysaccharides was established. Mucopolysaccharide accounts for about one-fifth of the dry weight of the non-dialyzable material in the extrapallial fluid of the clam *Mercenaria,* the remainder being protein (Crenshaw, 1971a).

De Waele (1929) found that the following inorganic ions were present in the extrapallial fluid of the freshwater clam *Anodonta cygnea:* sodium, potassium, calcium, magnesium, manganese, chloride, sulphate, and phosphate. The concentrations of the principal cations and anions in the extrapallial fluid of three marine bivalves are given in Table I (Crenshaw, 1971c). There were evident differences between extrapallial fluid and seawater; all major cations and total carbon dioxide were higher in the extrapallial fluid. The Donnan ratios of the cations between extrapallial fluid and seawater were similar within a species and fell between 1.03 and 1.05, except for calcium which was higher. The bound calcium may occur in a calcium–glycoprotein complex, which is found both in the extrapallial fluid (Crenshaw and Young, 1971) and within the shell (Crenshaw, 1971b). De Waele (1929) considered the extrapallial fluid of *Anodonta* to be equivalent to blood, with similar proportions of constituents. However, later work indicates that the two fluids are distinct. The extrapallial fluid of *Anodonta* lacks catalase present in blood (Florkin and Bosson, 1935); and in *Mercenaria,* the ionic concentrations of extrapallial fluid and blood, though similar, are not identical (Crenshaw, 1971c). The extrapallial fluids of *Mercenaria, Crassostrea,*

TABLE I
INORGANIC COMPOSITION[a] OF EXTRAPALLIAL FLUIDS

Animal	No. of determinations	Na (m*N*)	K (m*N*)	Ca (m*N*)	Mg (m*N*)	Cl (m*N*)	SO_4 (m*N*)	CO_2 (m*M*)	pH
Mercenaria	24	444	9.6	23.6	120	472	46.1	5.2	7.33
mercenaria		±9	±0.8	±2.0	±10	±8	±5.1	±1.9	±0.15
Crassostrea	13	441	9.4	21.5	114	480	48.3	5.0	7.41
virginica		±9	±0.5	±1.7	±6	±9	±2.3	±0.8	±0.16
Mytilus	10	442	9.5	21.3	116	477	47.3	4.2	7.39
edulis		±10	±0.5	±1.2	±6	±8	±2.3	±0.5	±0.17
Seawater	47	427	9.0	18.5	106	496	51.1	2.5	7.91
		±9	±0.1	±0.4	±5	±6	±2.6	±0.1	±0.11

[a] Figures show means and standard deviations. m*N* = millinormal; m*M* = millimolar

and *Mytilus* were slightly alkaline, the mean pH range being 7.33–7.41 (Table I), in good agreement with the values given by Watabe and Kobayashi (see Wilbur, 1964) on *Mercenaria, Crassostrea,* and two other marine bivalves. When the valves were closed, the pH decreased slightly to 7.0–7.2.

From analyses of calcium and carbonate in *Mercenaria,* it appears that the extrapallial fluid is probably saturated with respect to calcite and aragonite (Crenshaw, 1971a). Potts (1954) has calculated that the blood of *Mytilus* and *Anodonta* is also probably saturated or supersaturated with respect to aragonite, which is somewhat more soluble than calcite. The blood of *Helix aspersa* has been found by Burton (1970) to be supersaturated with respect to both aragonite and calcite. Exposure to air results in some precipitation of calcium carbonate in the extrapallial fluid of *Mercenaria* (Crenshaw, 1971a), the blood of *Anodonta* (de Waele, 1929), and the blood of the snail *Otala* (Speeg and Campbell, 1968), a result to be expected following the loss of carbon dioxide and an increase in pH in a saturated solution. Such observations might be taken to indicate that precipitation also may occur *in vivo* through a change in carbon dioxide concentration or pH. However, a marked precipitation of this kind does not occur normally in extrapallial fluid. Instead, crystals develop from nuclei on the growing shell surface, as indicated by the regularity of crystal orientation and the well ordered formation of crystal layers.

The extrapallial fluid and the shell matrix both derive from the mantle and are in intimate contact; however, their relative amino acid composition is strikingly different, as shown in analyses of several species of bivalves (Wada, 1967a,b; Hare, cited in Wilbur and Watabe, 1967). Possible reasons for the difference are considered in Section VII,A.

In concluding the discussion of the extrapallial fluid, four observations may be made relative to its analysis: (1) Since the extrapallial space is commonly a double compartment bounded by differing types of epithelial cells (Fig. 1), the extrapallial fluid secreted into each compartment may be different, as Taylor *et al.* (1969) have pointed out. Analyses to date have concerned only the central compartment. (2) The composition of the fluid in the outer compartment may be influenced by the access of the external medium to this compartment. (3) Since the deposition of shell probably does not proceed continuously or at a uniform rate, adequacy of sampling becomes difficult. (4) Any disturbance to the mantle in sampling is likely to cause reactions which may alter the composition of the fluid.

V. Calcium Carbonate Deposition

Calcification of the shell has most commonly been viewed as the result of two closely interrelated series of processes. The first is the secretion of the extrapallial fluid by the mantle and the crystallization of calcium carbonate from this fluid. The second involves the organic matrix, also secreted by the mantle, and nucleation, crystal orientation, and growth. These aspects of shell formation parallel those proposed for the formation of calcium phosphate in bone (Posner, 1969). A third sequence of events should also be recognized. Shell growth is not continuous, but proceeds in small increments which are individual crystal layers. We now consider these three phases of shell formation.

The probable sources of shell carbonate are the bicarbonate of the outer medium and metabolic carbon dioxide. The utilization of bicarbonate from seawater was shown by placing *Crassostrea* in seawater containing sodium bicarbonate-^{14}C and recovering ^{14}C in shell carbonate and organic matrix (Hammen and Wilbur, 1959). The contribution of metabolic carbon dioxide was demonstrated by injecting bicarbonate-^{14}C and organic compounds into the snail *Otala* and measuring the ^{14}C of the shell carbonate (Campbell and Speeg, 1969). The reactions relating carbon dioxide and carbonate are shown in Eq. I and II.

$$\begin{array}{c} 2CO_2 + 2H_2O \underset{(1)}{\rightleftharpoons} 2H_2CO_3 \underset{(2)}{\rightleftharpoons} 2H^+ + 2HCO_3^- \underset{(3)}{\rightleftharpoons} 4H^+ + 2CO_3^{2-} \\ (4) \upharpoonleft\downharpoonright \\ CO_3^{2-} + CO_2 + H_2O \end{array} \qquad \text{(I)}$$

$$CO_2 + OH^- \rightleftharpoons HCO_3^- \qquad \text{(II)}$$

Carbonic anhydrase, present in the mantle of many species of mollusks, may accelerate calcium carbonate deposition through the catalysis of reactions of Eq. I (1), or II (Maren, 1967) (see Section VI). The formation of CO_3^{2-} from HCO_3^- will be a function of pCO_2 (reaction 4) and pH (reaction 3); and carbonate formation will be favored by the reduction in concentration of carbon dioxide and by removing protons. Hypotheses of calcification involving both of these mechanisms have been proposed and may be briefly mentioned.

De Waele (1929) considered that calcification occurred independently of cellular participation and was due to the loss of carbon dioxide from the extrapallial fluid, resulting in the formation of an insoluble protein–calcium carbonate complex. The carbon dioxide could be eliminated from the peripheral portion of the pallial cavity by diffusion through the thin curtain of periostracum connecting the mantle and the shell edge. In the region of the cavity central to the pallial attachment, carbon dioxide was thought to pass through pores in the shell (de Waele, 1929). To what extent molluscan shells generally may be permeable to carbon dioxide is unknown. However, Speeg and Campbell (1968) observed that a significant amount of carbon dioxide passed outward through the shell of *Otala*.

Two methods of forming carbonate by the removal of protons have been proposed. One is through the action of urease on urea resulting in ammonia production (Speeg and Campbell, 1969). The ammonia then accepts a proton from bicarbonate, and calcium carbonate would be deposited as shown in Eq. III.

$$NH_3 + HCO_3^- + Ca^{2+} \rightarrow CaCO_3 + NH_4^+ \qquad \text{(III)}$$

Ammonia required for the reaction is produced in gastropods (Speeg and Campbell, 1968), bivalves, and cephalopods (Campbell and Bishop, 1970). The contribution of urea to the formation of shell carbonate has been shown by injecting urea-^{14}C into *Otala* (Campbell and Speeg, 1969). In support of this mechanism is the inhibition of shell regeneration in *Helix aspersa* (Campbell and Speeg, 1969) by acetohydroxamic acid which inhibits urease. However, acetohydroxamic acid may also inhibit protein synthesis and thus retard calcification by interfering with shell matrix formation. The formation of ammonia is not limited to the action of urease on urea. Snails lacking urease may form ammonia through adenosine deaminase activity (Tramell and Campbell, 1971). Other possible sources of ammonia are glutamine and taurine (see Campbell and Bishop, 1970).

Another method proposed for the formation of calcium carbonate is the removal of protons by a streaming potential in the extrapallial fluid (Digby, 1968). The source of the potential would be mantle movements which are presumed to pull salt solution through the periostracum. Because cations permeate more readily than anions, the region inside becomes positive relative to the outside. Then, through an electrode effect or a Donnan effect, the inner solution becomes alkaline and the outer solution acid. Near the shell edge a pH of 9.4 is reported in *Mytilus edulis,* with decreasing alkalinity in the extrapallial fluid more distant from the edge. Deposition of calcium carbonate is considered to occur during conditions of elevated pH at sites where calcium is bound to organic matrix. Model experiments in which seawater was forced through the periostracum produced potentials and changes in pH as required by the hypothesis. There are several molluscan systems of calcium carbonate deposition which are not included within the proposed mechanism: (1) mollusks lacking a periostracum; (2) shell repair in which a periostracum is absent (Meenakshi, 1971); (3) terrestrial mollusks not living in an aqueous medium; and (4) cultured pearls in which the pearl nucleus is tightly enveloped by mantle epithelium within the gonad (Watabe, 1971).

More data are needed in support of the hypotheses relating calcium carbonate to ammonia production and streaming potential, including measurement of pH of the extrapallial fluid during calcification and in its absence. Such data are difficult to obtain, however, because the making of measurements may well interfere with the normal processes of calcification. We do not know the pH during the course of deposition of calcium carbonate, but measurements of the extrapallial fluid commonly show the pH to be between pH 7.0 and 7.6 (see Wilbur, 1964; Crenshaw and Neff, 1969). This is not to say that the pH may not be elevated during short periods, as one might suppose from the diurnal variation in ammonia production observed by Speeg and Campbell (1968) in *Otala lactea.* There may also be local areas of higher alkalinity as Digby (1968) reported. Any hypothesis of calcium carbonate deposition must take into account the variation in rate over the shell surface. Deposition in the oyster *Crassostrea* and the snail *Ampullarius* was found to be high at or near the shell edge and to decrease sharply with increasing distance inward (Wilbur and Jodrey, 1952; Zischke *et al.,* 1970). These differences in calcification rate can be explained, on the basis of the streaming potential hypothesis, by a gradient of pH. Considering de Waele's hypothesis (1929), the differences in rate would depend on a gradient of carbon dioxide. Still, a third possibility would be local cellular differences controlling the concentration of calcium or ammonia or of the amount of secreted matrix.

VI. Carbonic Anhydrase and Calcium Deposition

Carbonic anhydrase, an enzyme present in the mantle and blood, appears to play a role in calcification. It is widely distributed in the phylum Mollusca (for references, see van Goor, 1948; Wilbur, 1964; Polya and Wirtz, 1965). In freshwater bivalves, the carbonic anhydrase of the mantle is largely in an insoluble form (Istin and Girard, 1971b), in contrast to the soluble form found in vertebrates (Maren, 1967). Istin and Girard suggested that the enzyme may be located on the surface of the calcium spherites. Attempts to solubilize the enzyme through the use of pepsin, chymotrypsin, collagenase, β-mercaptoethenol and dodecyl sulfate were unsuccessful. Carbonic anhydrase has been detected in the mantle epithelium of the snail *Lymnaea stagnalis* (Timmermans, 1969) by using a histochemical test, the specificity of which has been questioned (Maren, 1967).

Carbonic anhydrase is presumed to increase the rate of calcium deposition in shell formation through the catalysis of Eq. I (1) or II, as we have mentioned. Both reactions would have the effect of increasing the rate of carbonate formation. The evidence for a role of carbonic anhydrase in shell formation is based on the presence of the enzyme in many, but not all, species of mollusks (Freeman and Wilbur, 1948) and the action of sulfonamide compounds, which are strong inhibitors of the enzyme. These compounds retard calcium deposition in normal shell (Wilbur and Jodrey, 1955; Freeman, 1960, Timmermans, 1969) and in regenerating shell (Stolkowski, 1951; Abolins-Krogis, 1968). Calcium carbonate deposition in the shell of the hen's egg is also decreased by a sulfonamide inhibitor of carbonic anhydrase (see Schraer and Schraer, 1970).

It is expected that an increase in carbon dioxide and the resulting higher hydrogen ion concentration would cause a decrease in insoluble carbonate of the mantle and thereby increase ionic calcium. The addition of 5% carbon dioxide to a mantle homogenate did increase ionic calcium (Istin and Girard, 1971a). However, the increase did not follow the expected relation and was repressed by a carbonic anhydrase inhibitor. The results have been interpreted as indicating that the equilibrium between the ionized and combined forms of calcium are controlled by carbonic anhydrase which may be localized on the calcium spherites (Istin and Girard, 1971b).

The carbonic anhydrase of the accessory boring organ of the gastropods *Thais* and *Urosalpinx* (Chétail and Fournié, 1969; Smarsh *et al.*, 1968) is thought to be involved in producing the acid which permits these animals to penetrate the shells of bivalves [see Eq. I (1) and (2)]. The hydrogen ions are released to the exterior, as indicated by a decrease in pH (Carriker *et al.*, 1967), and dissolve shell calcium carbonate. Calcium was found in increased concentration in the accessory boring organ during boring and is thought to pass into the cells in exchange for hydrogen ions (Chétail and Fournié, 1970).

VII. Shell Matrix and Crystallization

A. General Features

Two aspects of shell formation presenting challenging problems of analysis are the mechanisms involved in the highly ordered growth of calcium carbonate crystals in a variety of arrangements and the formation of individual crystal layers. The organic matrix is probably a key factor in both.

The major portion of the shell matrix is protein, some of which is tanned (Beedham, 1958). Mucopolysaccharide, lipid, and glycoprotein are also present (Wilbur and Simkiss, 1968; Crenshaw, 1971b). Details of matrix ultrastructure are discussed in Chapter 2. We will consider here those features of the matrix which relate to deposition of calcium carbonate crystals.

Shell protein can be separated into a water-soluble and a water-insoluble fraction (Meenakshi, in Wilbur and Simkiss, 1968; Voss-Foucart *et al.*, 1969; Crenshaw, 1971b). The soluble fraction is intracrystalline and presumably corresponds to the intracrystalline matrix described by Watabe (1965). Its composition varies in different species (Meenakshi *et al.*, 1971). The insoluble fraction is primarily intercrystalline, but its occurrence within crystals is not excluded. Analyses of the entire shell matrix (see Wilbur and Simkiss, 1968) and of the soluble and insoluble fractions (Meenakshi *et al.*, 1971) show that the matrix represents more than one protein rather than different states of polymerization of a single protein.

The mechanisms by which organic material secreted by the mantle becomes matrix have not been examined experimentally, but indirect evidence indicates at least three processes:

1. Matrix may be secreted by the mantle directly on the inner shell surface. Such matrix would appear in the extrapallial fluid only secondarily. Direct deposition is supported by the consideration that the extrapallial space peripheral to the pallial attachment in most Bivalvia adjoins two shell layers differing in structure, the more peripheral prismatic layer and the more central foliate layer (Fig. 1). The crystal type may be different as well, the prismatic layer being calcite and the foliate layer aragonite. The formation of these two kinds of shell could scarcely take place within the single common pool that bathes both, but this could be brought about by distinct mantle regions secreting matrix material differing in composition. De Waele (1929), on the other hand, proposed that differences in crystals and matrix could be the result of differences in dilution of the extrapallial fluid by water

entering through the periostracum at the mantle edge and from differences in the rate of loss of carbon dioxide.

2. Polymerization or hardening of proteins may take place subsequent to secretion (Wada, 1968). This follows from the sheetlike nature of the insoluble matrix separating crystal layers. Since the extrapallial fluid and the shell matrix differ markedly in amino acid composition (Wada, 1967a,b; Hare, cited in Wilbur and Watabe, 1967), any polymerization of protein and polypeptides would be selective, or would involve compounds secreted directly on the inner shell surface.

3. Soluble organic material is secreted into the extrapallial fluid and incorporated into the matrix as a soluble fraction. This was shown by the presence of a soluble calcium-binding glycoprotein both in the extrapallial fluid and within the shell crystals (Crenshaw and Young, 1971; Crenshaw, 1971b).

B. Crystal Nucleation

The relation of the composition and structure of the organic matrix to the crystalline phase of shell formation has not been extensively studied; and the few data available are inadequate to provide a picture of this basic aspect of shell formation. Of necessity, our remarks will be largely limited to some of the ideas advanced concerning possible relationships.

The organic matrix may play a role in crystal nucleation. Two mechanisms have been advanced. In the first, Ca^{2+} is thought to attach to carboxyl groups of aspartic and glutamic residues, and CO_3^{2-} is presumed to attach to amino groups of lysine or glucosamine (Matheja and Degens, 1968; see also Wilbur and Simkiss, 1968). Ca^{2+} and oxygen would form coordination polyhedra and provide the crystallographic structure of calcite (CaO_6) or aragonite (CaO_9). These polyhedra would then serve for nucleation of calcium carbonate crystals. Analyses show relatively high concentrations of aspartic and glutamic residues in matrix, including the soluble matrix (Matheja and Degens, 1968; Crenshaw, 1971b; Meenakshi *et al.*, 1971). However, if these residues occur as glutamine and asparagine, as indicated by Crenshaw's analyses of *Mercenaria*, Ca^{2+} could not be bound to them.

In the second mechanism, Ca^{2+} would be selectively bound to a highly sulfated glycoprotein reported in the soluble matrix of *Mercenaria* (Crenshaw, 1971b). It is suggested that Ca^{2+} could be chelated by ester sulfates located on adjacent polysaccharide chains (Marler and Davidson, 1965; Crenshaw, 1971b). A change in conformation of the glycoprotein could release calcium, causing a local increase in calcium and the initiation of crystal nuclei formation (Crenshaw, 1971b). Wada and Furuhashi (1970) suggested that in shell matrix an acid polysaccharide which appears to be part of a glycoprotein with sulfuric acid ester groups may be involved in mineralization. The possible role of acid mucopolysaccharide in mineralization is indicated by crystal formation around centers giving histochemical tests for these compounds on surfaces inserted between the mantle and shell (Saleuddin and Chan, 1969; Wada, 1970a).

C. Crystal Type

The crystal type—whether calcite, aragonite, or vaterite—may depend in part on the nature of the matrix on which the crystal is formed. The evidence comes from experiments in which crystals were deposited on different types of matrix and from regeneration studies in which both the crystal type and the matrix became altered (Wilbur and Watabe, 1963; Saleuddin and Hare, 1970). Because of the difficulty of controlling all factors, the evidence at present must be considered suggestive rather than firm.

D. Crystal Orientation

One of the striking, but not invariable, characteristics of molluscan shell is the uniformity of crystal orientation in any given region. The molecular orientation of the matrix may be one of several factors influencing crystal orientation. A study of the fine structure of the matrix in species of bivalves, gastropods, and a cephalopod has shown the presence of oriented fibrils (Mutvei, 1969); and in the bivalves *Pinctada* and *Pteria* the direction of matrix fibrils was observed to correspond with the b-axis of developing crystals (Wada, 1970a). However, the evidence for the role of the matrix in this respect is not compelling (Wilbur and Simkiss, 1968).

E. Inhibition of Crystal Growth

It appears probable that the deposition of matrix on the surface of crystals inhibits their growth and thus limits the thickness of individual crystal layers. This aspect will be considered in the discussion of crystal formation in the following section.

VIII. Formation of Crystal Layers

In considering crystallographic aspects of shell formation, we shall give attention to the foliate shell of the nacreous layer present in bivalves, gastropods, monoplacophorans, and cephalopods. The basic structure is a series of layers of tabular crystals, one crystal in thickness and separated by interlamellar organic matrix (Fig. 2). The unit of structure may be considered to be a single layer of crystals bounded above and below by matrix. Each crystal is also enveloped in a sheath of intercrystalline matrix. Organic matrix is also present within the crystals (Watabe, 1965; Mutvei, 1970) and is termed intracrystalline matrix. Calcium carbonate crystals form within the organic matrix and in contact with the extrapallial fluid, which is probably saturated with respect to aragonite and calcite.

Crystalline layers or lamellae of the nacreous region can be initiated in at least three ways, as described by Wada (1970a) from electron microscopic observation. (1) Microcrystals may develop on a flat crystal surface covered with organic matrix and, by growth and coalescence, produce a new layer in a localized area (Fig. 5). By a repetition of this process successive layers can be formed. (2) Spiral growth steps may form on the 001 plane of crystals (Fig. 6). (3) Steps forming an angle

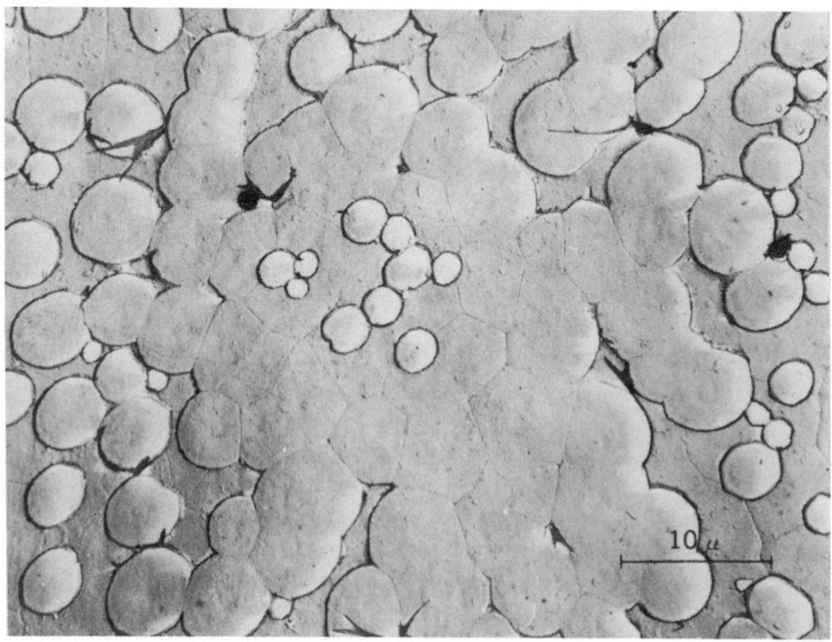

FIG. 5. Deposition of aragonite microcrystals at the center of a concentric growth hillock in *Pinctada*. (Wada, 1970a)

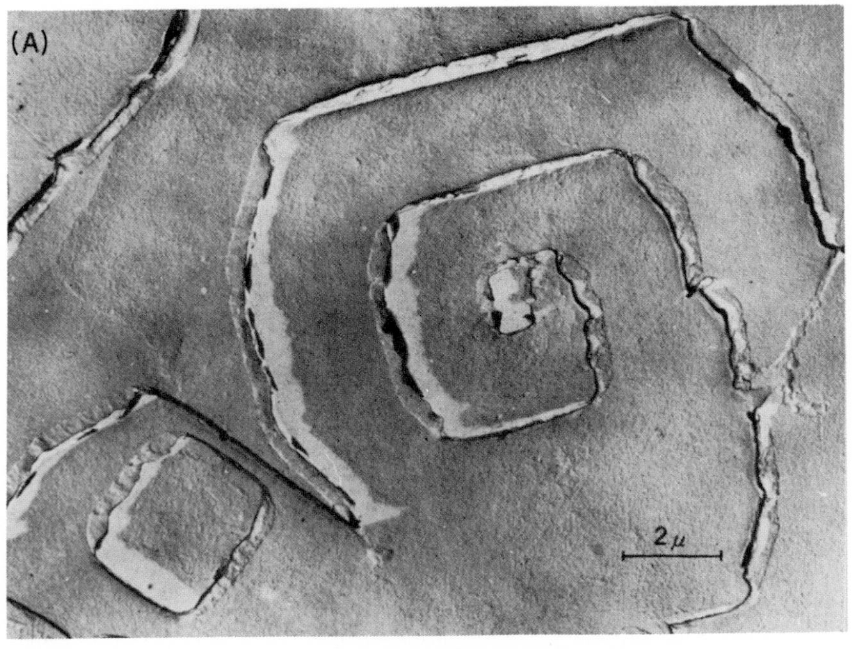

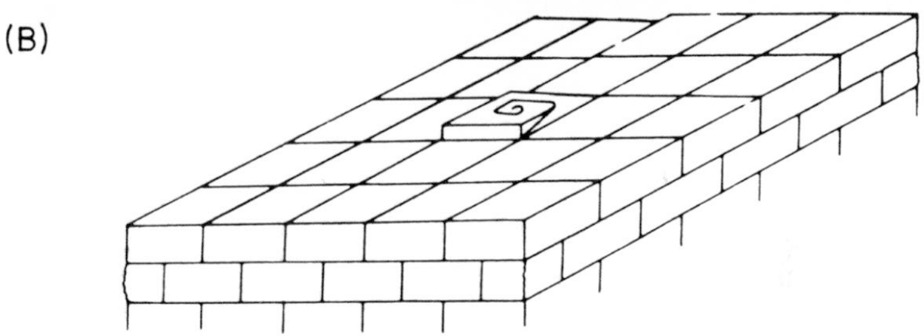

FIG. 6. A. Spiral growth step with a large step height on an aragonite crystal from the shell nacre of *Pinna attenuata*. B. Diagram of a spiral growth step. (Wada, 1970a)

may originate from differences in inclination between adjacent crystals or groups of crystals (Fig. 7).

We now turn to details of formation and growth of crystals in layer formation. Our discussion of the processes which come into play must, of necessity, lean heavily on speculation, since there is a paucity of solid information. As a first step, we will assume that the mantle has secreted organic material which forms a sheet of insoluble interlamellar matrix. Soluble matrix, differing in composition and probably secreted sepa-

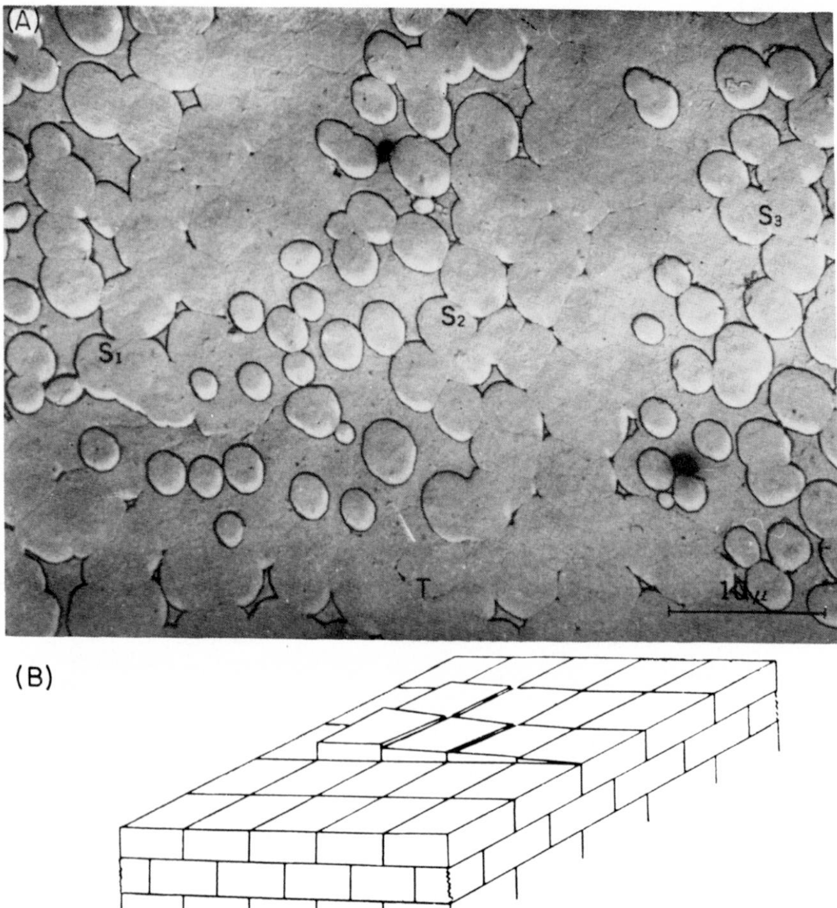

Fig. 7. A. Inclined steps (S_1, S_2, S_3) rising above a terrace (T) in the nacre of *Pinctada*. B. Diagram of an inclined step. (Wada, 1970a)

rately, will form a second phase. Then another sheet of interlamellar matrix is secreted (Bevelander and Nakahara, 1969). Between the two sheets of interlamellar matrix, and perhaps before the second sheet is deposited, crystal nuclei form within the soluble matrix. The crystal develops by the formation and growth of dendrites. Since the dendrites develop in the soluble matrix, this material becomes enclosed within the crystal as the crystal grows (Watabe, 1965). In this way, the soluble matrix comes to be intracrystalline. Nucleation and crystallite formation may result in the formation of small crystals in close proximity. Some of these coalesce and continue to grow; others are dissolved. The crystal will grow in thickness until it reaches the upper (Wada, 1970a) and lower interlamellar membranes or until matrix secretion on the crystal surface inhibits further increase in thickness (Wilbur, 1964; Watabe, 1965; Wada, 1968). However, lateral growth continues until neighboring crystals in the same plane have made contact, filling in all the space and so completing the layer (Wada, 1968). During growth of the crystal, the matrix that surrounds it is displaced (Bevelander and Nakahara, 1969) and forms the intercrystalline matrix sheath. This sheath becomes contiguous with the interlamellar matrix above and below the crystal and with neighboring crystals laterally. If crystallization does proceed in this manner, then the intercrystalline matrix would appear to be derived from the soluble matrix. Yet when the intercrystalline matrix is extracted, only insoluble protein is obtained, suggesting that any soluble matrix between crystals may perhaps become insoluble as in the case of the interlamellar matrix.

The structure of shell, with its alternating layers of crystals and sheets of organic matrix, indicates a well controlled periodic process. In the sequence described, Ca^{2+} and CO_3^{2-} would be in continuous supply from the mantle and the periodicity would be governed by the formation of sheets of insoluble matrix which, through inhibition of crystal growth, would determine crystal thickness (Watabe, 1965; Wada, 1968). An alternating secretion of insoluble and soluble matrix would also account for the layered structure, but periodic formation of insoluble matrix alone would be adequate (Meenakshi *et al.*, 1971). Another possible sequence giving the layered structure would be one in which growth of the calcium carbonate crystals would be controlled by periodic changes in concentrations of Ca^{2+} and CO_3^{2-} in the extrapallial fluid. Another variation would be a double alternation of calcium carbonate deposition and organic matrix secretion. Of these various mechanisms for producing crystal layers, the periodic secretion of matrix resulting in the inhibition of crystal growth would seem simplest. But data at hand are insufficient to provide a probable answer. Further, it is well to recognize the possibility that all species may not employ the same mechanism of crystal layer formation.

A study of the growing shell surface by transmission and scanning electron microscopy has shown that several crystal layers may form at the same time (Fig. 8) (Grégoire, 1962; Wise and Hay, 1968; Mutvei, 1970; Erben, 1971). Concurrent crystal growth in overlying layers requires movement of calcium and carbonate ions through the interlamellar membranes. The thin areas in the membranes described by Mutvei (1969) would presumably facilitate ionic movement. Fenestrations de-

FIG. 8. Scanning electron micrograph of the growth surface of the nacreous layer of *Turbo castanea,* a marine snail. New crystals are deposited at the tops of the stacks. Some crystals have been removed by acetate peels ($\times$3200) (With permission of Dr. Wise).

scribed by Grégoire *et al.* (1955), which would allow for ionic interchange, are thought to be the result of the method of membrane preparation (Mutevi, 1969).

Wada (1970b) has suggested that fenestrations in the interlamellar membranes that he finds in shell sections would also permit continuity of crystal growth through the membrane in an overlying layer (Bevelander and Nakahara, 1969). The new crystal, then, would grow laterally and vertically until it reached the next interlamellar membrane, when its continued vertical growth could initiate a third crystal layer, and so on. Because crystal thickness in tabular crystals is small as compared to surface area, several layers of crystals may form before lateral growth has been completed (Wise and Hay, 1968; Erben, 1971). The result may be a stack of superimposed growing crystals of equal thickness but of decreasing surface area, as shown in Fig. 8. Crystal continuity from layer to layer would also provide uniformity of crystal orientation, as Barker (1964b) has pointed out. However, information presently available does not permit an assessment of the importance of crystal continuity to orientation in shell generally. Moreover, consecutive layers may have different crystal orientation (Mutvei, 1970).

The individual crystals of molluscan shell are made up of units. From observations of thin sections of bivalve nacreous crystals, Watabe (1965) described the units as component blocks 0.2–0.5 μ in width and 1 μ in length. These presumably are formed by dendritic growth. Mutvei (1970) found the nacreous crystals of gastropods, bivalves, and *Nautilus* to consist of somewhat smaller mineral units termed laths which were made up of acicular elements, all oriented parallel to the crystallographic a-axis. The structures were revealed by removing organic material with hypochlorite and ion beam irradiation and by etching with chromium sulfate. The relation between the component blocks of Watabe and the crystalline laths of Mutvei remains to be clarified. Both studies, however, provide evidence of the crystalline components with an intracrystalline organic matrix. Towe and Hamilton (1968), on the other hand, do not support the view that the crystal is made up of subunits each surrounded by matrix.

Foliate shell represents one of the simplest of molluscan shell structures. The intimate relation between growing crystals and organic matrix described in this case is undoubtedly found in all molluscan shells. However, it should be appreciated that the structural characteristics of shell types are strikingly different. A discussion of the various types—foliate, prismatic, cross-lamellar, homogeneous, and myostracal—together with diagrams, photographs, and references, will be found in Grégoire (Chapter 2, this volume) and in Taylor *et al.* (1969).

IX. Growth Increments

One of the striking features of molluscan shell is its growth by increments. We have discussed one kind of increment, the individual crystal layer, which is a unit of shell thickness. Another kind of increment of shell growth is produced as the mantle grows in area and length. This growth unit is seen as bands or surface markings.

Bands 5–10 μ in width are a common feature of prismatic (Grégoire, Chapter 2, this volume), cross-lamellar, complex cross-lamellar, and homogeneous layers of bivalve shells (Taylor *et al.*, 1969). Other markings, regularly or irregularly spaced, may be of microscopic or macroscopic dimensions. Growth periods reflected in the banding have been shown to be annual, monthly, bidaily, daily, and less than one day. Daily growth markings have been established for the bivalves *Mercenaria* (Pannella and MacClintock, 1968), *Pecten* (Clark, 1968), *Cardium* (House and Farrow, 1968), the limpet *Acmaea* (Kenny, 1968), and the cuttlefish *Sepia* and *Siepiella* (Choe, 1963). To the biologist, such markings are of interest as indices of metabolic events and growth rates. To the geologist, the markings in fossil shells serve as indicators of environmental changes and help determine the length of synodic months in past ages (Pannella *et al.*, 1968). Periodic markings are not peculiar to molluscan shells but occur in several other calcifying systems, including teeth (Saxon and Higham, 1968), fish otoliths (Pannella, 1971) corals (Runcorn, 1966), insect cuticle, and crayfish gastroliths (Neville, 1967).

Studies on marked specimens have provided measures of linear growth rate under various natural environmental conditions. It is noteworthy that the daily pattern and machinery of shell formation continue near 0°C in nature and over several months in the laboratory without food (House and Farrow, 1968; Rhoads and Pannella, 1970). Light is one of the factors affecting the rate of growth of certain species (*Ampullarius, Acmaea*), as shown by the inhibition of linear growth in darkness (Zischke *et al.*, 1970; Kenny, 1969). Calcium deposition in the marine gastropod *Purpura,* as measured by ^{45}Ca, drops off sharply at night (Zischke, 1969) and is correlated with reduced feeding. On the other hand, light had no effect on calcium deposition in *Mytilus, Equipecten* (Dodd, 1969), or *Ampullarius* (Zischke *et al.*, 1970). Tidal action has been thought to cause the fine layering seen in cross sections of shell (Barker, 1964). However, the fine surface markings cannot be so explained, since their frequency of formation does not conform to tidal periods. In *Acmaea,* four to six striations are formed each day (Kenny, 1968), and in *Mercenaria* the number is larger (Fig. 9). Subdaily markings have been thought to be associated with periods of movement or valve closure (Pannella and MacClintock, 1968), but critical experimental data are needed for confirmation. Interestingly, the microscopic markings on the bivalve *Nucula cancellata* taken from 4970 meters were uniform with no apparent seasonal or tidal effects (Rhoads and Pannella, 1970). Regular surface markings were also found on several species of *Malletia* and *Abra* from 2000 meters to 5000 meters (Kilham, 1971).

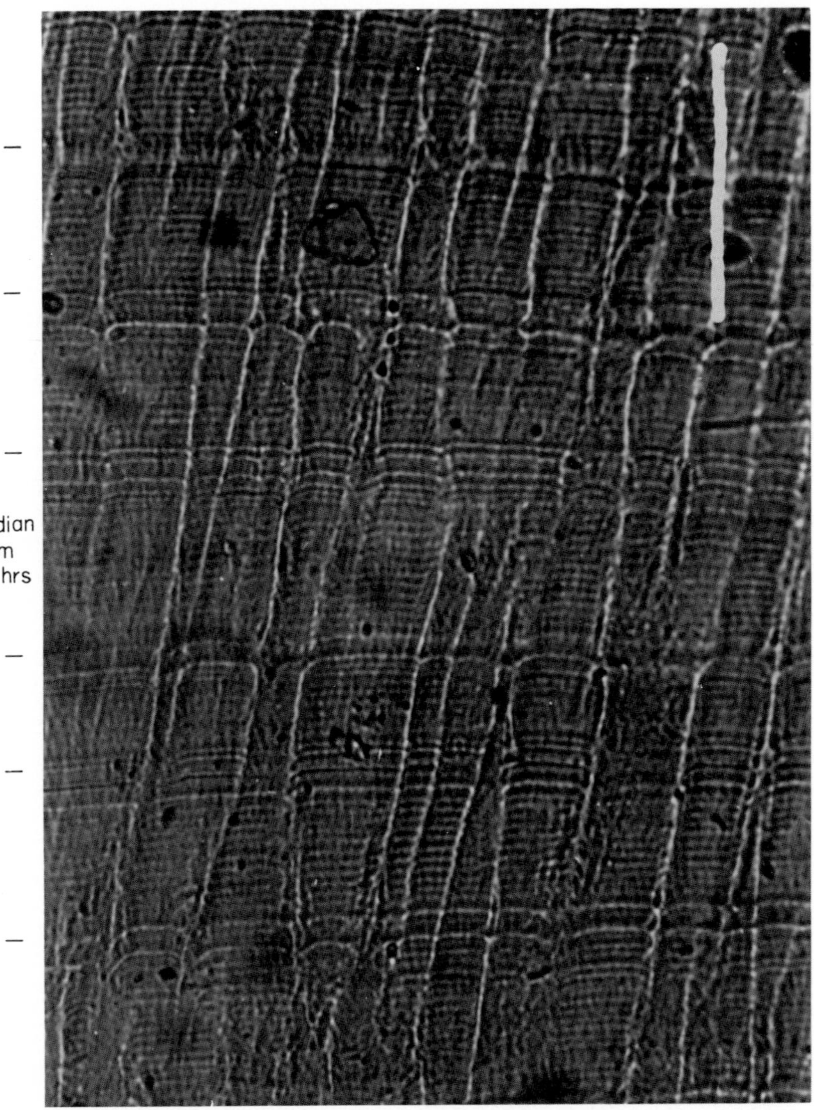

FIG. 9. Acetate replica of *Mercenaria campechensis*. The seven bands are presumed to represent growth increments of 24 hours. The white bar represents 100 μ. (With permission of Dr. Pannella.)

The presence of markings on shells from the deep sea where light is absent and the environmental factors are relatively constant points to an endogeneous rhythm of shell deposition. Such a rhythm also appears in the limpet *Acmaea* (Kenny, 1968) and the gastropod *Ampullarius*

(Zischke *et al.*, 1970) maintained under constant illumination in the laboratory. The latter species has a most interesting system of periodic shell deposition in its linear growth. In a rapidly growing specimen, thirty-six bands may be deposited each day, an average of one every 40 minutes. Each band is made up of some two hundred rows of crystallites 500–600 Å in thickness. Since the shell is of cross-lamellar structure, there is an alternation of crystal orientation with the formation of each band.

The rate at which crystalline lamellae are formed in bivalves has not been investigated extensively. However, in the pearl oyster it appears that as many as 9 to 15 lamellae may be formed daily (Wilbur and Watabe, 1971), as estimated from the rate of growth of shell thickness (Nakahara, 1961) and the thickness of lamellae (Wada, 1961). From an examination of pearls of known age, Watabe (1952) determined that the tissue surrounding the pearl formed two to eight lamellae per day.

It will be apparent that incremental growth of shell in forming crystal layers could occur in more than one way. We will mention two, realizing that other schemata might be proposed to account for the periodicity.

1. Let us assume that there is continuous movement of calcium and organic matrix from the mantle to the inner shell surface, with growth of calcium carbonate crystals in the matrix. If, at intervals, the composition of the extrapallial fluid were altered temporarily, bringing about a change in the matrix (e.g., hardening) such that crystal growth was limited, then that crystal layer would obviously no longer increase in thickness. With resumption of the previous condition, a new layer would form. A periodic change in the extrapallial fluid such as that mentioned could involve a feedback mechanism governed by pH or other factors at the surface of the mantle epithelium.

2. Rhythmic increases in the pH of the extrapallial fluid with a consequent precipitation of calcium carbonate has been suggested as a mechanism of producing crystal layers (Campbell, 1970). A possible source of the increased pH is ammonia production, as explained earlier (Speeg and Campbell, 1968). The precipitation of calcium carbonate would be accompanied by a decrease in pH. As more calcium passed into the extrapallial fluid and ammonia production continued, the cycle would repeat.

Final Comments

Growth of the shell proceeds as units of thickness and, commonly, as units of area and length. Stated differently, shell growth is incremental, not continous.

The mechanism of producing growth units remains to be investigated. However, it is evident that the growth units have their metabolic counterparts in the mantle or other organs. The metabolic events associated with periodic deposition of shell are completely obscure.

Unit deposition of crystal layers may rest with the control of shell matrix deposition or changes in the composition of the extrapallial fluid. The availability of calcium appears to be a less likely cause of unit deposition since it is probably supplied to the extrapallial fluid by diffusion from the mantle.

Although the formation of bands and shell markings results from growth of the mantle, unit growth of the mantle has not been established.

X. Shell Decalcification

Shell growth is commonly considered a unidirectional increase in mass of calcium carbonate and organic matrix. However, in some bivalves at least, turnover of the shell mineral probably takes place when the valves are closed and may be looked upon as a normal aspect of the metabolism of the organism (Crenshaw and Neff, 1969). Evidence of turnover is seen in the etching of crystals on the inner surface of mollusk shells (Watabe, 1971) and on the surface of growing pearls (Watabe, 1954). The changes in the extrapallial fluid during closure of the valves have been described for the clam *Mercenaria mercenaria* (Crenshaw and Neff, 1969): (1) the oxygen tension quickly dropped to a few mm. Hg, (2) the pH decreased a few tenths of a unit, (3) carbon dioxide and organic acids increased, and (4) the calcium content increased, due presumably to dissolution of the shell. In animals with calcium carbonate deposits in the mantle, calcium at this site might also be dissolved. Succinic acid made up 97% and lactic acid 2% of the organic acids of *Mercenaria;* and succinic acid accounted for 80% of the dissolved calcium. The predominance of succinic acid is a feature of interest in that it is weaker than lactic acid and would have less erosive action, as Hammen (1971) has pointed out. Succinic acid appears to be formed in some mollusks by the degradation of glucose to phosphoenolpyruvate (PEP), which is carboxylated by PEP carboxykinase to oxaloacetate and converted to succinate (Simpson and Awapara, 1966). Since two species had little or no PEP carboxykinase (Simpson and Awapara, 1964), the pathway may not be the same in all species. In the mantle fluid of the oyster *Crassostrea virginica,* an acid which chromatographs like propionic acid was present, but succinic, lactic, glyceric, and citric acids were absent (Wegener, 1971). The relatively small change in pH of the extrapallial fluid occurring during acid production is due to buffering by shell calcium carbonate (and perhaps by mantle calcium carbonate). The calcium carbonate of the shell together with that of the various tissues (see p. 108) undoubtedly serves as an alkali reserve which buffers against acids, including CO_2, in the organism generally (Dugal, 1939; Burton, 1970).

Under conditions of active growth, an alternation of calcium deposition and shell decalcification may occur several times each day as the valves open and close. During long periods of adverse environmental conditions, when calcium deposition is slow at best, shell decalcification may well predominate over growth (Davies and Crenshaw, 1971; Wada, 1971). This was shown in a study by Wada in which the crystals of the inner shell surface and shell weight of the pearl oyster *Pinctada martensii* were followed over an annual cycle (Fig. 10). Large crystals were present during November and December, when the water temperature and shell growth rate were declining. Then, in February, March,

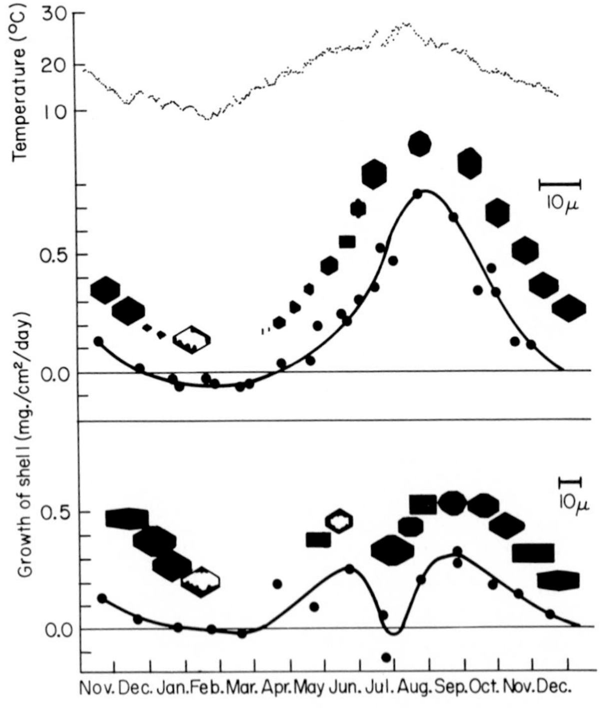

Fig. 10. Shell growth of *Pinctada martensii* (upper curve) and *Pinna attenuata* (lower curve) during a yearly cycle. Uppermost line shows temperature variation. Diagrams above curves indicate relative crystal size and shape during the year (with permission of Dr. Wada).

and April, the crystal size was small and dissolution of crystals occurred. During this period, the shell lost weight. From May through August growth rate, and crystal size increased. De Waele (1929) recognized that calcium may be removed from the shell when the uptake of calcium is deficient. The extreme case of shell dissolution is the macroscopic erosion of the inner shell surface evident in marine and freshwater bivalves removed from the water for periods of days (Dugal, 1939).

Studies on the extrapallial fluid relating to shell decalcification and to ion movements across the mantle indicate that measurement of shell growth by the incorporation of ^{45}Ca (Wilbur and Jodrey, 1952) may not be valid in all species. This method assumes isotopic equilibrium between the external medium and the extrapallial fluid. However, equilibrium was not reached in *Mercenaria* over several hours, as shown by sampling the extrapallial fluid by means of a catheter inserted through a hole in the shell (Goddard, 1966; Crenshaw and Neff, 1969). A further consideration is the finding by Crenshaw and Neff that when the valves were closed, shell dissolution was large compared with net mineral deposition during an equivalent period. Such dissolution will, of course, complicate the measurement of shell deposition.

The complications present in *Mercenaria* probably do not obtain in all mollusks. In the scallop *Argopecten irradians,* the valves normally remain open and sea water is constantly moved into the animal, presumably avoiding acid accumulation. Equilibrium of ^{45}Ca between the external medium and the mantle was attained within 2 hours and deposition of ^{45}Ca in the shell was linear after this time (Wheeler, 1971; see also Zischke *et al.*, 1970; Greenaway, 1971b). Deposition of $H^{14}CO_3$ added to the medium with ^{45}Ca followed the same course as ^{45}Ca.

XI. Shell Regeneration

The removal of a portion of shell of a mollusk commonly initiates a series of reactions which results in its replacement. The changes are clearly complex, and our understanding of the several mechanisms involved is fragmentary. We will briefly consider recent investigations of regeneration which have relevance to mechanisms of shell formation and its chemistry. References to earlier studies will be found in Abolins-Krogis (1960) and Wilbur (1964).

The sequence of changes during regeneration originates with stimulation of the mantle. The removal of part of the shell edge is a sufficient stimulus, apparently because of the altered inner shell surface with which the mantle comes in contact (Saleuddin *et al.*, 1970). Here the stimulus is clearly mechanical. In those cases in which removal of shell exposes a portion of the outer mantle surface not normally in contact with the external environment, the mantle may receive a chemical stimulus as well. The mantle, in turn, brings about metabolic changes within the animal, but whether through nerve or chemical transmission is not known. The effect is widespread as shown by histochemical changes in several tissues (Abolins-Krogis, 1960). We shall consider changes in the tissues, the processes of repair, and alterations in regenerated shell as compared to normal shell.

A. Mantle, Hepatopancreas, and Blood in Shell Regeneration

During regeneration, the mantle shows marked changes. These alterations, observed by histochemical procedures, are complex, differing with the mantle region and the stage of regeneration. For example, RNA in the outer epithelium of freshwater bivalves decreased and then increased above normal, while glycogen decreased temporarily in some cells and not in others (Saleuddin, 1967). Synthesis of RNA and protein in *Helix* was found to increase to a maximum 12 hours following shell injury. Also, alkaline phosphatase, cytochrome oxidase, and ATPase showed marked increases during shell regeneration (Chan, 1971). Other changes in the mantle include alteration in cell shape, accumulation of secretory granules, and increase in number of mitochondria. The form of the mitochondria may be altered, and unidentified crystals may appear within them (Tsujii, 1960; Saleuddin, 1970). In the mantle of the bivalve *Anodonta grandis* the alkaline phosphatase activity increased and an additional isoenzyme appeared during shell regeneration (Saleuddin, 1969). It may be presumed that the various deviations from the normal state of the mantle will be correlated with compositional changes in its secretion, and that the altered secretion, in turn, may contribute to the observed departure from the normal processes of shell formation described later.

The hepatopancreas also appears to play a role in shell repair. The calcium cells found in this organ are noteworthy in two respects: (1) the presence of spherites of calcium carbonate indicates that these cells have an intracellular system of calcification, and (2) the cells apparently provide a source of calcium in shell regeneration. The spherites of the calcium cells of *Helix* are associated with fibrous material arranged in concentric lamellae and having a spacing of 208–220 Å (Abolins-Krogis, 1960, 1965, 1970). Formation of the spherites occurs contiguous with a well developed Golgi complex. The relationship with the Golgi is especially interesting in view of the fact that in certain algae calcium carbonate structures (coccoliths) with an organic matrix are formed within Golgi vacuoles (Manton and Leedale, 1969). When shell regeneration is initiated by removing a piece of shell, the calcium spherites begin to be released from the cells within 1 hour (Abolins-Krogis, 1960). Amebocytes may then transfer them to the mantle and site of repair (Wagge, 1951; Abolins-Krogis, 1960; Tsujii, 1960).

During the first day of repair, blood calcium may remain at the normal level (Sioli, 1935; Saleuddin *et al.*, 1970), even though calcium is being released from the hepatopancreas and an increased rate of calcium deposition is occurring (Emerson, 1965). An increase in blood calcium after 7 days has been reported (Holtz and von Brand, 1940). During shell regeneration, a protein appears in the mantle of *Ampullarius* which is electrophoretically similar to hepatopancreatic protein (Saleuddin *et al.*, 1970). The finding is suggestive, but not proof, that the pancreas may release protein during regeneration. Another protein, normally detectable in the blood of a low percentage of animals, is increased during regeneration and may perhaps be due to an increased number of amebocytes in the blood.

B. Calcification during Shell Regeneration

Regenerating shell may be similar or distinctly different from normal shell, especially in its early repair. The organic matrix composition (Wilbur and Watabe, 1963; Saleuddin and Chan, 1969; Saleuddin and Hare, 1970), crystal type (Wilbur and Watabe, 1963; Abolins-Krogis, 1968; Saleuddin and Chan, 1969), and crystalline structure (Wilbur, 1964) may all be changed. We will summarize recent studies on these various aspects, beginning with the first stages of repair.

The early stages of calcification during shell repair of *Helix* have been observed by electron microscopy (Saleuddin and Wilbur, 1969; Saleuddin and Chan, 1969). After removal of an area of shell, electron microscope grids were placed directly on the mantle and removed after various periods for direct observation in the electron microscope. Organic matrix and crystals were present within 30 minutes. The matrix appeared as a homogeneous sheet or as a reticulum of fine fibers. Crystallization began with the formation of minute bodies in electron dense areas which stain for acid mucopolysaccharide. Calcified disk-shaped structures and spherulites, which stained for protein and mucopolysaccharide, were also found (Abolins-Krogis, 1968; Saleuddin and Chan, 1969). The layers of crystals in regenerated shell of *Helix* resembled one or more of the normal layers, although differences of crystal form were evident. Intracrystalline matrix was present as it is in normal shell (Saleuddin, 1971).

Amebocytes are present at the site of shell repair, as many investigators have observed. The importance of these cells in forming regenerating shell has been a matter of dispute (e.g., see Wagge, 1951; Durning,

1957; Wilbur, 1964; Abolins-Krogis, 1968). In the terrestrial snail *Euplecta,* amebocytes containing calcium granules were observed to rupture and the cytoplasm, containing protein, muco- and acid polysaccharides, and other substances, served as a substratum for fan-shaped crystallites (Kapur and Sen Gupta, 1971). These then developed into tabular crystals. As more amebocytes came to the area, the process continued until several crystalline layers were built up. Although the amebocytes may move some calcium to the regeneration area, the main source of calcium for crystal growth is almost certainly the extrapallial fluid.

The sequence of changes which produce a complete prismatic layer have also been observed on pieces of plastic or glass coverslips placed between the mantle and shell of the oyster *Crassostrea virginica* (Galtsoff, 1964).

C. Crystal Type in Regenerated Shell

The crystal type of the regenerated portions of the shell may be the same or different from the normal shell (Wilbur, 1964). For example, the snail *Nassarius obsoletus* and the clam *Mercenaria mercenaria,* both normally aragonitic, deposit only aragonite on regeneration, whereas the calcitic oyster *Crassostrea virginica* forms both calcite and aragonite crystals. *Helix pomatia* (Saleuddin and Wilbur, 1969; Saleuddin, 1970, 1971) and *Elliptio complanatus* (Wilbur and Watabe, 1963), both aragonitic, may lay down calcite and vaterite as well as aragonite during repair.

The factors controlling crystal type in molluscan shell may include both matrix protein, which may bring about nucleation (Degens *et al.,* 1967; Wilbur and Simkiss, 1968), and dissolved inorganic and organic substances (Kitano *et al.,* 1969). The crystal type which forms in association with the shell matrix under experimental conditions is correlated with the structural form of the matrix and the species from which the matrix is taken (Wilbur and Watabe, 1963). The organic matrix formed during regeneration may be structurally different from normal, and the crystal type may be changed as well. An examination of the composition of normal and regenerated matrix in *Helix pomatia* demonstrated that the regenerated matrix had a lower concentration of glycine and an increased glucosamine content (Saleuddin and Hare, 1970). The aragonite–calcite ratio of crystals formed *in vitro* can be influenced by the concentration of glycoprotein and the metabolic substrates succinate, citrate, pyruvate, malate, and lactate, as shown by the precipitation of calcium carbonate from a solution of calcium bicarbonate in the presence of these compounds (Kitano *et al.,* 1969). Since succinic and lactic acids and glycoprotein are known to be present in extrapallial fluid (Crenshaw, 1971a; Crenshaw and Neff, 1969), the possible effect of such compounds on the crystal type during shell regeneration must be recognized.

XII. Minor Elements and Trace Elements

The study of elements in addition to calcium in invertebrate skeletons has been given attention, particularly by geochemists, following the comprehensive work of Clarke and Wheeler (1917, 1922). Of the many elements found in molluscan shells, magnesium and strontium are frequently present in concentrations exceeding 1% and are appropriately termed minor elements (Dodd, 1967). Elements occurring in lesser amounts are designated as trace elements. Early studies of trace elements in molluscan shell were summarized in Vinogradov's (1953) extensive compilation. Although some of the techniques employed are now open to question, these studies clearly demonstrated that shell contained many trace elements. With the use of spectrographic analysis (e.g., Fox and Ramage, 1931; Brooks and Rumsby, 1965), activation analysis, and radioisotopes, the list of trace elements recognized in shell has become an extended one. The trace elements may be associated with crystalline material, bound to the several organic compounds of shell, or occur in certain organic molecules as in the case of the iron and bromine of shell pigments (Fox, 1966).

The composition of shells with respect to minor and trace elements depends upon a complexity of factors (Odum, 1957; Turekian and Armstrong, 1960; Dodd, 1967) which can be grouped into three major categories. One concerns crystallographic aspects of the incorporation of minor and trace elements into calcium carbonate. A second category comprises physiological and biochemical processes, both in the milieu in which shell formation takes place, and in the transfer of materials from the environment to the site of crystal deposition. The third division encompasses environmental factors, which of course affect physiological mechanisms and accordingly influence shell composition. Our discussion will consider these three categories as they apply to the composition of molluscan shells. For further discussion and review of the literature, the articles by Odum (1957), Waskowiak (1962), and Dodd (1967) will be found especially useful.

A. Crystallographic Factors

The type of crystals that form within organisms depends upon physical and chemical conditions. If the cationic radii are between 0.78 and 1.00

Å, hexagonal crystals will be deposited; and if the radii lie between 1.00 and 1.43 Å, the crystals will be orthorhombic. The ionic radius of calcium is 0.99 Å (Heslop and Robinson, 1967); and calcium carbonate occurs in the hexagonal form as calcite and in the orthorhombic form as aragonite. Both crystal types are found in mollusks. Rarely, a third form, vaterite, which is less stable, is deposited.

The influence of various factors on the polymorphic form has been shown by precipitating calcium carbonate from a calcium bicarbonate solution (Kitano *et al.*, 1969). It was found that the proportion of aragonite was increased by increasing the magnesium concentration and by increasing temperature. Copper, zinc, nickel, and manganese were more effective than magnesium in bringing about aragonite formation. Calcite formation was favored by an increase in concentration of certain organic compounds that complex calcium. The magnesium content of calcite increased with (1) the magnesium concentration of the solution; (2) the concentration of organic acids, and (3) temperature. The findings in these *in vitro* experiments are pertinent to the biological temperature range (5°–30°C). However, quantitative application of the results to the molluscan system can be made only when more complete analyses of the extrapallial fluid are carried out, since this is the site of calcium carbonate crystal formation.

Kitano *et al.* (1968) have provided data on the distribution coefficients of zinc and copper between solution and solid calcium carbonate precipitated in the presence of various organic compounds found in organisms. Here too, the data provide a pattern for the biological system, but calculation of the distribution coefficients for the organism requires data on body fluid composition, as the authors point out.

The distribution of minor and trace elements in shell has been frequently examined for correlations with mineralogy. Harriss (1965) has evaluated the relation of crystal type to trace element content by assembling data of several investigators in the form of frequency diagrams comprising ninety-five species of mollusks. Figure 11 shows the distribution for magnesium and strontium in shells which are either wholly aragonitic or calcitic. Calcitic shells have a magnesium content about tenfold that of aragonitic shells, with no overlap in the concentration ranges in the species tabulated, although there are exceptions (Kilham, 1970). In the case of strontium, aragonitic shells have a higher content generally, but Odum (1957) has shown that the two crystal types may have similar strontium concentrations. The difference between aragonitic and calcitic shells with respect to iron and manganese is less than with magnesium and strontium.

Dodd (1967), in reviewing the literature on magnesium and strontium

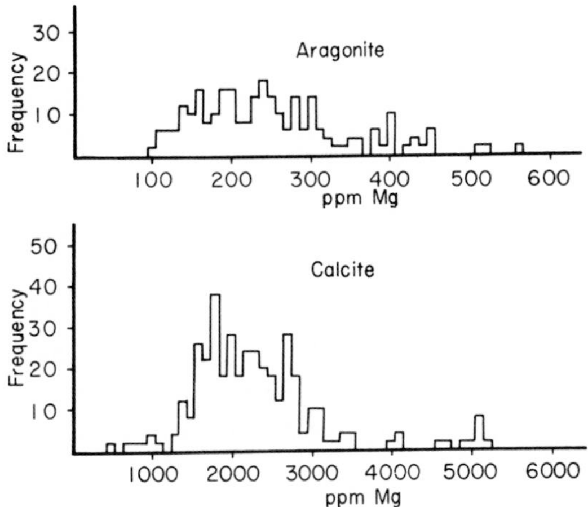

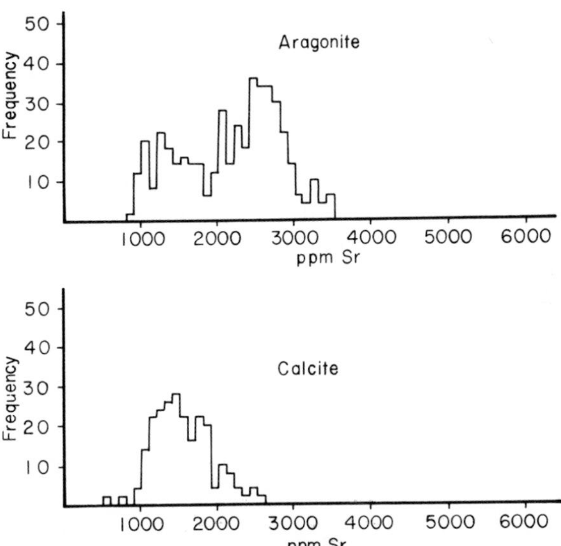

FIG. 11. Distribution of magnesium and strontium in skeletal material of ninety-five species of mollusks (Harriss, 1965). *Top.* Magnesium distribution in aragonite and calcite shell. *Bottom.* Strontium distribution in aragonite and calcite shell.

in calcareous skeletons, presents interesting comparisons between classes of mollusks and other invertebrates. The magnesium content of calcitic bivalve shell is very low as compared with the skeletons of most invertebrate groups, and the magnesium of aragonitic bivalves is lowest of all. Bivalve calcite has the least strontium found in invertebrates, and the strontium content of bivalve aragonite is also low. Gastropod shells, which are never entirely calcitic, are relatively low in both magnesium and strontium. In comparison with the aragonite of bivalves, gastropods, and scaphopods, the aragonite of cephalopods is somewhat higher in magnesium and considerably higher in strontium.

The higher concentration of magnesium in calcite as compared with aragonite has been explained on the basis of the easier substitution of the relatively small magnesium ion (0.66 Å) for the larger calcium ion (0.99 Å) in the calcite lattice, which is isostructural with magnesite ($MgCO_3$) (Dodd, 1967). The higher concentration of strontium observed in aragonite has been explained on a similar basis, aragonite being isostructural with strontianite ($SrCO_3$). However, the distinction between calcite and aragonite with respect to strontium content is much less sharp than with magnesium, as mentioned above.

In view of the other factors concerned in the distribution of minor and trace elements, it is not surprising that the correlations with crystal type are not especially strong.

Crystal growth rate is commonly thought to influence the extent of incorporation of trace and minor elements into the crystal lattice, substitution being greater as the rate of crystal growth is increased. One might then expect, for example, the magnesium content of a calcitic shell to increase as shell formation proceeded more rapidly. However, we shall see below that this relationship does not necessarily hold true in the living animal.

Our discussion has not considered crystal surface adsorption (Dodd, 1967) or the binding of ions by organic components of the shell. Both are undoubtedly involved in the uptake of minor and trace elements by molluscan shell, but their importance cannot yet be evaluated.

B. Physiological and Biochemical Processes

1. *The Compartments*

Earlier it was pointed out that ions which become a part of the shell pass from an external compartment consisting of the external medium, sediment, and food to the tissue and blood compartment, and into a third compartment, the extrapallial fluid, in contact with the growing shell surface (Fig. 2). The concentration of trace elements incorporated into shell will clearly depend upon their concentration in the extrapallial fluid. This concentration will be a function of the concentration in the tissue and blood and the rate of movement inward from the external environment.

The capacity of the tissues to concentrate certain trace elements may be very marked, the concentration being orders of magnitude above that of seawater (Table II). The tissue/shell ratio of trace elements may be either greater or less than unity, depending upon the element and the species (Table II). The significance of the ratio is uncertain because data on the extrapallial fluid concentration are lacking.

Comparisons of the ratios of minor and trace elements to calcium in the shell and in the seawater will indicate whether there has been a diminution or an enrichment relative to calcium in passing through the tissue-blood and the extrapallial fluid compartments. With strontium, a diminution occurs, and with magnesium, an extremely marked exclusion takes place (Table III) (Odum, 1957; Turekian and Armstrong, 1960). Barium, on the other hand, is enriched in shell (Turekian and Armstrong, 1960).

2. *Differences between Species*

The regulation by the mantle of the composition of the extrapallial fluid may be expected to vary with the species, resulting in differences

TABLE II

TISSUE AND SHELL CONCENTRATIONS OF TRACE ELEMENTS IN SOME BIVALVES[a]

Site of element	Element concentration (ppm dry weight)			
	Mn	V	Cu	Fe
Pecten novae-zelandiae				
Tissue	111	9	9	2,915
Shell	1	130	2	2,000
Tissue enrichment	55,500	4,500	3,000	291,500
Ostrea sinuata				
Tissue	8	3	41	682
Shell	1	49	<1	570
Tissue enrichment	4,000	1,500	13,700	68,200
Mytilus edulis				
Tissue	27	5	9	1,960
Shell	<1	110	3	<10
Tissue enrichment	13,500	2,500	3,000	196,000
Seawater[b,c]	2	2	3	10

[a] Brooks and Rumsby (1965).
[b] Seawater concentrations in ppb.
[c] Goldberg (1957).

TABLE III
RELATIVE CONCENTRATION OF ELEMENTS IN SHELL TO SEAWATER[a]

	(%Mg/%Ca) × 10^3	(%Sr/%Ca) × 10^3	(%Ba/%Ca) × 10^3
A. Molluscan shells—pelecypods (51 species) and gastropods (48 species)	1.33	3.7	0.025
B. Seawater[b]	5230	9.2	0.004
Fractionation (A/B)	0.00025	0.402	6.3

[a] Calculated from the data of Turekian and Armstrong (1960).
[b] Average element concentrations in seawater based on atomic weight of each.

in shell composition. The compositional differences in the shell among species or among higher taxa are referred to as generic control (Turekian and Armstrong, 1960), biochemical control (Lowenstam, 1963), or species effect. In order to evaluate this factor, other factors affecting composition should be constant or the extent of their influence on composition known. The factors concerned include temperature and other ecological conditions, mineralogy (i.e., whether calcite, aragonite, or both), and composition of food. Because of the lack of such data, evaluation of the biochemical factor from analyses in the literature may be difficult (Lowenstam, 1963). Ideally, analyses would be carried out on animals maintained in a controlled environment, a condition both difficult and impractical, especially when several species are to be studied. An alternative is the analysis of skeletons of different taxa having the same mineralogy living in a similar environment. Such analyses have been made for the magnesium and strontium content of representatives of several phyla having aragonitic skeletons from Bermuda waters (Lowenstam, 1963; see also Blackmon and Todd, 1959). The spread of magnesium and strontium values for the Polyplacophora, the Bivalvia, the Gastropoda, and the Cephalopoda was relatively small. Interestingly, the values of the Polyplacophora were higher than those of the other mollusks. The magnesium and strontium content of Bivalvia, Gastropoda, Cephalopoda, and fishes were generally lower than for the other animal phyla.

3. *Differences between Shell Layers*

The minor element content of individual shell layers may be quite different, as it is governed by the secretion of the mantle cells which form each layer. In *Mytilus*, for example, the magnesium content in the calcitic prismatic layer is about fourfold that of the aragonitic nacreous layer (Dodd, 1965). The Sr/Ca ratio also differs, the aragonitic layer being higher. But there may also be differences between layers in the shell of a single mineral species, as in the cephalopods *Sepia* and *Nautilus,* which have aragonitic shell (Price and Hallam, 1967). In *Sepia,* the strontium content in the phragmocone was found to be consistently higher than in the periostracum. Within the venter of *Nautilus,* the nacreous layer had a somewhat higher strontium content than the spherulitic and semiprismatic layers taken together. However, in *Mercenaria mercenaria* the aragonitic prismatic and nacreous layers have a similar Sr/Ca ratio (Odum, 1957).

4. *Effect of Size*

The rate of incorporation of minor elements in the shell may change with size of the animal. The content of both strontium and magnesium was found to decrease with size in *Mytilus* (Dodd, 1965). Measurements of strontium content of septa of *Nautilus* from the first formed to the more recently formed showed gradual changes. The pattern of change in the venter was quite different (Price and Hallam, 1967). Environmental factors and the effects of age may both be involved in the changes in *Nautilus.*

C. Environmental Factors

A method of assessing the influence of environmental factors on trace element distribution has been to collect animals from diverse localities and to examine the correlation between mean annual temperature or salinity values and the abundance of the element. Using this approach, the incorporation of several trace elements in the shell was found to increase with a decrease in salinity (Rucker and Valentine, 1961; Pilkey and Goodell, 1963). Temperature had a varied effect, apparently causing both increases and decreases in trace element concentration, depending upon the element, the species, and the shell layer (Pilkey and Goodell, 1963; Dodd, 1965). The use of mean annual figures for environmental factors and the analysis of samples from whole shells may obscure interesting relationships, as illustrated by studies on the uptake of strontium and magnesium by the bivalves *Crassostrea* and *Mytilus* (Dodd, 1965; Lerman, 1965). *Mytilus* shows a seasonal variation in the uptake of strontium, the rate being faster in the summer months (Fig. 12). However, the Sr/Ca ratios of animals living in waters of quite different temperatures may be similar (Odum, 1957). The *Mytilus* studies show further that in the calcitic prismatic layer there is a positive correlation of strontium content with environmental temperature whereas the temperature correlation in the argonitic nacreous layer is inverse (Fig. 13).

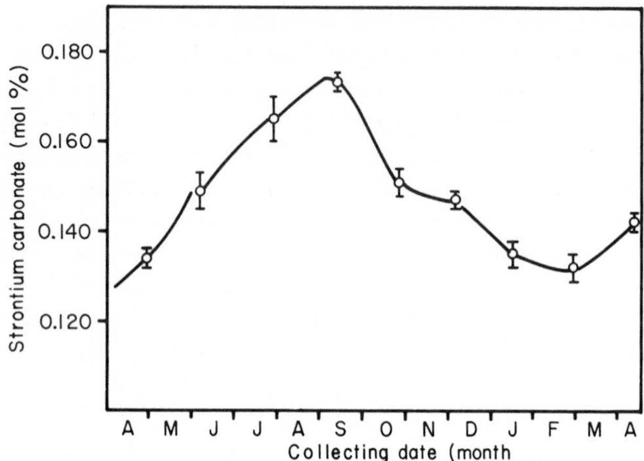

FIG. 12. Mol percent strontium carbonate in the most recently formed portion of the outer prismatic layer of *Mytilus edulis diegensis,* according to collection date in a 1 year period (Dodd, 1965).

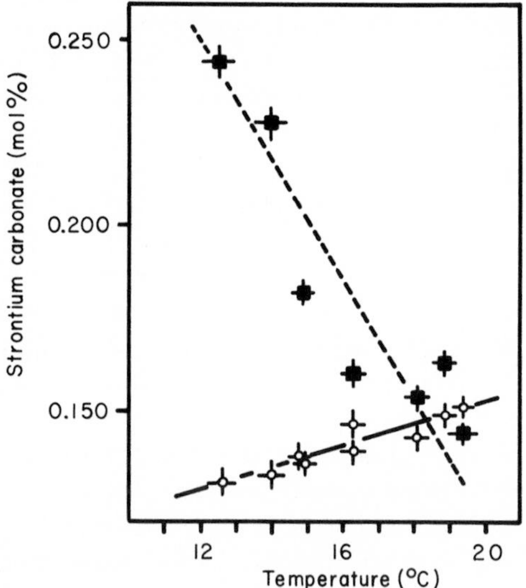

FIG. 13. Mol percent strontium carbonate in the most recently formed portion of the calcitic outer prismatic layer (open circles) and aragonitic nacreous layer (solid squares) of *Mytilus,* for the mean temperature for 2 weeks preceding the collecting date (Dodd, 1965).

The trace element concentration of the medium will be one of the factors determining the trace element composition of shell, as we have mentioned. The relationship between the composition of medium and shell is seen in the Sr/Ca ratios of shells of marine and fresh water mollusks. Analyses of a large number of species demonstrated a lower ratio generally in fresh water as compared to sea water (Odum, 1957; Dodd, 1967). A general proportionality between the concentrations of an element in the external medium and shell has been shown experimentally in snails using strontium (Odum, 1951; Likins *et al.*, 1963). However, the medium/shell ratios may vary with species.

It will be evident that the mineral content of individual crystal layers within each main layer will reflect environmental changes. By means of the electron microprobe, which permits analyses of areas of a few square microns, microvariations of composition can be measured (Moberly, 1968). The application of this method should permit the detection of changes in shell resulting from environmental variations of very short duration.

The strontium and magnesium content of aragonitic shells of a small number of representatives of the bivalve superfamilies Nuculanacea, Tellinacea, Limopsacea, Pectinacea, and Poromyacea has been determined as a function of depth to 5000 meters (Kilham, 1970). The Mg/Ca and Sr/Ca atom ratios were found to decrease with increasing depth. It will be interesting to examine this relationship further as more specimens of deep sea mollusks become available.

D. Experimental Approaches

The studies of minor and trace elements in molluscan shell have given attention to several factors involved in the incorporation of elements into the shell structure—conditions of crystal growth, crystal type, differences in shell layers, species differences, age, and conditions of the natural environment. With these studies serving as a base, experimental approaches in which mollusks are cultured under controlled conditions are needed to provide a more precise understanding of the rules governing shell composition. Laboratory studies would obviously simplify the analyses by eliminating environmental variation and by permitting the experimental manipulation of a single factor without the usual complication of other factors which may be changing simultaneously, as Lowenstam (1963) has pointed out. Selected marine, freshwater, and terrestrial species, including young stages, could be cultured for the investigation of age, temperature, salinity, and environmental concentrations of minor and trace elements. A critical gap is the paucity of information on minor and trace elements of the extrapallial fluid bathing the growing shell surface. The large bivalve species provide experimental opportunities for this phase of the problem (Crenshaw and Neff, 1969).

ACKNOWLEDGMENTS

I am grateful to Suzanne C. Ward, Lydia H. Brandon, Patricia L. Blackwelder, and Malvina M. Markman for assistance in the preparation of the manuscript; to Dr. M. A. Crenshaw and Dr. J. R. Dodd for review and helpful criticism; and to Dr. M. A. Crenshaw, Dr. M. Istin, Dr. S. P. Kapur, Dr. S. W. Wise, Jr., and Dr. K. Wada for kind permission to include unpublished data and photographs. Certain of the studies reported were supported by the Office of Naval Research, Oceanic Biology Program [(Grant Nonr 1181(06)] and the National Institute of Dental Research, National Institutes of Health (Grant DE-01382-06), and the Ford Foundation.

REFERENCES

Abolinš-Krogis, A. (1960). *Ark. Zool.* **13**, 159–201.
Abolinš-Krogis, A. (1965). *Ark. Zool.* **18**, 85–91.
Abolinš-Krogis, A. (1968). *Symp. Zool. Soc. London* **22**, 75–92.
Abolinš-Krogis, A. (1970). *Z. Zellforsch. Mikrosk. Anat.* **108**, 501–515.
Barker, R. M. (1964a). *Malacologia* **2**, 69–86.
Barker, R. M. (1964b). Personal Communication.
Beedham, G. E. (1958). *Quart. J. Microsc. Sci.* **99**, 341–357.
Bevelander, G., and Nakahara, H. (1969). *Calc. Tiss. Res.* **3**, 84–92.
Blackmon, P. D., and Todd, R. (1959). *J. Paleontol.* **33**, 1–15.
Bøggild, O. B. (1930). *Mem. Acad. Roy. Sci. Lett. Danemark, Sect. Sci.* [9] **2**, 232–326.
Brooks, R. R., and Rumsby, M. G. (1965). *Limnol. Oceanogr.* **10**, 521–527.
Burton, R. F. (1970). *Comp. Biochem. Physiol.* **37**, 193–203.
Campbell, J. W. (1970). Personal Communication.
Campbell, J. W., and Bishop, S. H. (1970). *In* "Comparative Biochemistry of Nitrogen Metabolism" (J. W. Campbell, ed.), Vol. 1, pp. 103–206. Academic Press, New York.
Campbell, J. W., and Speeg, K. V., Jr. (1969). *Nature* (*London*) **224**, 725–726.
Carriker, M. R., Charlton, G., and Van Zandt, D. (1967). *Science* **158**, 920–922.
Chan, W. (1971). Unpublished data.
Chétail, M., and Fournié, J. (1969). *Amer. Zool.* **9**, 983–990.
Chétail, M., and Fournié, J. (1970). *C. R. Acad. Sci.* **271**, 118–121.
Choe, S. (1963). *Nature* (*London*) **197**, 306–307.
Clark, G. R. (1968). *Science* **161**, 800–802.
Clarke, F. W., and Wheeler, W. C. (1917). *U.S., Geol. Surv., Prof. Pap.* **102**, 1–56.
Clarke, F. W., and Wheeler, W. C. (1922). *U.S., Geol. Surv., Prof. Pap.* **124**, 1–62.
Crenshaw, M. A. (1971a). Personal communication.
Crenshaw, M. A. (1971b). *Biomineralization* (in press).
Crenshaw, M. A. (1971c). Unpublished data (Table 1).
Crenshaw, M. A., and Neff, J. M. (1969). *Amer. Zool.* **9**, 881–885.
Crenshaw, M. A., and Young, S. D. (1971). Personal communication.
Davies, T. T., and Crenshaw, M. A. (1971). Personal communication.
Degens, E. T., Spencer, D. W., and Parker, R. H. (1967). *Comp. Biochem. Physiol.* **20**, 553–579.
de Waele, A. (1929). *Mem. Acad. Roy. Belg. Cl. Sci.* **10**, 1–52.
Digby, P. S. B. (1968). *Symp. Zool. Soc. London* **22**, 93–107.
Dodd, J. R. (1965). *Geochim. Cosmochim. Acta* **29**, 385–398.

Dodd, J. R. (1967). *J. Paleontol.* **41**, 1313–1329.
Dodd, J. R. (1969). *Nature (London)* **224**, 617–618.
Dugal, L. (1939). *J. Cell. Comp. Physiol.* **13**, 235–251.
Durning, W. C. (1957). *J. Bone Joint Surg.*, **39A**, 377–393.
Emerson, D. N. (1965). *Proc. S. Dak. Acad. Sci.* **44**, 109–112.
Erben, H. K. (1971). *Biomineralization* **3**, 51–64.
Florkin, M., and Bosson, G. (1935). *C. R. Soc. Biol.* **118**, 1222–1224.
Fox, D. L. (1966). *In* "Physiology of Mollusca," K. M. Wilbur and C. M. Yonge. Academic Press, New York, pp. 249–274.
Fox, H. M., and Ramage, H. (1931). *Proc. Roy. Soc. London.* **B 108**, 157–173.
Freeman, J. A. (1960). *Biol. Bull.* **118**, 412–418.
Freeman, J. A., and Wilbur, K. M. (1948). *Biol. Bull.* **94**, 55–59.
Galtsoff, P. S. (1964). *U.S. Fish Wildl. Serv., Fish. Bull.* **64**, 93–100.
Goddard, C. K. (1966). Personal communication.
Goldberg, E. D. (1957). *Geol. Soc. Amer., Mem.* **67**, 345–358.
Greenaway, P. (1971a). *J. Exp. Biol.* **54**, 199–214.
Greenaway, P. (1971b). *J. Exp. Biol.* **54**, 609–620.
Grégoire, C. (1962). *Bull. Inst. Roy. Soc. Natur. Belgique* **38**, 1–71.
Grégoire, C., Duchâteau, G., and Florkin, M. (1955). *Ann. Inst. Oceanogr. (Paris)* **30**, 1–36.
Haas, F. (1935). *In* "Klassen und Ordnungen des Tierreichs." **3**, Mollusca, III, Abt., Bivalvia, Teil I, (H. G. Bronn, ed.), Akad. Verlagsges, Leipzig.
Hammen, C. S. (1971). Personal communication.
Hammen, C. S., and Wilbur, K. M. (1959). *J. Biol. Chem.* **234**, 1268–1271.
Harriss, R. C. (1965). *Bull. Mar. Sci.* **15**, 265–273.
Heslop, R. B., and Robinson, P. L. (1967). "Inorganic Chemistry," 3rd ed. Elsevier, Amsterdam.
Holtz, F., and von Brand, T. (1940). *Biol. Bull.* **79**, 423–431.
House, M. W., and Farrow, G. E. (1968). *Nature (London)* **219**, 1384–1386.
Istin, M., and Girard, J. P. (1971a). *Calcif. Tissue Res.* **5**, 196–205.
Istin, M., and Girard, J. P. (1971b). *Calcif. Tissue Res.* **5**, 247–260.
Istin, M., and Kirschner, L. B. (1968). *J. Gen. Physiol.* **51**, 478–496.
Istin, M., and Maetz, J. (1964). *Biochim. Biophys. Acta* **88**, 225–227.
Kapur, S. P., and Gibson, M. A. (1968). *Can. J. Zool.* **46**, 987–990.
Kapur, S. P., and Sen Gupta, A. (1971). Personal communication.
Kenny, R. P. (1968). Personal communication.
Kilham, S. S. (1970). Ph.D. Thesis, Duke University.
Kilham, S. S. (1971). Personal communication.
Kitano, Y., Tokuyama, A., and Kanamori, N. (1968). *J. Earth Sci., Nagoya Univ.* **16**, 1–102.
Kitano, Y., Kanamori, N., and Tokuyama, A. (1969). *Amer. Zool.* **9**, 681–688.
Kobayashi, S. (1964a). *Biol. Bull.* **126**, 414–422.
Kobayashi, S. (1964b). *Bull. Jap. Soc. Sci. Fish.* **30**, 893–907.
Lehninger, A. L. (1970). *Biochem. J.* **119**, 129–138.
Lehninger, A. L., Carafoli, E., and Rossi, C. S. (1967). *Advan. Enzymol.* **29**, 259–320.
Lerman, A. (1965). *Science* **150**, 745–751.
Likins, R. C., Berry, E. G., and Posner, A. S. (1963). *Ann. N.Y. Acad. Sci.* **109**, 269–277.
Lowenstam, H. A. (1963). *In* "The Earth Sciences" (T. W. Donnelly, ed.), pp. 137–195. Univ. of Chicago Press, Chicago.

MacClintock, C. (1967). *Peabody Mus. Nat. Hist. Yale Univ. Bull.* **22**, 1–140.
Manton, I., and Leedale, G. F. (1969). *J. Mar. Biol. Ass. U.K.* **49**, 1–16.
Maren, T. H. (1967). *Physiol. Rev.* **47**, 595–781.
Marler, E., and Davidson, E. A. (1965). *Proc. Nat. Acad. Sci. U.S.* **54**, 648–656.
Matheja, J., and Degens, E. T. (1968). *N. Jahrb. Geol. Palaeont. Monatsh.* **4**, 215–229.
Meenakshi, V. R., Hare, P. E., and Wilbur, K. M. (1971). *Comp. Biochem. Physiol.* **40B**, 1037–1044.
Meenakshi, V. R. (1971). Personal communication.
Moberly, R. (1968). *Sedimentology* **11**, 61–82.
Mutvei, H. (1964). *Ark. Zool.* **16**, 221–278.
Mutvei, H. (1969). *Stockholm Contrib. Geol.* **20**, 1–17.
Mutvei, H. (1970). *Biomineralization* **2**, 49–72.
Nakahara, H. (1961). *Bull. Nat. Pearl Res. Lab.* **6**, 607–614.
Neville, A. C. (1967). *Biol. Rev.* **42**, 421–441.
Odum, H. T. (1951). *Science* **114**, 211–213.
Odum, H. T. (1957). *Inst. Mar. Sci.* **4**, 38–114.
Pannella, G. (1971). *Science* **173**, 1124–1127.
Pannella, G., and MacClintock, C. (1968). *Paleontol. Soc. Mem.* **2**, 64–80.
Pannella, G., MacClintock, C., and Thompson, M. N. (1968). *Science* **162**, 792–796.
Petitjean, M. (1965). Thesis, University of Paris.
Pilkey, O. H., and Goodell, H. G. (1963). *Limnol. Oceanogr.* **8**, 137–148.
Polya, J. B., and Wirtz, A. J. (1965). *Enzymologia* **29**, 27–37.
Posner, A. S. (1969). *Physiol. Rev.* **49**, 760–792.
Potts, W. T. W. (1954). *J. Exp. Biol.* **31**, 376–385.
Price, N. B., and Hallam, A. (1967). *Nature (London)* **215**, 1272–1274.
Rhoads, D. C., and Pannella, G. (1970). *Lethaia* 3, 143–161.
Rucker, J. B., and Valentine, J. W. (1961). *Nature (London)* **190**, 1099–1100.
Runcorn, S. K. (1966). *Sci. Amer.* **215**, 26–33.
Saleuddin, A. S. M. (1967). *Proc. Malacol. Soc. London* **37**, 371–380.
Saleuddin, A. S. M. (1969). *Malacologia* **9**, 501–507.
Saleuddin, A. S. M. (1970). *Can. J. Zool.* **48**, 409–416.
Saleuddin, A. S. M. (1971). *Can. J. Zool.* **49**, 37–41.
Saleuddin, A. S. M., and Chan, W. (1969). *Can. J. Zool.* **47**, 1107–1111.
Saleuddin, A. S. M., and Hare, P. E. (1970). *Can. J. Zool.* **48**, 886–888.
Saleuddin, A. S. M., and Wilbur, K. M. (1969). *Can. J. Zool.* **47**, 51–53.
Saleuddin, A. S. M., Miranda, E., Losada, F., and Wilbur, K. M. (1970). *Can. J. Zool.* **48**, 495–499.
Saxon, A., and Higham, C. F. W. (1968). *Nature (London)* **219**, 634–635.
Schmidt, W. J. (1921). *Zool. Jahrb., Abt. Anat. Ontog. Tiere* **45**, 1–148.
Schraer, R., and Schraer, H. (1970). *In* "Biological Calcification" (H. Schraer, ed.), pp. 347–373. Appleton, New York.
Simpson, J. W., and Awapara, J. (1964). *Comp. Biochem. Physiol.* **12**, 457–464.
Simpson, J. W., and Awapara, J. (1966). *Comp. Biochem. Physiol.* **18**, 537–548.
Sioli, H. (1935). *Zool. Jahrb., Abt. Allg. Zool. Physiol. Tiere* **55**, 507–534.
Smarsh, A., Chauncey, H. H., Carriker, M. R., and Person, P. (1968). *Amer. Zool.* **9**, 967–982.
Speeg, K. V., Jr., and Campbell, J. W. (1968). *Amer. J. Physiol.* **214**, 1392–1402.
Speeg, K. V., Jr., and Campbell, J. W. (1969). *Amer. J. Physiol.* **216**, 1003–1012.
Stolkowski, J. (1951). *Ann. Inst. Océanogr. (Paris)* **26**, 1–113.

Tanaka, S., and Hatano, H. (1955). *J. Chem. Soc. Jap., Pure Chem. Sect.* **72**, 602–605.

Taylor, J. D., Kennedy, W. J., and Hall, A. (1969). *Bull. Brit. Mus. (Nat. Hist.), Zool., Suppl.* **3**, 1–125.

Timmermans, L. P. M. (1969). *Neth. J. Zool.* **19**, 417–523.

Towe, K. M., and Hamilton, G. H. (1968). *Calcif. Tissue Res.* **1**, 306–318.

Tramell, P. R., and Campbell, J. W. (1971). Unpublished data.

Tsujii, T. (1960). *J. Fac. Fish., Prefect. Univ. Mie* **5**, 1–70.

Turekian, K. K., and Armstrong, R. L. (1960). *J. Mar. Res.* **18**, 133–151.

van Goor, H. (1948). *Enzymologia* **13**, 73–164.

van der Borght, O. (1963). *Arch. Internat. Physiol. Biochem.* **71**, 46–50.

van der Borght, O., and van Puymbroek, S. (1967). *In* "Radioecological Concentration Processes," (B. Åberg and F. P. Hungate, eds.) pp. 925–930, Pergamon Press, London.

Vinogradov, A. P. (1953). Sears Foundation Marine Research, No. II. Yale University, New Haven.

Voss-Foucart, M. F., Laurent, C., and Grégoire, C. (1969). *Arch. Int. Physiol. Biochim.* **77**, 901–915.

Wada, K. (1961). *Bull. Nat. Pearl Res. Lab.* **7**, 703–828.

Wada, K. (1967a). *Bull. Jap. Soc. Sci. Fish.* **33**, 613–617.

Wada, K. (1967b). *Bull. Jap. Soc. Sci. Fish.* **33**, 1007–1012.

Wada, K. (1968). *Bull. Nat. Pearl Res. Lab.* **13**, 1561–1596.

Wada, K. (1970a). *In* "Profiles of Japanese Science and Scientists" (H. Yukawa, ed.), pp. 227–244, Kodansha Ltd., Tokyo.

Wada, K., (1970b). Personal communication.

Wada, K. (1971). Personal communication.

Wada, K., and Furuhashi, T. (1970). *Bull. Jap. Soc. Sci. Fish.* **36**, 1122–1126.

Wagge, L. E. (1951). *Quart. J. Microsc. Sci.* **92**, 307–321.

Waskowiak, R. (1962). *Freiberg. Forschungsh.*, **136**, 1–155.

Watabe, N. (1952). *J. Fuji Pearl Inst.* **2**, 21–26.

Watabe, N. (1954). *Rep. Fac. Fish., Prefect. Univ. Mie* **1**, 449–454.

Watabe, N. (1965). *J. Ultrastruct. Res.* **12**, 351–370.

Watabe, N., (1971). Personal communication.

Wegener, B. A. (1971). Ph.D. Dissertation, University of Rhode Island.

Wheeler, A. P., (1971). Personal communication.

Wilbur, K. M. (1964). *In* "Physiology of Mollusca," (K. M. Wilbur and C. M. Yonge, eds.), Vol. 1, Chapter 8, pp. 243–282. Academic Press, New York.

Wilbur, K. M., and Jodrey, L. H. (1952). *Biol. Bull.* **103**, 269–276.

Wilbur, K. M., and Jodrey, L. H. (1955). *Biol. Bull.* **108**, 359–365.

Wilbur, K. M., and Simkiss, K. (1968). *In* "Comprehensive Biochemistry," (M. Florkin and E. H. Stotz, eds.), Vol. **26A**, pp. 229–295, Elsevier, New York.

Wilbur, K. M., and Watabe, N. (1970). Unpublished data.

Wilbur, K. M., and Watabe, N. (1963). *Ann. N.Y. Acad. Sci.* **109**, 82–112.

Wilbur, K. M., and Watabe, N. (1967). *Stud. Trop. Oceanogr. Miami* **5**, 133–154.

Wilbur, K. M., and Watabe, N. (1971). Unpublished observations.

Wise, S. W., Jr. (1970). Personal communication.

Wise, S. W., Jr., and Hay, W. W. (1968). *Trans. Amer. Microsc. Soc.* **87**, 419–430.

Zischke, J. A. (1969). Personal communication.

Zischke, J. A., Wilbur, K. M., and Watabe, N. (1970). *Malacologia* **10**, 423–439.

CHAPTER 4

Byssus Fiber—Mollusca

E. H. Mercer

I. Description of the Byssus

A tuft of short, straight, elastic, hairlike threads called the byssus is secreted by some bivalves and serves to fasten the animals to supports and to assist their limited movements. The byssus is the product of the foot and its associated structures. Of the many adaptive activities of the foot among mollusks, byssus is unique and the most remarkable.

Each component thread of the bundle of fibers is attached at its distal end by a small rounded attachment disk to a firm surface (usually a rock), and at its proximal end, it is secured by means of a small collar encircling the stout byssus stem as is shown in Fig. 1a. The byssus stem is embedded in and secreted by glands opening within the byssus cavity or pit at the base of the foot; the byssus threads themselves are secreted by the glands of the foot. The whole equipment may be referred to as the byssal apparatus. The animal may from time to time add new threads to its byssus to anchor itself more firmly as it increases in size or conditions change, or the whole byssus may be shed by pulling the stem from the pit, and a new byssus may be secreted to attach the animal in a new location.

The common edible mussel *Mytilus edulis* L. occurs in vast numbers on seashore rocks throughout the world and is a common article of diet. Since the byssus is almost completely resistant to the digestive enzymes, it is normally removed before the mussels are eaten, so that when mussels are used in large quantities large amounts of byssal thread are left over. This circumstance may have led to the use of byssus fibers as a textile material in earlier times. An early reference to its use as a textile occurs in the Bible (I Chronicles 4:21 and 15:27). It

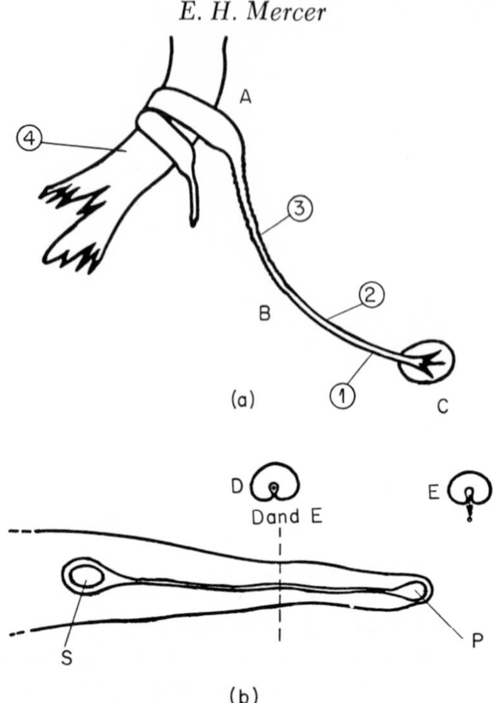

FIG. 1. The relationship between the byssus and the byssus foot in *Mytilus edulis* L. (Mercer, 1952). (a) *The byssus stem and thread.* At A is the collar-like attachment of the thread to the byssus stem; C is the attachment disk. The numbers indicate positions from which X-ray diffraction patterns were obtained. Position 1 yields an oriented fiber-type pattern; position 2 a less well oriented pattern. From 3 and 4 (the stem), only powder-type patterns result. (b) *The long extensible foot.* S is the aperture of the byssogenic cavity (pit) from which the stem, S, emerges. The groove leads to the sucker, P, which secretes the attachment disk, C. Byssal threads form in the groove between S and P. D and E are cross sections of the foot to show the position of the thread in the groove and the way in which it slips out of the groove when completed.

was spun from a bundle of threads like wool on to a spinning wheel to form a loose yarn. The yarns may be dyed—an indication of the proteinaceous character of the secretion. There are thus resemblances both to silk and to wool, and byssus has been referred to as "mussel silk." Scientific interest has been directed toward the elucidation of the molecular structure of the fibers and of the unusual method of their secretion and fashioning by the foot.

II. Molecular Structure of Byssus Fiber

Several authors (Trogus and Hess, 1933; Centola, 1936; Astbury, 1947) have sought by means of X-ray diffraction methods to elucidate the molecular structure of byssus and thus to find whether or not it bears any relationship to the more familiar natural fibers whose structure has been determined by the same method. Investigators continue their interest in this question in the hope that a knowledge of the distribution of the several fiber types will reveal phylogenetic affinities.

The approach to the problem was pioneered by Astbury (1939) whose comparative survey of numerous fibers revealed that in their small-scale molecular features, which contribute to the wide-angle diffraction pattern, the natural protein fibers appear to comprise only three classes: the α-proteins (type example: wool keratin) to which belong the proteins of the keratinized mammalian appendages, muscle myosin, etc.; the β-proteins (type example: silk fibroin); and the collagens (rat tail tendon), the common intercellular fiber of the vertebrates. Since these fibers are proteins, it may be assumed that their amino acid sequences are directly coded in the genome and that the distribution of the fiber types among the phyla will indicate the distribution of the genes coding them.

X-ray diffraction methods still offer a simple, direct nondestructive approach whose value is illustrated by the work of Astbury (1947) and Rudall (1950, 1968). The method is applicable even to an intractable material, such as byssus, which cannot be dissolved without destroying it. Centola (1936), who believed that byssus would prove to be a kind of silk related to the silks secreted by insect larvae and spiders, obtained powder-type diffraction patterns from byssus which were too vague and ambiguous to be decisive. Some features could be read as indicating either an α- or a β-protein. Trogus and Hess (1933), in a more extensive survey by diffraction methods of silks and similar fibers, obtained a sharp diffraction pattern from *Pinna* byssus, which appeared to have novel features.

Astbury apparently obtained better patterns from *Mytilus* threads which indicated a collagen fiber structure type. Mercer (1952) obtained an unambiguous collagen pattern from *Mytilus*. The crystallinity and orientation of the material in other byssus threads (e.g., *Pinna*), which are finer and more elastic than those of *Mytilus*, was too poor to give rise to an identifiable fiber pattern. However, in his comparative survey of the collagens, Rudall (1968) did not include the byssus.

Brown (1947, 1950, 1952), using chemical methods to examine a number of resistant natural materials, also concluded that byssus was related most closely to the collagens and that, to judge from its insolubility and brown color, it was tanned like the proteins of the insect cuticle.

Vertebrate collagens and some invertebrate collagens are characterized also by macromolecular regularities demonstrable by low-angle X-ray diffraction photographs (Bear, 1944) or by electron microscopy (Schmitt,

1944). Such collagens usually split into fairly distinct fine filaments, displaying a very characteristic set of longitudinal bands having a master periodicity of the order of 640 Å. Byssus fibers, probably because of their marked degree of tanning (see below), are difficult to split into fibrillar elements, and the fragments produced by prolonged mechanical and enzymatic disintegration do not reveal a periodicity of this type which would class them among the collagens (Mercer, 1952).

III. Elastic and Optical Properties of Byssus Fibers

No extensive study of the elastic properties of byssus has been published. A superficial examination is sufficient to show that even fibers taken from the same animal are not homogeneous in these properties and that the fibers of different species differ in diameter, extensibility, color, and transparency.

Mytilus threads may possess a reversible extensibility when wet of the order of 50–70% of their initial length. This long-range extensibility is similar to that of the keratin fibers (α-proteins) and is far greater than that of silk and collagen. Collagen in such well-oriented fibrous materials as tendon is virtually inextensible, and this property correlates with its molecular structure (see below). Older byssus threads are darker in color, less transparent, more brittle, and less extensible, which suggests that the training reaction continues for some time after the fibers are formed. *Mytilus* fibers (Mercer, 1952) are more extensible in their proximal sections nearer their attachment to the byssus stem than they are adjacent to the attachment disk. The same parts of the fibers are also somewhat thicker and display surface wrinkles. The less extensible portions nearer the attachment disk yield the best oriented X-ray diffraction patterns.

Close examination shows that the wrinkles visible on the surface of the portions of *Mytilus* threads adjacent to the stem smooth out on extension. They appear to be caused by a difference between the extensibility of a surface skin and a core. The surface, perhaps because of greater exposure to water and air or a richer admixture of the cross-linking component of the tanning system, might become more tanned and therefore less extensible than the core.

Byssus threads are birefringent, and the degree of birefringence is greater in the less extensible distal sections than in those adjacent to the byssus stem. When the threads stretch, the birefringence increases and the increase is greatest in the more extensible proximal portions. The birefringence (assuming this is of the intrinsic type) may be taken as a measure of the amount of oriented material in a fiber, whether this is included in a crystallite or is in a noncrystalline form, whereas the oriented fiber-type X-ray diffraction pattern arises from the oriented crystallites exclusively. In byssus, an increase in birefringence on extension occurs mainly in the proximal sections of the thread but is not accompanied by an improvement of the X-ray pattern from these parts. No improvement in the X-ray patterns accompanies the smaller increase in birefringence in the distal sections, which are already well oriented.

These observations seem to show that the very considerable amount of noncrystalline material in byssus threads is poorly oriented and that the orientation can be improved by stretching. Thus, extension is effected probably by the straightening of poorly oriented molecular chains without, however, crystallization of the aligned molecules occurring as happens, for example, in rubbers. The behavior resembles that of mammalian hairs in the earlier (less than 20%) range of their extensibility. It is possible that crystallization of byssus molecules is impeded during formation by the onset of the tanning reaction which, by establishing cross linkages between the polypeptide chains, would limit their free movement. These same cross linkages could prevent further crystallization of the molecules even when their alignment is improved by stretching.

The picture of the structure of byssus fiber as a heterogeneous mixture of oriented crystallites and poorly oriented chains would resolve the apparent paradox that we have in byssus a protein of the collagen type which apparently is capable of long-range extensibility. The molecular structure of collagen, as determined from its X-ray pattern, consists of three tightly fitted polypeptide chains twisted into a steep helix, a structure which permits little extension. Well-crystalline and well-oriented fibers composed of crystallites having this structure will also be inextensible. Byssus protein (or part of it) can assume this configuration and crystallize to yield crystallites capable of giving the definitive collagen pattern. However, if the greater part of the fiber consists of less oriented chains, the extensibility of the fiber as a whole could be yielded by these parts. The better crystalline regions are mostly located in the distal portion of the threads, and this could be due to their being stretched more during and immediately after formation before cross linkages are established. The differences observed between the byssus of different species could be attributed to differences between the degrees of tanning and crystallinity possible in such a system, but there is no evidence bearing on this point.

IV. The Formation of Byssus Fibers

The formation of byssus has been described by several authors (Williamson, 1906; Field, 1922; White, 1937) and we are concerned with it here only insofar as it offers an explanation of the mechanical properties of the threads.

The production of the byssus substance and the mechanical fashioning of the threads are activities of the muscular and glandular foot exclusively, whereas the byssus stem is a product of the byssal pit. Each thread is molded individually in a groove that extends along the ventral surface of the extensible tongue-like foot as shown in Fig. 1. The glands of the foot (probably of three kinds) secrete the precursors of the fibrous material directly into the groove. To fashion a thread, the foot is extended by the contraction of the circular muscles and the relaxation of the retractor muscles, and in this position the lips of the groove become tightly compressed together, converting the open groove into a closed tube (Fig. 1b). The tip of the extended foot is then placed against a firm surface such as a rock and a small rounded attachment disk is secreted from glands located at the extreme tip. The material of the disk appears to be distinct from that of the elastic byssus but has not been investigated. Meanwhile, the groove, acting as a mold, has filled with the plastic precursor of the threads secreted from the main glands, and a long thin rodlet is cast. When the thread has become sufficiently coherent, the foot muscles relax, the groove opens, and the still-plastic thread is released. Its distal end is embedded in the attachment disk and its proximal end, secreted from ducts opening from the foot glands at the mouth of the pit, is wrapped in the form of a collar around the byssus stem (Fig. 1a). The movements of the animal immediately after the release of a thread probably stretch it before it hardens fully and thus create or improve the orientation. The byssus body embedded in the byssogenous cavity or pit is a separate production secreted by glands opening into the pit and molded by the opening. It appears to be composed of a tanned protein similar to the threads but, as shown by X-rays and optical methods, is poorly oriented.

At first, the newly formed threads are creamy white in color; they slowly turn yellow then brown, a color change suggestive of tanning, which recalls the more adequately studied events associated with the formation of insect cuticle (Brown, 1947). The onset of the tanning reaction is probably the factor that sets a limit on the remodeling at the molecular level of the material of the threads. All the chemical components of the tanning system appear to be present in the precursors secreted by the foot glands. In insects, where tanning systems have been most adequately studied, a protein produced by one set of glands is cross-linked and stabilized by a polyphenol produced by a second set of glands. The polyphenol is secreted initially as an inactive precursor (a glucoside) from which it is split off by the action of an enzyme secreted with it. The material usually darkens, probably owing to the formation of melanin-like materials from the polyphenols (Jeuniaux, 1971).

The histology of the foot glands was described by the authors cited above, and more recently their histochemistry was reexamined by Gerzeli (1961), Oldfield (1961), Lim (1965), and Sullivan (1961). The arborescent branches of several glands are confusedly mixed in the body of the foot, rendering their distinction difficult. Nevertheless, three types of glandular tissue are demonstrable by histochemical methods, each presumably synthesizing an element of the tanning system. The largest gland, referred to as the white gland, produces the bulk of the protein; a purple gland (polyphenol gland) seems to contribute a quinone, the cross-linking component; and an enzyme gland, whose function (not yet demonstrated in this instance) is probably the activation of the polyphenol from an inert precursor. Mercer (1969) noted that the cells of the white gland were rich in ribosome-studded membranes enclosing secretion granules (diameter about 0.1μ) similar to those found in the left colleterial glands of the cockroach, which produce the protein of the tanned complex of the egg case. All recent investigators, using histochemical methods, recognize the presence of the three kinds of gland in a number of different species among the bivalvia.

From these findings, it is clear that the byssal system possess all three components of the classical tanning system, although corroborative evidence such as might be obtained were the various secretions extracted from the foot tissue, is still needed.

V. Summary

The byssus, by means of which numbers of bivalves attach themselves to rocks, consists of two parts—a thick stem secreted by glands of the byssus cavity and numerous threads secreted and fashioned by the foot. Both stem and threads are composed of an insoluble, tanned protein. Histochemical examination of the foot demonstrates the presence of a typical tanning system—a large white gland produces the protein forming the bulk of the secretion; a purple, or polyphenol gland secretes the cross-linking component, probably in a masked form; and an enzyme gland, whose product probably activates the polyphenol. These components are secreted from ducts opening into a groove running lengthwise along the extensible foot, where they mix and react to form a thread.

X-ray diffraction photographs of the better crystalline examples of byssus thread show that part of the protein has the molecular configuration of collagen. Most of the material is noncrystalline and poorly oriented. Fibers taken from *Mytilus* byssus show a long range reversible extensibility of about 50%, most of which is yielded by the more extensible less oriented distal portion of the thread. Mammalian collagen in well-oriented samples is essentially inextensible, and this feature finds its explanation in the current molecular model. The extension of byssus fiber appears to arise from its poorly oriented fraction, which, as is shown by birefringence measurements, becomes better oriented when stretched. No increase in crystallinity occurs on stretching; presumably the crosslinking introduced by tanning impedes crystallization.

REFERENCES

Astbury, W. T. (1939). *Annu. Rev. Biochem.* **8**, 113–132.

Astbury, W. T. (1947). *Proc. Roy. Soc., Ser B* **134**, 303–328.

Bear, R. S. (1944). *J. Amer. Chem. Soc.* **66**, 1297–1305.

Boutan, L. (1895). *Arch. Zool. Exp.* Gen. **23**, 295–338.

Brown, C. H. (1947). *6th Internat. Cong. Exp. Cytol., Stockholm,* 351.

Brown, C. H. (1950). *Quart. J. Microsc. Sci.* **91**, 331–339.

Brown, C. H. (1952). *Quart. J. Microsc. Sci.* **93**, 487–502.

Centola, G. (1936). *Gazz. Chim. Ital.* **66**, 71.

Field, I. A. (1922). *Bull. U.S. Fish. Bur.* **38**, 125–259.

Gerzeli, G. (1961). *Publ. Staz. Zool. Napoli* **32**, 88–103.

Jeuniaux, C. (1971). *Compr. Biochem.* **26C.**

Lim, C. F. (1965). *J. Anim. Morphol. Physiol.* **12**, 113–131.

Mercer, E. H. (1952). *Aust. J. Mar. Freshwater Res.* **3**, 199–205.

Mercer, E. H. (1969). Unpublished observations.

Oldfield, E. (1961). *Proc. Malacol. Soc. London* **31**, 226–249.

Rudall, K. M. (1950). *Progr. Biophys. Biophys. Chem.* **1**, 39–72.

Rudall, K. M. (1968). *In* "Biology of Collagen" Vol. 2, Part A, pp. 83–137. (B. S. Gould, ed.), Academic Press, New York.

Schmitt, F. O. (1944). *Advan. Protein Chem.* **1**, 25–68.

Sullivan, C. E. (1961). *Aust. J. Zool.* **9**, 219–257.

Trogus, C., and Hess, K. (1933). *Biochem. Z.* **260**, 376.

White, K. M. (1937). *Mytilus* L.M.B.C. Mem. No. 31 (Liverpool), pp. 2–117.

Williamson, H. C. (1906). *25th Annu. Rep. Fish. Board, Scotl.* Part III, pp. 221–253.

CHAPTER 5

Chemical Embryology of Mollusca

C. P. Raven

I. Oogenesis and Vitellogenesis

The egg cells of mollusks are formed in a hollow pouchlike or branched gonad. Egg formation is either of the solitary type, the egg cells protruding singly into the central cavity of the gonad (bivalves and scaphopods) or of the follicular type, in which the growing oocytes are partly surrounded by follicle cells (Raven, 1961, 1966).

During the early stages of oocyte development, while the premeiotic phenomena are taking place, the chromosomes of the egg cell nucleus are Feulgen-positive. With the beginning of oocyte growth, the Feulgen reaction of the chromatin becomes negative and remains so until the end of oogenesis, when the tetrads become visible preceding maturation. Feulgen-positivity of oocyte nucleoli has been described by various authors (Gabe and Prenant, 1949; Arvy, 1950a; Franc, 1951; Bolognari, 1954, 1956a, 1957). However, according to Bolognari (1960a) this apparent Feulgen-positivity may be due to refraction caused by granular inclusions of the nucleolus.

Collier and McCann-Collier (1962) found that the mature ovarian egg of *Ilyanassa* contains about thirty-two times the amount of DNA expected on the basis of its chromosomal complement. No attempt was made to determine whether the excessive amount of DNA is partially localized in the cytoplasm. According to Brown and Dawid (1968), *Spisula* oocytes contain extra copies of DNA for the synthesis of rRNA.

The nucleolus is rich in RNA in early oocytes. At later stages, amphinucleoli are formed as a rule, consisting of acidophilic and basophilic parts; only the latter contain RNA. If nucleolar buds are formed, they are either poor or rich in RNA (Berthier, 1948; Arvy, 1949, 1950a). During the second half of oogenesis, both the size and the RNA content of the nucleoli decreases as a rule.

Apart from the RNA content in oocyte nucleoli, the following substances have been identified by means of cytochemical methods: general proteins (Fautrez, 1959; Ubbels, 1968), basic proteins (Serra and Queiroz Lopes, 1945; Ranzoli, 1953; Bolognari, 1959, 1961; Ubbels, 1968), protein-bound sulfhydryl and disulfide groups (Serra and Queiroz Lopes, 1945; Bretschneider and Raven, 1951; Bolognari, 1959, 1961; Albanese and Bolognari, 1961; Ubbels, 1968), phospholipids (Ubbels, 1968), and polysaccharides (Albanese and Bolognari, 1961; Cowden, 1962; Bolognari and Cannata, 1963; Bedford, 1966a). A positive alkaline phosphatase reaction in the nucleoli has been described several times (Arvy, 1949, 1950a; Pelluet and Watts, 1951; Kobayashi, 1960; Bedford, 1966a) but should be judged with great caution (Laviolette, 1954). The same holds for a positive iron reaction of the nucleolus that has been described in *Planorbis* and *Limnaea* (Berthier, 1948; Bretschneider and Raven, 1951) but could not be confirmed with more reliable methods (Ubbels, 1968).

The following observations point to an intense metabolic activity of the nucleoli during oogenesis: changes in shape, structure, and composition of the nucleolus; formation and constriction of nucleolar buds; and extrusion of nucleolar vacuoles into the karyolymph (Raven, 1961, 1966). Rhythmic fluctuations in size of the nucleolus (Bretschneider and Raven, 1951; Ranzoli, 1953; Bolognari, 1956a; Cannata, 1962) may also be due to a periodic extrusion of substances.

The nucleoplasm may show positive cytochemical reactions of sulfhydryl and disulfide compounds (Serra and Queiroz Lopes, 1945; Ubbels, 1968), glycogen (Vasconcelos Frazao, 1957), and alkaline phosphatase (Bedford, 1966a). Extrusion of sulfhydryl compounds from nucleolar vacuoles into the nucleoplasm in *Limnaea* was described by Bretschneider and Raven (1951) and Ubbels (1968). Protein granules in the nucleus, giving positive reactions for general proteins, basic proteins, protein-bound sulfhydryl groups, and iron, were observed in oocytes of *Limnaea* by Ubbels (1968).

As a rule, the earliest oocytes have a cytoplasm that is very rich in RNA. During growth and vitellogenesis, a considerable decrease in

RNA concentration takes place. Often, the RNA content of the cytoplasm is greatest when the relative amount of nucleolar substance is maximal; increase and decrease of nucleolar volume and cytoplasmic RNA go hand in hand (Ranzoli, 1953; Bolognari, 1956b; Bolognari and de Raco, 1956; Cowden, 1961; Selwood, 1968). Part of the cytoplasmic RNA is bound to ribosomes. They are only partly connected with cytomembranes of the endoplasmic reticulum; most ribosomes occur freely in the cytoplasm. Toward the end of oogenesis, strings or whorls of connected ribosomes, probably representing polyribosomes, may occur (Recourt, 1961; Pasteels and de Harven, 1962b, 1963; Bedford, 1966b). Bolognari (1961) and Reverberi (1967) describe the passage of ribosomes through pores of the nuclear membrane.

The cytoplasm of the oocytes generally gives a moderate reaction for general proteins. Basic proteins show quantitative and qualitative variations with the species and the stage of development (Serra and Queiroz Lopes, 1945; Ranzoli, 1953; Cowden, 1962, 1966; Davenport and Davenport, 1965; Ubbels, 1968).

The cytoplasm of early oocytes of *Limnaea* initially contains a rather high amount of sulfhydryl- and disulfide-containing proteins, which decreases, however, with further growth. Small sulfhydryl-positive granules, probably mitochondria, are concentrated around the nucleus in previtellogenetic oocytes. Golgi bodies are probably stained too (Ubbels, 1968). The concentration of sulfhydryl proteins in oocytes of *Gryphaea* and *Ostrea* is higher on one side of the cell, where the density of mitochondria is greatest (Kobayashi, 1956).

Early oocytes of *Limnaea* have a considerable amount of phospholipid in the cytoplasm. At first the reaction is diffuse, but soon the cytoplasm becomes filled with small positive-staining granules, probably both Golgi bodies (see Malhotra, 1961) and mitochondria. After the beginning of vitellogenesis most of the phospholipid is bound to yolk granules (Ubbels, 1968). Glycogen in *Limnaea* is not formed in connection with visible cell structures, but diffusely in the cytoplasm (Bretschneider and Raven, 1951). Ubbels (1968) observed small metachromatic granules in the cytoplasm of *Limnaea.*

Ascorbic acid has been found in the eggs of various mollusks (Ries, 1937; Bretschneider and Raven, 1951; Pelluet and Watts, 1951), either diffusely in the cytoplasm or bound to Golgi bodies. The same is true for ferric iron (Berthier, 1948; Bretschneider and Raven, 1951; Ubbels, 1968). The benzidine peroxidase reaction is mainly bound to the mitochondria (Prenant, 1924; Ries, 1938; Raven and Bretschneider, 1942). A positive indophenol oxidase reaction may be either due to cytochrome oxidase or to a stable M-nadi oxidase, while some lipid substances also may give the reaction (Mancuso, 1954, 1955). In the first case, the reaction, which is inhibited by sodium azide, is bound to the mitochondria (Raven and Bretschneider, 1942; Bretschneider and Raven, 1951; Mancuso 1954, 1955; Attardo, 1955, 1957; Kobayashi, 1960; Pucci, 1961).

According to Bedford (1966b) in *Sypharochiton* new mitochondria are produced from a granular precursor substance, rich in RNA, lipoproteins, and phospholipids, which is apparently secreted by the nucleus and accumulated on the outer side of the nuclear membrane.

In bivalves, a special kind of granule is formed, which assembles toward the end of oogenesis in one layer beneath the surface. These cortical granules contain neutral mucopolysaccharides in *Mactra* (Sawada and Murakami, 1959) and acid mucopolysaccharides in *Barnea* (Pasteels and de Harven, 1962b). Davenport (1967) observed a layer of basic ribonucleoprotein granules beneath the surface of the *Haliotis* oocyte.

In *Limnaea* oocytes, a peripheral layer of hyaline cytoplasm is formed, which gives positive cytochemical reactions for general proteins, basic proteins, protein-bound sulfhydryl, RNA, and weakly acid polysaccharides, but a negative phospholipid reaction. Its outer lamella differs somewhat in its staining reactions; it possibly contains sialic acid as its main carbohydrate component. After ovulation and fertilization most of these reactions become negative (Ubbels, 1968). The surface layer of the ovulated egg cell is very rich in RNA, while a narrow subcortical layer beneath it contains somewhat more RNA than the rest of the cytoplasm (Raven, 1966).

Rebhun (1956a,b) described basophilic bodies, rich in RNA, in oocytes of *Spisula* and *Otala*. They consist of stacks of annulate lamellae, are supposed to arise by delamination from the nuclear membrane, and may give rise to "yolk nuclei" or ergastoplasm playing a part in the synthesis of proteid yolk. The yolk nucleus in *Helix* and *Tachea* is very rich in basic proteins (Serra and Queiroz Lopes, 1945).

The reserve substances (yolk or deutoplasm) in the egg cells of the mollusks can be divided into fatty yolk and proteid yolk. The fatty yolk consists of droplets or globules of lipid substances. They can be accumulated by centrifugal force into one part of the egg. The fat content, determined by this method, is about 5% in *Limnaea* (Raven, 1945), 2–5% in various nudibranchs, 10% in *Cumingia*, and 14% in *Mytilus* (Costello, 1939); these values have not been corrected for the cytoplasm occupying the interstices between oil drops, however. In general, the eggs of lamellibranchs have a high fat content. The eggs of *Pecten* have the highest iodine number known in all eggs (Needham, 1942).

The globules of the fatty yolk arise in most mollusks independently of visible cell structures in the cytoplasm (Parat, 1928; Yung, 1930; Bretschneider and Raven, 1951; Ubbels, 1968; Selwood, 1968); only for *Mytilus* has fat formation in complex Golgi systems during later phases of oogenesis been described (Worley, 1944). In addition to neutral fats, the globules may contain phospholipids—e.g., *Sypharochiton* (Selwood, 1968).

The proteid yolk consists of droplets or platelets of various sizes, shapes, and compositions. Its amount may vary considerably. In the eggs of some bivalves, no formed proteid yolk is said to be present (Worley, 1944). In centrifuged *Limnaea* eggs, the proteid yolk occupies about 50% of the egg volume (Raven, 1945; Bretschneider and Raven, 1951). In the eggs of nudibranchs, this amounts to 66–75% (Costello, 1939); in *Crepidula,* it is more than 75% (Conklin, 1917) (values not corrected for interstitial cytoplasm). Very yolk-rich eggs are found in the cephalopods.

Various cell structures may play a part in the formation of the yolk platelets: Golgi systems (Parat, 1928; Fahmy, 1949; Arvy, 1950b; Bretschneider and Raven, 1951; Yasuzumi and Tanaka, 1957; Bolognari, 1960b,c; Recourt, 1961; Beams and Sekhon, 1966; Bedford, 1966a), endoplasmic reticulum (Favard and Carasso, 1958; Bolognari, 1960b; Recourt, 1961; Bedford, 1966b), or mitochondria (Favard and Carasso, 1958; Albanese and Bolognari, 1964). In Polyplacophora (Gabe and Prenant, 1949) and in *Dentalium* (Arvy, 1950a), clear interstices appear in the basophilic cytoplasm. In *Sypharochiton,* they contain lipids, neutral and acid mucopolysaccharides, and alkaline phosphatase. These interstices become surrounded by a membrane and so transform into vesicles, in which the proteid yolk is formed (Bedford, 1966b).

It seems probable that these observations concern the accumulation and condensation of the yolk substances rather than their primary synthesis. Some of these substances may be synthesized elsewhere in the body and taken up in finished form by the oocyte from the blood (e.g., by micropinocytosis) (cf. Selwood, 1968). Other yolk substances may be synthesized in the oocyte from smaller precursors. Evidently, the ribosomes, either free or associated with "rough" endoplasmic reticulum, may be supposed to play an important part in this synthesis. According to Bolognari (1961), ribosomes in the cytoplasm increase to a size of 300 Å by the uptake of amino acids, and then aggregate to yolk granules. This aggregation could take place on various organelles or inclusions of the cytoplasm.

The proteid yolk of *Limnaea* (Bretschneider and Raven, 1951), *Myxas,* and *Succinea* (Jura, 1960) consists of two kinds of platelets, called β-

and γ-granules. They differ in size, shape, and probably in amino acid composition (Ubbels, 1968).

Electron microscopy has shown that the yolk granules of *Limnaea* (Elbers, 1957, 1959; Recourt, 1961) and *Planorbis* (Favard and Carasso, 1958) contain numerous electron-dense particles, for the greater part arranged in a regular crystalline pattern. The particles, probably globular macromolecules of an iron-containing protein (possibly ferritin), have a diameter of about 60–70 Å. Cytochemically, ferric iron can be demonstrated in the yolk granules of *Limnaea* (Bretschneider and Raven, 1951; Ubbels, 1968), *Planorbis* (Berthier, 1948) and *Bembicium* (Bedford, 1966a).

It is obvious that the proteid yolk platelets in most cases exhibit positive reactions for general proteins and basic proteins (Malhotra, 1960; Bedford, 1966b; Cowden, 1966; Ubbels, 1968), for protein-bound sulfhydryl groups (Ubbels, 1968), and for particular amino acids; in some species, the yolk is especially rich in tryptophan (Cowden, 1961, 1962). However, it by no means consists exclusively of proteins. In *Limnaea*, the yolk granules exhibit carbohydrate reactions from their first appearance; presumably, they consist at least partly of glyco- or mucoproteins (Ubbels, 1968). In *Barnea* and *Gryphaea* (Pasteels and Mulnard, 1957) and in *Bembicium* (Bedford, 1966a), likewise, carbohydrates are associated with the yolk platelets. Phospholipids may also be bound to the proteid yolk granules (Malhotra, 1960; Bedford, 1966b; Ubbels, 1968). In *Planorbis* (Berthier, 1948) and *Limnaea* (Bretschneider and Raven, 1951; Ubbels, 1968), the yolk granules become pyroninophilic toward the end of oogenesis or after ovulation; apparently, they become associated with ribonucleoprotein at this time. A positive alkaline phosphatase reaction of yolk granules has been described by Bedford (1966a,b). The yolk of *Limnaea* contains calcium, probably in the form of calcium phosphate (Malhotra, 1960). Finally, pigments may be bound to the yolk; in *Mytilus* it is a carotenoid (Worley, 1944), in *Limnaea* a melanin (Raven and Bretschneider, 1942).

It is evident that the growth of the oocyte and the synthesis of the yolk call for a regular supply of nutritive substances. In *Sypharochiton* vitellogenesis does not commence until the blood supply of the oocytes has been established. Part of the plasma membrane of each oocyte is bathed by the blood (Selwood, 1968). In cephalopods acidophilic substances, probably representing strongly hydrated simple proteins (histones and protamines), are secreted by the follicle cells into the oocyte. At the same time, the formation of secondary yolk begins, probably by condensation of the previtelline substances supplied by the follicle cells (Yung, 1930; Konopacki, 1933). In *Limnaea*, the follicle cells contain large fat drops, consisting of triglycerides and free fatty acids; their cytoplasm is rich in phospholipid and sulfhydryl-rich granules (Ubbels, 1968).

Apart from their function in the nutrition of the growing oocyte, the follicle cells may also take part in the formation of the egg envelopes. In *Limnaea* they contain metachromatic granules which may have to do with the formation of the vitelline membrane; the latter after ovulation shows a strong metachromatic staining with azure A at alkaline pH (Ubbels, 1968). In *Sypharochiton,* the chorion has three layers. The inner one, consisting of an acid mucopolysaccharide, is secreted by the Golgi bodies of the oocyte; the middle (protein) and outer (lipid) layers are secreted by the follicle cells, mainly through the activity of their Golgi bodies (Bedford, 1966b).

II. Fertilization

As in other groups of animals, substances produced by the germ cells play an important part in the fertilization of mollusks. The substances secreted by the eggs (fertilizin) activate and agglutinate the sperms (Tyler and Fox, 1939; Southwick, 1939; Tyler, 1940, 1949a; von Medem, 1942, 1945; Metz and Donovan, 1949; Galtsoff, 1964). The sperm agglutinin of *Megathura* is thought to be a protein (Tyler and Fox, 1940). Antifertilizin from sperm inactivates other sperm and neutralizes the sperm agglutinins from eggs (Tyler, 1939; von Medem, 1942, 1945).

Egg membrane lysins have been demonstrated in sperm extracts of various species (Tyler, 1939; von Medem, 1942, 1945; Berg, 1950; Krauss, 1950; Dan, 1962; Humphreys, 1962a,b). The lysin of *Megathura* is apparently a protein (Tyler, 1949b). The lysin of the archaeogastropod *Tegula pfeifferi* is a globulin that dissolves most readily at approximately the ionic strength of seawater and possesses polysaccharase activity (Dan, 1967). The egg membrane lysin of bivalves is probably contained in the acrosome of the sperm and is liberated when the latter breaks down and is transformed into an acrosome filament upon contact with the vitelline membrane of the egg (Dan and Wada, 1955; Wada *et al.*, 1956; Pasteels and de Harven, 1962a; Niijima and Dan, 1965).

Although the fertilized egg generally becomes refractory to the entrance of further sperms within seconds, this fertilization reaction is not attended by marked morphological changes in mollusk eggs. Some diminution of the number of cortical granules may take place (Sawada and Murakami, 1959; Sawada, 1964), but this is not generally the case. According to Humphreys (1967), extrusion of cortical granules in *Mytilus* occurs gradually at later stages of development, probably adding mucopolysaccharide material to the vitelline coat. If there is any elevation of the vitelline membrane at fertilization, it is only slight.

When the sperm has entered the egg of *Barnea,* a positive ATPase reaction becomes visible around the sperm head. During the subsequent maturation divisions and copulation of the pronuclei, the reaction gradually increases in the rest of the cytoplasm (Dalcq and Pasteels, 1963). Increase in ATPase activity with fertilization also occurs in *Crassostrea.* The same holds for succinate dehydrogenase and reduced NAD dehydrogenase in this species (Kobayashi, 1968a,b). In *Spisula,* the activity of ATPase and adenylate kinase does not change at fertilization, however (Horwitz and Nelson, 1964). Protein synthesis increases significantly immediately after fertilization in *Spisula;* permeability to amino acids increases sharply after meiosis is completed about 50 minutes after fertilization (Horwitz and Nelson, 1964; Monroy and Tolis, 1964; Bell and Reeder, 1967).

The changes in respiratory activity of the egg that occur at the moment of fertilization vary in different species. In *Mactra lateralis,* respiration increases with fertilization by a factor 1.8, in *Ostrea virginica* by a factor 1.4 (Ballentine, 1940). In *Ostrea commercialis* and *Spisula,* however, no change in respiratory rate takes place at fertilization (Cleland, 1950; Horwitz, 1965). In *Cumingia,* it decreases by a factor 0.45 (Whitaker, 1933). The respiratory quotient does not change at fertilization in *O. commercialis* and *Spisula.* In *Ostrea,* it is about 0.8 both in unfertilized and recently fertilized eggs, indicating a combustion of carbohydrates and fat. In *Spisula,* a respiratory quotient in the vicinity of 0.7 was found (Sclufer, 1955).

III. Ooplasmic Segregation

Immediately after ovulation and fertilization, in the eggs of most mollusks a process begins whereby various components and inclusions of the cytoplasm, originally more or less evenly distributed throughout the egg, are accumulated or concentrated at certain places. This process is called ooplasmic segregation. It brings about a situation in which different parts of the egg differ in their cytoplasmic composition. The ooplasmic segregation continues during cleavage. The segregated egg substances are distributed unequally among the cleavage cells, which are endowed from the outset with a different chemical composition of their cytoplasm.

It was Spek who in the early thirties drew attention to the fact that when the eggs of various animals are treated with vital stains having the properties of pH indicators, color differences between the animal and vegetative sides appear at the time of maturation or early cleavage. This he explained by assuming that this bipolar differentiation was due to a segregation of positively and negatively charged colloid particles, a "kataphoresis in the living cell" (Spek, 1934). Later investigations showed, however, that the observed differences in staining are due to a concentration of the proteid yolk in the vegetative part of the egg, leaving at the animal pole an extensive area consisting only of hyaline protoplasm, or at most containing part of the fatty yolk (Ries and Gersch, 1936; Raven, 1938). This process is most clearly expressed in the eggs of Opisthobranchiata, but it occurs to a lesser extent also in some Prosobranchiata. Apart from the proteid yolk, other inclusions of the cytoplasm also take part in the process, which therefore results in a kind of stratification of egg components along the main axis of the egg.

The eggs of *Aplysia* will be treated as an example (Ries and Gersch, 1936; Ries, 1937; Attardo, 1957). Ooplasmic segregation here begins immediately before the extrusion of the polar bodies and leads to a bipolar differentiation in which the proteid yolk occupies somewhat more than the vegetative half of the egg. Then follows a narrow band of "*Speichergranula*" (probably Golgi bodies), a zone of lipid yolk droplets, a ring of mitochondria, and finally an area of clear cytoplasm at the animal pole. The latter is the carrier of a leukomethylene blue oxido-reductase, whereas the mitochondria contain cytochrome oxidase, the *Speichergranula* ascorbic acid, while the benzidine peroxidase reaction is strongest in the zone of proteid yolk, though the enzyme is not bound to yolk granules (Ries, 1938). At cleavage, this array of substance is parceled out in a definite way among the cells (Fig. 1). The eggs of the opisthobranchiate *Navanax* (Worley and Worley, 1943) and the prosobranchiate *Columbella* (Spek, 1934) behave in a somewhat similar way.

In other instances, there is likewise a polar segregation of cytoplasmic substances. By the repeated formation of a polar lobe at the vegetative pole and its secondary fusion with certain cleavage cells, some of these substances are shunted into special blastomeres. In *Ilyanassa,* the polar lobe is rich in proteid yolk but poor in fatty yolk and in mitochondria, compared with the rest of the egg (Clement and Lehmann, 1956; Crowell, 1964). It also contains a rich supply of polynucleotides (Berg and Kato, 1959), of total phosphorus, acid-soluble phosphorus, and phospholipid phosphorus (Collier, 1960a), but is rather poor in RNA (Collier, 1960b). In *Dentalium,* on the other hand, the polar lobe does not contain any proteid yolk, but is rich in mitochondria (Reverberi, 1958). It has, moreover, an accumulation of DNA granules immediately

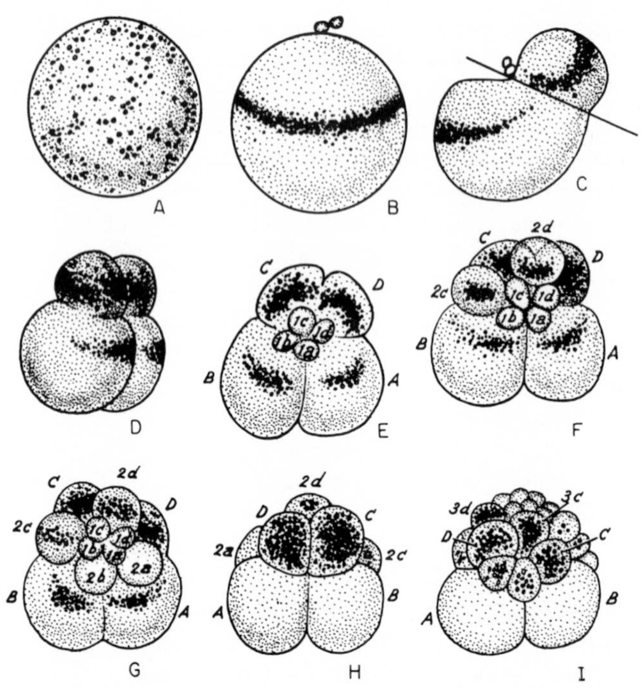

FIG. 1. Distribution of ascorbic acid (black granules) during early development of *Aplysia limacina.* (A) Immature egg shortly after laying; (B) mature egg after completion of ooplasmic segregation; (C) beginning first cleavage; (D) four-cell stage; (E) Eight-cell stage; (F–I) later cleavage stages. (After Ries, 1937.)

beneath the egg cortex at the vegetative pole (Timmermans *et al.*, 1970). In *Gryphaea,* the polar lobe lacks yolk platelets and polysaccharides, but is very rich in RNA (Pasteels and Mulnard, 1957). Finally, in *Mytilus,* the polar lobe does not differ in its composition from the rest of the egg with regard to any of the components tested—proteid yolk, lipid yolk, mitochondria, or cytochrome oxidase (Berg, 1957; Pucci, 1961; Reverberi and Mancuso, 1961; Humphreys, 1962a,b, 1964).

It would be erroneous to assume that a mere bipolar differentiation along one axis, whether combined with polar lobe formation or not, suffices to explain ooplasmic segregation in all cases. In many instances, a more precise localization of substances is attained by secondary accumulations and substance displacements superimposed on the former. The relationships in *Limnaea* may serve as an example.

In the eggs of *Limnaea stagnalis* (Raven, 1945, 1946, 1963, 1967, 1970), a vegetative pole plasm is formed in the 3–5 hours between ovulation and oviposition, by the accumulation of certain components of the ooplasm beneath a special area of the plasmalemma. It is very rich in β-granules of the proteid yolk, exhibiting positive iron (Arendsen de Wolff-Exalto, 1947) and RNA reactions. The rest of the ooplasm contains fewer β-granules, but is rich in γ-granules, fat droplets, mitochondria, and glycogen. Moreover, in the equatorial region of the egg there is a ring of six lenticular subcortical accumulations (SCA), which contain a special kind of granule and are rather rich in RNA and in lipids. During the maturation divisions, the vegetative pole plasm spreads beneath the surface, finally surrounding the whole egg as a layer of nearly equal thickness. Meanwhile, a new cytoplasmic differentiation appears—the animal pole plasm. It is very rich in mitochondria, but does not contain any fat droplets. Shortly before the third cleavage, the vegetative pole plasm substance concentrates markedly toward the animal side of the cells and unites with the animal pole plasm. At the subsequent divisions, this common pole plasm substance is differentially distributed among the cleavage cells, so that the relative amount of pole plasm substance in the cells of the blastula decreases from the animal towards the vegetative pole.

Meanwhile, the SCA have been displaced toward the most vegetative part of the embryo. Their substance is distributed in a regular way among the macromeres and second and third micromeres. In the former, coarse dark granules are formed in the SCA plasm at the vegetative pole, which are very rich in RNA. These granules move during the twenty-four cell stage toward the inner ends of the macromeres, where they condense into compact dark bodies, very rich in RNA. Glycogen granules accumulate in the adjacent parts of the micromeres. It seems that a transfer of RNA-containing substances from the vegetative to the animal blastomeres takes place at later development (Minganti, 1950; Raven, 1970).

More fragmentary cytological and cytochemical observations suggest that ooplasmic segregation takes a similar course in the related basommatophore pulmonates *Myxas* (Raven, 1964), *Planorbis, Physa* (Wierzejski, 1905; Mancuso, 1953), and in the stylommatophores *Succinea* (Jura, 1960), *Arion* (Lams, 1910), and *Limax* (Guerrier, 1968). The same holds, to a lesser extent, for the prosobranch snails *Bithynia* (Attardo, 1955) and *Bembicium* (Bedford, 1966a).

IV. Osmotic Regulation

The osmotic relations of marine eggs with the surrounding medium generally raise no special difficulties. As a rule, the eggs are in osmotic equilibrium with seawater; they swell upon dilution of the medium and shrink when it is made hypertonic.

In eggs laid in fresh water, a great discrepancy exists between the inner and outer osmotic pressure. The envelopes and capsular membranes of *Limnaea* and *Biomphalaria* are freely permeable to water and inorganic ions (Beadle, 1969a; Raven, 1970). We may expect to find in such eggs special conditions serving to prevent excessive swelling.

In *Physa* (Clement, 1938) and *Limnaea* (Raven, 1945), a slow swelling of the eggs occurs during the uncleaved stage. In both cases, the egg volume increases by about 40–50% until first cleavage. This swelling is an osmotic phenomenon; its rate is related to the osmotic pressure of the medium (Raven and Klomp, 1946). The recently laid *Limnaea* egg is in osmotic equilibrium with a 0.093 *M* nonelectrolyte solution, having an osmotic pressure of about 2.1 atm. The nonsolvent volume of the *Limnaea* egg has been calculated at 54–57% of the original egg volume. Compared with a calculated nonsolvent volume of 44% in *Ostrea* (Lucké and Ricca, 1941) and of 12–39% in *Spisula* (Schechter, 1956), this is an excessively high value. The difference may be linked up with special conditions of freshwater eggs, but it is very probable that other factors, e.g., the resistance of the egg membrane against swelling, influence the results (Raven and Klomp, 1946; Schechter, 1956).

The surface membrane of the *Limnaea* egg probably has a very low ion permeability (Elbers, 1959, 1966, 1969; Raven, 1970). Its permeability constant for water (expressed in moles per second per square centimeter per mole per liter) lies in the neighborhood of 2×10^{-7} (Raven and Klomp, 1946). From the figures given by Lucké and Ricca (1941), one can calculate that the permeability constant in the eggs of *Cumingia* is about four times, in those of *Ostrea* six times, as large. Water permeability is therefore evidently restricted in the freshwater mollusk in comparison with marine species.

Uncleaved *Limnaea* eggs chilled to 3°C show a significant increase in volume of about 5% over eggs kept at room temperature; a similar though not fully significant swelling was observed in eggs treated with 0.001 *M* potassium cyanide (Raven *et al.*, 1953). Although these results might point to the existence of an active water excretion mechanism, it seems that such a mechanism plays, at best, only a minor part during the uncleaved stage.

The swelling of the eggs of *Limnaea* continues after the onset of cleavage, even at a greatly increased rate (Raven *et al.*, 1952; Raven, 1966). This cannot be explained by the increase in surface area or a rise in osmotic pressure of the cells, but must be due to an increase in water permeability of the egg surface. Only at later cleavage stages does water extrusion by means of the cleavage cavity appear to match or even to exceed the osmotic water intake. This is accompanied by a decrease in the vacuolization of the cytoplasm.

In many freshwater mollusks and land pulmonates, there is, from the two-cell stage on, a wide cleavage cavity, which opens periodically to the exterior and ejects its contents to the medium. This recurrent cleavage cavity has first been described in detail by Kofoid (1895) in *Limax*. It has since been found in various land and freshwater snails and in freshwater bivalves (Raven, 1966). A special study of its physiology in various gastropods has been made by Comandon and de Fonbrune (1935).

The cleavage cavity of *Biomphalaria* shows a great swelling after subjecting the eggs to anoxia, cyanide, or low temperature. Since the cleavage cells are not obviously swollen after such treatment, it is concluded that water is not actively pumped into the cavity from the cells, but the active process is concerned with removing fluid from the cavity (Beadle, 1969b).

V. Breakdown and Utilization of the Yolk

The proteid yolk of *Limnaea* and some other pulmonates consists of two kinds of platelets, β- and γ-granules, differing in various aspects (see above p. 159). During the passage of the ovulated egg through the oviduct, the γ-granules of *Limnaea* show the first indications of a peculiar swelling process (Bretschneider and Raven, 1951). This swelling becomes much more pronounced after oviposition, when clear, watery vacuoles form around the γ-granules and gradually increase in size (Raven, 1945). Since these vacuoles are visible in the living egg, they are not fixation artifacts. Under the influence of centrifugal force, the granules move together with their vacuoles and cannot be separated from them (Andrew, 1959).

Apart from various *Limnaea* species, formation of vacuoles around γ-granules has also been observed in *Myxas* and in *Succinea* (Jura, 1960). Similar vacuoles, but without distinct γ-granules in them, can be seen in the eggs of *Planorbis* and *Physa*. Clement (1938) has adduced evidence that they also arise by a swelling of granules.

Electron microscopic observations have shown that the vacuoles in *Limnaea* are surrounded by a triple-layered membrane. The γ-granule resembles a β-granule in its structure, but has a less homogeneous matrix, which sometimes entirely disintegrates, whereby the crystalline aggregates of protein molecules disperse throughout the vacuole (Elbers,

1959). Recourt (1961) suggests that there is no essential difference between β- and γ-granules, the latter representing stages in the breakdown of the β-granules. This is disproved by the fact that in recently-laid eggs the γ-granules are not only larger, but also heavier than the β-granules, accumulating at the centrifugal end of the egg during centrifugation. Only when vacuoles begin to form around them do they become lighter (Raven and Bretschneider, 1942).

Cytochemically, the β- and γ-granules show clear differences in composition. The β-granules of *Limnaea* contain ribonucleoproteins; in the γ-granules, the RNA test is entirely negative (Raven, 1945). The β-granules are PAS-positive, the γ-granules negative. Moreover, β-granules stain with Millon's reagent (for tyrosine) and with Serra's arginine reaction; both reactions are negative in γ-granules (Andrew, 1959). On the other hand, the γ-granules exhibit a strong positive reaction for acid phosphatase, whereas neither the β-granules nor any other egg component shows any acid phosphatase activity (Bluemink, 1967). The watery contents of the vacuoles exhibits weak reactions for polysaccharides, tyrosine, and possibly arginine (Andrew, 1959).

At later cleavage stages, the vacuolization of the cytoplasm in *Limnaea* diminishes strongly. The γ-granules are probably entirely broken down and digested during early cleavage. The proteid yolk granules of the blastula are probably derived from the β-granules of the uncleaved egg. Breakdown of the yolk continues at the blastula stage. It is attended with swelling of the granules, segregation of their matrix, decomposition of the yolk crystallites into their elementary particles and further breakdown of these particles, increase in membrane complexes, and the appearance of acid phosphatase and organophosphate-resistant esterase activity (Bluemink, 1967). Golgi bodies probably play a part in the breakdown and transformation of the yolk (Raven, 1946; Malhotra, 1960). Hydrolases (acid phosphatase) produced in Golgi cisternae may be transported by Golgi vesicles to the yolk granules (Bluemink, 1967).

Pasteels and Mulnard (1957) observed in uncleaved eggs of *Barnea* and *Gryphaea* stained with toluidine blue the presence of numerous small granules (α-granules) exhibiting a metachromatic staining. During early cleavage, larger (β) granules are formed in the neighborhood of the nuclei and asters; they do not stain directly, but their strong metachromatic staining arises at the expense of the α-granules. The β-granules show an active motility, moving to and fro between the astral rays.

Similar granules were observed by Rebhun (1958, 1959, 1960) in *Spisula*. They show the same intimate relationships to the asters. However, in *Spisula*, the β-granules are already present prior to fertilization, and they can be directly stained. Pasteels and Mulnard (1963) maintain that in *Barnea* only an indirect staining of the β-granules (via α-granules) is possible.

Cytochemically, the metachromatic granules of *Barnea* contain mucopolysaccharides and show a strong activity of acid phosphatase (Pasteels and Mulnard, 1957). They are further assumed to be identical with granules containing AMPase, ATPase, and still other dephosphorylating enzymes (Dalcq and Pasteels, 1962, 1963). As to their nature, Pasteels and Mulnard (1957) first considered them to be mitochondria, but later (1960) inclined toward the view that the metachromatic β-granules were Golgi bodies. Rebhun (1959, 1960) first expressed the view, based on electron-microscopic observations, that the β-granules are multivesicular bodies. This opinion has been adopted by Pasteels and Mulnard (1963), who suggest that the α-granules of *Barnea* are yolk platelets, whereas the β-granules represent multivesicular bodies arising by transformation of the former. In view of their content of hydrolytic enzymes, they show an analogy to lysosomes. Pasteels (1966a,b), moreover, stresses their resemblance to the "vacuome" in the sense of Parat; with neutral red, the multivesicular bodies of *Barnea* swell to vacuoles.

The utilization of the yolk in Prosobranchia, and its relation to embryonic nutrition, has been extensively reviewed by Fioroni (1966). In cephalopods special relationships occur. Cleavage is meroblastic and discoidal. Gastrulation begins with the formation of a flat yolk epithelium or perivitelline membrane, originating from the peripheral cells of the blastoderm, and extending around the yolk. Later, this becomes a yolk syncytium, which begins at an early stage with the digestion of the yolk. The blastoderm then extends peripherally over the perivitelline membrane, thereby forming the external yolk sac. As the embryo develops, yolk is transported by muscular contractions toward its interior, where it forms an inner yolk sac.

The perivitelline membrane plays an important part in yolk resorption (Portmann and Bidder, 1928; Yung, 1930; Sacarrao, 1945). Liquefaction of the yolk begins in the intraembryonic part of the yolk sac. The yolk syncytium penetrates deeply into the yolk, especially in the region of the sinus posterior, where digested food substances pass directly into the circulation.

According to Konopacki (1933) accumulations of lipids occur already at early stages near the nuclei of the perivitelline membrane; presumably a degradation of lipoproteins takes place under the influence of enzymes secreted by the cells. The liberated lipids are passed on to the embryo, in which they are first found in the intercellular spaces. At a later stage, lipids and proteins are found in the blood sinuses. Glycogen also appears in the perivitelline membrane, especially near the vena cava; it is passed on to the blood in this vein. Near the shell gland, glycogen passes directly from the yolk into the neighboring tissue. Toward the end of the embryonic period, the breakdown of the yolk is greatly increased in the regressing outer yolk sac.

According to Portmann and Bidder (1928) and Sacarrao (1945), the liver plays an important part in the passage of the digested yolk substances to the embryo. The liver cells come in direct contact with the perivitelline membrane. This shows considerable activity, and an active digestion of the yolk occurs in this region. Presumably breakdown products are passed through the liver cells and taken up in the lumen; they are probably transmitted to the blood in the pancreas, which is completely surrounded by a sinus pancreaticus.

VI. Embryonic Nutrition

The eggs of many gastropods are surrounded by capsules and are bathed by an egg capsule fluid or albumen, which as a rule contains both polysaccharides and proteins. Egg clutches of *Helix, Limnaea,* and *Australorbis* contain galactogen, those of *Lanistes* and *Pomacea* contain a galactose-fucose polysaccharide (May and Weinland, 1953; McMahon *et al.*, 1957); in none of the species is glycogen present at first. The egg capsule fluid of *Limnaea stagnalis* contains about 14.3% dry matter. Half of this is galactogen; the other half mainly protein. Three globulins and two albumins can be distinguished. No lipids in detectable quantity are present (Raven, 1966). Electrophoretically, several bands of proteins, aromatic esterases, and polysaccharides are produced, with species-specific differences between various species of *Limnaea* (Morrill *et al.*, 1964). According to George and Jura (1958), the egg capsule fluid of *Succinea* consists of various parts. The outer transparent insoluble part consists of polysaccharide and protein, giving the reactions of arginine, tyrosine, and disulfide groups. The inner part can be separated by centrifuging into a centripetal fraction of soluble protein with mainly tyrosine, α-amino acids, and disulfide compounds, and a centrifugal fraction mainly consisting of polysaccharides.

The eggs of pulmonates (*Limnaea, Planorbis, Succinea, Limax*) at cleavage stages begin to ingest albumen in all cells; this albumen is laid down in the ectoplasmic part of the cells in special albumen vacuoles (Raven, 1946). The contents of these vacuoles exhibit strong polysaccharide reactions. The uptake of the albumen takes place partly by a peculiar process of pinocytosis in large crater-like pits at the surface of the cells, partly by micropinocytosis. After their formation, the albumen vacuoles coalesce. The vacuole membrane may rupture within 15 minutes after vacuole formation, releasing the albumen into the cytoplasm. Coalescence of albumen vacuoles with yolk granules could also be demonstrated. Older albumen vacuoles exhibit positive reactions for acid phosphatase and organophosphorus-resistant esterase, whereas neither extracellular capsule fluid nor freshly ingested albumen vacuoles show such hydrolytic enzyme activity. Therefore, it is believed that the albumen vacuoles acquire their hydrolytic enzymes by coalescence with yolk granules. Apparently it is the capsule fluid, not the intracellular yolk that is the main nutrient reserve for the developing embryo. The function of the yolk granules is thought to be mainly enzymatic; they supply the hydrolytic enzymes necessary for the digestion of the albumen (Bluemink, 1967).

While originally in pulmonates all superficial cells of the embryo take part in the uptake of the albumen, after gastrulation this is more and more restricted to the endoderm. At later stages, part of the endoderm cells develop into the larval liver, which in many gastropods plays an important part in the uptake and digestion of the albumen. The ingested albumen is first accumulated in smaller vacuoles in the apical part of the cells, then in bigger irregular vacuoles in the basal part. A change in stainability of the albumen in the vacuoles may indicate that chemical transformations are taking place. In *Limnaea,* the contents of the larger vacuoles show an acid reaction (Raven, 1946). The intracellular digestion may be preceded by an extracellular digestion in the gut lumen by means of enzymes secreted by the cells. Extrusion of secretion droplets into the lumen by the cells has sometimes been observed; in *Limnaea,* these cells temporarily exhibit a distinct apocrine secretion (Raven, 1952).

The dry weight of *Limnaea* embryos remains practically the same throughout cleavage. Apparently, the uptake of albumen just compensates for the weight loss by respiration. During development the embryos ingest a great deal of the galactogen present in the egg capsule fluid. They contain a measurable quantity of β-galactosidase from the trochophore stage. Galactogen breakdown can only be demonstrated after blood circulation has begun. Freshly hatched snails show a rapid combustion of galactogen during starvation (Horstmann, 1956a, 1958, 1964).

A great part of the ingested galactogen is not combusted, however, but utilized in fat synthesis. The lipid content per embryo increases from 0.15 μg. to 3.85 μg. during development. A rapid rise in fat synthesis after the fourth day coincides with an increased uptake of galactogen (Horstmann, 1956b). In *Helix,* galactogen from the egg capsule fluid is taken up in the stomach, but its breakdown occurs only after hatching (Hunger and Horstmann, 1968).

In addition to carbohydrate, the proteins of the egg capsule fluid are likewise used for the growth of the embryo. In *Limnaea,* the protein content of the embryo begins to rise after gastrulation, with a concomitant decrease in protein of the egg capsule fluid (Morrill, 1964). In *Physa,* the total nitrogen increases ninetyfold from oviposition to hatching. A first increase occurs from the third to the seventh day; then, after a plateau, there is a strong rise from the ninth day to hatching (d'Ancona Lunetta and Minganti, 1964). A protein hydrolyzate of the egg capsule fluid of *Limnaea* contains eighteen amino acids, of which glutamic acid, glycine, alanine, leucine–isoleucine, and phenylalanine preponderate. The same eighteen amino acids, in similar proportions, are found in free form in the 7-day-old embryo; it seems probable that they are derived from ingested albumen (Morrill, 1963).

Other relationships are found in those Prosobranchiata, where the embryos develop in a cocoon and feed on abortive eggs or retarded embryos. These cases have been reviewed by Fioroni (1966).

VII. Nucleic Acids

Collier and McCann-Collier (1962) have shown that the *Ilyanassa* embryo by the twenty-five cell stage has begun to synthesize new DNA; it has approximately twice the DNA content of the ovarian egg. Each cell of the twenty-five-cell embryo contains about 34.6 pg. of DNA, which is approximately five times the diploid amount. The authors conclude that once the embryo begins to make new DNA, it synthesizes too much. In view of the fact that half the amount of DNA found per cell is possibly cytoplasmic DNA already present in the uncleaved egg, that the S phase in these embryonic cells is likely to be situated early in the cycle, and that the cell number was not accurately established, this conclusion may appear premature.

DNA reduplication preceding first cleavage takes place in *Spisula* both in the sperm and the egg nucleus at the time of pronucleus formation (Ito and Leuchtenberger, 1955).

In *Limnaea,* during cleavage, DNA replication in each cell begins when it is still in telophase. A G_1-phase is lacking up to the forty-eight-cell stage. Cleavage becomes asynchronous after the eight-cell stage. Lengthening of the cell cycle is at first only due to a prolongation of the G_2 phase; after the twenty-four-cell stage, the S phase is also lengthened in certain cells of the animal hemisphere, but not in vegetative cells. Moreover, from this stage on, corresponding cells in different quadrants begin to divide asynchronically, as a first indication of dorsoventrality of the embryo (Van den Biggelaar, 1971a,b).

In *Biomphalaria* the nuclei of the blastomeres of early stages have a DNA content varying between 2c and 4c, corresponding to a regular reduplication cycle. After gastrulation in the nuclei of the large-celled endoderm cells the DNA content increases to 16c by endomitosis (Schreiber and Camey, 1966).

Collier (1963) has shown that the *Ilyanassa* embryo between days 1 and 6 of embryogenesis utilizes 97.6 $\mu\mu$moles of thymidine-^{14}C and 104.6 $\mu\mu$moles of uridine-^{14}C for DNA synthesis. Therefore, uridine utilization for DNA synthesis is a major metabolic pathway in this embryo.

The RNA content of the *Ilyanassa* egg remains more or less constant during the first 3 days of development. The first major synthesis of RNA occurs during the fourth day, and this period of active RNA synthesis continues throughout the fifth day. Incorporation of phosphorus-32 into RNA shows a sharp increase during the third day of development; hence, 24 hours before any net synthesis of RNA takes place. It is concluded that RNA is metabolically inactive during the first 2 days, hence during cleavage and gastrulation, begins to play some functional role on the third day, while production of new RNA begins just before the onset of histological differentiation (Collier, 1961b). Later experiments have shown that there is some minimal synthesis of RNA even during the first 2 days, though it cannot be decided whether this is "turnover" or net gain (Collier, 1965b).

According to Brachmachary *et al.* (1968), phosphorus-32 incorporation in *Limnaea* begins in the uncleaved eggs, rises to a peak at the late trochophore stage, then declines throughout the veliger stage. Incorporation of uridine-^{3}H into RNA is demonstrable from the sixteen-cell stage on (Van den Biggelaar, 1971c). The RNA-rich granules in the macromeres of the twenty-four-cell stage and later stages remain unlabeled after incubation of the eggs with uridine-^{3}H. Apparently, their RNA is of maternal origin.

A high molecular weight RNA having a base composition similar to that of DNA, and therefore judged to be a messenger RNA, has been isolated from 4- to 5-day *Ilyanassa* embryos (Collier, 1965a) and also from 3-day embryos (Collier, 1965b). Both "light" and "heavy" dRNA is formed in the 5-day embryo; at earlier stages, the "heavy" dRNA may predominate (Collier and Yuyama, 1969). Removal of the polar lobe results in decreased protein and DNA synthesis and a decrease in total RNA synthesis, but a per cell RNA content equivalent to that of the normal embryo (Collier, 1965b). Since a major part of the pool of nucleic acid precursors is probably localized in the polar lobe (Collier,

1960a), it is suggested that the lobeless embryo fails to differentiate because it is deficient in nucleic acid precursors, and that this represses informational RNA synthesis (Collier, 1965b). Davidson *et al.* (1965) found that there was no difference in the incorporation of uridine into RNA between normal and lobeless embryos during the first hours of development, but after 6 hours the rate of RNA synthesis in lobeless embryos became smaller; in 30-hour embryos it was from 1.5–2 times smaller than in the normal. They conclude that gene activation and the consequent synthesis of informational RNA is repressed in lobeless embryos. On the other hand, Clement and Tyler (1967) observed that the isolated polar lobe continues to incorporate labeled amino acids into protein during 24 hours at a constant rate, and conclude that it contains a long-lived messenger RNA.

Actinomycin D does not block cleavage, gastrulation, and ciliation in *Ilyanassa,* but it depresses further development. From experiments in which the embryos are exposed to actinomycin during restricted periods of development, it is concluded that transcription precedes cell differentiation by 1–2 days, and lasts several hours (Collier, 1966). *Planorbis* embryos treated with barbituric acid (Sherbet and Lakshmi, 1964a) or diethylbarbituric acid (Rao and Mulherkar, 1964) at higher concentrations are blocked at the gastrula stage.

VIII. Protein Synthesis

The uncleaved fertilized eggs of *Aplysia* contain a pool of twelve free amino acids. Quantitative and qualitative changes occur during development. At gastrulation, asparagine and hydroxyproline disappear, while arginine and proline appear in sensible quantities. Cysteine and methionine were not found at all; probably their place is taken by taurine (Reverberi *et al.*, 1964).

Jokusch (1968) studied the incorporation of leucine-^{3}H in early stages of *Limnaea* by means of autoradiography. Incorporation increases during the first three cleavage divisions; it fluctuates with the cell cycle, incorporation at metaphase being about 30% of that during the next interphase. Incorporation is partially inhibited by 5×10^{-3} *M* puromycine. Van den Biggelaar (1972) found that protein synthesis is demonstrable starting at the second maturation division. Synthesis is minimal during mitosis, while there is a marked peak in synthetic activity during the latter part of the S phase and the beginning of the G_2 phase, especially at the two-cell stage. The activity of the nuclei and mitotic apparatus is higher than that of the cytoplasm, but the chromosomes are unlabeled. Van der Wal (1972) observed by means of electron microscope autoradiography of *Limnaea* eggs incubated with leucine-^{3}H at the uncleaved stage or during early cleavage that the labeled proteins are especially localized in the mitochondria, the disintegrating yolk granules, the lipid bodies, and the nucleus. In the yolk granules, incorporation is restricted to the surface of intact granules, the borders of surrounding vacuoles, and the diffuse matrix of disintegrated yolk granule substance in the vacuoles. This localization suggests that the newly synthesized proteins represent mainly enzymes playing a part in the breakdown of the yolk.

The rate of protein synthesis in *Spisula* after fertilization increases steadily for at least 3 hours. There is at the same time an increase in the number and specific activity of polysomes (Bell and Reeder, 1967). In *Ilyanassa* protein synthesis, measured by the incorporation of leucine-^{14}C, begins at the third day, and reaches a maximum during the fourth day, just preceding visible organ differentiation (Collier, 1961a,b).

Changes in activity of different enzymes during development in various species were recorded by Collier (1957a,b,c), Morrill (1964), and d'Ancona Lunetta and Minganti (1964). In early stages of development of *Helix* the chromatin of the nuclei contains a weakly basic histone, differing in its coloration with fast green from that of later stages. During gastrulation, this is replaced by the adult histone (Bloch and Hew, 1960).

With electrophoretic methods, Norris and Morrill (1964) found in adult organs of *Limnaea palustris* forty-two hydrolytic enzymes. In early embryos, five of these enzymes can be detected. From the third day, the number of enzymes gradually increases, and in the recently hatched young snail, twenty-nine of the forty-two enzymes can be demonstrated. No bands were found that did not occur in the adult. In *Ilyanassa,* however, some hydrolytic enzyme bands are only present before the fourth day, other ones appear at or after this day, while a further group of enzymes is present throughout development (Morrill and Norris, 1965). In *Argobuccinum,* in uncleaved and cleaving eggs, five isozymes of lactic dehydrogenase are present. Two of them disappear at the blastula stage. In the early veliger, a new band appears, in the mid-veliger, a second one, so that there are finally five bands at about equal electrophoretic distances. It is assumed that the two early disappearing bands belong to ovarian enzymes, whereas the two new ones represent new synthesis in connection with differentiation (Goldberg and Cather, 1963).

Removal of the polar lobe in *Ilyanassa* results in decreased protein synthesis (Collier, 1961a, 1965b). The isolated polar lobe incorporates labeled amino acids into proteins, but its rate of incorporation per unit volume is only about half that of the whole egg (Clement and Tyler, 1967). In *Mytilus,* the incorporation rate of labeled amino acids is also much lower in the isolated polar lobe than in the rest of the egg, but no differences were observed prior to isolation (Abd-el-Wahab and Pantelouris, 1957).

Puromycin blocks cleavage reversibly in *Ilyanassa* (Collier, 1966). Chloramphenicol in low concentrations blocks development at an early gastrula stage in *Planorbis* (Sherbet and Lakshmi, 1964b).

IX. Respiration

Cleland (1950) studied the respiration of developing *Ostrea* eggs. Respiration begins to rise with the onset of cleavage. It exhibits rhythmic fluctuations with the cleavage cycles, the rate of respiration being high from prophase to metaphase or perhaps anaphase, and low during telophase and interphase. Geilenkirchen (1961) studied the respiration of single cleaving *Limnaea* eggs by means of the Cartesian diver technique. Oxygen uptake drops from prophase to telophase, and rises from telophase to prophase, with perhaps an interruption in midinterphase of the third cleavage cycle.

Generally, respiration continues to increase during development, but the rise is not always regular and may be interrupted by plateaus or even drops at certain stages (Cleland, 1950; Berg and Kutsky, 1951; Sclufer, 1955; Perlogawora-Szumlewicz and von Brand, 1957; Horstmann, 1958; Black, 1962a; Hunger and Horstmann, 1968).

In *Ostrea,* the respiratory quotient remains constant during cleavage at a value of about 0.8. Presumably, a mixture of carbohydrate and fat is combusted during cleavage (Cleland, 1950). In *Spisula,* the respiratory quotient, which is about 0.7 immediately after fertilization, shows a gradual increase toward unity (Sclufer, 1955). A respiratory quotient of about 0.8–0.85 has been found during cleavage in *Aplysia* (Buglia, 1908). Meyerhof (1911), studying the calorific quotients in addition to the respiratory exchanges, concluded that fat is the principal fuel throughout development in this species. On the contrary, the mean value of the respiratory quotient during early stages of development of *Limnaea* was found to be 1.05. This is connected with the fact that fat is being synthesized at the expense of carbohydrate (Baldwin, 1935; Horstmann, 1956a,b).

Berg and Kutsky (1951) studied the respiration of isolated blastomeres in *Mytilus.* The oxygen consumption is higher in CD than in AB, but when it is calculated per unit of volume, it is 13% lower in CD. The oxygen consumption per unit volume of the isolated polar lobe is 25% less than that of the whole egg.

X. Cell Metabolism

The terminal electron transport system in unfertilized eggs of *Spisula* agrees with that in mammalian tissues, chick embryos, and in other marine eggs. It is mainly bound to mitochondria. Succinate cytochrome c reductase, DPNH cytochrome c reductase, and cytochrome oxidase exhibit no clear quantitative changes during the first 22 hours of development to a swimming larva. Apart from the first one, these enzymes are present in excess with respect to the actual respiratory rate. There are no differences in enzyme concentration between blastomeres A, B, C, and D (Strittmatter and Strittmatter, 1961). According to Krane and Crane (1960), TPNH increases three- to sevenfold during the first hours after fertilization. TPN shows no significant change. It is concluded that glucose utilization takes place mainly via the hexose monophosphate shunt. In *Crassostrea*, cytochrome oxidase, DPNH oxidase, succinic dehydrogenase, malic dehydrogenase, and aconitase exhibit no changes until the trochophore stage; then the activity of cytochrome oxidase decreases to the veliger stage. Isocitric dehydrogenase and α-ketoglutaric dehydrogenase increase two- to fourfold from the blastula to the trochophore (Black, 1962a). Succinic dehydrogenase and DPNH oxidase are almost completely localized in the large granule fraction of the fertilized egg, but in the trochophore about one-third of each enzyme is found in the supernatant fraction. High percentages of aconitase, isocitric dehydrogenase, and α-ketoglutaric dehydrogenase are found in the granules of both stages; the latter enzyme appears to be completely localized in the granules of the trochophore. Fumarase, malic dehydrogenase, and DPNH cytochrome c reductase are almost absent from granules of the egg, but considerable proportions of these enzymes are found in trochophore granules. All of the enzymes except DPNH oxidase increase in activity, relative to succinic dehydrogenase, in the granules during development to the trochophore stage. This indicates that differentiation of the population of respiratory granules occurs during development in *Crassostrea* (Black, 1962b). α-Amylase is lacking prior to the veliger stage; then its activity increases with the development of the gut. After 37 hours there is a sharp drop, probably due to starvation. The larval enzyme has the same pH and substrate specificity as the adult one (Black and Pengelley, 1964).

In *Limnaea*, anaerobic production of lactic acid by the breakdown of galactogen begins at the fifth day. The following glycolytic enzymes were found: hexokinase, phosphoglucomutase, glucose phosphate isomerase, aldolase, enolase, and lactic acid dehydrogenase. Glucose breakdown apparently takes place via the Embden–Meyerhof pathway. Moreover, the enzymes of the pentose phosphate cycle are present in the eggs (Horstmann, 1960a,b). In *Helix*, the activity of aldolase and β-galactosidase, relative to dry weight, remains unchanged after the third to fourth day of development. Lactate dehydrogenase rises until the eighth to eleventh day, then remains constant. Glucose-6-phosphate dehydrogenase first increases, but shows a continuous drop in activity after the ninth day (Hunger and Horstmann, 1968).

XI. Cell Differentiation

In general, histological differentiation of cells begins shortly after gastrulation. It is attended with a "chemical differentiation," as the cells in various organs begin to differ in their chemical composition, e.g., by the synthesis of special proteins essential for their function. In most mollusks, where development is indirect, cell differentiation is at first restricted to some larval organs (e.g. velum, larval livers, larval kidneys), whereas in large parts of the body the cells remain more or less undifferentiated and continue dividing in preparation for the development of adult organs at some later stage. In those cases, often a marked difference in composition may be seen between the cells of larval organs and the undifferentiated cells elsewhere. The larval organs consist of large cells with big nuclei, which stain only weakly with Feulgen. They have lost the capacity to divide. Their cytoplasm contains many Golgi bodies and mitochondria, often big albumen vacuoles, and a great amount of fat, glycogen, and iron, but it is poor in RNA. The primordia of adult structures, on the other hand, have small cells which are actively dividing, have small strongly Feulgen-positive nuclei, possess no or only small albumen vacuoles, few Golgi bodies and mitochondria, little fat, glycogen, and iron, but are rich in RNA (Raven, 1946; Bedford, 1966a). This suggests that the two types represent opposite trends of cell life—on the one hand, functional elements containing a great amount of fuel and having a predominantly catabolic metabolism; on the other hand, formative cells in which anabolism preponderates. However, Van den Biggelaar (1972) found that the cells of the velum and larval kidney at the trochophore stage of *Limnaea* exhibit a very strong incorporation both of uridine-^{3}H in their nucleoli and of leucine-^{3}H.

In addition to the common characteristics of larval cells enumerated above, the larval ciliary cells of velum, apical plate, foot, and stomodaeum are also rich in cytochrome oxidase (Ries, 1937; Raven,

1946; Attardo, 1955; Reverberi, 1958) and in alkaline phosphatase (Minganti, 1950; Bedford, 1966a) in various species. Occasionally, they also show positive reactions of peroxidase (Prenant, 1924; Raven, 1946), leukomethylene blue oxidoreductase (Ries and Gersch, 1936), cholinesterase (Buznikov, 1960), and ascorbic acid (Peltrera, 1940).

In some cases, the distribution of a special component among the embryonic organs is suggestive of its functional role in the cell. For instance, alkaline phosphatase, apart from its occurrence in ciliary cells, is especially found at places where secretory or excretory processes are taking place—the shell gland and mantle fold (Minganti, 1950; Cather, 1967; Timmermans, 1969), the statocyst during the secretion of the statolith (Minganti, 1950; Bedford, 1966a), the wall of the nephric channel in the larval kidney, the wall of the foregut in the region of the chitinous jaws, the odontoblasts (Minganti, 1950), and the supraradular epithelium (Gabe and Prenant, 1952b) (Fig. 2).

The distribution of substances, as revealed by cytochemical methods may show a pattern from which conclusions on the role of the cells can be drawn. In the mantle fold of *Limnaea,* a clear zonation exists in the region which is active in shell secretion. Nearest the edge of the mantle there is a band of cells rich in RNA and peroxidase, which is probably involved in the synthesis, secretion, and tanning of the proteins of the periostracum. In the next zone the cells are also rich in RNA, but peroxidase is lacking; presumably it forms the inner part of the periostracum or the proteinaceous matrix of the calcareous shell layers. This, in turn, is succeeded by a region, where the cells are rich

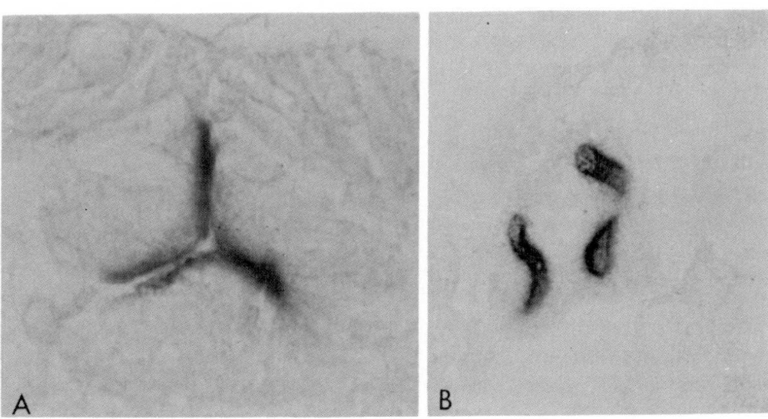

FIG. 2. Alkaline phosphatase reaction (azo dye method) in (A) shell gland and (B) larval kidney of 3-day-old embryo of *Limnaea stagnalis.* (Courtesy of Miss Dr. L. P. M. Timmermans.)

in glycogen, alkaline phosphatase, carbonic anhydrase, ATPase, cytochrome oxidase, and the enzymes of the tricarboxylic acid cycle. Apparently, these are the cells providing for the segregation, transfer, and secretion of the calcium and carbonate ions for building the calcareous substance of the shell (Timmermans, 1969).

A somewhat similar case is that of the radular sac, where the different parts involved in the secretion of the radula differ in their cytochemical composition. The odontoblasts, which produce the chitinous teeth, are generally rich in carbohydrates, but have only a small amount of ribonucleoproteins and sulfhydryl proteins, and are poor in mineral substances. Alkaline phosphatase may either be present (*Limnaea*) or absent (*Actaeon*) (Minganti, 1950; Gabe and Prenant, 1952a,b, 1957, 1959). The subradular epithelium, which takes part in the secretion of both the radular and the subradular membrane, is rich in carbohydrates, both glycogen and mucopolysaccharides; its cells further contain a large amount of RNA and of sulfhydryl proteins (Gabe and Prenant, 1952b, 1959). Finally, the supraradular epithelium, which takes part in hardening of the teeth by their impregnation with either calcium or iron salts and with proteins, is rich in carbohydrates, RNA, sulfhydryl proteins, alkaline phosphatase, and either calcium or iron (Spek, 1921; Gabe and Prenant, 1952a,b, 1959). Runham (1963) supposes that the first hardening of the radular teeth takes place by the formation of covalent bonds between chitin end groups and protein carboxyl.

REFERENCES

Abd-el-Wahab, A., and Pantelouris, E. M. (1957). *Exp. Cell Res.* **13**, 78.
Albanese, M. P., and Bolognari, A. (1961). *Caryologia* **14**, 329.
Albanese, M. P., and Bolognari, A. (1964). *Experientia* **20**, 29.
Andrew, A. (1959). *Proc., Kon. Ned. Akad. Wetensch., Ser. C* **62**, 68.
Arendsen de Wolff-Exalto, E. (1947). *Proc., Kon. Ned. Akad. Wetensch.* **50**, 315.
Arvy, L. (1949). *C.R. Acad. Sci.* **228**, 1983.
Arvy, L. (1950a). *Arch. Biol.* **61**, 187.
Arvy, L. (1950b). *Bull. Soc. Zool. Fr.* **75**, 159.
Attardo, C. (1955). *Ric. Sci.* **25**, 2797.
Attardo, C. (1957). *Acta Embryol. Morphol. Exp.* **1**, 65.
Baldwin, E. (1935). *J. Exp. Biol.* **12**, 27.
Ballentine, R. (1940). *J. Cell. Comp. Physiol.* **15**, 217.
Beadle, L. C. (1969a). *J. Exp. Biol.* **50**, 473.
Beadle, L. C. (1969b). *J. Exp. Biol.* **50**, 491.
Beams, H. W., and Sekhon, S. S. (1966). *J. Morphol.* **119**, 477.
Bedford, L. (1966a). *J. Embryol. Exp. Morphol.* **15**, 15.
Bedford, L. (1966b). Thesis, University of Sydney.
Bell, E., and Reeder, R. (1967). *Biochim. Biophys. Acta* **142**, 500.
Berg, W. E. (1950). *Biol. Bull.* **98**, 128.
Berg, W. E. (1957). *Biol. Bull.* **113**, 365.

Berg, W. E., and Kato, Y. (1959). *Acta Embryol. Morphol. Exp.* **2**, 227.
Berg, W. E., and Kutsky, P. B. (1951). *Biol. Bull.* **101**, 47.
Berthier, J. (1948). *Bull. Biol. Fr. Belg.* **82**, 61.
Black, R. E. (1962a). *Biol. Bull.* **123**, 58.
Black, R. E. (1962b). *Biol. Bull.* **123**, 71.
Black, R. E., and Pengelley, E .T. (1964). *Biol. Bull.* **126**, 199.
Bloch, D. P., and Hew, H. Y. C. (1960). *J. Biophys. Biochem. Cytol.* **8**, 69.
Bluemink, J. G. (1967). "The subcellular Structure of the Blastula of Limnaea stagnalis L. (Mollusca) and the Mobilization of the Nutrient Reserve." Ph.D. Thesis, Univ. of Utrecht.
Bolognari, A. (1954). *Boll. Zool.* **21**, 185.
Bolognari, A. (1956a). *Acta Histochem.* **2**, 229.
Bolognari, A. (1956b). *Arch. Zool. Ital.* **41**, 241.
Bolognari, A. (1957). *Boll. Soc. Ital. Biol. Sper.* **33**, 46.
Bolognari, A. (1959). *Acta Histochem.* **8**, 504.
Bolognari, A. (1960a). *Boll. Soc. Ital. Biol. Sper.* **36**, 575.
Bolognari, A. (1960b). *Nature (London)* **186**, 490.
Bolognari, A. (1960c). *Nature (London)* **186**, 565.
Bolognari, A. (1961). *Atti Soc. Peloritana Sci. Fis. Mat. Natur.* **7**, 1.
Bolognari, A., and Cannata, F. (1963). *Atti Soc. Peloritana Sci. Fis. Mat. Natur.* **9**, 273.
Bolognari, A., and de Raco, H. (1956). *Atti Soc. Peloritana Sci. Fis. Mat. Natur.* **2**, 131.
Brahmachary, R. L., Banerjee, K. P., and Basu, T. K. (1968). *Exp. Cell Res.* **51**, 177.
Bretschneider, L. H., and Raven, C. P. (1951). *Arch. Neer. Zool.* **10**, 1.
Brown, D. D., and Dawid, I. B. (1968). *Science* **160**, 272.
Buglia, G. (1908). *Arch. Fisiol.* **5**, 455.
Buznikov, G. A. (1960). *Dokl. Biol. Sci.* **132**, 373.
Cannata, F. (1962). *Atti Soc. Peloritana Sci. Fis. Mat. Natur.* **8**, 335.
Cather, J. N. (1967). *J. Exp. Zool.* **166**, 205.
Cleland, K. W. (1950). *Proc. Linn. Soc. N.S. Wales* **75**, 282.
Clement, A. C. (1938). *J. Exp. Zool.* **79**, 435.
Clement, A. C., and Lehmann, F. E. (1956). *Naturwissenschaften* **43**, 478.
Clement, A. C., and Tyler, A. (1967). *Science* **158**, 1457.
Collier, J. R. (1957a). *Embryologia* **3**, 243.
Collier, J. R. (1957b). *Biol. Bull.* **113**, 340 (abstr.).
Collier, J. R. (1957c). *Exp. Cell Res.* **13**, 122.
Collier, J. R. (1960a). *Exp. Cell Res.* **21**, 548.
Collier, J. R. (1960b). *Exp. Cell Res.* **21**, 126.
Collier, J. R. (1961a). *Acta Embryol. Morphol. Exp.* **4**, 70.
Collier, J. R. (1961b). *Exp. Cell Res.* **24**, 320.
Collier, J. R. (1963). *Exp. Cell Res.* **32**, 442.
Collier, J. R. (1965a). *Science* **147**, 150.
Collier, J. R. (1965b). *In* "The Biochemistry of Animal Development" (R. Weber, ed.), Vol. 1, pp. 203–244. Academic Press, New York.
Collier, J. R. (1966). *Curr. Top. Develop. Biol.* **1**, 39.
Collier, J. R., and McCann-Collier, M. (1962). *Exp. Cell Res.* **27**, 553.
Collier, J. R., and Yuyama, S. (1969). *Exp. Cell Res.* **56**, 281.
Comandon, J., and de Fonbrune, P. (1935). *Arch. Anat. Microsc.* **31**, 79.

Conklin, E. G. (1917). *J. Exp. Zool.* **22**, 311.
Costello, D. P. (1939). *Physiol. Zool.* **12**, 13.
Cowden, R. R. (1961). *Biol. Bull.* **120**, 313.
Cowden, R. R. (1962). *Growth* **26**, 209.
Cowden, R. R. (1966). *Histochemie* **6**, 226.
Crowell, J. (1964). *Acta Embryol. Morphol. Exp.* **7**, 225.
Dalcq, A., and Pasteels, J. J. (1962). *Arch. Anat. Histol. Embryol.* **44**, Suppl., 75.
Dalcq, A., and Pasteels, J. J. (1963). *Develop. Biol.* **7**, 457.
Dan, J. C. (1962). *Biol. Bull.* **123**, 531.
Dan, J. C. (1967). *In* "Fertilization: Comparative Morphology, Biochemistry and Immunology" (C. B. Metz and A. Monroy, eds.), Vol. 1, pp. 237–295. Academic Press, New York.
Dan, J. C., and Wada, S. K. (1955). *Biol. Bull.* **109**, 40.
d'Ancona Lunetta, G., and Minganti, A. (1964). *Acta Embryol. Morphol. Exp.* **7**, 249.
Davenport, R. (1967). *Exp. Cell Res.* **47**, 606.
Davenport, R., and Davenport, J. C. (1965). *Exp. Cell Res.* **39**, 74.
Davidson, E. H., Haslett, G. W., Finney, R. J., Allfrey, V. G., and Mirsky, A. E. (1965). *Proc. Nat. Acad. Sci. U.S.* **54**, 696.
Elbers, P. F. (1957). *Proc., Kon. Ned. Akad. Wetensch., Ser. C* **60**, 96.
Elbers, P. F. (1959). "Over de beginoorzaak van het Li-effect in de morphogenese." Ph.D. Thesis, Univ. of Utrecht.
Elbers, P. F. (1966). *Biochim. Biophys. Acta* **112**, 318.
Elbers, P. F. (1969). *J. Embryol. Exp. Morphol.* **22**, 449.
Fahmy, O. G. (1949). *Quart. J. Microsc. Sci.* **90**, 159.
Fautrez, J. (1959). *Biol. Jaarb.* **27**, 17.
Favard, P., and Carasso, N. (1958). *Arch. Anat. Microsc. Morphol. Exp.* **47**, 211.
Fioroni, P. (1966). *Rev. Suisse Zool.* **73**, 621.
Franc, A. (1951). *Ann. Sci. Nat. Zool. Biol. Anim.* [11] **13**, 135.
Gabe, M., and Prenant, M. (1949). *Cellule* **53**, 99.
Gabe, M., and Prenant, M. (1952a). *C.R. Acad. Sci.* **235**, 1050.
Gabe, M., and Prenant, M. (1952b). *Arch. Zool. Exp. Gen.* **89**, N. 15.
Gabe, M., and Prenant, M. (1957). *Ann. Sci. Nat. Zool. Biol. Anim.* [11] **19**, 587.
Gabe, M., and Prenant, M. (1959). *Ann. Histochim.* **3**, 95.
Galtsoff, P. S. (1964). *U.S., Fish. Wildl. Serv., Fish. Bull.* **64**.
Geilenkirchen, W. L. M. (1961). "Effects of Mono- and Divalent Cations on Viability and Oxygen Uptake of Eggs of Limnaea stagnalis." Ph.D. Thesis, Univ. of Utrecht.
George, J. C., and Jura, Cz. (1958). *Proc., Kon. Ned. Akad. Wetensch., Ser. C* **61**, 598.
Goldberg, E., and Cather, J. N. (1963). *J. Cell. Comp. Physiol.* **61**, 31.
Guerrier, P. (1968). *Ann. Embryol. Morphol.* **1**, 119.
Horstmann, H. J. (1956a). *Biochem. Z.* **328**, 342.
Horstmann, H. J. (1956b). *Biochem. Z.* **328**, 348.
Horstmann, H. J. (1958). *Z. Vergl. Physiol.* **41**, 390.
Horstmann, H. J. (1960a). *Hoppe-Seyler's Z. Physiol. Chem.* **319**, 110.
Horstmann, H. J. (1960b). *Hoppe-Seyler's Z. Physiol. Chem.* **319**, 120.
Horstmann, H. J. (1964). *Hoppe-Seyler's Z. Physiol. Chem.* **337**, 57.
Horwitz, B. A. (1965). *Exp. Cell Res.* **38**, 620.
Horwitz, B. A., and Nelson, L. (1964). *Biol. Bull.* **127**, 374 (abstr.).

Humphreys, W. J. (1962a). *Proc. Int. Congr. Electron Microsc., 5th, 1962* Vol. II, Art. NN-10.
Humphreys, W. J. (1962b). *J. Ultrastruct. Res.* **7**, 467.
Humphreys, W. J. (1964). *J. Ultrastruct. Res.* **10**, 244.
Humphreys, W. J. (1967). *J. Ultrastruct. Res.* **17**, 314.
Hunger, J., and Horstmann, H. J. (1968). *Z. Biol. (Munich)* **116**, 90.
Ito, S., and Leuchtenberger, C. (1955). *Chromosoma* **7**, 328.
Jokusch, B. (1968). *Z. Naturforsch. B* **23**, 1512.
Jura, Cz. (1960). *Zool. Pol.* **10**, 95.
Kobayashi, H. (1956). *J. Sci. Hiroshima Univ., Ser. B, Div. 1* **16**, 1.
Kobayashi, H. (1960). *Jap. J. Zool.* **12**, 497.
Kobayashi, H. (1968a). *Cytologia* **33**, 112.
Kobayashi, H. (1968b). *Cytologia* **33**, 118.
Kofoid, C. A. (1895). *Bull. Mus. Comp. Zool., Harvard Coll.* **27**, 35.
Konopacki, M. (1933). *Bull. Int. Acad. Pol. Sci. Lett., Cl. Sci. Math. Natur., Ser. B* **2**, p. 51.
Krane, S. M., and Crane, R. K. (1960). *Biochim. Biophys. Acta* **43**, 369.
Krauss, M. (1950). *J. Exp. Zool.* **114**, 239.
Lams, H. (1910). *Mem. Acad. Roy. Belg.*, Cl. Sci. (In 4°), *Ser. II* **2**.
Laviolette, P. (1954). *Ann. Sci. Nat. Zool. Biol. Anim.* [11] **11**, 427.
Lucké, B., and Ricca, R. A. (1941). *J. Gen. Physiol.* **25**, 215.
McMahon, P., von Brand, T., and Nolan, M. O. (1957). *J. Cell. Comp. Physiol.* **50**, 219.
Malhotra, S. K. (1960). *Cellule* **61**, 109.
Malhotra, S. K. (1961). *Quart. J. Microsc. Sci.* **102**, 83.
Mancuso, V. (1953). *Rend. Ist. Super. Sanita. (Ital. Ed.)* **16**, 367.
Mancuso, V. (1954). *Ric. Sci.* **24**, 1886.
Mancuso, V. (1955). *Ric. Sci.* **25**, 2843.
May, F., and Weinland, H. (1953). *Z. Biol. (Munich)* **105**, 339.
Metz, C. B., and Donovan, J. E. (1949). *Biol. Bull.* **97**, 257.
Meyerhof, O. (1911). *Biochem. Z.* **35**, 246.
Minganti, A. (1950). *Riv. Biol.* **42**, 295.
Monroy, A., and Tolis, H. (1964). *Biol. Bull.* **126**, 456.
Morrill, J. B. (1963). *Acta Embryol. Morphol. Exp.* **6**, 339.
Morrill, J. B. (1964). *Acta Embryol. Morphol. Exp.* **7**, 131.
Morrill, J. B., and Norris, E. (1965). *Acta Embryol. Morphol. Exp.* **8**, 232.
Morrill, J. B., Norris, E., and Smith, S. D. (1964). *Acta Embryol. Morphol. Exp.* **7**, 11.
Needham, J. (1942). "Biochemistry and Morphogenesis." Cambridge Univ. Press, London and New York.
Niijima, L., and Dan, J. (1965). *J. Cell Biol.* **25**, 243 and 249.
Norris, E., and Morrill, J. B. (1964). *Acta Embryol. Morphol. Exp.* **7**, 29.
Parat, M. (1928). *Arch. Anat. Microsc.* **24**, 73.
Pasteels, J. J. (1966a). *J. Embryol. Exp. Morphol.* **16**, 301.
Pasteels, J. J. (1966b). *Bull. Ass. Anat.* **51**, 763.
Pasteels, J. J., and de Harven, E. (1962a). *Arch. Biol.* **73**, 445.
Pasteels, J. J., and de Harven, E. (1962b). *Arch. Biol.* **73**, 465.
Pasteels, J. J., and de Harven, E. (1963). *Arch. Biol.* **74**, 415.
Pasteels, J. J., and Mulnard, J. (1957). *Arch. Biol.* **68**, 115.
Pasteels, J. J., and Mulnard, J. (1960). *C.R. Acad. Sci.* **250**, 190.

Pasteels, J. J., and Mulnard, J. (1963). *Arch. Biol.* **74**, 319.
Pelluet, D., and Watts, A. H. G. (1951). *Quart. J. Microsc. Sci.* **92**, 453.
Peltrera, A. (1940). *Pubbl. Staz. Zool. Napoli* **18**, 20.
Perlogawora-Szumlewicz, A., and von Brand, T. (1957). *J. Wash. Acad. Sci.* **47**, 11.
Portmann, A., and Bidder, A. M. (1928). *Quart. J. Microsc. Sci.* **72**, 301.
Prenant, M. (1924). *Arch. Morphol. Gen. Exp.* **21**, 1.
Pucci, I. (1961). *Acta Embryol. Morphol. Exp.* **4**, 96.
Ranzoli, F. (1953). *Caryologia* **5**, 137.
Rao, K. V., and Mulherkar, L. (1964). *Proc., Kon. Ned. Akad. Wetensch., Ser. C* **67**, 273.
Raven, C. P. (1938). *Acta Neer. Morphol.* **1**, 337.
Raven, C. P. (1945). *Arch. Neer. Zool.* **7**, 91.
Raven, C. P. (1946). *Arch. Neer. Zool.* **7**, 353.
Raven, C. P. (1952). *J. Exp. Zool.* **121**, 1.
Raven, C. P. (1961). "Oogenesis: The Storage of Developmental Information." Pergamon Press, Oxford.
Raven, C. P. (1963). *Develop. Biol.* **7**, 130.
Raven, C. P. (1964). *J. Embryol. Exp. Morphol.* **12**, 805.
Raven, C. P. (1966). "Morphogenesis. The Analysis of Molluscan Development," 2nd ed. Pergamon Press, Oxford.
Raven, C. P. (1967). *Develop. Biol.* **16**, 407.
Raven, C. P. (1970). *Int. Rev. Cytol.* **28**, 1.
Raven, C. P., and Bretschneider, L. H. (1942). *Arch. Neer. Zool.* **6**, 255.
Raven, C. P., and Klomp, H. (1946). *Proc., Kon. Ned. Akad. Wetensch.* **49**, 101.
Raven, C. P., Bezem, J. J., and Isings, J. (1952). *Proc., Kon. Ned. Akad. Wetensch., Ser. C* **55**, 248.
Raven, C. P., Bezem, J. J., and Geelen, J. F. M. (1953). *Proc., Kon. Ned. Akad. Wetensch., Ser. C* **56**, 409.
Rebhun, L. I. (1956a). *J. Biophys. Biochem. Cytol.* **2**, 93.
Rebhun, L. I. (1956b). *J. Biophys. Biochem. Cytol.* **2**, 159.
Rebhun, L. I. (1958). *Biol. Bull.* **115**, 325 (abstr.).
Rebhun, L. I. (1959). *Biol. Bull.* **117**, 518.
Rebhun, L. I. (1960). *Ann. N.Y. Acad. Sci.* **90**, 357.
Recourt, A. (1961). "An Electron Microscopic Study of Oogenesis in Limnaea stagnalis L." Ph.D. Thesis, Univ. of Utrecht.
Reverberi, G. (1958). *Acta Embryol. Morphol. Exp.* **2**, 79.
Reverberi, G. (1967). *Acta Embryol. Morphol. Exp.* **10**, 1.
Reverberi, G., and Mancuso, V. (1961). *Acta Embryol. Morphol. Exp.* **4**, 102.
Reverberi, G., Molinaro, M., and Metafora, S. (1964). *Acta Embryol. Morphol. Exp.* **7**, 101.
Ries, E. (1937). *Pubbl. Staz. Zool. Napoli* **16**, 363.
Ries, E. (1938). *Biodynamica* **40**, 1.
Ries, E., and Gersch, M. (1936). *Pubbl. Staz. Zool. Napoli* **15**, 223.
Runham, N. W. (1963). *Ann. Histochim.* **8**, 433.
Sacarrao, G. F. (1945). *Arq. Mus. Bocage* **16**, 33.
Sawada, N. (1964). *Mem. Ehime Univ., Sect. II B* **5**, 53.
Sawada, N., and Murakami, T. H. (1959). *Mem. Ehime Univ., Sect. II* 3, 235.
Schechter, V. (1956). *Exp. Cell Res.* **10**, 619.
Schreiber, G., and Camey, T. (1966). *Ann. Histochim.* **11**, 313.

Sclufer, E. (1955). *Biol. Bull.* **109**, 113.
Selwood, L. (1968). *J. Morphol.* **125**, 71.
Serra, J. A., and Queiroz Lopes, A. (1945). *Port. Acta Biol.* **1**, 51.
Sherbet, G. V., and Lakshmi, M. S. (1964a). *Wilhelm Roux' Arch. Entwicklungsmech. Organismen* **155**, 144.
Sherbet, G. V., and Lakshmi, M. S. (1964b). *Wilhelm Roux' Arch. Entwicklungsmech. Organismen* **155**, 429.
Southwick, W. E. (1939). *Biol. Bull.* **77**, 157.
Spek, J. (1921). *Z. Wiss. Zool.* **118**, 313.
Spek, J. (1934). *Wilhelm Roux' Arch. Entwicklungsmech. Organismen* **131**, 362.
Strittmatter, P., and Strittmatter, C. F. (1961). *J. Cell. Comp. Physiol.* **57**, 87.
Timmermans, L. P. M. (1969). *Neth. J. Zool.* **19**, 417.
Timmermans, L. P. M., Geilenkirchen, W. L. M., and Verdonk, N. H. (1970). *J. Embryol. Exp. Morphol.* **23**, 245.
Tyler, A. (1939). *Proc. Nat. Acad. Sci. U.S.* **25**, 317.
Tyler, A. (1940). *Biol. Bull.* **78**, 159.
Tyler, A. (1949a). *Physiol. Rev.* **28**, 180.
Tyler, A. (1949b). *Amer. Natur.* **83**, 195.
Tyler, A., and Fox, S. W. (1939). *Science* **90**, 516.
Tyler, A., and Fox, S. W. (1940). *Biol. Bull.* **79**, 153.
Ubbels, G. A. (1968). "A Cytochemical Study of Oogenesis in the Pond Snail Limnaea stagnalis." Ph.D. Thesis, Univ. of Utrecht.
Van den Biggelaar, J. A. M. (1971a). *J. Embryol. Exp. Morphol.* **26**, 351.
Van den Biggelaar, J. A. M. (1971b). *J. Embryol. Exp. Morphol.* **26**, 367.
Van den Biggelaar, J. A. M. (1971c). *Exp. Cell Res.* **67**, 207.
Van den Biggelaar, J. A. M. (1972). In preparation.
Van der Wal, U. P. (1972). In preparation.
Vasconcelos Frazao, J. (1957). *C.R. Soc. Biol.* **151**, 1487.
von Medem, F., Graf (1942). *Biol. Zentralbl.* **62**, 431.
von Medem, F., Graf (1945). *Zool. Jahrb., Abt. Allg. Zool. Physiol. Tiere* **61**, 1.
Wada, S. K., Collier, J. R., and Dan, J. C. (1956). *Exp. Cell Res.* **10**, 168.
Whitaker, D. M. (1933). *J. Gen. Physiol.* **16**, 475 and 497.
Wierzejski, A. (1905). *Z. Wiss. Zool.* **83**, 502.
Worley, L. G. (1944). *J. Morphol.* **75**, 77.
Worley, L. G., and Worley, E. K. (1943). *J. Morphol.* **73**, 365.
Yasuzumi, G., and Tanaka, H. (1957). *Exp. Cell Res.* **12**, 681.
Yung, K. C. (1930). *Ann. Inst. Oceanogr. (Paris)* **7**, 301.

CHAPTER 6

Pigments of Mollusca

T. W. Goodwin

I. Carotenoids

A. Lamellibranchia

The unique pigment pectenoxanthin, isolated some 35 years ago by Lederer (1934) from the gonads of the scallop *Pecten maximus,* has now been shown (Campbell *et al.*, 1967) to be identical with the acetylenic carotenoid, alloxanthin (I), first isolated from various cryptomonad algae (Chapman, 1966; Mallams *et al.*, 1967). In addition, a new acetylenic carotenoid, pectenolone (II), was isolated from *P. maximus* (Campbell *et al.*, 1967). Pectenoxanthin is also present in *Pecten jacobaeus* (Lederer, 1934), the common mussel (*Mytilus edulis*) (Campbell *et al.*, 1967), and probably *Volsella modiolus* (von Euler *et al.*, 1934). *Mytilus californianus,* the California sea mussel, is said to contain zeaxanthin (III) and an acidic pigment of unknown structure, mytiloxanthin (m.p. 140°–144°C; λ_{max} (CS_2) 500 nm) (Scheer, 1940). The techniques available in 1940 would almost certainly not have distinguished between alloxanthin and zeaxanthin. Another carotenoid of unknown structure is glycymerin (a neutral xanthophyll, m.p. 148°–153°C, λ_{max} (CS_2) 495 nm) from the gonads of *Pectunculus glycymeris* (Lederer, 1933).

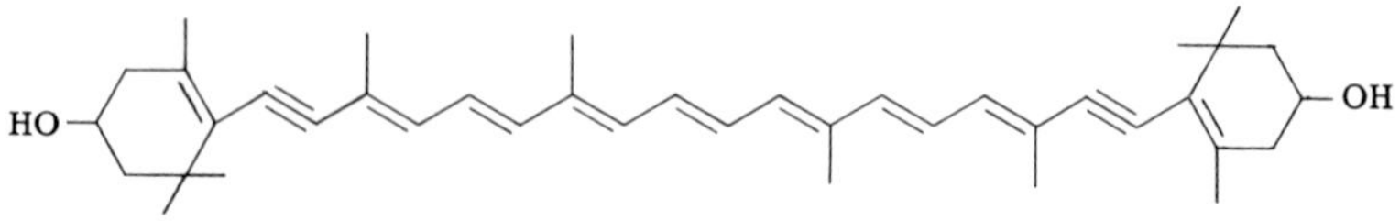

(I)

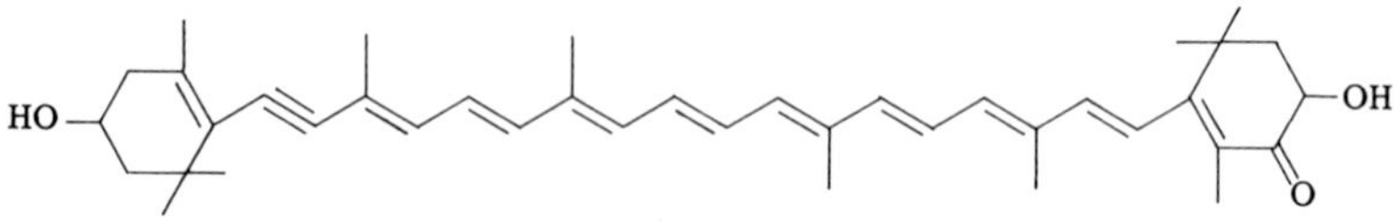

(II)

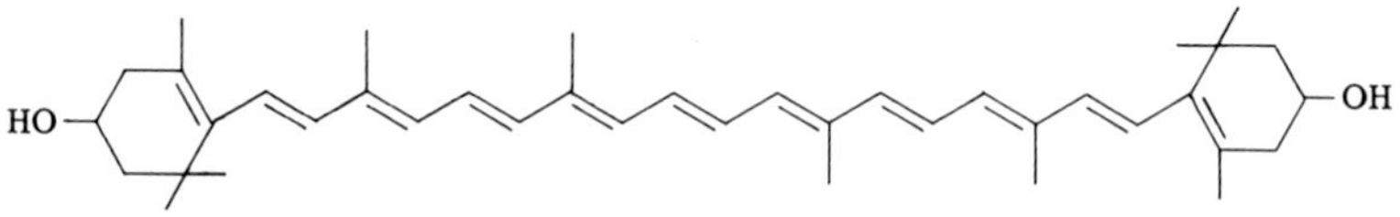

(III)

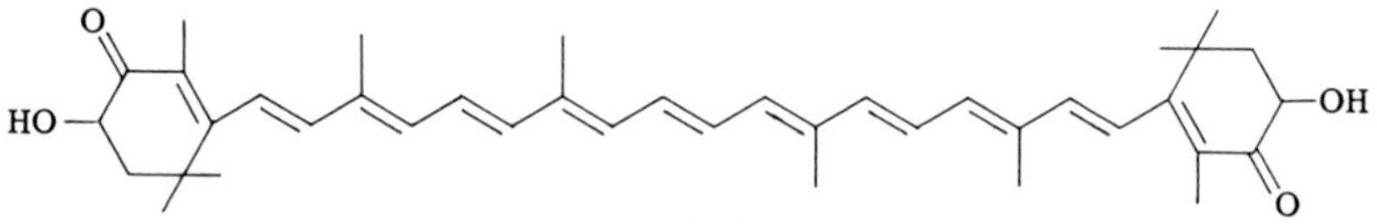

(IV)

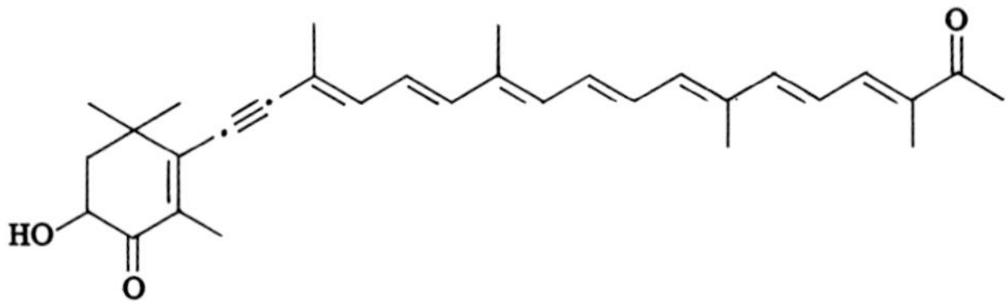

(IVa)

Astaxanthin (IV) is reported in the feet of *Lima excavata* (Sörensen, 1937). Fisher *et al.* (1956a) examined a large group of lamellibranchs for the quantitative distribution of β-carotene and xanthophylls; they did not identify the xanthophylls. The lamellibranchs in which carotenoids have been noted are listed in Table I.

Formation and Metabolism

The only detailed investigation reported is that of Scheer (1940) on *Mytilus californianus*. There was no marked seasonal variation in the

TABLE I
LAMELLIBRANCHS THAT HAVE BEEN REPORTED TO CONTAIN CAROTENOIDS

Species	Pigments	Reference
Anomia ephippium	β-Carotene (V)	Lönnberg (1931)
Astarte sulcata	β-Carotene (V)	Lönnberg (1931)
Cardium echinatum	β-Carotene (V), lutein (VI)(?)	Lönnberg (1931)
Cardium edule	β-Carotene (V), mixture of unidentified xanthophylls	L. R. Fisher *et al.* (1956a)
Cardium tuberculatum	Lutein (VI)(?)	Lederer (1938), Lönnberg (1931)
Chlamys septemradiatus	Mixture of unidentified xanthophylls	L. R. Fisher *et al.* (1956a)
Cochleodesma praetense	β-Carotene (V)	Lönnberg (1931)
Cultellus pellucidus	β-Carotene (V), lutein (VI)	Lönnberg (1931)
Dosinia exoleta	Lutein (VI)	Lönnberg (1931), D. L. Fox and Crane (1942)
Gryphaea angulata	β-Carotene (V), mixture of unidentified xanthophylls	Lönnberg (1933)
Leda parvula	β-Carotene (V), lutein (VI)(?)	L. R. Fisher *et al.* (1956a), Lönnberg (1931)
Lima excavata	Astaxanthin (IV)	Sörensen (1937)
Lima hians	Mixture of unidentified xanthophylls	L. R. Fisher *et al.* (1956a)
Lima doscombei	β-Carotene (V), lutein (VI)	Lönnberg (1931)
Lucina borealis	β-Carotene (V)(?)	Lönnberg (1931)
Modiolaria marmorata	β-Carotene (V)(?), lutein (VI)(?)	Lönnberg (1931)
Modiolus modiolus	β-Carotene (V), lutein (VI)(?)	Lönnberg (1931)
Mya truncata	β-Carotene (V)(?), lutein (VI)(?)	Lönnberg (1931)
Mya arenaria	β-Carotene (V), mixture of unidentified xanthophylls	L. R. Fisher *et al.* (1956a)
Mytilus californicus	Zeaxanthin (III)(?), Mytiloxanthin	Scheer (1940)
Mytilus edulis	β-Carotene (V), alloxanthin (= pectenoxanthin)(I)	Campbell *et al.* (1967), L. R. Fisher *et al.* (1956a)
Meretrix eusoria	β-Carotene (V), lutein (VI), zeaxanthin (III)(?)	Shimizu and Monura (1968)
Nucula sulcata	β-Carotene (V)(?), lutein (VI)	Lönnberg (1931)
Pecten jacobaeus	Alloxanthin (= pectenoxanthin)(I)	Karrer and Solmssen (1935), Lederer (1934), Campbell *et al.* (1967)
Pecten maximus	β-Carotene (V), alloxanthin (= pectenoxanthin)(I), pectenolone (II)	L. R. Fisher *et al.* (1956a)
Pecten opercularis	β-Carotene (V), lutein (VI)(?)	Lönnberg (1931)
Pecten strictus	β-Carotene (V)	Lönnberg (1931)
Psammobia ferroensis	β-Carotene (V), lutein (VI)	Lönnberg (1931)
Pectunculus glycymeris	Glycymerin (unknown structure)	Lederer (1938)
Ostrea edulis	β-Carotene (V), mixture of unidentified xanthophylls	L. R. Fisher *et al.* (1956a)

TABLE I (*Continued*)

Species	Pigments	Reference
Solen ensis	β-Carotene (V), lutein (VI)	Lönnberg (1931)
Scrobicularia plana	β-Carotene (V), mixture of unidentified xanthophylls	L. R. Fisher *et al.* (1956a)
Saxicava rugosa	β-Carotene (V), lutein (VI)	Lönnberg (1931)
Spisula solida	β-Carotene (V)(?), lutein (VI)(?)	Lönnberg (1931)
Spisula subtruncata	β-Carotene (V), lutein (VI)(?)	Lönnberg (1931)
Tapes pullastra	β-Carotene (V), lutein (VI)	Lönnberg (1931)
Tellina crassa	β-Carotene (V)	Lönnberg (1931)
Thiacia convexa	β-Carotene (V)	Lönnberg (1931)
Venus fasciata	β-Carotene (V)	Lönnberg (1931)
Venus gallina	β-Carotene (V)	Lönnberg (1931)
Venus japonica	β-Carotene (V), zeaxanthin (III), flavoxanthin (VII)	Shimizu and Uchida (1968)
Venus ovata	β-Carotene (V)	Lönnberg (1931)
Volsella barbata	β-Carotene (V), lutein (VI)(?)	Lönnberg (1931)

carotenoid levels, and prolonged fasting (196 days) did not affect the levels, although the amounts, especially in the gonads, fell considerably. The levels were also independent of the tissue lipid levels. On a carotenoid-rich diet of *Nitzschia closterium*, carotenoid levels increased, but on a carotenoid-free diet, the levels dropped below that of the starved animals.

During spawning, the carotenoids in the spent ova are the same, qualitatively and quantitatively, as in the female gonads, but the total amount does not account for the difference between ripe and spent females. Spermatozoa do not contain carotenoids, but there is a suggestion of a loss of carotenoids from males during spawning.

The characteristic of mussels is the high concentration of xanthophylls relative to carotenes and there is evidence from Scheer's work that the animals preferentially absorb xanthophylls from the intestinal tract.

B. Gastropoda

1. *Marine Gastropods*

Hopkinsiaxanthin first described in the nudibranch mollusk, *Hopkinsia rosacea* (Strain, 1949) has recently been shown to have structure IVa (McBeth quoted by Isler, 1971).

The limpets *Patella vulgata* and *Patella depressa* contain α-carotene (VIII) and β-carotene (V), echinenone (IX), β-cryptoxanthin (3-hydroxy-β-carotene) (X), and zeaxanthin (III), which are mainly concen-

TABLE II
MARINE GASTROPODS KNOWN TO CONTAIN CAROTENOIDS

Species	Pigment	Reference
Aplysia depilans	β-Carotene (V), unknown xanthophylls	L. R. Fisher *et al.* (1956a)
Aporrhais pes-pelecani	Unknown xanthophylls	L. R. Fisher *et al.* (1956a)
Astea undosa	β-Carotene (V), unknown xanthophylls	L. R. Fisher *et al.* (1956a)
Buccinum undatum	β-Carotene (V), unknown xanthophylls, lutein (VI)	Lönnberg (1931), Fisher *et al.* (1956a)
Capulus hungaricus	β-Carotene (V), lutein (VI)	Lönnberg (1931)
Clione limacina	β-Carotene (V), unknown xanthophylls	L. R. Fisher *et al.* (1956a)
Crepidula fornicata	β-Carotene (V), unknown xanthophylls	L. R. Fisher *et al.* (1956a)
Cypraea spadicea	β-Carotene (V), unknown xanthophylls	L. R. Fisher *et al.* (1956a)
Cyprina islandica	β-Carotene (V)	Lönnberg (1931), Fabre and Lederer (1934)
Gibbula cineraria	β-Carotene (V), unknown xanthophylls	Karrer and Solmssen (1935)
Gibbula tumida	β-Carotene (V), lutein (VI)(?)	Lönnberg (1931)
Haliotis fulgens	β-Carotene (V), unknown xanthophylls	L. R. Fisher *et al.* (1956a)
Hopkinsia rosacea	Hopkinsiaxanthin (IVa)	Strain (1949), Isler (1971)
Lima excavata	Astaxanthin (IV)	Sörensen (1937)
Lima loscombi	β-Carotene (V), lutein (VI)	Lönnberg (1931)
Limacina retroversa	Unknown xanthophylls	L. R. Fisher *et al.* (1956a)
Littorina littoralis	β-Carotene (V), unknown xanthophylls	L. R. Fisher *et al.* (1956a)
Littorina littorea	β-Carotene (V), lutein (VI)	Lönnberg (1933), L. R. Fisher *et al.*, (1956a)
Littorina rudis	β-Carotene (V), unknown xanthophylls	L. R. Fisher *et al.* (1956a)
Megathura crenulata	β-Carotene (V), unknown xanthophylls	L. R. Fisher *et al.* (1956a)
Nassa incrassata	β-Carotene (V)	Lönnberg (1933)
Nassa reticulata	β-Carotene (V)	Lönnberg (1931)
Natica nitida	β-Carotene (V), lutein (VI)	Lönnberg (1931)
Osilinus lineatus	β-Carotene (V), unknown xanthophylls	L. R. Fisher *et al.* (1956a)
Patella depressa	β-Carotene (V), echinenone (IX), zeaxanthin (III), cryptoxanthin (X)	Goodwin and Taha (1950)
Patella vulgata	β-Carotene (V), echinenone (IX), zeaxanthin (III), cryptoxanthin (X)	Goodwin and Taha (1950), L. R. Fisher *et al.*, (1956a)
Philine aperta	β-Carotene (V) (?), lutein (VI)	Lönnberg (1931)
Pleurobranchus sp.	Astaxanthin (IV)	Karrer and Solmssen (1935)
Purpura lapillus	β-Carotene (V), lutein (VI)	Lönnberg (1931)
Rissoa spp.	β-Carotene (V), lutein (VI)	Lönnberg (1931)
Trivia europaea	β-Carotene (V)	Lönnberg (1931)
Trochus zizyphinus	β-Carotene (V), lutein (VI)	Lönnberg (1931)

trated in the testes and ovaries in the ratio 1:5:3:3:3 (Goodwin and Taha, 1950). Carotenoids have also been observed in *Littorina planaxis* (North, 1953), *Cerithidea californica* (Nadakal, 1960), and *Tugalia gigas* (Nishibori, 1956–1957). The last named contained an astaxanthin ester and an unidentified xanthophyll [m.p. 178°–9°C, λ_{max} (ethanol) 430.5, 451, 480 nm]. The known distribution of carotenoids in marine gastropods is given in Table II.

2. *Freshwater and Land Gastropods*

The eggs of *Pomacea* spp. (*Pila* spp.) were first shown to contain carotenoids by Comfort (1947), and later Villela (1956) found

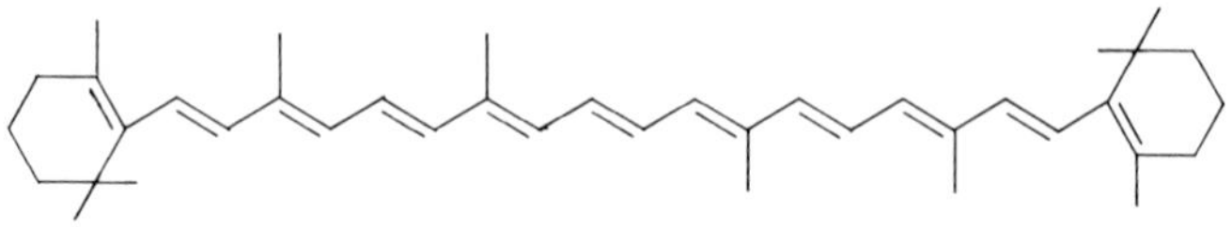

(V)

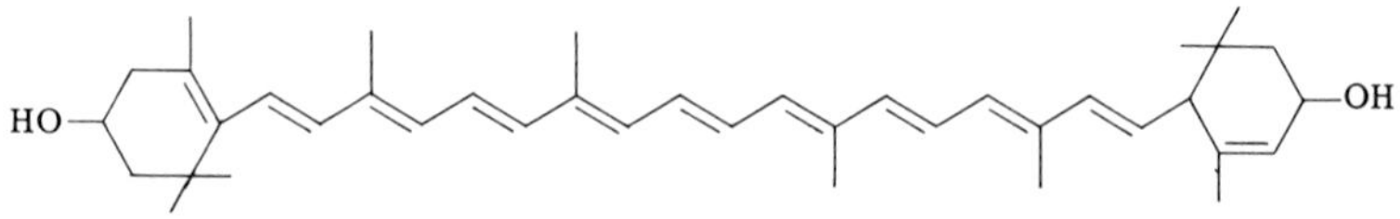

HO OH

(VI)

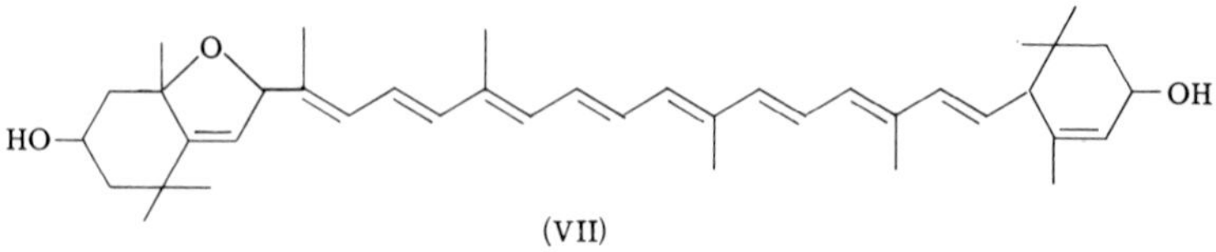

O
HO OH

(VII)

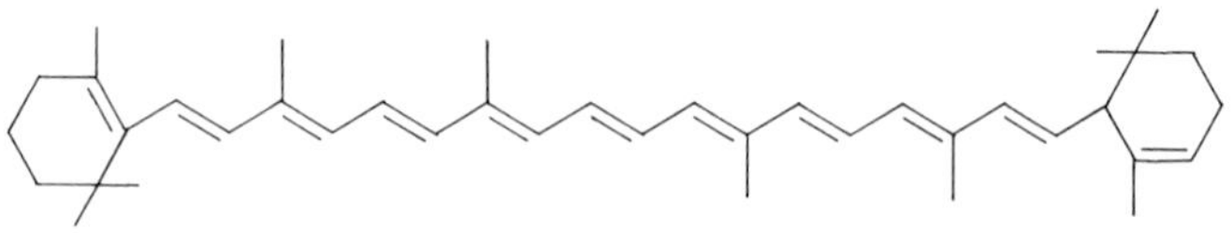

(VIII)

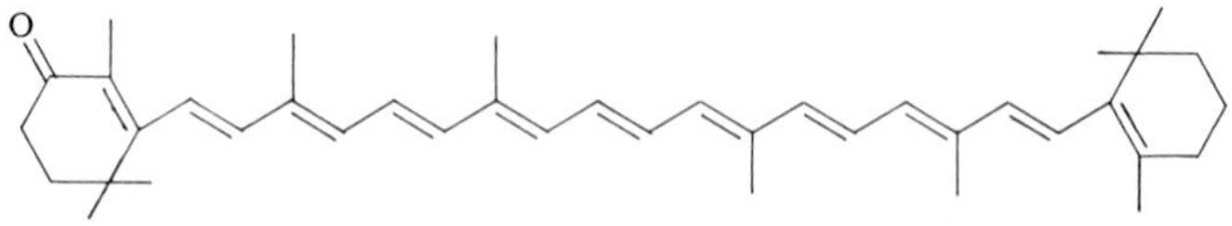

O

(IX)

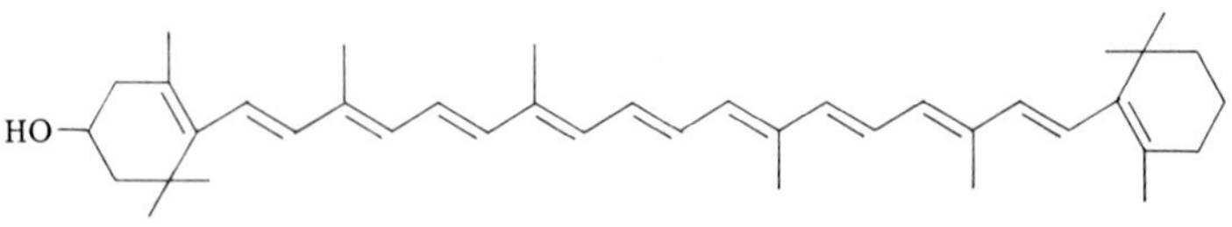

HO

(X)

that the pigments of the omnivores *P. haustrum* and *P. dolioides* were exclusively xanthophylls, while *P. sordida,* which is mainly herbivorous, stored only α- and β-carotene. The red glycochromoprotein from the eggs of *P. canaliculatus australis* contains astaxanthin as its prosthetic group (Cheesman, 1958). The pigment and protein can be reversibly dissociated.

Planorbis corneus var. *rubra* contains traces of β-carotene with unidentified xanthophylls, as does the snail *Helix aspersa* (Fisher *et al.*, 1956a). The accumulation of carotenoids in the neurones of *H. aspersa, P. corneus,* and *Limnaea stagnalis* can be demonstrated histochemically (Cain, 1948).

C. Other Classes

i. Loricata. Stenoplax conspicua contains β-carotene and an unidentified mixture of xanthophylls, which possibly includes astaxanthin (L. R. Fisher *et al.*, 1956a).

ii. Amphineura. Chaetoderma nitidulum contains β-carotene (Lönnberg, 1931).

iii. Scaphopoda. Dentalium entale is said to contain lutein (Lönnberg, 1931).

iv. Cephalopoda. The characteristic of the class Cephalopoda is the comparative absence of carotenoids. Trace amounts have been detected in the animals listed in Table III.

TABLE III

Cephalopods in Which Traces of Carotenoids Have Been Detected

Species	Reference
Decapoda	
Loligo forbesi	L. R. Fisher *et al.* (1956a)
Loligo opalescens	D. L. Fox and Crane (1942)
Ommastrephes pteropus	L. R. Fisher *et al.* (1956b)
Todaropsis eblanae	L. R. Fisher *et al.* (1956b)
Parasepia elegans	L. R. Fisher *et al.* (1956b)
Rossia macrosoma	Lönnberg (1935), L. R. Fisher *et al.* (1956b)
Sepia officinalis	L. R. Fisher *et al.* (1956b)
Sepiola scandica	Lönnberg (1935)
Sepiola sp.	L. R. Fisher *et al.* (1956b)
Octopoda	
Eledone cirrosa	Lönnberg (1935), L. R. Fisher *et al.* (1956b)
(Par) *Octopus bimaculatus*	Lönnberg (1935), D. L. Fox and Crane (1942)

II. Indigoids

The extirpated hypobranchial glands of certain marine gastropods turn blue or red on standing. The pigment Tyrian purple, from *Murex brandaris,* is the 6,6′-dibromo derivative of indigotin (XI). Other sources of these pigments are *Nucella* (*Purpura*) *lapillus, Murex trunculus,* and *Mitra* spp. (see, e.g., D. L. Fox, 1966).

Two prochromogens were obtained from the glands of *M. trunculus* and one from *M. brandaris.* All release sulfate on treatment with the enzyme sulfatase. Prochromogen I from *M. trunculus* is apparently indoxyl, which is oxidizable to indigotin. Prochromogen II from *M. trunculus* and the prochromogen of *M. brandaris* are both 6-bromoindigotin (Bouchilloux and Roche, 1955).

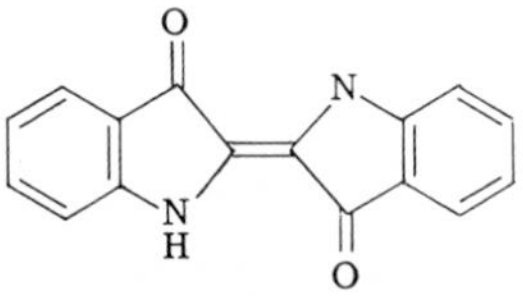

(XI)

III. Melanins

One of the classic sources of melanin is cephalopod ink, particularly that from *Sepia officinalis* (Gortner, 1910; Nicolaus, 1962; Thomson, 1962). Melanin some 150 million years old has been found in the ink sacs of the fossil *Geotenthis* (W. Buckland, 1827, quoted by H. M. Fox and Vevers, 1960). The integument of some gastropod molluscs (e.g., *Limnaea*) also contains melanin (André, 1896). Other possible unconfirmed sources of melanin are discussed by D. L. Fox (1966).

IV. Sclerotins

The hardening (tanning) of the cuticle of invertebrates is brought about by cross-linking of the cuticular proteins with quinones derived from dihydroxyphenols. The simultaneous darkening frequently observed is due to the formation of aminoquinones (see Dennell, 1958). These chromoproteins are frequently called sclerotins; they are ill defined chemically but can be clearly distinguished from melanins and ommochromes. In the Mollusca, such pigments have been observed in the byssus and periostracum of *Mytilus* spp. (H. M. Fox and Vevers, 1960; D. L. Fox, 1966) and the ligament of the fresh water *Anodonta* (H. M. Fox and Vevers, 1960).

V. Ommochromes

The yellow, orange, and brown pigments in the skin chromatophores of *Sepia officinalis* are said to be ommochromes. The different colors may be ommochromes in different redox states or attached to protein; ommochromes are also present in cephalopod eyes (Schwinck, 1953).

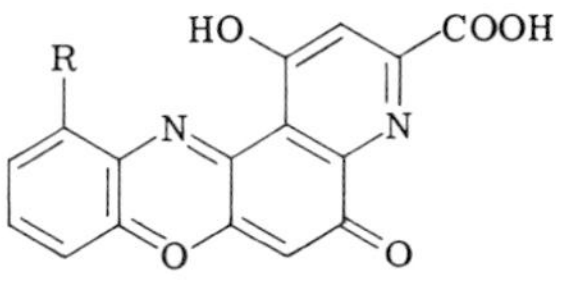

(XII)

The nature of the cephalopod ommochromes is not known, but the characteristic structure of these pigments is exemplified by xanthommatin (XII).

VI. Porphyrins (Excluding Hemoglobins)

Uroporphyrin I (XIII) is the major porphyrin in the shells of forty-two genera of mollusks examined by Comfort (1951), who also noted porphyrins in some fossil species and found that they do not occur in the shells of any land or freshwater Mollusca except certain Neretinidae. An elusive pigment, conchoporphyrin (Fischer and Orth, 1934; Lederer, 1940), reported in a number of molluscan shells, is probably uroporphyrin (Tixier, 1952; Nicholas and Comfort, 1949).

Uroporphyrin I is also present in the integument of the common black garden slug (*Arion ater* ≡ *A. empiricorum* (Kennedy, 1959), *Akera bullata, Duvaucelia plebia* (≡ *Tritonia*) and *Aplysia punctata* (Kennedy and Vevers, 1953). In *Arion ater* the amount of uroporphyrin I present is directly proportional to the melanin content of the animal. The red-orange integumentary freckles on the siphon of *Bankia setacea* contain solid aggregations of protoporphyrin IX (XIV) (P. M. Townsley, 1964, quoted by D. L. Fox, 1966). It is interesting that this is a derivative of uroporphyrin III (XV), which has not been noted elsewhere in this phylum.

Free and metalloporphyrins in pearls give them their characteristic green (metallo) and pink (free) colors (Kosaki, 1947; Takagi and Tanaka, 1955; Takagi, 1956).

VII. Bile Pigments (Bilichromes)

A number of bile pigments have been well characterized in Mollusca, but their exact structures are still unknown. These compounds are rufine from the epidermis of *Arion rufus* and rufescine from the shell of *Haliotis rufescens* (Dhéré and Baumeler, 1928; Dhéré *et al.*, 1928; Tixier, 1945), haliotiviolin from the shell of *Haliotis cracherodii* (Tixier and Lederer, 1949; Tixier, 1952), and turboglaucobilin from the shells of *Turbo regenfussi, Turbo marmoratus,* and *Turbo elegans* (Tixier, 1952).

The ink of the sea hare *Aplysia* sp. contains a purple pigment recently identified as the monomethyl ester of phytoerythrobilin (XVI) (Rüdiger, 1967; Siegelman *et al.*, 1968), the prosthetic group of the red pigment phycoerythrin of red algae (Chapman *et al.*, 1971). In *Aplysia californica,* this compound represents 75% of the total pigment present, the remainder is mainly the free acid with traces of the dimethyl ester (Siegelman *et al.*, 1968).

Aplysia spp. are strict herbivores, and *A. californica* clearly obtains its pigment from the phycoerythrobilin of red algae. However, the European species tend to graze on the green alga *Ulva,* so the immediate source of their pigment is not yet known.

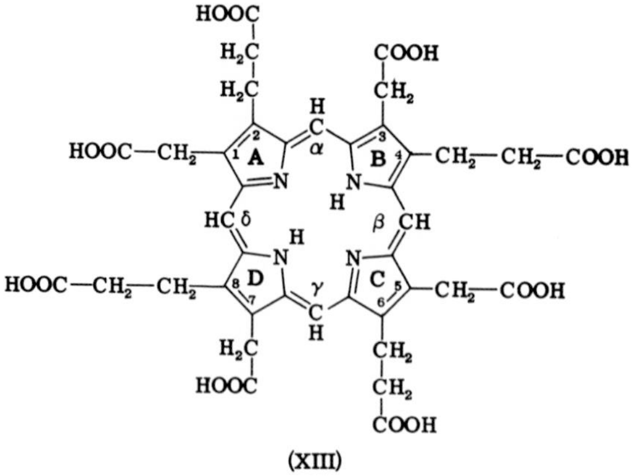

(XIII)

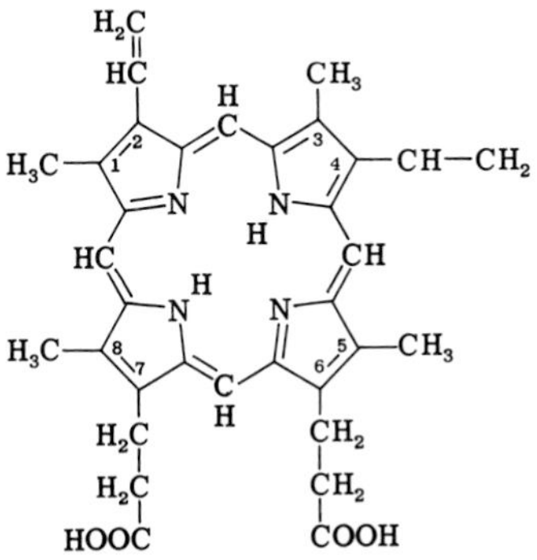

(XIV)

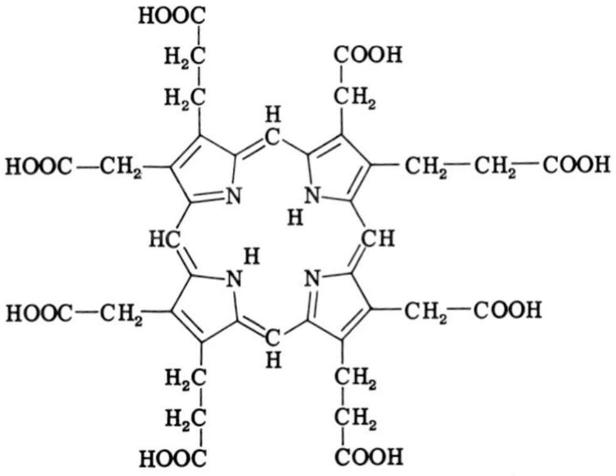

(XV)

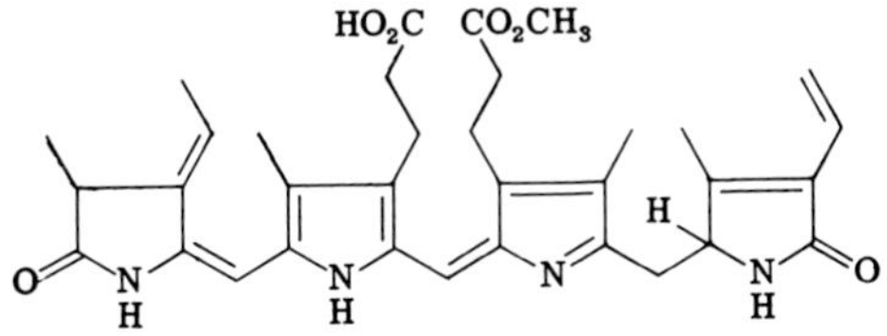

(XVI)

VIII. Hemochromogens (Other Than Hemoglobins)

Helicorubin is a chromoprotein found in the gut of various gastropods (but not for example in *Planorbis*) and in the cephalopod *Loligo*. Its best known source is the snail *Helix aspersa* (Sorby, 1876; Krunkenberg, 1882). The prosthetic group is heme, as in hemoglobin, but the protein part is different from that in hemoglobin (Anson and Mirsky, 1925; Roche and Morena, 1936). In helicorubin from *Helix pomatia,* the minimum molecular weight is 18,500, based on the presence of one iron atom per molecule (Keilin, 1956). A compound similar to helicorubin was found in the digestive gland of *H. pomatia* and called cytochrome h (Keilin, 1956).

IX. Miscellaneous Pigments

Adenochrome, a red pigment of unknown structure, is present in the branchial hearts of *Octopus bimaculatus* (D. L. Fox and Updegraff, 1944).

REFERENCES

André, J. (1896). *Rev. Suisse Zool.* 3, 429.
Anson, M. L., and Mirsky, A. E. (1925). *J. Physiol.* (*London*) **60**, 50.
Bouchilloux, S., and Roche, J. (1955). *Bull. Inst. Oceanogr.* **52**, 1.
Cain, A. J. (1948). *Quart. J. Microsc. Sci.* **89**, 421.
Campbell, S. A., Mallams, A. K., Waight, E. S., Weedon, B. C. L., Barbier, M., Lederer, E., and Salaque, A. (1967). *Chem. Commun.* p. 941.
Chapman, D. J. (1966). *Phytochemistry* **5**, 1331.
Chapman, D. J., Cole, W. J., and Siegelman, H. W. (1971). In press.
Cheesman, D. F. (1958). *Proc. Roy. Soc., Ser. B* **149**, 571.
Comfort, A. (1947). *Nature* (*London*) **175**, 310.
Comfort, A. (1951). *Biol. Rev. Cambridge Phil. Soc.* **26**, 285.
Dennell, R. (1958). *Biol. Rev. Cambridge Phil. Soc.* **33**, 178.
Dhéré, C., and Baumeler, C. (1928). *C. R. Soc. Biol.* **99**, 492 and 726.
Dhéré, C., Baumeler, C., and Schneider, A. (1928). *C. R. Soc. Biol.* **99**, 722.
Fabre, R., and Lederer, E. (1934). *Bull. Soc. Chim. Biol.* **16**, 105.
Fischer, H., and Orth, H. (1934). *Annu. Rev. Biochem.* **3**, 410.
Fisher, L. R., Kon, S. K., and Thompson, S. Y. (1956a). *J. Mar. Biol. Ass. U.K.* **35**, 41.
Fisher, L. R., Kon, S. K., and Thompson, S. Y. (1956b). *J. Mar. Biol. Ass. U.K.* **35**, 63.
Fox, D. L. (1966). *In* "Physiology of Mollusca" (K. M. Wilbur and C. M. Yonge, eds.), Vol. 2, p. 249. Academic Press, New York.
Fox, D. L., and Crane, S. C. (1942). *Biol. Bull.* **82**, 284.
Fox, D. L., and Updegraff, D. M. (1944). *Arch. Biochem.* **1**, 339.
Fox, H. M., and Vevers, G. (1960). "The Nature of Animal Colours." Sidgwick & Jackson, London.
Goodwin, T. W., and Taha, M. M. (1950). *Biochem. J.* **47**, 244.
Gortner, R. A. (1910). *J. Biol. Chem.* **8**, 341.
Isler, O. (1971). Carotenoids, Birkhauser, Basel.
Karrer, P., and Solmssen, U. (1935). *Helv. Chim. Acta* **18**, 915.
Keilin, J. (1956). *Biochem. J.* **64**, 663.
Kennedy, G. Y. (1959). *J. Mar. Biol. Ass. U.K.* **38**, 27.
Kennedy, G. Y., and Vevers, H. G. (1953). *Nature* (*London*) **171**, 81.

Kosaki, T. (1947). *Seiri Seitai* **1**, 247.
Krukenberg, C. F. W. (1882). "*Vergl.-physiol. Studien,*" Vol. 2. Carl Winter, Heidelberg.
Lederer, E. (1933). *C. R. Soc. Biol.* **113**, 1015.
Lederer, E. (1934). *C. R. Soc. Biol.* **116**, 150.
Lederer, E. (1938). *Bull. Soc. Chim. Biol.* **20**, 567.
Lederer, E. (1940). *Biol. Rev. Cambridge Phil. Soc.* **15**, 273.
Lönnberg, E. (1931). *Ark. Zool. Ser. A* [1] **22**, No. 14; **23**, No. 14.
Lönnberg, E. (1933). *Ark. Zool., Ser. A* [1] **26**, No. 7.
Lönnberg, E. (1935). *Ark. Zool., Ser. B* [1] **28**, No. 8.
Mallams, A. K., Waight, E. S., Weedon, B. C. L., Chapman, D. J., Haxo, F. T., Goodwin, T. W., and Thomas, D. M. (1967). *Chem. Commun.* p. 301.
Nadakal, A. N. (1960). *Biol. Bull.* **119**, 90.
Nicholas, R. E. H., and Comfort, A. (1949). *Biochem. J.* **45**, 208.
Nicolaus, R. A. (1962). *Conf. Lette VII Corso Estivo Accad. Naz. Lincei.*
Nishibori, K. (1956–1957). *Nippon Suisan Gakkaishi* **22**, 715.
North, W. J. (1953). Doctoral Dissertation, University of California, Los Angeles, California.
Roche, J., and Morena, J. (1936). *C. R. Soc. Biol.* **123**, 1215.
Rüdiger, W. (1967). *Hoppe-Seyler's Z. Physiol. Chem.* **348**, 129 and 1554.
Scheer, B. T. (1940). *J. Biol. Chem.* **136**, 275.
Schwinck, I. (1953). *Naturwissenschaften* **40**, 365.
Shimizu, T., and Uchida, K. (1968). *Nippon Suisan Gakkaishi* **34**, 154.
Shimizu, T., and Monura, R. (1968). *Nippon Suisan Gakkaishi* **34**, 159.
Siegelman, H. W., Chapman, D. J., and Cole, W. J. (1968). *Biochem. Soc. Symp.* **28**, 107.
Sörensen, N. A. (1937). *Kgl. Nor. Vidensk. Selsk. Skr.* No. 1.
Sorby, H. C. (1876). *Quart. J. Microsc. Sci.* **16**, 76.
Strain, H. H. (1949). *Biol. Bull.* **97**, 206.
Takagi, Y. (1956). *Nippon Kagaku Zasshi* **77**, 1717.
Takagi, Y., and Tanaka, S. (1955). *Nippon Kagaku Zasshi* **76**, 406 and 602.
Thomson, R. H. (1962). *Comp. Biochem.* **3**, Part A, p. 727.
Tixier, R. (1945). *Ann. Inst. Oceanogr.* **22**, 343.
Tixier, R. (1952). *Mem. Mus. Hist. Natur. Paris* **5**, 41.
Tixier, R., and Lederer, E. (1949). *C. R. Acad. Sci.* **228**, 1669.
Villela, G. G. (1956). *Nature (London)* **178**, 93.
von Euler, H., Hellström, H., and Klussman, E. (1934). *Hoppe-Seyler's Z. Physiol. Chem.* **228**, 77.

CHAPTER 7

Respiratory Proteins in Mollusks

F. Ghiretti and A. Ghiretti-Magaldi

I. Introduction

Distilled water, saturated with oxygen, contains about 10 ml. of oxygen per liter at 0°C. This value decreases with increasing temperature and salinity. At 20°C, seawater contains only 5.3–5.6 ml. oxygen per liter. If compared with air (210 ml. oxygen per liter), the aquatic medium appears undoubtedly less favorable for life than the terrestrial environment, at least as far as oxygen is concerned. At first sight, therefore, species that live in water appear to be in less advantageous respiratory conditions than terrestrial forms. Aquatic mollusks, however, have developed efficient respiratory mechanisms that allow a sufficient uptake of oxygen, even where this gas is present at low tensions. Moreover, they are able to reduce their metabolic needs according to the oxygen supply of the environment so that many of them can survive when the oxygen concentration of the medium drops to very low levels. On the other hand, there are many forms which require fully oxygenated water, and the amount of oxygen available may be the limiting factor for their physiological activities.

During evolution, the increase in size and activity of the animals has required a parallel increase of the efficiency of the circulatory and respiratory mechanisms. With the development of a circulatory system, the transport of gases from the external medium to the internal environment and vice versa is no longer dependent on diffusion only; moreover, the presence of large respiratory surfaces, as in gills and lungs, allows the circulating fluids to come into intimate contact with the external medium, thus facilitating the gas exchanges through the separation surfaces.

In mollusks, many types of circulatory apparatus are present, from a very simple lacunar or open system in which the movement of the blood depends entirely on the body muscle contractions to a fully closed system in which the circulation is maintained by the heart. On one side, we find the Monoplacophora with a rudimentary heart and a lacunar system; on the other, the Cephalopoda, which are provided with well developed contractile organs and with arteries, veins, and capillaries.

As for the respiratory mechanisms, they are strictly related to the habitats in which the forms have radiated in the course of evolution. The great majority of mollusks are aquatic and have developed highly specialized respiratory organs (the ctenidia) in the mantle cavity; others live temporarily or permanently on land and the mantle cavity is converted to a respiratory chamber or to highly vascularized air-breathing lungs, where large amounts of blood come in close proximity with air. There are several species, however, that lack a definite respiratory system; this is then represented by variable structures, sometimes extremely reduced, sometimes so exaggerated that no relation can be found between structure and function. Finally, in the absence of any organ, either specific or accessory, respiration is accomplished by diffusion through the undifferentiated surface of the body. Gas exchanges through the body surface generally occur in all aquatic mollusks, even when differentiated organs are present, and under certain conditions, they may become of vital importance.

The amount of oxygen that dissolves in the blood is determined by the absorption coefficient and by the partial pressure (pO_2) of the gas. Since pO_2 in the external medium cannot rise beyond a certain value (normally 160 mm Hg in the most favorable habitats), the oxygen content of the blood can be augmented by increasing the absorption coefficient. The respiratory pigments, for their ability to combine reversibly with oxygen, have exactly the function of increasing the absorption capacity of the internal medium. An oxygen-carrying pigment can be considered, in fact, either as oxygen storage for intermittent use, or as an oxygen transporting device for carrying oxygen from the respiratory organs to the tissues. Both instances require that the pigment be a compound that combines loosely with oxygen.

The main property of a respiratory pigment, therefore, is its ability to combine with oxygen at high gas pressure and to release it when the pressure decreases. Accordingly, the pigment combines with oxygen at the respiratory surface, carries it through the circulatory system, and gives up the bound oxygen to the tissues. The pigment, therefore, behaves as an oxygen buffer, minimizing the changes of oxygen tension due to the absorption and the liberation of the gas. The dissociation curve of the blood is much higher than that of a fluid in which oxygen is physically dissolved, and its shape is more or less sigmoid. It can be seen from Fig. 1 that the amount of oxygen liberated when the pO_2 drops from T_a (tension in the arterial blood) to T_v (tension in the venous blood) is much higher than would be if a straight line relationship were obtained. While the amount of oxygen physically dissolved rises linearly as long as the pO_2 increases, the amount of gas that combines with the pigment approaches a maximum, and at a certain pO_2 value, most of the pigment becomes saturated. The oxygen capacity is the effective transporting ability of the pigment and is measured in milliliters per hundred (percent) or volumes per hundred of combined oxygen. It is a function of the pigment concentration in the blood. The oxygen affinity—measured as the amount in percent of the oxygenated pigment at different pO_2—is characterized by the slope of the dissociation curve. A steep slope indicates a high affinity for oxygen (i.e., the strength by which the pigment binds oxygen), whereas a flat curve is found when the pigment has a low affinity for the gas. Two parameters are commonly used for indicating the oxygen affinity of the blood or of

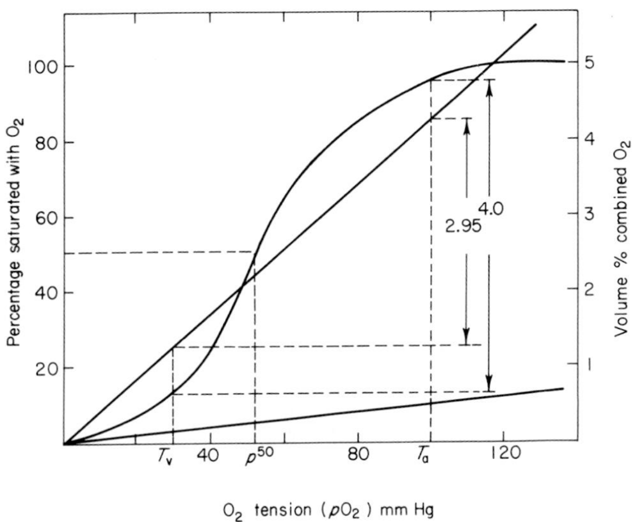

FIG. 1. Hypothetical oxygen dissociation curve (from Jones, 1963).

a respiratory pigment: the loading tension (t_L), which is the partial pressure at which the blood becomes 95% saturated with oxygen, and the unloading tension (t_U), which corresponds to the half-saturation value. The t_U indicates the facility with which oxygen is transferred from the blood to the tissues; a high value means high availability of oxygen to the tissues; a low value means that oxygen is released only when it is almost completely exhausted in the tissues and its tension is very low.

The oxygen dissociation curve of a respiratory pigment is affected by several factors. Lowering the pH generally reduces the affinity for oxygen, and the curve shifts to the right (high t_U values). A high carbon dioxide pressure, therefore, facilitates deoxygenation (Bohr effect). The presence of neutral salts also lowers the affinity for oxygen. This decreases also by increasing the temperature. The heat of oxygenation (ΔH) can be calculated by measuring the oxygenation reaction at different temperatures or by direct calorimetry.

Of the four respiratory pigments known in nature—hemoglobin, hemocyanin, hemerythrin, and chlorocruorin—only hemoglobin and hemocyanin are present in mollusks. Several species, however, lack respiratory pigment in the blood completely; others have both of them, hemocyanin in the blood and hemoglobin in the tissues. The physiology and the chemistry of these pigments have been reviewed by Redfield (1952), Manwell (1960a, 1964), Jones (1963), Read (1966a), Ghiretti (1962, 1966).

II. Hemoglobin

Hemoglobins are found in mollusks in the body tissues or in the circulating fluids, dissolved in hemolymph or contained in the hemocoelic erythrocytes. While the molecular weight of tissue hemoglobin (myoglobin) is about 17.000 and that of blood cell hemoglobin is never higher than 73.000, the pigments in solution in the hemolymph have molecular weights up to several millions. They are also called erythrocruorin. The physicochemical properties of these extracellular hemoglobins, however, are not unequivocal criteria for differentiating them, and Keilin and Hartree (1951) and Florkin (1969) have advocated that the name erythrocruorin be abandoned.

Hemoglobins circulating in the blood have the function of oxygen carriers; myoglobins function as oxygen storage, and by facilitating the diffusion of the gas, they maintain a partial pressure gradient in the tissues.

A. Tissue Hemoglobins

The distribution of myoglobin in mollusks is very irregular and haphazard. When present, however, the pigment is located especially in those structures that are engaged in prolonged intermittent activity such as the stomach and the radular muscles. A complete list of molluscan species containing tissue hemoglobin has been compiled by Read (1966a).

The carnivorous Tectibranch *Navanax inermis*, however, has no myoglobin in its powerful pharyngeal muscles (Manwell, 1958). In gastropods, myoglobin has also been found in the nervous tissue of *Planorbis, Limnaea* (Berthier, 1947), *Aplysia* (Chalazonitis and Arvanitaki, 1951, 1963; Arvanitaki and Chalazonitis, 1952), and in the odontophore "carlage" of *Busycon canaliculatum* (Lash, 1959; Person *et al.*, 1959). The pigment found by Strittmatter and Busch (1963) in the nervous tissue of *Spisula solidissima* has properties that resemble those of cytochromes as well as of myoglobin. No myoglobin is present in the tissues of terrestrial pulmonata (Manwell, 1958). Several gastropod species that possess tissue hemoglobin have hemocyanin in the blood. Amphineura also have hemocyanin as the blood oxygen-carrying pigment, and here too myoglobin is present in the radular muscles (Giese, 1952; Manwell, 1958, 1960a). In several species of bivalves, myoglobin is present in the adductor muscle and the heart (Ball and Meyerhof, 1940; Manwell, 1963); the pismo clam *Tivela stultorum* owes the pink color of the mantle, gills and adductor muscle to this pigment (D. L. Fox, 1953). The myoglobin content of the buccal muscle of *Aplysia depilans* ranges from 4.85 to 6.15% dry weight (Rossi-Fanelli and Antonini, 1957); this value is higher than that found in the radular muscles of *Cryptochiton stelleri* (1–2%) and of *B. canaliculatum* (3–4%) by Manwell (1963).

The myoglobin of *Aplysia* has been subjected to extensive investigation and is perhaps the best known among the variety of tissue hemoglobins described in mollusks. The pigment has been isolated and crystallized from the buccal muscles of the mediterranean species *Aplysia limacina* and *A. depilans*, and its chemical and physicochemical properties (the equilibrium and kinetic constants for the reaction with oxygen and carbon monoxide, the amino acid composition, the tryptic peptide pattern, and the partial amino acid sequence) have been studied (Rossi-Fanelli and Antonini, 1957; Rossi-Fanelli *et al.*, 1958; Wittenberg *et al.*, 1965a; Tentori *et al.*, 1971).

Aplysia myoglobin is similar to that of mammals (horse, whale, seal) with respect to several properties such as molecular weight (18.000), heme content (one heme per molecule), shape of the oxygen dissociation curve, and the absence of Bohr effect. It can be recalled that none of the chiton myoglobin has a Bohr effect, with the exception of the pigment of *C. stelleri* which occasionally shows a small transient heme–heme interaction (Manwell, 1958). Several features, however, strongly differentiate the myoglobin of *Aplysia* from that of other sources. In fact, the dissociaton constant of oxymyoglobin is six to seven times greater in *Aplysia* than in horse myoglobin, which accounts for the much lower oxygen affinity observed in the molluscan protein. *Aplysia* myoglobin shows also different spectral properties in the ferric form and has a low isoelectric point (4.75) in accordance with the high amount of dicarboxylic amino acids that have been found in the protein. Moreover, the pigment contains only one histidine residue per molecule and has no tyrosine or cysteine.

A peculiar behavior is shown by *Aplysia* myoglobin upon heating. When heated to about 90°C in the presence of salts, the ferrimyoglobin undergoes a rapidly reversible denaturation which is associated with large changes in the spectral properties, in the optical rotatory dispersion, and in fluorescence. The transition equilibrium curve and the kinetics of the transition have been studied recently by Brunori *et al.* (1968). The primary structure of *Aplysia* myoglobin has confirmed the unique chemical feature of this protein, i.e., the presence of a single histidine per molecule. In the partial sequence determined by Tentori *et al.* (1971), this histidine corresponds to residue 102, a position compatible with that normally occupied by the proximal histidine in human hemoglobin (87 and 92 in the α and β chains, respectively) and sperm whale myoglobin (93).

The myoglobin of the radular muscle of *B. canaliculatum* has one heme group per 31,000 molecular weight unit; its histidine content is similar to that of vertebrate myoglobin and distinguishes the pigment from all invertebrate myoglobins so far investigated (Read, 1966b).

As already mentioned, hemoglobin has been identified in the nerves and gangliar cells of *Aplysia* (Chalazonitis and Arvanitaki, 1951). Nerve fragments of *Aplysia californica,* when sealed in a small chamber, exhaust the oxygen dissolved in their substance. J. B. Wittenberg *et al.* (1965) have found that, under these conditions, hemoglobin is converted to an unfamiliar derivative exhibiting an absorption spectrum characterized by maxima at 423, 548, and 580 mμ. The reaction is reversible, since the spectrum of oxyhemoglobin reappears when the tissue is returned to oxygen. The same occurs with slices of the red muscle. The suggestion has been advanced that the new derivative, characterized by an absorption maximum at 423 mμ might be a higher oxidation state of hemoglobin and might be involved as an electron carrier in cyanide-insensitive respiration.

Nerve hemoglobin has been partially purified and its heme identified as iron porphyrin IX. Its molecular weight, as determined by sedimentation equilibrium, is equal to 16,400, and the oxygen dissociation curve is hyperbolic.

B. Circulating Hemoglobins

The distribution of circulating hemoglobins in mollusks is as scattered and haphazard as that of myoglobins. A number of bivalve species have hemoglobin in the blood, whereas among gastropods, it has been found only in *Planorbidae*. No hemoglobin has been observed in the hemolymph of Amphineura and Cephalopoda.

In planorbid gastropods, as well as in two bivalve species *Cardita floridana* and *Astarte alaskensis*, hemoglobin is dissolved in the hemolymph (Manwell, 1963); in all other bivalves, the pigment is contained in the hemocoelic erythrocytes. As mentioned, dissolved hemoglobins have a much higher molecular weight than intracellular hemoglobins, ranging from 1,634,000 for *Planorbis corneus* to 3,000,000 for C. *floridana*. (Svedberg, 1933; Svedberg and Hedenius, 1934). No electron microscope observations have been made on hemoglobins of high molecular weight from mollusks. We don't know, therefore, whether or not the giant molecules of *C. floridana* resemble those of *Lumbricus* or *Arenicola* studied by Levin (1963a).

Bivalve hemoglobins show quite different dissociation curves, and their oxygen affinity is in general lower than that of myoglobin. In fact, whereas muscle myoglobin of *Mercenaria mercenaria* has a very high oxygen affinity (t_U = 0.55 mm Hg), the hemocoelic hemoglobins, either dissolved or contained in the erythrocytes, show considerably lower oxygen affinity (11 mm Hg in *C. floridana,* 13.5 mm Hg in *Noetia ponderosa*) (Manwell, 1963). None of them has the sigmoid oxygen dissociation curve typical of many vertebrate hemoglobins nor shows a Bohr effect, the oxygen equilibrium being independent of pH. The significance of negative heme–heme interactions in clam hemoglobins has been discussed by Manwell (1960b).

According to Kruger (1958), vascular hemoglobins are less important to mollusks than to vertebrates. This idea is supported by the finding that the pigment is lacking even in forms that live in habitats poor in oxygen. It has also been observed that individuals of the same species and living in the same environment (as, for instance, *Glycymeris nummaria*) have a very wide range of hemoglobin content, some specimens actually lacking the pigment. Finally, there is no relationship between metabolic activity and amount of hemoglobin present; specimens with the lowest respiratory rate sometimes have the highest hemoglobin content.

On the other hand, it has been shown that several species are able to gain or lose hemoglobin in the blood according to the oxygen content of the waters in which they live. In adult *P. corneus,* oxygen has no effect on the hemoglobin concentration; the animals live as well in water with 4% as with 21% of oxygen; newly hatched snails, however, live longer in fully aerated water, and a small increase in the synthesis of hemoglobin occurs when they are kept at low oxygen concentration (H. M. Fox, 1945, 1955). It would be interesting to follow the oxygen dissociation curve during the development of the animals.

The physiological significance of the circulating hemoglobin in the planorbid gastropods has been the object of extensive investigations (Leitch, 1916; Borden, 1931; Wolvekamp, 1938; Hazelhof, 1939; Precht, 1939; Florkin *et al.,* 1941; Lindroth, 1943; H. M. Fox, 1945; Zaajier and Wolvekamp, 1958; Manwell, 1964; Jones, 1963, 1964). The results obtained, however, were rather discordant. It has been found that the affinity for oxygen is very high as shown by a steep dissociation curve, the t_U ranging from 2.7 to 3.0 mm Hg at 20°C. At lower temperatures, the dissociation curve becomes steeper (the oxygen affinity increases) and the sigmoidal shape in its lower part tends to disappear. The effect of hydrogen concentration is very complex, and considerable variations in the shape of the curve are observed at different pHs. At low carbon dioxide tensions (in the pH range 7.67–7.52), a normal Bohr effect is found which is reversed in the pH range 7.67–8.12. The principal function of the pigment seems to be its participation in the oxygen transport in media where the oxygen tension is low. According to Hazelhof (1939), the snails go to the water surface when the oxygen tension in the lung has dropped to 30 mm Hg or even to lower values. In the presence of carbon dioxide, the animals swim to the surface when the oxygen tension in the lung is much higher. Compared to *Limnaea stagnalis,* which lacks a vascular respiratory pigment, *P. corneus* is able to exploit its pulmonary oxygen storage to a greater extent and roam longer in water in search of food (Jones, 1964).

In many invertebrate species, it is rather difficult to demonstrate conclusively whether or not a respiratory pigment is absolutely necessary for the transport of oxygen. Direct measurements of oxygen in arterial and venous blood may be not an easy task. However, it is reasonable to believe that at least in some circumstances, the pigment can function *in vivo* as it behaves *in vitro.* The primary function of a respiratory protein is, as already mentioned, to increase the absorption coefficient of oxygen in circulating fluids, thus augmenting the oxygen content of the blood. It has also been shown that any oxygen-binding protein of relatively low molecular weight enhances the rate of the steady state diffusion of oxygen into the tissues (J. B. Wittenberg, 1966; Scholander, 1960; Hemmingsen, 1962, 1963).

It is impossible to draw any conclusion on the evolutionary aspects of hemoglobin from the available data on the presence and the distribution of the pigment in mollusks. Most species have been studied at random, and the presence of the pigment was stated sometime on the basis of few and incomplete observations. In very few cases, the physical and chemical properties of the protein have been studied in an exhaustive fashion. On the other hand, it is the absence more than the presence of hemoglobin in mollusks that raises important genetic, ecological, and evolutionary problems. Mollusks are provided with a complete electron transport system made up by flavoproteins and iron porphyrin compounds (Ghiretti-Magaldi *et al.*, 1958; Ghiretti *et al.*, 1959). They should be able, therefore, to synthesize hemoglobin as well. It would be interesting to know why and how many species, disregarding the habitat in which they live and their position in the zoological scale, lack hemoglobin in the blood or have it only in the tissues or, alternatively, synthesize an entirely different pigment such as the hemocyanin.

The hemoglobin of the hemocoelic erythrocytes of the bivalve *Anadara inflata* has received much attention by the Japanese authors Kawamoto (1928), Sato (1931), Kobayashi (1935), Kubo *et al.* (1956), Ui (1957), Yagi *et al.* (1957), and Sumita *et al.* (1964). Based on the sedimentation and diffusion constants, a molecular weight of 73,000 has ben established for this protein. Since the minimal molecular weight calculated from the iron content is 19,000. the pigment contains four hemes per molecule. No evidence of dissociation has been found, and the molecule, therefore, is probably tetrameric. According to Read (1965), the hemoglobin of the very closely related species *Anadara ovalis* (*Arca pexata*), for which a molecular weight of 33,000 had been reported (Svedberg and Hedenius, 1934; Svedberg and Pedersen, 1940), may also form tetrameric aggregates. The prosthetic group is very similar, if not identical, to that of vertebrate hemoglobin (Kubo *et al.*, 1956). The protein moiety, however, is remarkably different, as shown by the rather acid isoelectric point (between pH 5 and 6), its electrophoretic mobility, the amino acid composition, and the N- and C-terminal amino acids (Yagi *et al.*, 1957). Another peculiarity of this hemoglobin (common, however, to *Aplysia* myoglobin) is the very low content of histidine (2.3%).

Read (1962) has found that the hemoglobin of *Phacoides pectinatus* consists of at least three components Hb_1, Hb_2, and Hb_3, which form aggregates of different size. The molecular weight of Hb_1, calculated from its amino acid composition, is equal to 14,682; in concentrated solutions, this component forms tetrameric, octameric, and even higher aggregates which dissociate upon dilution.

III. Hemocyanin

Among all copper proteins known to occur in nature, only hemocyanin is able to combine reversibly with oxygen and function as an oxygen-carrier pigment. Its distribution in animals is much more restricted than that of hemoglobin; hemocyanin, in fact, is present in two phyla only: Mollusca and Arthropoda. In mollusks, this protein has been found in Amphineura, Cephalopoda, and many Gastropoda. Bivalvia and Gastropoda Opistobranchia lack hemocyanin. As with hemoglobins, the distribution of hemocyanins shows no ecological or evolutionary correlation. Among Amphineura and Gastropoda (but not Cephalopoda), there are species that have hemocyanin in the blood and hemoglobin in the tissues. Hemocyanin is always present in solution in the blood.

A. Function

Evidence indicates that the presence of copper confers to the protein its function as respiratory pigment. Pure hemocyanin shows phenolase and catalase activities (Ghiretti, 1956; Felsenfeld and Printz, 1959); these accessory functions (demonstrated *in vitro* only) are also due to the presence of copper.

The respiratory function of hemocyanin has been largely investigated by many authors (for references, see Redfield, 1934, 1952; Manwell, 1960a, 1964; Ghiretti, 1962, 1966; Redmond, 1968), and the relation of the blue color of the blood to oxygen transport was recognized by P. Bert since 1867, before the nature of the pigment had been discovered. The oxygen transport by hemocyanin was unequivocally demonstrated by measuring directly the oxygen concentration of the prebranchial and postbranchial blood in *Octopus vulgaris* (Winterstein, 1909), *Loligo pealii* (Redfield and Goodkind, 1929), and *Sepia officinalis* (Wolvekamp, 1938), and the ability of the blood to combine reversibly with oxygen has been observed in all mollusks containing hemocyanin. In *Helix pomatia,* the blood taken from the vena magna is colorless, while the postbranchial blood drawn from the heart is blue (Spoek *et al.*, 1964). Although investigations in many mollusks (see, for instance, *Busycon canaliculatum* among gastropods) are limited to *in vitro* studies of the oxygen equilibrium curve, there is no reason to doubt, unless a case is discovered to the contrary, that hemocyanin has a respiratory function in all animals that have it. "Hemocyanin definitely serves as normal transporter of oxygen in molluscs. The extent, however, to which it is essential for this activity probably varies from that in the Cephalopods where it is necessary for life, to that in the more sluggish forms, which, especially when not under stress of greater than usual activity, could possibly do without its aid in oxygen transport" (Redmond, 1968).

The oxygen capacity of the molluscan blood is rather low, ranging from 1 to 5 volumes per hundred. In Cephalopods, up to 90% of the oxygen carried by hemocyanin is removed when passing through the tissues. Loading and unloading tensions depend on salt concentration, pH, and temperature. The oxygen equilibrium curve varies from hyperbolic to sigmoid and the kind and extent of the Bohr effect are very variable. The hemocyanins of *O. vulgaris, L. pealei,* and *S. officinalis* show a high sensitivity to pH and an extreme Bohr effect, whereas the hemocyanins of *H. pomatia* and *B. canaliculatum* have a reverse Bohr effect. According to Redmond (1968), this effect is a peculiar characteristic of the shelled snails and is of functional significance, facilitating the oxygen distribution in the body when the animal retracts into the shell.

B. Chemistry

Most of our knowledge of the chemical properties of hemocyanin come from studies that have been carried out in the last ten to fifteen years (for earlier literature, see Redfield, 1952; Ghiretti, 1962, 1966). Some information is now available on the amino acid composition of the protein, the mode of attachment of copper, the nature of oxygen binding, the quaternary structure of the molecule, and conformational modifications upon oxygenation. Nevertheless, considerably more investigation is required before the mechanisms correlating structure with function in hemocyanin will be clearly understood.

In hemocyanin, copper appears to be attached to the protein directly. It can be removed by cyanide (Kubowitz, 1938), but in molluscan hemocyanin, part of the metal is resistant to cyanide treatment (Ghiretti-Magaldi *et al.*, 1966; Witters and Lontie, 1968; Moore *et al.*, 1968). The apoprotein obtained by Konings *et al.* (1969) from *Helix pomatia* still contained 0.24–0.48 μg. of copper per milligram of protein. Copper can be restored to apohemocyanin when added as chloride (Kubowitz, 1938), oxide (Ghiretti-Magaldi and Nardi, 1964), or acetonitrile complex (Lontie *et al.*, 1965) of monovalent copper.

The minimum molecular weight of molluscan hemocyanin, as calculated from analytical data (0.25% of copper) is 25,000. Since it has been unequivocally established that the stoichiometry of the oxygen binding is 2 Cu per 1 O_2, the minimum molecular weight of the function unit is equal to 50,000. Hemocyanins, however, exist in the blood of mollusks as aggregates of very high molecular weight, up to several millions, and *in vitro,* they undergo characteristic reversible dissociation depending on protein concentration, ionic environment, pH, etc. (Svedberg and Pedersen, 1940; Eriksson-Quensel and Svedberg, 1936; Brohult, 1940). The hemocyanin of *H. pomatia* is a mixture of two components known as α- and β-hemocyanin in a 3:11 ratio (Lontie *et al.*, 1962). Both have a molecular weight 9×10^6 and dissociate successively into halves, tenths, and twentieths when the pH increases from 7 to 10 in the absence of calcium or magnesium ions (Brohult, 1940; Heirweg *et al.*, 1961). *Loligo pealei* hemocyanin has been reinvestigated by van Holde and Cohen (1964), who found that the 59 S protein (3.8×10^6) dissociates in alkaline solution into particles one-fifth and one-tenth its weight.

The structure at the quaternary level of hemocyanin has been the object of extensive studies using high-resolution electron microscopy (van Bruggen *et al.*, 1962a,b, 1963; Levin, 1963b; Fernández-Morán *et al.*, 1966; Eskeland, 1967). Molluscan hemocyanins have a cylindrical structure with the same diameter (340 Å) and increasing height (140–680 Å), built up by an increasing number of rows: three in *Octopus* and *Loligo;* six in *Helix;* nine or twelve in *Busycon.* A higher degree of organization has been observed in the gastropod *Kelletia kelletia;* electron micrographs show linear polymers formed by end-to-end aggregation of the protein molecules (Condie and Langer, 1964).

van Bruggen *et al.* (1962a,b) could observe no structural differences between hemocyanin and apohemocyanin from *H. pomatia.* By the criterion of sedimentation velocity, Cohen and van Holde (1964) also found that native hemocyanin of *L. pealei* and its subunits are not modified when copper has been removed. It seems, therefore, that the metal is not directly involved in the binding together of subunits.

The absorption spectrum of oxygenated hemocyanin shows three bands with maxima at 278 mμ (the protein band), in the near ultraviolet around 346 mμ (the so-called copper band), and in the visible region at about 580 mμ. The last two bands disappear during deoxygenation. van Holde (1967) has found that the absorption band in the visible spectrum is not simple but involves at least three components centered at 440, 570, and about 700 mμ, which correspond closely to the structures of spectra of a number of proteins containing cupric copper.

On the basis that (1) cuprous compounds are required for the regeneration of active hemocyanin, (2) the metal is released as monovalent copper when is detached from the deoxygenated protein (Klotz and

Klotz, 1955; Felsenfeld, 1960), and (3) no electron paramagnetic resonance signal is shown by hemocyanin in the oxygenated state (Manwell, 1964; see also Nakamura and Mason, 1960, for crustacean hemocyanin), it was generally accepted that copper in both oxy- and deoxyhemocyanin is entirely cuprous. According to van Holde (1967), this hypothesis must be rejected. He argues that the blue color and the very intense copper band which appear by addition of oxygen might correspond either to $Cu^{+} \rightarrow Cu^{2+}$ charge transfer or to charge transfer between copper and oxygen.

The effect of aging has been studied on *Helix* and *Octopus* hemocyanins. When stored in the cold for a few months, the copper band decreases (Ghiretti-Magaldi *et al.*, 1962; Lontie and Witters, 1966) and an electron spin resonance signal is observed, indicating that the metal has changed to Cu^{2+}. The copper band (and the oxygen-binding capacity) can be partially regenerated by reducing agents such as hydrazine, sulfite, thiosulfate, and sulfhydryl compounds (Lontie and Witters, 1966). Aged hemocyanin, therefore, resembles methemocyanin. According to Konings *et al.* (1969), the aging process does not change some of the molecules completely, but all or most molecules partly.

The amino acid composition is very similar in all the molluscan hemocyanins analyzed so far and is characterized by a high content of dicarboxylic acids (Ghiretti-Magaldi *et al.*, 1966; Gruber, 1968; Witters and Lontie, 1968). The most striking dissimilarity lies in the content of half cystine residues. No free sulfhydryl groups can be determined both in native and apohemocyanin, a finding which does not support the common belief that copper is linked to sulfhydryl groups in hemocyanin (Ghiretti-Magaldi and Nuzzolo, 1965).

Ghiretti-Magaldi *et al.* (1971) have investigated the N-terminal amino acids of *Octopus vulgaris* hemocyanin. At least one residue of aspartic acid is present at the N-terminal position. In both α- and β-hemocyanin of *H. pomatia* Gruber (1968) also found one N-terminal amino acid, arginine.

Molecular subunits have been recently obtained by gel filtration of SDS (sodium dodecyl sulfate)-treated *Octopus* hemocyanin. Two subunits are obtained having molecular weights of 25,000 and 100,000 (Salvato *et al.*, 1972). Albergoni *et al.* (1972) have analyzed the carbohydrate content of the hemocyanin of *O. vulgaris* and found mannose, fucose, and glucosamine amounting to 4% of the protein.

Ten years ago, in a review of the state of knowledge on hemocyanin, special attention was given to the fundamental problems that remained unsolved or had never been attacked before (Ghiretti, 1962). Since then, many aspects of the chemistry of hemocyanins have been clarified; yet the two questions of paramount importance—how copper is bound to the protein molecule and what is the mechanism of the oxygenation reaction—are still unsolved. Work is in progress in several laboratories, and at present, it is rather difficult to discriminate in the most recent literature tentative suggestions from certainly demonstrated facts.

Copper is undoubtedly bound to very alkaline groups of the protein molecule, as shown by potentiometric titration of hemocyanin of *O. vulgaris*. Salvato *et al.* (1971), working with native and copper-free hemocyanin, found four basic groups per atom of copper. In recent years, two main suggestions have been made: (1) that the copper is bound to sulfydryl groups, (Klotz and Klotz, 1955), and (2) imidazole groups are involved in the binding (Lontie, 1958). While the participation of cysteine can be excluded, the binding of copper to histidine would have received further support by Wood and Bannister (1968) if their method (photooxidation in the presence of methylene blue) were not so unspecific.

The results of spectrophotometric titration by Salvato *et al.* (1972) on oxy- and copper-free hemocyanin of *Octopus* indicate that two tyrosine residues are involved in the oxygenation reaction, although the fluorescence variation that has been observed by Shaklai and Daniel (1970) in the hemocyanin of the gastropod *Levantina hierosolima* should be due to tryptophan more than to tyrosine.

The kinetics of the reaction of hemocyanin with oxygen has been reinvestigated by Brunori (1969), who found that the binding of oxygen might proceed via a reaction mechanism involving essentially two coupled elementary steps, a faster bimolecular process and a slower monomolecular one.

van Holde (1967) and De Phillips *et al.* (1969, 1970) have inquired whether there are conformational changes of the hemocyanin molecule accompanying oxygenation. By circular dichroism and sedimentation studies, they found that oxygenated and deoxygenated hemocyanin of *L. pealei* and *Busycon canaliculatum* are conformationally distinct. Klotz and Heiney (1957) have also observed pronounced changes in optical rotation upon deoxygenation of *Busycon* hemocyanin. It has also been found that the dissociation of hemocyanin is a function of the percentage of oxygenation and that the dissociated subunits bind more oxygen than the undissociated molecule. This would indicate that "the binding of a small ligand by a multi-subunit protein results in changes of the bonding energy between the subunits" (De Phillips *et al.*, 1970).

REFERENCES

Albergoni, V., Cassini, A., and Salvato, B. (1972). *Comp. Biochem. Physiol.* (in press).
Arvanitaki, A., and Chalazonitis, N. (1952). *Arch. Sci. Physiol.* **6**, 213.

Ball, E. G., and Meyerhof, B. (1940). *J. Biol. Chem.* **134,** 483.
Bert, P. (1867). *C.R. Acad. Sci.* **65,** 300.
Berthier, J. (1947). *C.R. Acad. Sci.* **225,** 957.
Borden, M. A. (1931). *J. Mar. Biol. Ass. U.K.* **17,** 709.
Brohult, S. (1940). *Nova Acta Regiae Soc. Sci. Upsal.* **12,** No. 4, 1.
Brunori, M. (1969). *J. Mol. Biol.* **46,** 213.
Brunori, M., Antonini, E., Fasella, P., Wyman, J., and Rossi-Fanelli, A. (1968). *J. Mol. Biol.* **34,** 497.
Chalazonitis, N., and Arvanitaki, A. (1951). *Bull. Inst. Oceanogr.* **48,** 996.
Chalazonitis, N., and Arvanitaki, A. (1963). *Bull. Inst. Oceanogr.* **61,** No. 1282.
Cohen, L. B., and Van Holde, K. E. (1964). *Biochemistry* **3,** 1809.
Condie, R. M., and Langer, R. B. (1964). *Science* **144,** 1138.
De Phillips, H. A., Nickerson, K. W., Johnson, M., and Van Holde, K. E. (1969). *Biochemistry* **8,** 3665.
De Phillips, H. A., Nickerson, K. W., and Van Holde, K. E. (1970). *J. Mol. Biol.* **50,** 471.
Eriksson-Quensel, I. B., and Svedberg, T. (1936). *Biol. Bull.* **71,** 498.
Eskeland, T. (1967). *J. Ultrastruct. Res.* **17,** 544.
Felsenfeld, G. (1960). *Arch. Biochem. Biophys.* **87,** 247.
Felsenfeld, G., and Printz, M. P. (1959). *J. Amer. Chem. Soc.* **81,** 6259.
Fernandez-Morán, H., Van Bruggen, E. F. J., and Ohtsuki, M. (1966). *J. Mol. Biol.* **16,** 191.
Florkin, M. (1969). *Chem. Zool.* **4,** 111.
Florkin, M., Laurent, Y., Lefebre, L., and Scalais, G. (1941). *Acta Biol. Belg.* **1,** 227.
Fox, D. L. (1953). "Animal Biochromes and Structural Colours." Cambridge Univ. Press, London.
Fox, H. M. (1945). *J. Exp. Biol.* **21,** 161.
Fox, H. M. (1955). *Proc. Roy. Soc., Ser. B* **143,** 203.
Ghiretti, F. (1956). *Arch. Biochem. Biophys.* **63,** 165.
Ghiretti, F. (1962). *In* "Oxygenases" (O. Hayaishi, ed.), p. 517. Academic Press, New York.
Ghiretti, F. (1966). *In* "Physiology of Mollusca" (K. M. Wilbur and C. M. Yonge, eds.), Vol. 2, p. 233.
Ghiretti, F., Ghiretti-Magaldi, A., and Tosi, L. (1959). *J. Gen. Physiol.* **42,** 1185.
Ghiretti-Magaldi, A., and Nardi, G. (1964). *Protides Biol. Fluids, Proc. Colloq.* **11,** 507.
Ghiretti-Magaldi, A., and Nuzzolo, C. (1965). *Comp. Physiol. Biochem.* **16,** 249.
Ghiretti-Magaldi, A., Giuditta, A., and Ghiretti, F. (1958). *J. Cell. Comp. Physiol.* **52,** 389.
Ghiretti-Magaldi, A., Nardi, G., Ghiretti, F., and Zito, R. (1962). *Boll. Soc. Ital. Biol. Sper.* **38,** 1845.
Ghiretti-Magaldi, A., Nuzzolo, C., and Ghiretti, F. (1966). *Biochemistry* **5,** 1943.
Ghiretti-Magaldi, A., Salvato, B., and Albergoni, V. (1971). *Boll. Soc. Ital. Biol Sper.* **46,** 943.
Giese, A. C. (1952). *Anat. Rec.* **113,** 609.
Gruber, M. (1968). *In* "Physiology and Biochemistry of Haemocyanins" (F. Ghiretti, ed.), p. 49. Academic Press, New York.
Hazelhof, E. H. (1939). *Z. Vergl. Physiol.* **26,** 306.
Heirwegh, K., Borginon, H., and Lontie, R. (1961). *Biochim. Biophys. Acta* **48,** 517.

Hemmingsen, E. A. (1962). *Science* **135**, 733.

Hemmingsen, E. A. (1963). *Comp. Biochem. Physiol.* **10**, 239.

Jones, J. D. (1963). *Prob. Biol.* **1**, 9.

Jones, J. D. (1964). *Comp. Biochem. Physiol.* **12**, 283.

Kawamoto, N. (1928). *Sci. Rep. Tohoku Univ., Fourth Ser.* **3**, 561.

Keilin, D., and Hartree, E. F. (1951). *Nature (London)* **168**, 266.

Klotz, I. M., and Heiney, R. E. (1957). *Proc. Nat. Acad. Sci. U.S.* **43**, 717.

Klotz, I. M., and Klotz, T. A. (1955). *Science* **121**, 477.

Kobayashi, S. (1935). *Sci. Rep. Tohoku Univ., Ser. 4* **10**, 257.

Konings, W. N., Van Driel, R., Van Bruggen, E. F. J., and Gruber, M. (1969). *Biochim. Biophys. Acta* **194**, 55.

Kruger, F. (1958). *Zool. Jahrb., Abt. Allg. Zool. Physiol. Tiere* **67**, 311.

Kubo, M., Kobayashi, T., and Kishita, M. (1956). *Bull. Chem. Soc. Jap.* **29**, 767.

Kubowitz, F. (1938). *Biochem. Z.* **299**, 32.

Lash, J. W. (1959). *Science* **130**, 334.

Leitch, I. (1916). *J. Physiol. (London)* **50**, 370.

Levin, O. (1963a). *J. Mol. Biol.* **6**, 95.

Levin, O. (1963b). *Acta Univ. Upsal., Abstr. Uppsala Diss. Sci.* No. 24.

Lindroth, A. (1943). *Ergeb. Biol.* **19**, 324.

Lontie, R. (1958). *Clin. Chim. Acta* **3**, 68.

Lontie, R., and Witters, R. (1966). *In* "Biochemistry of Copper" (J. Peisach, P. Aisen, and W. E. Blumberg, eds.), p. 455. Academic Press, New York.

Lontie, R., Brauns, G., Cooreman, H., and Vanclef, A. (1962). *Arch. Biochem. Biophys., Suppl.* **1**, 295.

Lontie, R., Blaton, V., Albert, M., and Peeters, B. (1965). *Arch. Intern. Physiol. Biochim.* **73**, **150**.

Manwell, C. (1958). *J. Cell. Comp. Physiol.* **52**, 341.

Manwell, C. (1960a). *Annu. Rev. Physiol.* **22**, 191.

Manwell, C. (1960b). *Arch. Biochem. Biophys.* **89**, 194.

Manwell, C. (1963). *Comp. Biochem. Physiol.* **8**, 209.

Manwell, C. (1964). *In* "Oxygen in the Animal Organism" (F. Dickens and E. Neil, eds.), p. 49. Macmillan, New York.

Moore, C. H., Henderson, R. W., and Nichol, L. W. (1968). *Biochemistry* **7**, 4075.

Nakamura, T., and Mason, H. S. (1960). *Biochem. Biophys. Res. Commun.* **3**, 297.

Person, P., Lash, J. W., and Fine, A. (1959). *Biol. Bull.* **117**, 504.

Precht, H. (1939). *Z. Vergl. Physiol.* **26**, 696.

Read, K. R. H. (1962). *Biol. Bull.* **123**, 605.

Read, K. R. H. (1965). *Comp. Biochem. Physiol.* **15**, 137.

Read, K. R. H. (1966a). *In* "Physiology of Mollusca" (K. M. Wilbur and C. M. Yonge, eds.), Vol. 2, p. 209. Academic Press, New York.

Read, K. R. H. (1966b). *Comp. Biochem. Physiol.* **17**, 375.

Redfield, A. C. (1934). *Biol. Rev. Cambridge Phil. Soc.* **9**, 175.

Redfield, A. C. (1952). *In* "Symposium on Copper Metabolism" (W. D. McElroy and B. Glass, eds.), p. 174. Johns Hopkins Press, Baltimore, Maryland.

Redfield, A. C., and Goodkind, R. (1929). *J. Exp. Biol.* **6**, 340.

Redmond, J. R. (1968). *In* "Physiology and Biochemistry of Haemocyanins" (F. Ghiretti, ed.), p. 5. Academic Press, New York.

Rossi-Fanelli, A., and Antonini, E. (1957). *Biochemistry (USSR)* **22**, 312.

Rossi-Fanelli, A., Antonini, E., and Povoledo, D. (1958). *In* "Symposium on Protein Structure" (A. Neuberg, ed.), p. 144. Methuen, London.

Salvato, B., Sartore, S., Zaccaria, G., and Ghiretti-Magaldi, A. (1972). *Boll. Soc. Ital. Biol. Sper.* (in press).

Salvato, B., Ghiretti-Magaldi, A., and Tallandini, L. (1972). *Boll. Soc. Ital. Biol. Sper.* (in press).

Sato, T. (1931). *Z. Vergl. Physiol.* **14,** 763.

Scholander, P. E. (1960). *Science* **131,** 585.

Shaklai, N., and Daniel, E. (1970). *Biochemistry* **9,** 564.

Spoek, G. L., Bakker, H., and Wolvekamp, H. P. (1964). *Comp. Biochem. Physiol.* **12,** 209.

Strittmatter, P., and Burch, H. B. (1963). *Biochim. Biophys. Acta* **78,** 562.

Sumita, N., Kajita, A., and Kaziro, K. (1964). *J. Biochem. (Tokyo)* **55,** 148.

Svedberg, T. (1933). *J. Biol. Chem.* **103,** 311.

Svedberg, T., and Hedenius, A. (1934). *Biol. Bull.* **66,** 191.

Svedberg, T., and Pedersen, K. O. (1940). "The Ultracentrifuge." Oxford Univ. Press, London and New York.

Tentori, L., Vivaldi, G., Carta, S., Marinucci, M., and Massa, A. (1971). *FEBS Lett.* **12,** 181.

Ui, N. (1957). *J. Biochem. (Tokyo)* **44,** 9.

van Bruggen, E. F. J., Wiebenga, E. H., and Gruber, M. (1962a). *J. Mol. Biol.* **4,** 1.

van Bruggen, E. F. J., Wiebenga, E. H., and Gruber, M. (1962b). *J. Mol. Biol.* **4,** 8.

van Bruggen, E. F. J., Schuiten, V., Wiebenga, E. H., and Gruber, M. (1963). *J. Mol. Biol.* **7,** 249.

van Holde, K. E. (1967). *Biochemistry* **6,** 93.

van Holde, H. E., and Cohen, L. B. (1964). *Biochemistry* **3,** 1803.

Winterstein, H. (1909). *Biochem. Z.* **19,** 384.

Wittenberg, B. A., Brunori, M., Antonini, E., Wittenberg, J. B., and Wyman, J. (1965a). *Arch. Biochem. Biophys.* **111,** 576.

Wittenberg, B. A., Wittenberg, J. B., Stolzberg, S., and Valenstein, E. (1965b). *Biochim. Biophys. Acta* **109,** 530.

Wittenberg, J. B. (1966). *J. Biol. Chem.* **241,** 104.

Wittenberg, J. B., Brown, P. K., and Wittenberg, B. A. (1965). *Biochim. Biophys. Acta* **109,** 518.

Witters, R., and Lontie, R. (1968). *In* "Physiology and Biochemistry of Haemocyanins" (F. Ghiretti, ed.), p. 61. Academic Press, New York.

Wolvekamp, H. P. (1938). *Z. Vergl. Physiol.* **25,** 541.

Wood, E. J., and Bannister, W. H. (1968). *Biochim. Biophys. Acta* **154,** 10.

Yagi, H., Mishima, T., Tsujimura, T., Sato, K., and Egami, F. (1957). *J. Biochem. (Tokyo)* **44,** 1.

Zaaijer, J. J. P., and Wolvekamp, H. P. (1958). *Acta Physiol. Pharmacol. Neer.* **7,** 56.

CHAPTER 8

Carbohydrates and Carbohydrate Metabolism in Mollusca

Esther M. Goudsmit

I. Introduction

This chapter emphasizes the biochemical aspects of molluscan carbohydrates. The wealth of histochemical data has been recently reviewed by Libbie Hyman (1967). Her book, which covers four classes of mollusks, provides both a background and a stimulus for research on any facet of the phylum.

II. Blood Sugar

The blood sugar of all classes of mollusks thus far investigated appears to be glucose. However, in their review, Goddard and Martin (1966) caution that most investigators measured total reducing compounds rather than glucose specifically.

The nonreducing disaccharide, trehalose, the principal sugar of insect hemolymph, has been found in nine species of marine and freshwater mollusks (Fairbairn, 1958; Badman, 1967). Since trehalose was extracted only from whole animals in these studies, the quantitative and metabolic relationship between trehalose and glucose in the molluscan hemolymph remains unknown.

The question of regulatory factors controlling blood glucose levels remains open. None of the vertebrate hormones, such as insulin, adrenaline, glucagon, noradrenalin, dopamine, or serotonin, seems to have physiologically significant effects (Martin, 1966; Goddard, 1968). Goddard (1969) has now begun to look for regulatory activity in molluscan extracts with promising success. Aqueous extracts of several tissues of *Octopus dofleini* were tested for glycemic activity *in vivo.* Catheterization into the vena cava made possible direct infusion of the test substances and hourly measurements of glucose. The test substances were prepared by homogenization of a specific tissue in filtered seawater and centrifugation at medium speed. The supernatant solution was used without further purification. Thus, extracts of optic gland, subesophageal ganglia, renopericardial canal tissue, and branchial heart appendages were active. Seven other tissue extracts such as supraesophageal ganglia, branchial heart, and hepatopancreas showed no activity and were used as control infusions. The meaning of these results awaits further purification of the active factors.

III. Sugar Nucleotides

Many of the important pathways of carbohydrate metabolism involve nucleoside diphosphate sugars. Leloir and Cardini (1957) found that UDP-D-glucose was the activated precursor in mammalian glycogen synthesis. Since that major discovery, the synthesis of other polysaccharides has been shown to proceed via sugar nucleotides as shown in Table I. The sugar nucleotide content of mollusks has been investigated in only a few cases. Wheat (1960) isolated and identified UDP-glucose, UDP-acetylglucosamine, and UDP-acetylgalactosamine in whole body extracts of the slug, *Limax maximus.* Horstmann (1965a) examined freshly hatched *Helix pomatia* and found UDP-glucose, UDP-galactose, and "UDP-*N*-acetylhexosamine" to be present.

While the enzymatic synthesis of glycogen has not been followed in mollusks, UDP-glucose is probably the glucosyl donor. UDP-D-galactose is utilized in the formation of galactogen, the galactose homopolymer found in snail albumen glands (Goudsmit and Ashwell, 1965). GDP-D-mannose and GDP-L-galactose are present in the albumen gland of *Helix pomatia* (Goudsmit and Neufeld, 1966). These two sugar

TABLE I
SACCHARIDE SYNTHESIS FROM SOME SUGAR NUCLEOTIDES[a]

Precursor	Product	Experimental source and reference[b]
UDP-glucose	Glycogen (amylose)	Liver, mammalian (1)
ADP-glucose	Glycogen (amylose)	Bacteria (2)
ADP-glucose	Starch (amylose)	Wheat germ (3)
GDP-glucose	Cellulose	Mung bean (4)
GDP-mannose	Mannan	Yeast (5)
UDP-galactose	Galactogen	*Helix pomatia* (6)
UDP-*N*-acetylgalactosamine	Chitin	*Neurospora* (7)
UDP-glucuronic acid + UDP-*N*-acetylglucosamine	Hyaluronic acid	Skin, fetal rat (8)

[a] Adapted from White *et al.* (1963).
[b] References: (1) Leloir and Cardini (1957); (2) Greenberg and Preiss (1964); (3) Recondo and Leloir (1961); (4) Elbein *et al.* (1964); (5) Algranati *et al.* (1963); (6) Goudsmit and Ashwell (1965); (7) Glaser and Brown (1957); (8) Schiller (1964).

nucleotides are interconvertible (Goudsmit and Neufeld, 1967). GDP-L-galactose may be the activated donor of the L-galactose residues, which are part of the galactogen molecule of *Helix*.

Presumably, the UDP-*N*-acetylhexosamines play a significant role in the synthesis of heteropolysaccharides. The great variety of mucopolysaccharides and glycoproteins present among the Mollusca suggests that other and unusual sugar nucleotide combinations remain to be discovered.

IV. Mucopolysaccharides and Glycoproteins

A. DEFINITIONS, OCCURRENCE, AND FUNCTIONS

Mucopolysaccharides are composed of alternate units of glycuronic acid and amino sugar, usually *N*-acetylated (Table II). Sometimes acidic, this trait may be heightened by the presence of sulfate groups. Recent research indicates the association of protein.

Glycoproteins are essentially proteins that contain covalently bonded carbohydrate portions which may contribute from 1% to 80% of the weight of the molecule. The sugar moiety may contain from two to six different sugars, including D-galactose, D-mannose, L-fucose, D-glucose, D-glucosamine, D-galactosamine, and sialic acids. Sulfate may also be present.

These complex oligo- and polysaccharides are important components of animal mucins, connective tissue, and skin. They provide mechanical

TABLE II
SOME MUCOPOLYSACCHARIDES AND THEIR REPEATING DISACCHARIDE UNITS[a]

Mucopolysaccharide	Monosaccharides in the disaccharide units	Linkages[b]
Hyaluronic acid	D-Glucuronic acid; *N*-acetyl-D-glucosamine	β, 1 → 3; β, 1 → 4
Chondroitin	D-Glucuronic acid; *N*-acetyl-D-galactosamine	β, 1 → 3; β, 1 → 4
Chondroitin sulfate A	D-Glucuronic acid; *N*-acetyl-D-galactosamine 4-sulfate	β, 1 → 3; β, 1 → 4
Chondroitin sulfate C	D-Glucuronic acid; *N*-acetyl-D-galactosamine 6-sulfate	β, 1 → 3; β, 1 → 4
Dermatan sulfate	L-Iduronic acid; *N*-acetyl-D-galactosamine 4-sulfate	α, 1 → 3; β, 1 → 4
Keratosulfate	D-Galactose; *N*-acetyl-D-glucosamine 6-sulfate	β, 1 → 4; β, 1 → 3

[a] From White *et al.* (1963).

[b] The linkages are given in the same order as the names of the monosaccharide units; e.g., for chondroitin the linkages, as given, the linkage of glucuronic acid to acetylgalactosamine is β, 1 → 3, and that of acetylgalactosamine to the next glucuronic acid is β, 1 → 4, etc.

or protective support and lubrication. In the oviparous mollusks, they are of further significance in the composition of egg jellies and capsules.

Protein-polysaccharide complexes in mollusks and other invertebrates are covered extensively by Hunt (1970).

B. MUCINS

The production of mucins by mollusks is a well known trait. Land snails crawl along a slime path. Freshwater pulmonates travel from substrate to surface along mucoid tightropes. Mucus figures dramatically in the courtship behavior of slugs (Hyman, 1967, p. 602). Among the Pelecypoda and many Gastropoda, a mucous thread entraps food particles and carries them to the gut. Salivary glands secrete a lubricatory slime. The prosobranch hypobranchial gland secretion catches, compacts, and extrudes particles that might otherwise hamper the flow of seawater over the gills and osphradium. These mucins, the secretion products of epithelial glands, contain a mixture of glycoproteins and mucopolysaccharides. Many histochemical studies of mucous glands have been carried out (reviewed by Hyman, 1967), but biochemical analyses are few.

Hypobranchial mucins have been the secretions studied most often. Hunt and Jevons (1963) detected a complex conjugate of glycoprotein and an acidic polysaccharide in the hypobranchial secretion of *Buccinum undatum.* Excised glands were agitated in a solution of cold 0.55 *M* sodium chloride, causing mucin to flow, whereupon it was wound around a glass rod and collected. The whole mucin contained 41.0% protein, 30.0% neutral carbohydrate, 2.0% hexosamine, 13.2% sulfate, 10% water, and less than 5% calcium. After analytical ultracentrifugation and moving boundary electrophoresis, the presence of two main components was indicated. Further phenol extraction proved these to be a glycoprotein and an acidic polysaccharide.

The glycoprotein contained 8% of neutral sugars including galactose, mannose, fucose, and glucose, as well as 4.5% of galactosamine and glucosamine. The peptide moiety was high in aspartic and glutamic acids, totalling 23% of the amino acid content (Hunt and Jevons, 1965b).

The acidic polysaccharide was also characterized. Its molecular weight was 1.7×10^5 (Hunt and Jevons, 1966). It was found to be a polyglucose sulfate with one ester–sulfate group per monosaccharide residue. A linear β $(1 \rightarrow 4)$ linkage indicated a cellulose-like structure (Hunt, 1967). The polyglucose sulfate was found to be loosely joined to the protein moiety of the glycoprotein through ionic linkages (Hunt and Jevons, 1963) and to be covalently bound to another separate peptide moiety (Hunt and Jevons, 1965a). Viscosity measurements of the whole mucin and of protease-treated mucin indicated the polyglucose sulfate to be responsible for the viscosity of the mucin in solution (Hunt and Jevons, 1966).

Analyses of hypobranchial mucins of other prosobranchs indicated that their chemical compositions were different in each case. That of *Busycon canaliculatum* contained polyhexosamine sulfate linked to an acidic protein moiety through calcium ions (Shashoua and Kwart, 1959; Kwart and Shashoua, 1958). The mucin of *Charonia lampas* included a mixture of β $(1 \rightarrow 4)$ and α $(1 \rightarrow 6)$ polyglucose sulfates as well as protein (Iida, 1963). The hypobranchial mucin of *Neptunea antiqua* consisted of a polyhexosamine sulfate with some glucose and fucose present (Doyle, 1964).

Pancake and Karnovsky (1971) analyzed mucus extruded from the foot of *Otala lactea*. The deproteinized mucin contained a single major component which differed from common sulfated acidic mucopolysaccharides in that the component sugars were L-iduronic acid and glucosamine. This combination has not previously been found together in a mucopolysaccharide.

Each of these secretions contain linear polysaccharides bearing strongly acidic sulfate groups. In contrast, the carbohydrate portion of vertebrate viscous secretions often achieve their acidic character from terminal sialic acid residues. Endogenous sialic acids have not been detected in mollusks (Warren, 1963).

C. Connective Tissue and Skin

Connective tissue is found throughout the body of mollusks and provides the supportive substance for internal organs.

1. *Collagen*

The highly structured protein collagen is the main component of vertebrate connective tissue, comprising one third of the total body protein by weight. Collagen is regarded as a glycoprotein by carbohydrate chemists because it contains 5% or more by weight of tightly bound glucose and galactose.

Invertebrate collagens are reported to contain a wide variety of hexoses as well as hexosamine and pentoses. However, it now appears that these collagens may have been incompletely purified. Thus, Katzman *et al.* (1969) isolated two collagen fractions from the body wall of the echinoderm *Thyone briareus* by gelatinization and subsequent chromatography on DEAE-cellulose. Both fractions contained covalently linked glucose and galactose as the sole carbohydrates.

The existence of molluscan collagens is uncertain. The byssus threads of *Mytilus* and other bivalves may be largely collagen (Rudall, 1955). Byssus is discussed in Chapter 4 of this volume.

2. *Cartilage*

Vertebrate cartilage contains the mucopolysaccharides and chondroitin sulfates A and C, as well as keratosulfate (Table II). Histologically it is identified by its hyaline metachromatic matrix. Molluscan cartilages show some interesting deviations in biochemical structure.

The radula of snails is supported by a chondroid apparatus, the odontophore. It is assumed on histological evidence to be a typical cartilage. However, Lash and Whitehouse (1960) found the odontophore of the whelk *Busycon caniculatum* contained polyglucose sulfate, rather than the usual chondroitin sulfates. Polyglucose sulfate thus far has been detected only in mollusks (Hunt and Jevons, 1966; Egami *et al.*, 1956; Iida, 1963).

Cartilages of squid also resemble vertebrate cartilage histologically. Biochemical analysis in one case has revealed the presence of a chondroitin sulfate (Mathews *et al.*, 1962). Both the cranial and nuchal cartilages of *Loligo* were found to contain a chondroitin sulfate which differed from that found in vertebrate cartilage in having 1.5 ester sulfate groups per monosaccharide residue and a slightly higher optical rotation.

A chrondroitin sulfate almost identical to that found in *Loligo* was isolated from the cartilage of the ancient arthropod *Limulus* (Mathews *et al.*, 1962). The authors were impressed by this example of "biochemical convergence in the evolution of connective tissue." Although they tentatively suggested the existence of a common acid polysaccharide precursor in the members of some ancestral phylum, they recognized the multiplicity of chemical evolution even within the Mollusca, as evidenced by the presence of polyglucose sulfate.

In a search for a common acid polysaccharide precursor in other phyla, Katzman and Jeanloz (1969) found chondroitin sulfate and polyfucose sulfate in the connective tissue of an echinoderm, *Thyone*, but not in a coelenterate, *Metridium dianthus*, nor in a poriferan, *Hippospongia gossypina*. They concluded that closely related acid polysaccharides may have had independent biogeneses. Yet, as is always the case with missing links, they are missing until they have been found.

3. *Skin*

Vertebrate skin contains primarily hyaluronic acid and dermatan sulfate (Table II). Molluscan skin has been investigated only in the case of two cephalopods. The skin of *Ommastrephes sloani* contains a chondroitin, which comprises 70% of the total mixture of mucopolysaccharides present (Anno *et al.*, 1964). *Loligo opalescens* skin may contain a highly sulfated chondroitin sulfate, since Srinivasan and coworkers (1969) found equal amounts of glucuronic acid and *N*-acetylgalactosamine in several fractions of that tissue. The molar ratio of sulfate to hexosamine was at least 2:1, compared with 1:1 for vertebrate chondroitin sulfates. The presence of large amounts of protein and of glucose, as well as some glucosamine, galactose, mannose, and xylose, indicates a complexity in the mucopolysaccharide and glycoprotein constituents which has yet to be resolved.

D. EGG JELLIES

Molluscan zygotes are commonly enclosed in nutritive and protective layers and perhaps by a gelatinous or hard capsule. The composition of these layers and their origin in various glands of the reproductive tract has been the subject of several investigations. Histochemical data suggests some complexity in the secretory products. Thus, several sulfated and nonsulfated mucopolysaccharides were detected within various cell types of the oothecal gland of *Limnaea stagnalis*, a freshwater pulmonate (Fantin and Vigo, 1968). When undertaking biochemical analyses of glandular secretions or of egg layers, these complexities should be resolved, if possible, by separation techniques such as analytical ultracentrifugation and moving boundary electrophoresis. In this way, homogeneity of each mucopolysaccharide and glycoprotein is more nearly assured prior to its chemical and structural analysis.

McMahon *et al.* (1957) analyzed the carbohydrate content of the reproductive tract and eggs of several gastropods. Only high molecular weight polysaccharides were preserved, since extracts were prepared by heating in 60% potassium hydroxide and precipitating with ethanol. The various polysaccharides were distinguished by their electrophoretic mobility as well as by carbohydrate content. Thus, the reproductive tract origin of several egg layer components could be determined. *L. stagnalis* muciparous gland and isolated egg mucus each contained electrophoretically identical polysaccharides, which upon hydrolysis yielded galactose, glucose, glucosamine, and an unknown hexose. The oothecal gland and isolated egg membranes each possessed the same galactose, fucose, glucosamine, and unknown hexose polymer. Extracts of isolated uteri and of the freshly laid eggs of *Lanistes bolteniamus,* a dioecious operculate, yielded identical galactose- and fucose-containing polysaccharides. The structure and linkage of these polymers is unknown. Whether or not they are components of glycoproteins also remains to be shown.

May (1932a,b) ascertained that the albumen gland of *Helix pomatia* contained a galactose homopolymer, galactogen. Concurrently he noticed that in the freshly laid egg of *Helix,* galactogen was the only detectable polysaccharide. Since that time, the albumen glands of many species of gastropods have been analyzed for the possible presence of this homopolymer. This subject is discussed later in this chapter.

Bayne (1966) has studied the amino acid, lipid, and carbohydrate content of the several layers surrounding the zygote of the slug, *Agriolimax reticulatus.* Thin-layer chromatography of hydrolyzed perivitelline fluid indicated the presence of galactose, glucosamine, and several amino acids. Attempts to demonstrate a protein–polysaccharide complex were not conclusive. Acid hydrolyzates of egg jelly contained only galactose and glucosamine. Neither protein nor lipid were detected. Egg shell hydrolyzates contained galactose, fucose, and amino acids. Bayne (1967) also examined the free and bound carbohydrate and amino acid composition of *Agriolimax* reproductive glands. Extracts of whole glands were analyzed rather than secretory material alone, leading to the recovery of some amino acids and of glucose which were characteristic of the gland tissue itself. In any case, biochemical analyses plus direct observation of dissected material confirmed that the perivitelline fluid was the product of the albumen gland, that the jelly layer originated in the upper oviducal gland, and that the eggshell originated in the lower oviducal gland.

E. Egg Capsules

Among higher prosobranchs such as *Buccinum*, the egg capsule or oötheca is secreted by the capsule gland within the distal portion of the pallial oviduct and then is molded and hardened when in contact with the pedal gland (Hyman, 1967, p. 305). Thus, the completed oötheca most certainly has a complex composition.

Thus far, the sclerotized egg capsules of marine prosobranchs have been analyzed only after acid hydrolysis, since they are resistant to all attempts at solubilization. Thus, Hunt (1966) found the egg capsule of *Buccinum undatum* insoluble in distilled water, 0.5 *M* sodium chloride, 4 *N* sodium hydroxide, 5% sodium sulfide, performic acid at 4°C, 45% aqueous phenol at 90°C, and 4% trichloroacetic acid at 60°C. The oötheca was also impervious to treatment with crystalline pepsin, trypsin, chymotrypsin, or pronase. Finally, acid hydrolysis, for which details are not given, permitted analysis of the carbohydrate and amino acid content. The neutral sugars galactose, glucose, mannose, xylose, fucose, and rhamnose were present in a 2:1:3:1:2:1 ratio and accounted for 5–6% of the total weight of dried material. Two percent of an unidentified hexosamine was found, while a mixture of fifteen amino acids comprised 77.5% of the dry weight.

V. Galactogen

A. Definition and Occurrence

Galactogen is a high molecular weight homopolymer of galactose. It is confined to the albumen gland of adult pulmonate snails and to the albumen gland region of the pallial oviduct of some prosobranchs. The albumen gland secretes the perivitelline fluid immediately bathing the zygote. This fluid, composed of galactogen and proteins, but not of lipid, is metabolized by the developing embryo (Horstmann, 1956a) or utilized after hatching (Horstmann, 1965b).

Galactogen itself represents 29.7% of the dry weight of the freshly laid egg of *Helix pomatia* (May, 1932b) and 36.1% of the dry weight of new *Limnaea stagnalis* eggs (Horstmann, 1956a). The large amount of galactogen in the egg and its utilization by the embryo or freshly hatched snail has led many to believe that this polysaccharide must have a considerable influence upon embryonic development.

Aside from *Helix* and *Limnaea*, galactogen has been looked for and found in the albumen glands of the following pulmonates: *Oxychilus cellarius* (Righby, 1963), *Achatina fulica* (Ghose, 1963), *Aplexa nitens* and *Otala lactea* (McMahon *et al.*, 1957), *Bulimnaea megasoma*

(Goudsmit, 1964), *Ariolimax columbianis* (Meenakshi and Scheer, 1968), *Ariophanta* (Meenakshi, 1954), *Biomphalaria glabrata* (Corrêa *et al.*, 1967), and *Strophocheilus oblongus* (Duarte and Jones 1971). Galactogen has also been identified in the prosobranchs *Pila virens* and in *Viviparus* (Meenakshi, 1954).

B. Chemical and Structural Analyses

The original discovery of an unusual polysaccharide in snails was made by Hammarsten (1885) while investigating the structure of muco- or glycoproteins prepared from the mantle, foot, and albumen gland of *Helix pomatia*. Upon mild alkaline hydrolysis of the mucin prepared from albumen gland, Hammarsten isolated a polysaccharide which differed from glycogen in several ways. It did not react with iodine to give the characteristic purple color, it was not destroyed by saliva, and it was levorotatory in solution. Hammarsten named the polysaccharide *tierisches Sinistrin.*

Acid hydrolysis of sinistrin yielded a dextrorotatory product whose identity was not determined until forty years later by Levene (1925). Mucic acid analysis and phenylosazone preparation established the hydrolyzed product to be D-galactose. This was the first time a galactan had been isolated from animal tissues. According to Levene (1925), the polysaccharide was not part of a glycoprotein complex, but occurred in a free state in the albumen gland, since the pure polysaccharide could be extracted from the gland with water.

May (1931, 1932a) rediscovered the albumen gland polysaccharide and renamed it galactogen. The specific rotation of the component galactose was +60.7°, lower than the +80.0° of an equilibrium mixture of pure α- and β-D-galactose; thus, an unknown, optically active substance was present. Bell and Baldwin (1941) determined that the low specific rotation was the result of the presence of L-galactose. They pointed out that this was the first observed instance of L-galactose in an animal polysaccharide. Since then, L-galactose has been identified in the jelly coat of one genus of sea urchin (Vasseur, 1950), in the galactan of one group of marine algae (Mori, 1953; Su and Hassid, 1962a), and in the galactogen of one other genus of snail, *Biomphalaria* (Corrêa *et al.*, 1967).

Methylation analysis of *H. pomatia* galactogen by Bell and Baldwin (1938) yielded equimolar proportions of 2,3,4,6-tetramethyl-D-galactose and 2,4-dimethyl-D-galactose. These products and other data suggested that the structure of galactogen was either a β (1 → 3) linked chain, each galactopyranose unit bearing a C-6 linked side chain, or a β (1 → 6) linked chain, with the side chains attached at C-3. The ratio of D- to L-galactose was thought to be either 6:1 or 7:1, and L-galactose was believed to be part of the side chains (Bell and Baldwin, 1941; May and Weinland, 1954, 1956).

Further work on the structure of *H. pomatia* albumen gland galactogen has been carried out by O'Colla (1953) using a modification of a periodate oxidation method (Barry, 1943), which is specifically suited to the structural analysis of $(1 \rightarrow 3)$ linked polysaccharides. O'Colla's results, together with those of Bell and Baldwin (1938) suggested a dichotomous structure, with each galactopyranose unit bearing a branch or side chain.

Similar results were obtained by Corrêa and his co-workers (1967) after analysis of *Biomphalaria glabrata* galactogen. In addition, this galactogen was found to contain a high proportion (36%) of L-galactose. The location of L-galactose in the galactogen molecule has not been conclusively demonstrated.

Molecular weight determinations have been performed with galactogen extracted from the eggs of *H. pomatia* with cold water. Ultracentrifugal sedimentation analysis showed an average molecular weight of 4×10^6 (Horstmann, 1964). The molecular weight of galactogen similarly extracted from eggs of *Limnaea stagnalis* is 2.2×10^6 (Fleitz and Horstmann, 1967). This polymer contains only D-galactose (Horstmann and Geldmacher-Mallinkrodt, 1961).

The albumen gland is a compound tubular organ composed of centrotubular cells and secretory cells. The latter contain large periodic acid–Schiff-positive, diastase resistant globules which are assumed to contain galactogen (Grainger and Shillitoe, 1952). A recent ultrastructural investigation of *H. pomatia* albumen glands has disclosed more information about the content of these globules (Nieland and Goudsmit, 1969). In the golgi zone of the secretory cells, small vesicles containing an amorphous matrix coalesce and enlarge to form the huge secretory globules, which then accumulate discrete granules 200 Å in diameter (Fig. 1). This particulate component was identified as galactogen on the basis of comparison with cold water extracted, negatively stained pure galactogen. Galactogen therefore resembles glycogen, since the particulate nature of the latter has been well documented (Revel *et al.*, 1960). The appearance of galactogen granules within the secretory globules suggests that this is the site of biosynthesis of the polysaccharide. By way of comparison, Neutra and Leblond (1966) followed the incorporation of galactose-^{3}H into various glycoprotein- and mucopolysaccharide-secreting cells. Their results indicated that carbohydrates destined for secretion are synthesized in "saccules" released by the golgi stacks. These complex carbohydrates are not particulate as is galactogen.

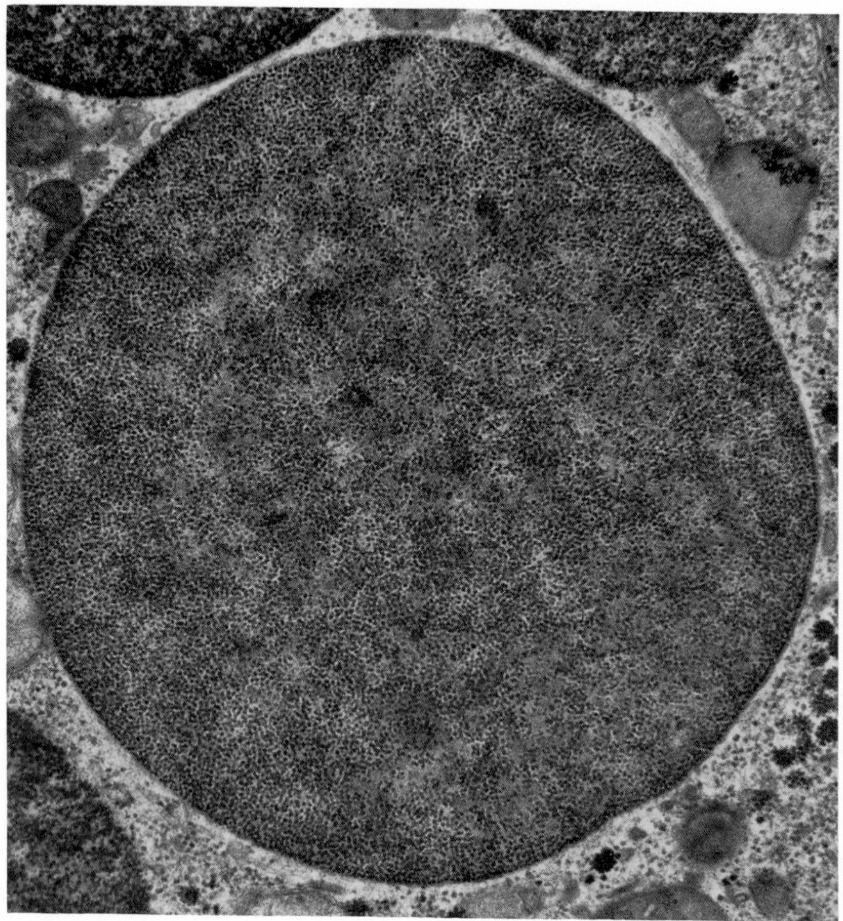

FIG. 1. A secretory globule composed of galactogen granules (200 Å diameter) embedded in a homogeneous matrix. The globules occur within the secretory cells of the albumen gland of *Helix pomatia*. Electron micrograph: Epon-Araldite embedded; stained with uranyl acetate and lead citrate; ×35,300. (Nieland and Goudsmit, 1969).

C. Galactogen Content of Adult and Embryonic Snails

May (1932a) ascertained that galactogen was confined to the albumen gland of mature *Helix pomatia*. Glands showed a large variation in galactogen content during the different seasons of the year (May, 1934b). This polysaccharide accumulated in the spring when the snails emerged from hibernation and began feeding. In summer, the amount dropped sharply as galactogen was transferred to the eggs as part of their albumen (perivitelline) layer. No more galactogen was built up until the next egg laying season. Glycogen, present in most tissues and organs, such as the hepatopancreas, accumulated prior to hibernation and was slowly utilized during the winter. May (1934c) also found that during starvation, although glycogen was depleted in about 10 days, the galactogen reserves remained untouched. Galactogen was utilized as an emergency food source only after the glycogen had been used up. It thus appeared that glycogen was the immediate energy source in the adult snail, while galactogen was not normally metabolized but was presumably held in reserve for its role in egg laying.

The fertilized egg cell with its albumen layer and surrounding albumen membrane is termed the ovum (Bondesen, 1950). The absorption of the albumen layer by cells of the developing embryos has been observed in several cases. According to Bondesen (1950) and Bloch (1938), during gastrulation of *Ancylus fluviatilis* the albumen is absorbed by cells lining the blastopore, and during later embryonic stages by cells lining the stomodaeum and "almost assumes the character of a spontaneous ingestion of food."

During development of *Limnaea stagnalis,* the cells of the stomodaeum become ciliated, and the albumen is conveyed to the midgut by ciliary movement. The albumen is engulfed by the gut endoderm so that numerous small vacuoles form. During organogenesis some endodermal cells develop into the "larval liver," (Raven, 1958). These large cells have so many albumen-containing vacuoles that the nucleus is displaced toward the cell membrane. The larval liver represents the main digestive organ of the larva as well as the site of albumen storage (Fraser, 1946). This organ may form the adult liver directly, or as in the case of *Limnaea* (Bloch, 1938), it may disappear at hatching and be replaced by the cells of the definitive adult liver. The young pulmonate or prosobranch at hatching possesses a rudimentary reproductive system and an albumen gland is entirely lacking. The above accounts strongly suggest that albumen and its component galactogen may be the nutritive substrate for the developing embryo. The utilization of galactogen by embryos or by freshly hatched snails has been documented.

May (1932b) noticed that in the freshly laid egg of *H. pomatia,* galactogen was the only detectable polysaccharide and that it represented 37.8% of the ash-free dry weight of the egg. Following the course of galactogen and glycogen content during embryonic development, May and Weinland (1953) found that while the galactogen content seemed to remain constant at 705 mg. per 100 eggs, the glycogen content rose from zero during the first 8 days to 36 mg. per 100 eggs at the time of hatching. If there was a decrease in galactogen content during development, it was too small to be detected. The possible role of galactogen as a nutritive substrate for the developing *Helix* embryo remained uncertain.

Horstmann (1965b) noticed that freshly hatched *Helix* remained in their earthen brood-hole for up to 2 weeks without eating, although food was made available. Guessing that galactogen reserves played an important role during this time, Horstmann repeated the measurements of May and Weinland (1953) extended to 3 weeks after hatching. As before, there was no detectable galactogen dimunition until birth, the value remained constant at 5 mg. per embryo. However, 5 days after hatching, the value dropped to 3 mg. per embryo. It continued to drop in the fasting young snails, to less than 0.5 mg. per embryo at 15 days after hatching. Interestingly, the glycogen values rose slightly during this period, but were never higher than 1 mg. per embryo. Thus, there appears to be a post-embryonic fasting period during which galactogen, stored in the albumen sac, is first utilized. After 2 weeks, the new snail must emerge from the nest and hunt for food, for then galactogen and the newly synthesized glycogen are used up. Horstmann (1965b) also found the respiratory quotient during the post-embryonic fasting period to average 0.85–0.90. Lipid values rose during embryonic and post-embryonic development. Protein, a major component of the albumen layer, was not measured.

The apparent metabolic inertia of *Helix* embryonic galactogen can be contrasted with its earlier catabolism by *L. stagnalis* embryos (Horstmann, 1956a). The eggs of this fresh water pulmonate began detectable utilization of the polymer by the eighth day of development. At hatching (day 12), galactogen had dropped 42% to 12 μg per embryo. This remainder was used up within 4 days if no exogenous food was available to the young snails. Horstmann (1956b) also investigated the lipid content of embryonic *L. stagnalis*. Freshly laid eggs each contained 0.15 μg. This value increased to 3.85 μg at the time of hatching. It was concluded that lipid was formed at the expense of galactogen.

The protein component of the albumen layer appears to be heterogeneous in three species of Limnaeid (Morrill *et al.*, 1964), since eleven distinct bands can be visualized after starch gel electrophoresis. The transfer of protein from albumen layer to embryo was measured in *L. palustris* (Morrill, 1964). The first detectable changes came after gastrulation with formation of major organ primordia, including the larval liver. Then protein in the perivitelline fluid dropped steadily from its initial value of 17 μg per embryo to 3 μg at hatching. Protein was correspondingly picked up by the embryo, which had started with only 2 μg and at hatching contained 19 μg. What fraction of this protein was stored in the gut and larval liver and what was transformed into the embryo's own protoplasm remains unknown.

D. METABOLISM

Egg galactogen diminishes during development (Horstmann, 1956a) or just after hatching (Horstmann, 1965b). The enzymes responsible for catabolism of the polysaccharide remain unknown as of this writing. Galactogen content increases in the albumen gland of adult snails prior to egg laying (May, 1934b) and it is subsequently deposited with the eggs. The enzyme system responsible for galactogen synthesis has begun to be successfully investigated.

A cell-free extract from the albumen gland of *Helix pomatia* utilized UDP-D-galactose-^{14}C for the transfer of galactose-^{14}C to acceptor galactogen (Goudsmit and Ashwell, 1965). The enzyme extract exhibited an absolute requirement for acceptor molecule. Furthermore, when *Helix* galactogen was replaced by galactogen from *Limnaea stagnalis*, incorporation of galactose-^{14}C was reduced 80%. This indicated some enzyme specificity as well as pointing up the chemical and structural differences in the two galactogens. Glycogen was completely unable to replace galactogen in the formation of radioactive polymer. The linkage of the newly incorporated D-galactose remains unknown.

Presumably, at least three enzymes are active in the synthesis of galactogen. One would be involved in the formation of β-D($1 \rightarrow 3$) linkages and a second in the formation of β-D($1 \rightarrow 6$) linkages. Specific sugar nucleotide precursors for each linkage may be needed, or the case might be analogous to that of glycogen synthesis. Here it has been shown that only the α-D($1 \rightarrow 4$) linkages of glycogen are formed from UDP-D-glucose (Leloir *et al.*, 1959). The α-D($1 \rightarrow 6$) bonds are synthesized by a branching enzyme, which removes sections of the glycogen α-1,4 chain and hooks them on again in an α-1,6 linkage (Larner, 1953).

The third enzyme system, for the attachment of L-galactose, in those galactogens containing this optical isomer may involve a new sugar nucleotide. GDP-L-galactose was isolated from *H. pomatia* albumen glands (Goudsmit and Neufeld, 1966). This sugar nucleotide has been previously found by Su and Hassid (1962b) in a red alga, where it was postulated to be an intermediate in the synthesis of the L-galactose residues of galactan. In *Helix*, an enzyme system capable of forming GDP-L-galactose was also present in albumen glands (Goudsmit and Neufeld, 1967). Thus, cell-free extracts converted GPD-D-mannose-^{14}C to GDP-L-galactose-^{14}C by a direct double epimerization. D-Mannose and L-galactose differ from each other in their configuration at about the third and fifth carbon atom. The role of GDP-L-galactose as a possible donor of the L-galactose residues of *H. pomatia* galactogen remains to be investigated.

Glycogen breakdown proceeds by hydrolytic fragmentation (amylases, maltase) or by phosphorylytic attack at the terminus of each chain, releasing glucose molecules as glucose 1-phosphate. In the latter case, hydrolytic cleavage is also necessary for removal of glucose present in α-1,6 linkage at the branch points. In animals, amylases and maltase are generally found in the digestive tract, while only the phosphorylase system is significant within cells. Other hydrolytic enzymes, the glucosidases, are often found within lysosomes.

If galactogen breakdown were analogous to glycogen degradation, one could hopefully look for and find galactogen phosphorylases and galactogen hydrolases. Thus far, instead of discovering new enzymes, investigators have limited themselves to determining the presence of the well known β-galactosidases. Although these enzymes are often present in snail tissue, their role in galactogen metabolism has not yet been shown, as will be clear from the cases outlined below.

Adult snails utilize galactogen only when they are starving (May, 1934c). A search for the enzyme involved led to a preliminary finding of two β-galactosidases in *Helix aspersa* (Barnett, 1965). One, with a pH optimum of 5.4 was confined to hepatopancreas, crop, salivary gland, and digestive juice. The other, with a pH optimum of 2.2 was widely distributed in tissues, including the albumen gland. Presumed to be lysosomal, it was implicated in the catabolism of galactogen. Unfortunately, the enzyme was not tested on galactogen as a substrate, so its relevance remains unproved. It is not unlikely that such an enzyme might be released from lysosomes during the severity of starvation.

Horstmann (1965b) found a β-galactosidase to be present in freshly hatched *H. pomatia*, but not in embryos. This enzyme was not tested on galactogen and yet was stated to be responsible for its degredation. Weinland and Nuchterlein (1954) maintained that *H. pomatia* galactogen was split by a commercial pectinase preparation containing β-galactosidase. The mixture was incubated for 17 days before a maximum 7% breakdown of galactogen was thought to have resulted. The same commercial preparation is said to have caused the complete cleavage of embryonic *L. stagnalis* galactogen after 22 hours at 45°C (Horstmann and Geldmacher-Mallinkrodt, 1961).

In another case, an electrophoretic analysis of seven hydrolytic enzymes in *Limnaea palustris* revealed the presence of β-galactosidase in albumen glands, in freshly laid eggs, and in embryos at all stages of development (Norris and Morrill, 1964). These authors were primarily interested in the distribution and heterogeneity of all seven enzymes.

β-galactosidase may well be important in the release of galactose from galactogen. It would be necessary to see what products are formed by the action of snail galactosidase on galactogen from the same species. On the other hand, some as yet unknown hydrolases or phosphorylases may be the main catalytic agents which make galactose available for its entrance into metabolic pathways. Since embryonic galactogen is stored in the gut or in the evanescent larval liver, extracellular or lysosomal enzymes would seem to be the prime candidates.

Weinland (1953a,b) searched for a galactogen-splitting enzyme in *H. pomatia.* An extract of acetone-powdered *Helix* intestinal juice catalyzed a 4% hydrolysis of the polysaccharide after prolonged incubation (94 hours, pH 4.7, 30°C). The extent of hydrolysis was measured by increase in reducing power or by titration with thiosulfate. The control solution contained enzyme alone and showed extremely high blank values. It would have been revealing to include two other control samples: (1) galactogen alone, and (2) galactogen plus heat-inactivated enzyme. In any case, it is not likely that albumen gland galactogen is degraded by intestinal juice enzymes *in vivo.* Hopefully, in the future investigators will look for the relevant enzyme systems in albumen glands or in embryos that catabolize galactogen and will test these enzymes on galactogen extracted from the same snail species.

In embryos and in fasting freshly hatched snails, galactogen is most likely an important donor (via galactose) of intermediates to the metabolic machinery. In feeding snails, the primary diet is starch and cellulose. These polymers are digested and contribute glucose to the pathways of intermediary metabolism. Moreover, this glucose is probably a significant source of the UDP-D-galactose necessary for galactogen synthesis in maturing snails.

The route for the conversion of glucose to UDP-D-galactose is shown in Fig. 2. All of the metabolites are present in young *H. pomatia* except galactose oligosaccharides (Horstmann, 1956b, 1965a,b). Furthermore, most of the enzymes listed are now known to be present in snail tissues. Thus, the following were found to be active in cell-free extracts of *H. pomatia* and *Bulimnaea megasoma* albumen glands: phosphoglucomutase, galactose 1-phosphate uridyl transferase, and glycogen phosphorylase (Goudsmit, 1964). UDP-glucose pyrophosphorylase and epimerase are also present in *Helix* albumen glands (Sawicka and Chojnacki, 1968; Goudsmit, 1965). Assays for UDP-galactose pyrophosphorylase were negative for the same tissue (Goudsmit, 1965) Phosphoglucomutase and hexokinase have been detected in 7-day-old *L. stagnalis* embryos (Horstmann, 1960a). Galactokinase has not been looked for, nor has glycogen synthetase. Although these Leloir pathway metabolites

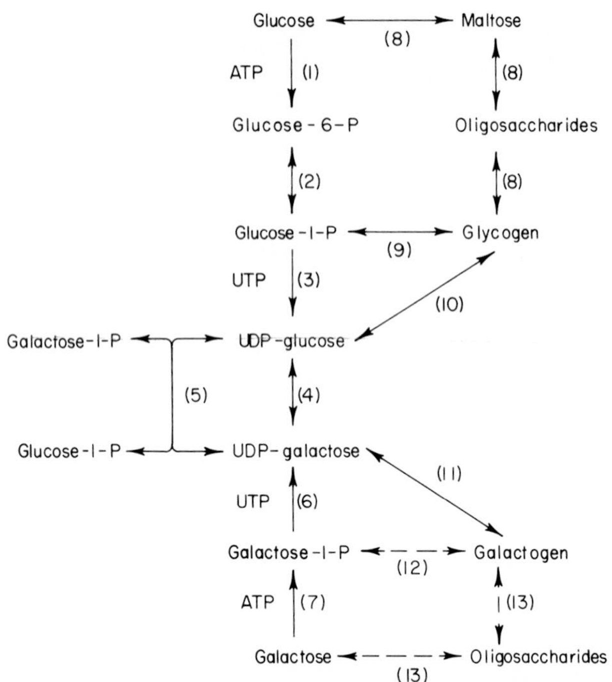

FIG. 2. Enzymatic pathways for the interconversion of glucose and galactose. Hypothetical enzymes are indicated by a broken arrow; (1) hexokinase; (2) phosphoglucomutase; (3) UDP-glucose pyrophosphorylase; (4) UDP-galactose-4-epimerase; (5) nucleotidyl transferase; (6) UDP-galactose pyrophosphorylase; (7) galactokinase; (8) glycogen amylases, maltases, glucosidases; (9) glycogen phosphorylase; (10) glycogen synthetase; (11) galactogen synthetase; (12) galactogen phosphorylase (?); (13) galactogen hydrolases (?).

are present in young snails and the enzymes are functional in adult albumen gland extracts, their physiological significance must await kinetic studies.

Galactogen and glycogen are found in the same snail (May, 1934a,b) and sometimes the same cell (Nieland and Goudsmit, 1969). A glance at Fig. 2 will reveal how closely tied their synthetic pathways are. UDP-glucose and UDP-galactose are precursors in glycogen and galactogen synthesis respectively. These two sugar nucleotides are interconvertible via two routes epimerase and nucleotidyl transferase. The factors governing biosynthesis of the parallel polysaccharides galactogen and glycogen within the snail would give a fascinating insight into control mechanisms.

VI. Glycogen

A. Detection

Glycogen is a branched chain homopolysaccharide, which upon hydrolysis yields D-glucose as the only product. The straight chain portions of the molecule are linked by $\alpha 1 \rightarrow 4$ bonds, while the branch points are joined by $1 \rightarrow 6$ α-glucosidic linkages. Approximately ten to eighteen glucose units occur between branch points (Kjölberg *et al.*, 1963). Glycogen is particulate (Lazarow, 1942) and can be distinguished from its surroundings in electron micrographs (Revel *et al.*, 1960; Revel, 1964; Fawcett, 1966). It occurs as simple spherical β particles from 150–350 Å in diameter or as a covalently bonded cluster of a variable number of simple particles, termed a rosette or α particle.

The variability in size of glycogen granules is reflected in its wide molecular weight spectrum. When obtained by cold water extraction procedures which prevent degradation, glycogens have molecular weights ranging from 5×10^6 for that of *Mytilus edulis* (Kjölberg *et al.*, 1963) to 500×10^6 for rabbit liver glycogen (Orrell and Bueding, 1964). Moreover, there is a large amount of polydispersity.

Histochemically, glycogen may be demonstrated by the periodic acid–Schiff (PAS) method when it is successfully combined with diastase or amylase treated control sections. In sectioned material prepared for electron microscopy, glycogen shows a marked and unexplained affinity for lead stains. It is difficult to distinguish the smaller β-glycogen from free ribosomes of similar size.

The biochemical isolation and identification of glycogen in most research publications has remained unchanged for a century, since the days of Claude Bernard and Pflüger. Generally, the tissue is dissolved in hot 30–60% potassium hydroxide for an hour or more, cooled, and the polysaccharides which resist this treatment are precipitated from solution with ethanol. They are purified by repeated precipitations. Subsequently, a solution of the polysaccharide(s) is completely hydrolyzed in acid at 100°C for 3 hours and the resultant sugars are determined by chromatographic, colorimetric, or enzymatic means. The presence of glucose is taken as the indication that glycogen was originally present. This method can be complicated by the presence of more than one polysaccharide, especially if they contained some glucose. Also, it may be difficult to detect the small changes in glycogen content which occur under various natural or experimental conditions.

A specific enzymatic method for the microdetermination of glycogen is now being used in several laboratories, although not on molluscan material. According to this procedure of Bueding and Hawkins (1964a,b), glycogen is completely degraded by the combined action of the crystalline enzymes, glycogen phosphorylase, and amylo-1,6-glucosidase to glucose 1-phosphate and glucose. These products are subsequently quantitated utilizing a spectrophotometric–enzymatic assay. The glycogen samples are usually portions of crude homogenates or whole tissues that have been heated in 5.6% potassium hydroxide for an hour, then have been cooled and neutralized. By their method, Hawkins, Bueding and co-workers routinely assay large numbers of portions each containing 2–20 μg of glycogen.

A further fluorometric modification of this method gives linearity within the range of 0.015–1.0 μg glycogen and has been used to determine amounts of that polysaccharide in cleaving mouse eggs (Stern and Biggers, 1968). Since physiologically important fluctuations in glycogen content are localized in certain organs, tissues, or even parts of cells, a combination of histochemical and microbiochemical analyses yields the most information. This may be especially true in gonadal tissues and developing embryos.

B. Metabolism

Glycogen of muscle, hepatopancreas, and other tissues serves as a source of immediately available or reserve energy via its component glucose. Glucose and its metabolic derivatives are used in the synthesis of other hexoses, of some amino acids, of neutral lipids, and of pentoses which are an integral portion of nucleic acids.

The immediate reactions concerning glycogen synthesis and catabolism have been neglected in research on mollusks. Glycogen synthetase has not been looked for. Glycogen phosphorylase is present in partially purified extracts of albumen glands and in crude homogenates of embryos of the pulmonate snails, *Helix pomatia* and *Bulimnaea megasoma* (Goudsmit, 1964). Phosphorylase activity was also detected in cell-free hepatopancreas preparations from the above named snails (Goudsmit, 1964), from the mussel *Mytilus californianus*, and the abalone *Haliotis rufescens* (Bennett and Nakada, 1968). The glucose 1-phosphate produced by the phosphorylase reaction can be converted to glucose 6-phosphate in a freely reversible reaction catalyzed by phosphoglucomutase. This enzyme has been detected in several mollusks (Bennett and Nakada, 1968; Hunger and Horstmann, 1968; Goudsmit, 1964).

Glucose 6-phosphate stands at the crossroads of several metabolic pathways. It may be dephosphorylated to free glucose and thus be available for transport to other parts of the body. It may be converted to fructose 6-phosphate and enter the glycolytic route and continue through the citric acid cycle giving rise to ATP, neutral lipids, and glycogenic amino acids. Glucose 6-phosphate may be oxidized to yield 6-phosphogluconic acid and enter the hexose monophosphate shunt, there to be converted to pentoses and to participate in the generation of NADPH.

All these possibilities are now theoretically open to mollusks. The metabolites and enzymes that participate in each route and cycle have been assayed directly or indirectly in several cases. All of the enzymes of glycolysis and the key enzymes in the pentose phosphate pathway have been detected in the mussel *M. californianus* and the abalone, *H. rufescens* (Bennett and Nakada, 1968). Slices of whole *Mytilus*, of hepatopancreas, and of gill tissue were incubated with glucose-1-^{14}C or glucose-6-^{14}C and the percentage of ^{14}C in respired carbon dioxide was measured. Glucose-1-^{14}C was oxidized at a faster rate, giving supportive evidence for the operation of the pentose cycle.

Glycogen synthesis could be shown at the same time as glycogenolysis in the experiments of Bennett and Nakada (1968). Slices of hepatopancreas from *Mytilus* and *Haliotis* were incubated in the presence of isotopically labeled glucose, galactose, or fructose and the amount of radioactivity incorporated into glycogen was measured. Fructose-^{14}C was the most efficient; its ^{14}C was incorporated into glycogen at three times the rate of similarly labeled glucose. Galactose lagged behind both the others.

Several glycolytic enzymes and pentose shunt enzymes have been reported for embryos of *Lymnaea stagnalis* (Horstmann, 1960a,b). Some glycolytic enzymes have also been shown for the freshwater snail, *Physa halai* (Beams, 1963) and the freshwater mussel, *Unio* (Karpiak, 1960), as well as for the Australian oyster, *Crassostrea commercialis* (Humphrey, 1950). Many more accounts demonstrate the presence of glycolytic intermediates [reviewed by Bennett and Nakada (1968) and by Hammen (1969)].

Simpson, Awapara, and their colleagues are continuing to explain how the brackish water clam *Rangia cuneata* produces so much succinate as an end product of anaerobic glycolysis. Succinate accounts for 40% of the glucose degraded. The proposed scheme involves a highly active phosphoenol pyruvate (PEP) carboxykinase (Simpson and Awapara, 1964, 1966). PEP is carboxylated to oxaloacetate, which is then successively transformed into malate, fumarate, and succinate in a reversal of the citric acid cycle. The problem arises in balancing the cofactor NAD formed and utilized during glycolysis and subsequent succinate formation. For each molecule of glucose, two trioses are produced during glycolysis. Each triose utilizes two molecules of NAD for the continuation of the reaction sequence, the reduction of PEP to succinate. Thus, in the whole scheme, for each molecule of glucose, four molecules of NAD are needed. Yet, during glycolysis only two molecules of NAD are made available. Therefore, only one molecule of triose can be transformed into succinate. What happens to the other triose? Stokes and Awapara (1968) may have solved this problem. During glycolysis in *Rangia,* alanine is produced in equimolar amounts with succinate. All the carbons of alanine are derived from glucose. The amino acid is presumably formed from pyruvate by transamination, a reaction known to occur in *Rangia* (Awapara and Campbell, 1964). The clam's ability to form enough pyruvate during glycolysis is not discussed. Formation of alanine balances the oxidation–reduction picture, as well as taking care of the disposition of carbons. For succinate, one carbon is picked up from carbon dioxide during the PEP carboxykinase reaction.

The American oyster, *Crassostrea virginica,* produces more succinic acid than lactic acid during anaerobic glycolysis (Simpson and Awapara, 1966). Mantle tissue of the oyster incorporated ^{14}C labeled bicarbonate into succinate, malate, and fumarate (Hammen and Wilbur, 1959). Carbon dioxide fixation appeared to take place by the carboxylation of pyruvate to form malate (Hammen, 1966). Oyster tissue was incubated in $NaH^{14}CO_3$ for various periods. The carboxylic acids were then extracted and identified. In the shortest incubation periods of less than 2 minutes, malic acid contained 50–65% of the radioactivity. Furthermore, malic dehydrogenase activity was present in homogenates. Thus, in the American oyster, continuation of glycolysis to pyruvate formation seems to be favored. Comparing key enzyme activities in both this species and in *Rangia,* pyruvic kinase is one hundred times more active in the oyster and there is no impediment to the conversion of PEP to pyruvate. The oyster has only 5% of the PEP carboxykinase activity of *Rangia* (Hammen, 1969).

Martin (Goddard and Martin, 1966) has reviewed the relevant literature concerning glycogen content of various mollusks. As much as is known about the relation of lipids and glycogen in the "economy" of marine invertebrates has been set down by Giese (1966). Mollusks are well represented. It should now be possible to continue and enlarge upon the exciting aspects of carbohydrate metabolism. Oysters, mussels, brackish water clams, snails, and squid each metabolize glucose according to its own needs. These needs are dependent upon nutritional or reproductive state and upon seasonal, tidal, latitudinal, and phylogenetic factors. For some few mollusks it is now possible to progress from a list of metabolites and enzymes to an investigation of pool sizes, kinetics, and functions.

REFERENCES

Algranati, I. D., Carminatti, H., and Cabib, E. (1963). *Biochem. Biophys. Res. Commun.* **12**, 504–509.

Anno, K., Kawai, Y., and Seno, N. (1964). *Biochim. Biophys. Acta* **83**, 348–349.

Awapara, J., and Campbell, J. W. (1964). *Comp. Biochem. Physiol.* **11**, 231–235.

Badman, D. G. (1967). *Comp. Biochem. Physiol.* **23**, 621–629.

Barnett, J. E. G. (1965). *Biochem. J.* **96**, 72p.

Barry, V. C. (1943). *Nature (London)* **152**, 537–538.

Bayne, C. (1966). *Comp. Biochem. Physiol.* **19**, 317–338.

Bayne, C. (1967). *Comp. Biochem. Physiol.* **23**, 761–773.

Beams, C. (1963). *Comp. Biochem. Physiol.* **8**, 109–114.

Bell, D. J., and Baldwin, E. (1938). *J. Chem. Soc., London* pp. 1461–1465.

Bell, D. J., and Baldwin, E. (1941). *J. Chem. Soc., London* pp. 125–132.

Bennett, R., and Nakada, H. I. (1968). *Comp. Biochem. Physiol.* **24**, 787–797.

Bloch, S. (1938). *Rev. Suisse Zool.* **45**, 157–220.

Bondesen, P. (1950). "A Comparative Morphological-Biological Analysis of the Egg Capsules of Freshwater Pulmonate Gastropods." Naturhistorisk Museum, Aarhus, Denmark.

Bueding, E., and Hawkins, J. T. (1964a). *Anal. Biochem.* **7**, 26–36.

Bueding, E., and Hawkins, J. T. (1964b). *Anal. Biochem.* **9**, 115–126.

Corrêa, J. B. C., Dmytraczenko, A., and Duarte, J. H. (1967). *Carbohyd. Res.* **3**, 445–452.

Doyle, J. (1964). *Biochem. J.* **91**, 6P.

Duarte, J. H., and Jones, J. K. N. (1971). *Carbohyd. Res.* **16**, 327–335.

Egami, F., Asahi, T., Takahashi, N., Suzuki, S., Shikata, S., and Nisizawa, K. (1955). *Bull. Chem. Soc. Jap.* **28**, 685–689.

Elbein, A. D., Barber, G. A., and Hassid, W. Z. (1964). *J. Amer. Chem. Soc.* **86**, 309–310.

Fairbairn, D. (1958). *Can. J. Zool.* **36**, 787–795.

Fantin, A. M. B., and Vigo, E. (1968). *Histochemie* **15**, 300–311.

Fawcett, D. W. (1966). "The Cell—An Atlas of Fine Structure." Saunders, Philadelphia, Pennsylvania.

Fleitz, H., and Horstmann, H. J. (1967). *Hoppe-Seyler's Z. Physiol. Chem.* **348**, 1301–1311.

Fraser, L. A. (1946). *Trans. Amer. Microsc. Soc.* **65**, 279–293.

Ghose, K. C. (1963). *Proc. Zool. Soc. London* **140**, 681–750.

Giese, A. C. (1966). *Physiol. Rev.* **46**, 244–298.

Glaser, L., and Brown, D. H. (1957). *Biochim. Biophys. Acta* **23**, 449–450.

Goddard, C. K. (1968). *Comp. Biochem. Physiol.* **27**, 275–285.

Goddard, C. K. (1969). *Comp. Biochem. Physiol.* **28**, 271–291.

Goddard, C. K., and Martin, A. W. (1966). *In* "Physiology of Mollusca" (K. M. Wilbur and C. M. Yonge, eds.), Vol. 2, pp. 275–308. Academic Press, New York.

Goudsmit, E. M. (1964). Ph.D. Thesis, University of Michigan.

Goudsmit, E. M. (1965). Unpublished experiments.

Goudsmit, E. M., and Ashwell, G. (1965). *Biochem. Biophys. Res. Commun.* **19**, 417–422.

Goudsmit, E. M., and Neufeld, E. F. (1966). *Biochim. Biophys. Acta* **121**, 192–195.

Goudsmit, E. M., and Neufeld, E. F. (1967). *Biochem. Biophys. Res. Commun.* **26**, 730–735.

Grainger, J. N. R., and Shillitoe, A. J. (1952). *Stain Technol.* **27**, 81–85.
Greenberg, E., and Preiss, J. (1964). *J. Biol. Chem.* **239**, PC4314.
Hammarsten, O. (1885). *Arch. Gesamte Physiol. Menschen Tiere* **36**, 373–456.
Hammen, C. S. (1966). *Comp. Biochem. Physiol.* **17**, 289–296.
Hammen, C. S. (1969). *Amer. Zool.* **9**, 309–318.
Hammen, C. S., and Wilbur, K. M. (1959). *J. Biol. Chem.* **234**, 1268–1271.
Horstmann, H. J. (1956a). *Biochem. Z.* **328**, 342–347.
Horstmann, H. J. (1956b). *Biochem. Z.* **328**, 348–351.
Horstmann, H. J. (1960a). *Hoppe-Seyler's Z. Physiol. Chem.* **319**, 110–119.
Horstmann, H. J. (1960b). *Hoppe-Seyler's Z. Physiol. Chem.* **319**, 120–125.
Horstmann, H. J. (1964). *Biochem. Z.* **340**, 548–551.
Horstmann, H. J. (1965a). *Biochem. Z.* **342**, 23–30.
Horstmann, H. J. (1965b). *Z. Biol. (Munich)* **115**, 133–155.
Horstmann, H. J., and Geldmacher-Mallinckrodt, M. (1961). *Hoppe-Seyler's Z. Physiol. Chem.* **325**, 251–259.
Humphrey, G. F. (1950). *Aust. J. Exp. Biol. Med. Sci.* **28**, 151–160.
Hunger, V. J., and Horstmann, H. J. (1968). *Z. Biol. (Munich)* **116**, 90–104.
Hunt, S. (1966). *Nature (London)* **210**, 436–437.
Hunt, S. (1967). *Carbohyd. Res.* **4**, 259–261.
Hunt, S. (1970). "Polysaccharide-Protein Complexes in Invertebrates." Academic Press, New York.
Hunt, S., and Jevons, F. R. (1963). *Biochim. Biophys. Acta* **78**, 376–378.
Hunt, S., and Jevons, F. R. (1965a). *Biochim. Biophys. Acta* **101**, 214–216.
Hunt, S., and Jevons, F. R. (1965b). *Biochem. J.* **97**, 701–709.
Hunt, S., and Jevons, F. R. (1966). *Biochem. J.* **98**, 522–529.
Hyman, L. H. (1967). "The Invertebrates," Vol. VI. McGraw-Hill, New York.
Iida, K. (1963). *J. Biochem. (Tokyo)* **54**, 181–188.
Karpiak, S. E. (1960). *Acta Physiol. Pol.* **11**, 153–155.
Katzman, R. L., and Jeanloz, R. W. (1969). *Science* **166**, 758–759.
Katzman, R. L., Bhattacharya, A. K., and Jeanloz, R. W. (1969). *Biochim. Biophys. Acta* **184**, 523–528.
Kjölberg, O., Manners, D. J., and Wright, A. (1963). *Comp. Biochem. Physiol.* **8**, 353–365.
Kwart, H., and Shashoua, V. E. (1958). *J. Amer. Chem. Soc.* **80**, 2230–2236.
Larner, J. (1953). *J. Biol. Chem.* **202**, 491–503.
Lash, J. W., and Whitehouse, M. W. (1960). *Biochem. J.* **74**, 351–355.
Lazarow, A. (1942). *Anat. Rec.* **84**, 31–50.
Leloir, L. F., and Cardini, C. E. (1957). *J. Amer. Chem. Soc.* **79**, 6340–6341.
Leloir, L. F., Olavarria, J. M., Goldemberg, S. H., and Carminatti, H. (1959). *Arch. Biochem. Biophys.* **81**, 508–520.
Levene, P. A. (1925). *J. Biol. Chem.* **65**, 683–700.
McMahon, P., von Brand, T., and Nolan, M. O. (1957). *J. Cell. Comp. Physiol.* **50**, 219–240.
Martin, A. W. (1966). *In* "Physiology of Mollusca" (K. M. Wilbur and C. M. Yonge, eds.), Vol. 2 p. 291. Academic Press, New York.
Mathews, M. B., Duh, J., and Person, P. (1962). *Nature (London)* **193**, 378–379.
May, F. (1931). *Z. Biol. (Munich)* **91**, 215–220.
May, F. (1932a). *Z. Biol. (Munich)* **92**, 319–324.
May, F. (1932b). *Z. Biol. (Munich)* **92**, 325–330.
May, F. (1934a). *Z. Biol. (Munich)* **95**, 277–297.

May, F. (1934b). *Z. Biol. (Munich)* **95**, 401–430.
May, F. (1934c). *Z. Biol. (Munich)* **95**, 606–613.
May, F., and Weinland, H. (1953). *Z. Biol. (Munich)* **105**, 339–347.
May, F., and Weinland, H. (1954). *Hoppe-Seyler's Z. Physiol. Chem.* **296**, 154–166.
May, F., and Weinland, H. (1956). *Hoppe-Seyler's Z. Physiol. Chem.* **305**, 219–225.
Meenakshi, V. R. (1954). *Curr. Sci.* **23**, 301–302.
Meenakshi, V. R., and Scheer, B. T. (1968). *Comp. Biochem. Physiol.* **26**, 1091–1097.
Mori, T. (1953). *Advan. Carbohyd. Chem.* **8**, 315–350.
Morrill, J. B. (1964). *Acta Embryol. Morphol. Exp.* **7**, 131–142.
Morrill, J. B., Norris, E., and Smith, S. D. (1964). *Acta Embryol. Morphol. Exp.* **7**, 155–166.
Neutra, M., and Leblond, C. P. (1966). *J. Cell Biol.* **30**, 137–150.
Nieland, M. L., and Goudsmit, E. M. (1969). *J. Ultrastruct. Res.* **29**, 119–140.
Norris, E., and Morrill, J. B. (1964). *Acta Embryol. Morphol. Exp.* **7**, 29–41.
O'Colla, P. (1953). *Proc. Roy. Irish Acad., Sect. B* **55**, 165–170.
Orrell, S. A., and Bueding, E. (1964). *J. Biol. Chem.* **239**, 4021–4026.
Pancake, S. J., and Karnovsky, M. L. (1971). *J. Biol. Chem.* **246**, 253–262.
Raven, C. P. (1958). "Morphogenesis: The Analysis of Molluscan Development," p. 162. Pergamon Press, Oxford.
Recondo, E., and Leloir, L. F. (1961). *Biochem. Biophys. Res. Commun.* **6**, 85–88.
Revel, J. P. (1964). *J. Histochem. Cytochem.* **12**, 104–114.
Revel, J. P., Napolitano, L., and Fawcett, D. W. (1960). *J. Biophys. Biochem. Cytol.* **8**, 575–589.
Rigby, J. (1963). *Proc. Zool. Soc. London* **141**, 311–360.
Rudall, K. M. (1955). *Symp. Soc. Exp. Biol.* **9**, 49–95.
Sawicka, T., and Chojnacki, T. (1968). *Comp. Biochem. Physiol.* **26**, 707–713.
Schiller, S. (1964). *Biochem. Biophys. Res. Commun.* **15**, 250–255.
Shashoua, V. E., and Kwart, H. (1959). *J. Amer. Chem. Soc.* **81**, 2899–2905.
Simpson, J. W., and Awapara, J. (1964). *Comp. Biochem. Physiol.* **12**, 457–464.
Simpson, J. W., and Awapara, J. (1966). *Comp. Biochem. Physiol.* **18**, 537–548.
Srinivasan, S. R., Radhakrishnamurthy, B., Dalferes, E. R., and Berenson, G. S. (1969). *Comp. Biochem. Physiol.* **28**, 169–176.
Stern, S., and Biggers, J. D. (1968). *J. Exp. Zool.* **168**, 61–66.
Stokes, T., and Awapara, J. (1968). *Comp. Biochem. Physiol.* **25**, 883–892.
Su, J. C., and Hassid, W. Z. (1962a). *Biochemistry* **1**, 468–474.
Su, J. C., and Hassid, W. Z. (1962b). *Biochemistry* **1**, 475–482.
Vasseur, E. (1950). *Acta Chem. Scand.* **4**, 1144–1145.
Warren, L. (1963). *Comp. Biochem. Physiol.* **10**, 153–171.
Weinland, H. (1953a). *Biochem. Z.* **324**, 19–31.
Weinland, H. (1953b). *Biochem. Z.* **324**, 74–82.
Weinland, H., and Nuchterlein, H. (1954). *Hoppe-Seyler's Z. Physiol. Chem.* **298**, 48–54.
Wheat, R. W. (1960). *Science* **132**, 1310–1311.
White, A., Handler, P., and Smith, E. L. (1963). "Principles of Biochemistry," 3rd ed. McGraw-Hill, New York.

CHAPTER 9

Lipid and Sterol Components and Metabolism in Mollusca

P. A. Voogt

I. Introduction

Many papers have been devoted to lipids and lipid composition in mollusks. Several of them, however, give but little information, as only overall values like lipid content, melting point, and optical rotation of total sterol contents; mean degree of unsaturation in fatty acids; or some of the predominant fatty acids have been determined and reported. Attention paid to each of the five classes within the phylum of Mollusca is quite different. Most papers are concerned with bivalves and gastropods, while only few deal with cephalopods and still fewer with Amphineura. Nothing seems to be known about lipid composition in scaphopods. A number of causes has brought about this remarkable discrimination. Without doubt bivalves and gastropods are the most successful classes of the mollusks as to the number of genera and the absolute number of specimens. Therefore, they are easily available, which stimulates investigation into these classes. A second reason for this unequal interest may be the economic importance of several representatives of the gastropods, bivalves, and cephalopods. Finally, study of scaphopods is hampered seriously by the minute dimensions of these animals.

A similar situation exists with respect to the investigation of the different classes of lipids. Whereas numerous papers, though several are rather preliminary, have appeared about sterols and have enabled Bergmann (1949, 1962) to give a picture of their occurrence and distribution within the phylum of Mollusca, our knowledge about occurrence and distribution of lipids belonging to other classes is still fragmentary, and is confined to a small number of mollusks investigated.

Part of the interest in sterols, roused earlier than that in other lipid classes, was directed to provitamin D contents (Bock and Wetter, 1938; Kind and Herman, 1948; Matsumoto *et al.*, 1955; Matsumoto and Tamura, 1955b; Petering and Waddell, 1951; Rosenberg and Waddell, 1951; Toyama and Takagi, 1954; Van der Vliet, 1948a,b). Later on, interest in sterols decreased, owing to lack of means of separating and identifying them. Important progress was made with the introduction of gas chromatography into sterol research (Beerthuis and Recourt, 1960; Brooks and Hanaineh, 1963; Clayton, 1962; Hamilton *et al.*, 1963; Horning *et al.*, 1959, 1961; Ikekawa *et al.*, 1968; Knights, 1966, 1967; F. A. Vanden Heuvel and Court, 1968, 1969; W. J. A. Vanden Heuvel *et al.*, 1960, 1966; W. J. A. Vanden Heuvel and Horning, 1962). Identification of sterols was improved by using infrared spectrometry and mass spectrometry (Brooks *et al.*, 1968). Application of these techniques yielded more details about sterol composition in recent years (Idler *et al.*, 1964; Tamura *et al.*, 1964a,b; Wainai *et al.*, 1964; Voogt, 1969).

The development of chromatographic procedures in several forms also animated the investigation into other classes of lipid. Complex mixtures could be separated on columns filled with silicic acid (Hirsch and Ahrens, 1958) or Florisil (Carroll, 1961). The same was obtained with thin-layer chromatography (Freeman and West, 1966). Fatty acids were separated according to degree of unsaturation on columns (Elovson, 1965; Goldfine and Bloch, 1961) or on thin-layer plates (Mangold and Kammereck, 1961; White and Williams, 1965; White, 1966). Gas–liquid chromatography has been applied with success to fatty acids (Ackman, 1963a,b; Ackman and Burgher, 1963; Haken, 1966, 1969a,b; Iverson *et al.*, 1965). With the exception of sterols no review has yet appeared of the lipid components and metabolism in mollusks.

Structure and Biosynthesis of Naturally Occurring Lipids

These subjects will be given only in outline.

1. *Hydrocarbons*

Among hydrocarbons present in animals, a very important place is occupied by squalene ($C_{30}H_{50}$) (I), though its concentration is usually rather low (Popják, 1954). Its biosynthesis has been the subject of many investigations. The first part, from acetate to mevalonate, is not well

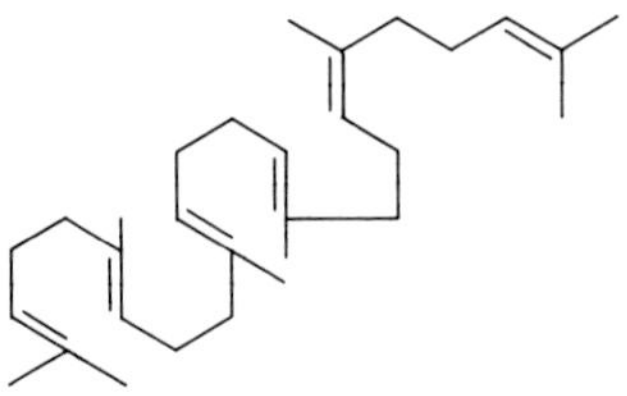

Squalene

(I)

understood yet (Danielsson and Tchen, 1968; Higgins and Kekwick, 1969), but the latter part, from mevalonate to squalene, is well established. Squalene itself is a precursor to terpenoids and sterols.

2. *Terpenoids*

Squalene is, via the intermediate 2,3-oxidosqualene (Corey *et al.*, 1966; Van Tamelen *et al.*, 1966), cyclized to form lanosterol (II), a triterpene.

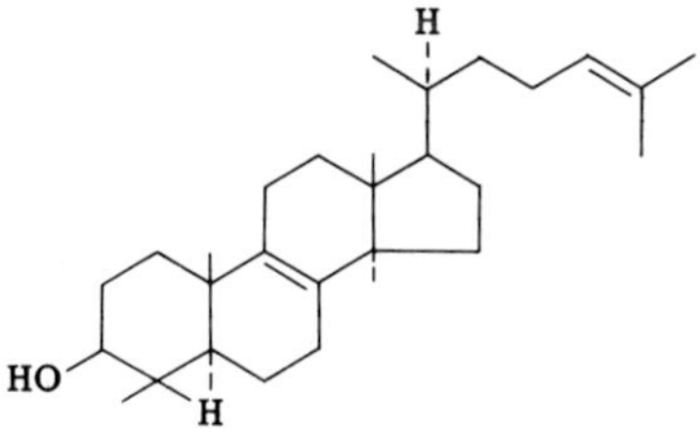

H

HO

H

Lanosterol

(II)

In plants, the first triterpene formed seems to be cycloartenol (von Ardenne *et al.*, 1965). Other precursors for terpenoids are mevalonate, farnesyl pyrophosphate, and geranyl pyrophosphate. In this way, all compounds built up from isoprenoid units are unified in the terpenoids group. These include quinones (Wiss and Gloor, 1969; Rudney, 1969), several alkaloids in plants, carotenoids, and rubber. Reviews on this subject are given by Sandermann (1962) and Thomson (1962).

3. *Sterols*

The triterpene lanosterol is transformed into zymosterol (III), which is believed to be the precursor of all other sterols. These sterols may occur in free form or esterified to fatty acids. Cholestane (IV), essen-

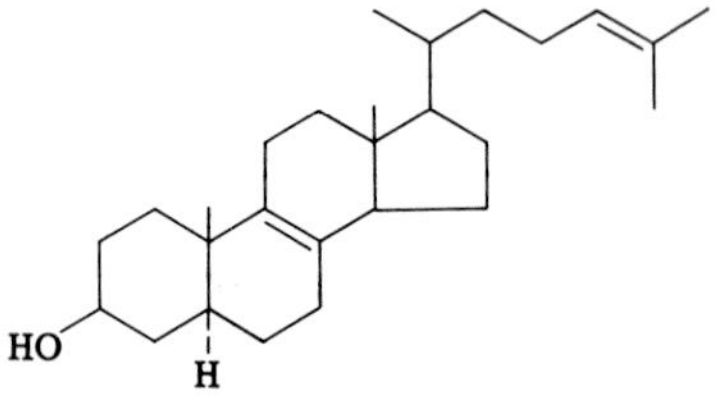

Zymosterol
(III)

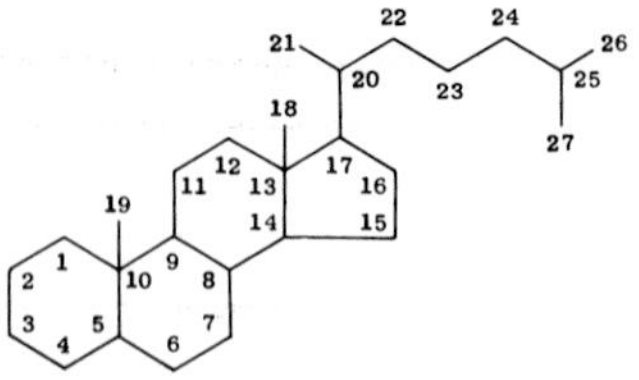

Cholestane
(IV)

tially a hydrocarbon, can be considered as the parent nucleus of most sterols. Cholesterol, one of the most important sterols, can be represented now as cholest-5-en-3β-ol or Δ^5-cholestenol and ergosterol as 24β-methyl-$\Delta^{5,7,22}$-cholestatrienol. The biosynthesis of sterols has been reviewed by Olson (1965) and by Danielsson and Tchen (1968), while reviews of the structure and distribution of sterols have been given by Bergmann (1949, 1962).

4. Fatty Acids

Fatty acids may occur as free fatty acids, but they are mostly encountered in esterified form. They are present in methyl esters, sterol esters, glycerides (mono-, di-, or triglycerides), and phospholipids. Chain length may vary from two to twenty-seven carbon atoms (Cerbulis and Wight Taylor, 1969; Sipos and Ackman, 1968), and even twenty-eight carbon atoms have been observed (Terpstra and Voogt, 1969; Van der Horst and Voogt, 1969b; Van Gennip and Voogt, 1969). Though most fatty acids possess straight chains, considerable amounts of branched-chain fatty acids (V, VI) may be present. In anteiso fatty acids, the

$$CH_3-\underset{\displaystyle CH_3}{\underset{|}{C}H}-(CH_2)_n-COOH$$

Isofatty acid
(V)

$$CH_3-CH_2-\underset{\displaystyle CH_3}{\underset{|}{C}H}-(CH_2)_n-COOH$$

Anteisofatty acid
(VI)

methyl side chain is in the antepenultimate position. Besides satured, there are also unsaturated fatty acids.

The fatty acid composition in organisms is influenced by several external conditions. Ackman (1964) points to the structural homogeneity in unsaturated fatty acids of marine lipids, but in another paper (Ackman, 1967), he indicates several differences in fatty acid composition of marine and freshwater fish oils. In a review by Lovern (1964), differences between the fatty acids of marine and terrestrial organisms are given. Lewis (1962) has indicated the effects of temperature and pressure on fatty acid composition. Eicosapentenoic acid (VII) and docosahexenoic acid (VIII) are considered to be characteristic for

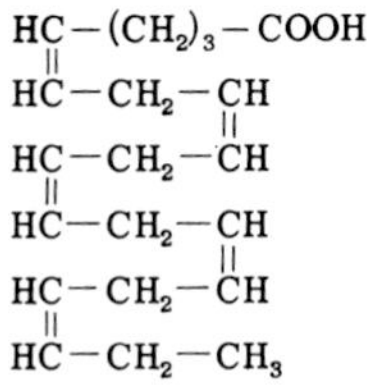

$$\begin{array}{l} HC-(CH_2)_3-COOH \\ \| \\ HC-CH_2-CH \\ \qquad\qquad \| \\ HC-CH_2-CH \\ \| \\ HC-CH_2-CH \\ \qquad\qquad \| \\ HC-CH_2-CH \\ \| \\ HC-CH_2-CH_3 \end{array}$$

Eicosapentenoic acid
20:5 (5, 8, 11, 14, 17)$^+$
(VII)

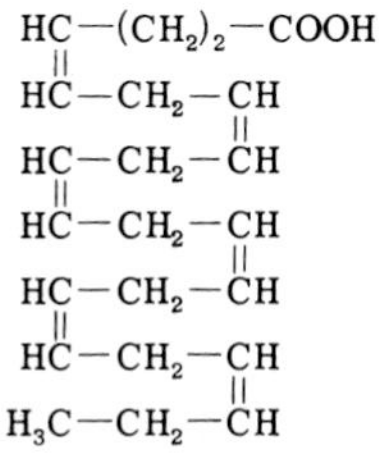

$$\begin{array}{l} HC-(CH_2)_2-COOH \\ \| \\ HC-CH_2-CH \\ \qquad\qquad \| \\ HC-CH_2-CH \\ \| \\ HC-CH_2-CH \\ \qquad\qquad \| \\ HC-CH_2-CH \\ \| \\ HC-CH_2-CH \\ \qquad\qquad \| \\ H_3C-CH_2-CH \end{array}$$

Docosahexenoic acid
22:6 (4, 7, 10, 13, 16, 19)
(VIII)

$^+$The first number indicates the length of the carbon chain, the second the number of double bonds, and the positions of the double bonds are given in parentheses.

marine organisms. A review on the comparative aspects of fatty acid occurrence and distribution has been given by Shorland (1962). For details about the biosynthesis of fatty acids, see Green and Allmann (1968).

5. *Phospholipids*

This group is very heterogenous, but all compounds have in common that they contain phosphorus. The most simple phospholipid is phosphatidic acid (IX) from which all other phospholipids can be thought

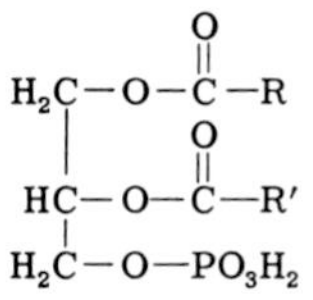

$$\begin{array}{l} H_2C-O-C(=O)-R \\ | \\ HC-O-C(=O)-R' \\ | \\ H_2C-O-PO_3H_2 \end{array}$$

Phosphatidic acid
(IX)

to be derived by esterification of the phosphate residue with choline (X), ethanolamine (XI), or serine (XII). As can be seen from the

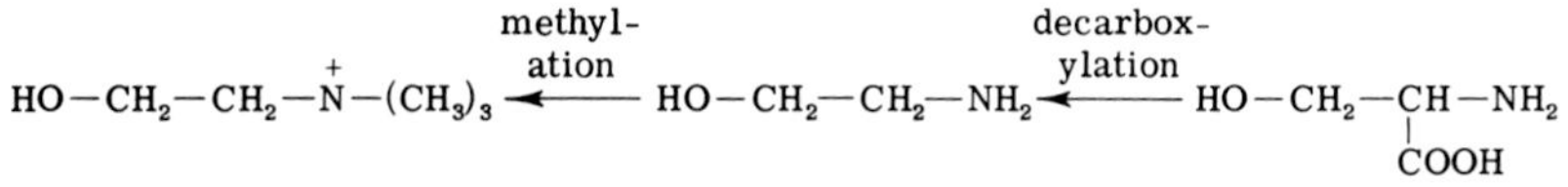

$$HO-CH_2-CH_2-\overset{+}{N}-(CH_3)_3 \xleftarrow{\text{methylation}} HO-CH_2-CH_2-NH_2 \xleftarrow{\text{decarboxylation}} HO-CH_2-\underset{\displaystyle COOH}{\underset{|}{CH}}-NH_2$$

Choline (X) Ethanolamine (XI) Serine (XII)

formulas there is a marked structural relationship between these substances. Phosphatidylcholine was formerly called lecithin, while the other ones were called cephalins. Lysophospholipids are obtained when the fatty acid in β-position is removed from the diacylphospholipids.

A very interesting group of phospholipids is constituted by the sphingolipids. The main constituent ceramide is built up from the amino alcohol sphingosine (XIII) and a fatty acid, forming a carbon–nitrogen bond. Ceramide can be esterified by its terminal hydroxyl group to phosphorylcholine forming sphingomyelin, or to phosphorylethanolamine (2-amino-ethylphosphate). Recently, a remarkable compound, 2-amino-ethyl-phosphonic acid (XIV), containing a carbon–phosphorus bond has

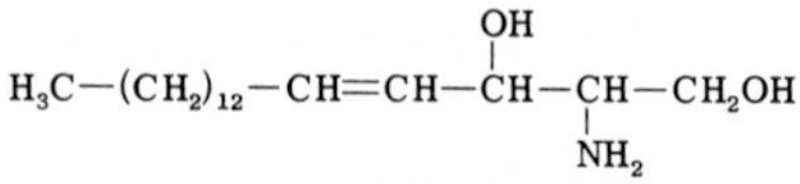

$$H_3C-(CH_2)_{12}-CH{=}CH-CH(OH)-CH(NH_2)-CH_2OH$$

Sphingosine

(XIII)

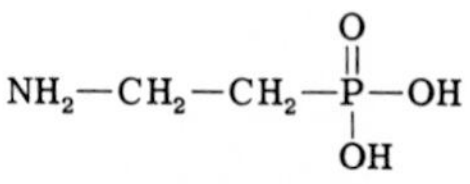

$$NH_2-CH_2-CH_2-P(=O)(OH)-OH$$

2-Aminoethyl-phosphonic acid

(XIV)

been observed (Quin, 1965). Phospholipids containing this component are called phosphonolipids. An interesting review on the distribution of phospholipids has been given by Dittmer (1962). The biochemistry and biosynthesis of phospholipids has been reviewed by Dawson (1962), Misra (1966), and Rossiter (1968).

6. Lipids Containing Ether Bonds

Usually these lipids are not considered as a separate category, but are classed with the triglycerides or phospholipids. The ether linkage, first observed by Tsujimoto and Toyama (1922), has been found in glycerides and phospholipids. Instead of a fatty acid in acyl linkage, a hydrocarbon in ether linkage is present. Usually, the alkyl chain is attached at the α-position of glycerol. These lipids are often indicated as α-glyceryl ethers or alkyl glyceryl ethers. Dialkyl ethers are also observed. A well known glyceryl ether is batyl alcohol (XV). Glyceryl

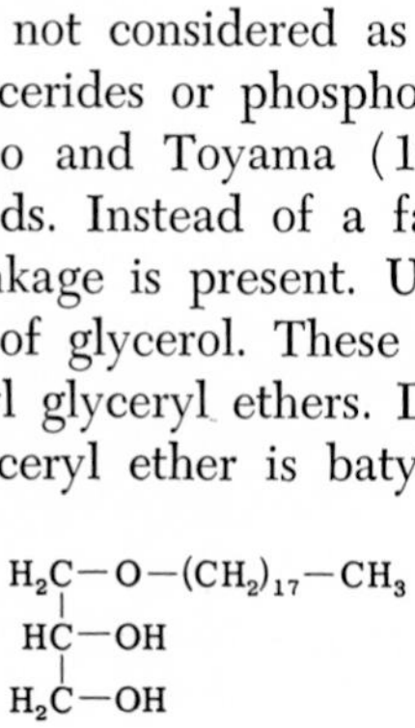

$$\begin{array}{l} H_2C-O-(CH_2)_{17}-CH_3 \\ \;\;| \\ HC-OH \\ \;\;| \\ H_2C-OH \end{array}$$

Batyl alcohol

(XV)

ethers in which the alkyl chain is α,β unsaturated are called plasmalogens or alk-1-enyl glyceryl ethers. Just like the alkyl glyceryl ethers, the alk-1-enyl glyceryl ethers are present both in triglycerides and phospholipids.

In a series of interesting experiments, Snyder *et al.* (1969a) showed that the alkylchain was derived from fatty alcohols instead of from fatty acids or aldehydes. In another paper (Snyder *et al.*, 1969b), they reported the biosynthesis of alkyl ether bonds from glyceraldehyde 3-phosphate and fatty alcohols by microsomal enzymes and suggested a possible metabolic pathway. The same was found by these authors (Snyder *et al.*, 1969c) for *Asterias forbesi.* However, alk-1-enyl glyceryl ethers were not synthesized. Wykle and Snyder (1969) demonstrated that the glyceryl moiety of the glyceryl ethers was provided by glyceraldehyde 3-phosphate and to a lesser degree by dihydroxyacetone phosphate. Thompson (1966) showed that plasmalogens were derived from glyceryl ethers by unsaturation of the alkyl chain and a possible metabolic pathway was presented. Till then it was believed that plasmalogens were the precursors for glyceryl ethers. In 1968, this author (Thompson, 1968) confirmed his former results. An excellent review on biochemistry, biological functions, and biosynthesis of alkyl glyceryl ethers and alk-1-enyl glyceryl ethers has been given by Snyder (1969).

7. *Glycolipids*

These lipids, often called cerebrosides, consist of ceramide and one or more sugars (neutral sugars or aminosugars). Another type of glycolipid has been identified by Nakazawa (1959).

II. Distribution of Lipids

A. Amphineura

This class has been the subject of but few investigations. The results have provided only scattered data, all related to the subclass Polyplacophora. No data are known about the other subclasses. Giese (1966), in studying the lipid contents in several organs of the chitons *Katharina tunicata* and *Mopalia hindsii,* found that the gonads in particular contained considerable amounts of lipids. In one case, the lipid content in other organs varied strongly with the sexual cycle. In general, gravid animals did contain more lipids than immature or spent ones.

1. *Sterols*

The chiton *Liolophura japonica* has been studied by several investigators. Lipid values from these studies have been summarized in Table I. Toyama and Shibano (1943), in studying the sterols of *L. japonica* could not obtain constant melting points of the sterols and their acetates.

TABLE I
LIPIDS OF *Liolophura japonica*

Lipid	Amount: Toyama and Tanaka, 1953 I	Toyama and Tanaka, 1953 II	Takagi and Toyama, 1956	Kita and Toyama, 1959
Total lipid (% dry weight)	1.1	2.0	1.5	1.0
Phospholipids (% total lipid)	20	14	—	—
Acetone-soluble lipids (% total lipid)	80	86	51	—
Unsaponifiable lipids (% acetone-soluble lipids)	22	11.43	24	14.2[a]
Sterols (% unsaponifiable lipids)	45	—	51.5	27

[a] Percent of total lipids.

When Toyama and Tanaka (1953) studied this chiton again—caught, however, at another locality—the melting point was constant after four recrystallizations. They proved the sterol to be identical with Δ^7-cholestenol. To explain the different results, chitons from the first locality were studied once more, and again the authors could not obtain a constant melting point. From this and from other indications, they concluded that in this case, besides Δ^7-cholestenol, another sterol must be present. Kind and Meigs (1955) showed that Δ^7-cholestenol was also the principal sterol in *Chiton tuberculatus* L. Takagi and Toyama (1956) studied the chitons *Acanthochiton rubrolineatus, Cryptoplax japonica,* and *Liolophura japonica.* Lipid values obtained for *Acanthochiton* and *Cryptoplax* were comparable with those of *Liolophura* recorded in Table I. In all cases, only Δ^7-cholestenol was found, while the presence of $\Delta^{5,7}$-sterols could not be demonstrated. Sterols of *A. rubrolineatus, C. japonica,* and *L. japonica* made up 43.8, 42.9, and 51.5% of the unsaponifiable matter, respectively. Kita and Toyama (1959) also studied the sterols of several chitons, using fresh instead of dried animals. There were some obvious differences; the percentages of nonsaponifiable lipids and sterols were much lower than those obtained formerly. These values are given in Table II. The most important difference, however, is the presence of considerable amounts of $\Delta^{5,7}$ sterols.

In *Lepidochiton* sp., five sterols could be observed (Voogt, 1969a), of which $\Delta^{5,7,22}$-cholestatrienol was only present in trace amount. The other sterols were identified as cholesterol (7.7%), $\Delta^{7,22}$-cholestadienol (2.2%),

TABLE II
UNSAPONIFIABLE LIPIDS AND STEROLS IN SOME CHITONS

Species	Unsaponifiable matter[a]	Sterols[b]	$\Delta^{5,7}$ sterols
Liolophura japonica	14.19	27.36	3.5
Onithochiton hirasei	13.74	60.57	4.7
Cryptoplax japonica	22.45	29.44	12.5

[a] Percent of total lipids.
[b] Percent of unsaponifiable matter.

Δ^7-cholestenol (88.6%), and $\Delta^{5,7}$-cholestadienol (1.5%). These results are rather different from those given above for other chitons in that cholesterol is the main sterol but one.

2. *Fatty Acids*

Takagi and Toyama (1956) reported that the polyunsaturated fatty acids of *Liolophura japonica* ranged from diunsaturated to hexaunsaturated, with different members in every class. The ratio of di- and triunsaturated fatty acids to tetra- and pentaunsaturated ones was found to be larger than for other aquatic animals.

3. *α-Glyceryl Ethers*

A small amount of saturated α-glyceryl ethers, possibly chimyl and batyl alcohol was found in the unsaponifiable matter of *Liolophura japonica* (Toyama and Tanaka, 1953). Thompson and Lee (1965) studied the alkyl glyceryl ethers in *Katherina tunicata* and found that these substances were present both in the triglycerides (4.1%) and in the phospholipids (7.9 weight percent or 24.5 avery mole percent). The neutral lipids contained 15:0 (4%)—16:0 (chimyl alcohol, 72%), 18:0 (batyl alcohol, 8%), and 18:1 (selachyl alcohol, 16%)—alkyl glyceryl ethers, while these values in the phospholipids were 8, 75, 6, and 10%, respectively. Besides these, the 14:0 alkyl glyceryl ether (2%) was present in the phospholipids. In comparison with the representatives of the other classes of mollusks investigated in this study, the glyceryl ether composition of *K. tunicata* was very simple as only four or five different side chains were present. It may be interesting that the content of selachyl alcohol in this chiton far exceeded that in the animals of the other classes.

B. Gastropoda

1. *Prosobranchia*

a. Archeogastropoda. Unlike Ampineura, which may have rather high lipid contents in several organs and generally a low glycogen level, *Archeogastropoda* store large amounts of glycogen (Giese, 1966).

i. Sterols. Most investigations of archeogastropods are concerned with sterols. However, most of them are old and give little information about the composition of sterols present in the animals investigated. Studying the sterols of *Haliotis gigantea,* Tsujimoto and Koyanagi (1934a,b) found the sterols (m.p. 147°C) to consist mainly of cholesterol and some sterols of the pelecypods. In a study by Toyama *et al.* (1955b), *Haliotis* was separated into flesh and viscera, and percentages of unsaponifiable matter (39.09, 6.19) sterols (52.8, 42.82) and provitamin D ($\Delta^{5,7}$ sterols, 0.21, 0.34), were determined.

Of the family Patellidae, the genus *Cellana* has been investigated several times. Tsujimoto and Koyanagi (1934b) reported that the sterols of *Cellana nigrolineata* melted at 144–145°C and contained some conchasterol, which is identical with 24-methylenecholesterol. Toyama *et al.* (1955a) studied *Cellana toreuma* and *Cellana nigrolineata* and reported the presence of about 24% unsaponifiable matter, melting points of the sterols of 142°C and 127°C, respectively, and the absence of $\Delta^{5,7}$ sterols. Kita and Toyama (1959) reinvestigated *Cellana toreuma,* using fresh instead of dried animals. The nonsaponifiable matter and sterols amounted to 17.38 and 52.38%, respectively. The sterols melted at 137–140°C and contained 0.7% $\Delta^{5,7}$ sterols. Moreover, from the ultraviolet absorption curve, they concluded the presence of some 24-methylenecholesterol.

The sterols of *Patella coerulea* (L.) were investigated by Voogt (1968b, 1971b) and consisted of cholesterol (69.5%), desmosterol (18%), 24-methylenecholesterol (2.5%), stigmasterol (0.5%), β-sitosterol (2.0%), and fucosterol (7.5%).

Three genera of the family Trochidae have been studied. Tsujimoto and Koyanagi (1935) reported that the sterols of *Tegula xanthostigma* consisted mainly of cholesterol and that probably some conchasterol (24-methylenecholesterol) was present. Toyama and Tanaka (1956a) came to the conclusion that cholesterol was the principal sterol in *Tegula argyrostoma sublaevis* with about 1% $\Delta^{5,7}$ sterols. Higher contents of $\Delta^{5,7}$ sterols were found by Kita and Toyama (1959) in *Tegula argyrostoma basilirata* (1.8%) and *Tegula pfeifferi* (2.1%). *T. pfeifferi* might contain some 24-methylenecholesterol. The sterols of *Calliostoma unicum* (Kita and Toyama, 1959) contained 3.9% $\Delta^{5,7}$ sterols with some 24-methylenecholesterol. According to Toyama *et al.* (1955b) and to Kita and Toyama (1959) the sterols of *Monodonta labio* consist substantially of cholesterol with 1.8% $\Delta^{5,7}$ sterols and again some 24-methylenecholesterol. *Monodonta turbinata* (Born) was studied by Voogt (1968b, 1971b). Its sterols consist in $\Delta^{5,7,22}$-cholestatrienol (2%), cholesterol (80%), desmosterol (8%), 7-dehydrocholesterol (2.5%), 24-methylenecholesterol (2.7%), stigmasterol (0.4%), β-sitosterol (1.7%), and fucosterol (2.3%).

The sterols of *Lunella coronata coreensis* and *Turbo cornutus*, belonging to the family Turbinidae, contained mainly cholesterol and about 1% $\Delta^{5,7}$ sterols (Toyama *et al.*, 1955b; Kita and Toyama, 1959). Tsujimoto and Koyanagi (1934a) reported the presence of some conchasterol besides cholesterol in *Turbo cornutus.* Of the genus *Nerita* (Neritidae), three species have been studied. In *N. peleronta* only cholesterol could be detected (Kind *et al.*, 1948). Toyama *et al.* (1955b) showed the presence of 0.11% $\Delta^{5,7}$ sterols in *N. albicilla.* Kita and Toyama (1959) using fresh animals found 5.3% $\Delta^{5,7}$ sterols in *N. japonica.*

In conclusion, one can say that the older investigations agree in that the sterols of Archeogastropoda consist exclusively or substantially of cholesterol, that $\Delta^{5,7}$ sterols are present in amounts of about 1%, and that in some cases 24-methylenecholesterol is present or suspected. Recent investigations have shown that cholesterol is indeed the principal sterol, but that other sterols may make up 20–30%. This indicates that the older data may be of little value.

ii. Fatty acids. Fatty acids in Prosobranchia have been little investigated. Analyses without gas–liquid chromatography give but scant information. Toyama *et al.* (1955b) studied the fatty acids of *Lunella coronata* and found di- (6.4%), tri- (5.0%), tetra- (11.0%), penta- (8.5%), and hexaunsaturated (2.5%) fatty acids. For *Monodonta labio* (same authors) these values were 6.0, 0.2, 8.6, 11.0, and 3.4%, respectively. Striking are the low values of the hexene acids in both animals and the low content of diunsaturated fatty acids in *M. labio.*

Shimma and Taguchi (1964) studied the fatty acids of *Haliotis discus, Haliotis japonica,* and *Turbo cornutus,* while Bannatyne and Thomas (1969) studied those of *Haliotis iris.* Fatty acids of *Patella coerulea* have been reported by Tibaldi (1966). The results of these investigations are summarized in Table III. This table shows good agreement in the fatty acid composition of the *Haliotis* species, apart from slight differences in that of *Turbo;* that of *Patella* is somewhat different, however, particularly with regard to 14:1, 20:2, which are present in *Patella,* and to 22:4 and 22:5, which are absent in this animal. This means that in *Patella* the fatty acids characteristic for marine animals, namely

TABLE III
FATTY ACIDS OF SOME ARCHEOGASTROPODA

Fatty acid (short hand designation)	Content (%)				
	Haliotis iris	*Haliotis discus*	*Haliotis japonica*	*Turbo cornutus*	*Patella coerulea*
14:0	5.1	3.9	4.9	1.4	1.6
14:1	0.3	—	—	—	6.9
15:0	0.7	2.9	3.2	4.9	—
16:0	22.8	20.9	19.8	13.6	16.3
16:1	1.8	3.3	4.4	—	1.7
16:2	0.2	—	—	—	—
17:0	1.0	—	Trace	—	—
18:0	6.7	5.1	3.9	4.9	1.4
18:1	15.7	16.4	17.1	6.6	19.4
18:2	0.5	1.2	1.6	3.5	—
18:3	1.2	—	—	—	—
18:4	—	—	0.9	Trace	—
20:1	3.7	4.8[a]	5.9[a]	3.3[a]	7.3
20:2	—	—	—	—	24.2
20:4	13.4	10.7[b]	12.3[b]	16.4[b]	—
20:5	8.0	8.8	10.0	6.7	4.4
22:1	5.3	—	—	—	—
22:2	—	5.5[c]	3.9[c]	7.3[c]	8.3
22:4	3.2	2.0[c]	2.6[c]	9.5[c]	—
22:5	10.4	7.3	8.4	8.7	—
22:6	—	—	—	—	1.5
24:4(?)	—	3.5	1.1	1.6	—

[a] Combined with 18:3.
[b] Combined with 22:1.
[c] Identification not sure.

20:5 and 22:6, are present only in small amounts. The shift in *Haliotis* and *Turbo* from 20:5 and 22:6 to 20:4 and 22:5, respectively, is obvious. The absence of 22:6 in these species is striking.

iii. Phospholipids. The discovery of a sphingolipid containing 2-aminoethylphosphonic acid (carbon–phosphorus bonded) by Rouser *et al.* (1963) has roused great interest in the study of phospholipids. Apart from those of Etienne-Petitfils (1956) and of Etienne and Kahane (1958) of the phosphatidylcholines of *Patella vulgata,* the other investigations are very recent. Vaskovsky and Kostetsky (1969) showed that phospholipids made up 27.2% of the lipids in *Tegula rustica.* By thin-layer chromatography, six phospholipids were detected, including phosphatidylethanolamine, phosphatidylcholine, and sphingomyelin.

De Koning (1966a,b) analyzed the phospholipids of the abalone *Haliotis midae* and reported many details. Phospholipids made up 70% of total lipids. Very interesting was the high plasmalogen content (23%). Most plasmalogens were of the ethanolamine type, the remainder was present in phosphatidylcholine and cardiolipin. The phospholipids consisted of 32% phosphatidylethanolamine and ethanolamine plasmalogen, 41% phosphatidylcholine, 5% phosphatidylserine, 5% phosphatidylinositol, 1% sphingomyelin with some lysophosphatidylcholine, and 6% ceramide-2-aminoethyl phosphonate (CAEP), leaving 10% unaccounted for. De Koning has also studied the fatty acid distribution in phospholipids, nonphosphorylated lipids, and CAEP of *H. midae*. His results are given in Table IV.

The fatty acid distribution in phospholipids and nonphosphorylated lipids were similar and in good agreement with those in Table III. As can be seen from Table IV, phospholipids contained more 20:4 and

TABLE IV

FATTY ACID DISTRIBUTION IN PHOSPHOLIPIDS, NONPHOSPHORYLATED LIPIDS, AND CERAMIDE-2-AMINOETHYL PHOSPHONATE OF *Haliotis midae*[a]

Fatty acids	Phospholipids	Nonphosphorylated lipids	Ceramide-2-aminoethyl phosphonate
14:0 iso	1	2	—
14:0	5	6	—
15:0	1	1	—
16:0	22	25	53
16:1	4	11	4
17:0	1	—	4
18:0	8	8	15
18:1	17	22	1
18:2 } 19:0 }	1	3	—
18:3	1	—	—
20:0	—	—	3
20:1	4	3	2
20:2	—	1	—
20:4	13	4	5
22:1(?)	7	3	—
20:5	5	6	11
22:4	—	1	—
22:5	4	2	—
22:6	9	1	—

[a] As percent of total fatty acid methyl esters.

22:6, but less 16:1 and 18:1 than nonphosphorylated lipids. Fatty acid composition in CAEP is quite different. The results suggest a preference for saturated fatty acids, as these make up 75% with 68% for 16:0 and 18:0.

Hori *et al.* (1967b) reported the presence of CAEP in *Turbo cornutus, Tegula argyrostoma, Monodonta labio, Tegula pfeifferi, Cellana eucosmia, Haliotis gurneri,* and *Stomatella lyrata.* In all cases, ceramide-aminoethyl phosphate (sphingoethanolamine) was absent. This might be characteristic. However, in another paper (Hori *et al.*, 1968c), its presence in *Monodonta labio* was reported. In addition, ceramide-2-monomethylaminoethyl phosphonate was found in *Monodonta labio* (Hori and Arakawa, 1969). In an interesting paper, Hori *et al.* (1967a) suggest that sphingomyelin and CAEP might replace each other.

In a preliminary note, Hayashi *et al.* (1969a) reported the isolation and characterization of a new sphingolipid from the viscera of *Turbo cornutus.* The sphingosine moiety turned out to contain sixteen carbon atoms instead of eighteen. The nitrogen base was the *N*-monomethyl derivative of 2-aminoethylphosphonic acid. Palmitic acid (16:0) made up 89.4% of total fatty acids in this phospholipid. Thus, the structure of this new sphingolipid seems to be *N*-palmitoylsphingosyl(C_{16})-2-*N*-methylaminoethyl phosphonate. The glycerophospholipids (Hayashi *et al.*, 1969b) were phosphatidylethanolamine and phosphatidylcholine. Besides the sphingolipid reported in the preliminary note, a second sphingolipid was present differing from the first one in the fatty acid moiety.

iv. Lipids containing ether bonds. Hardly anything is known about this group of lipids in Archeogastropoda. In his study of *Haliotis midae,* De Koning (1966b) found that the phosphatidylcholine fraction consisted of 92% diester phospholipids. In the remaining material (8%), some glyceryl ethers might have been present. Karnovsky *et al.*, (1946) reported that the unsaponifiable matter of *Patella granularis, Haliotis midae,* and *Turbo sarmaticus* contained 1.2%, 7.7%, and 1.4% glyceryl ethers, respectively.

v. Glycolipids. Hayashi *et al.* (1969b) have isolated a glycolipid from *Turbo cornutus* which contained sphingosine bases, fatty acids (palmitic acid and the same uncharacterized acid as in the second sphingolipid), and the sugars glucose, galactose, fucose, and glucosamine.

b. Mesogastropoda.

i. Sterols. Sterols in mesogastropods have been studied rather extensively. Of the family Viviparidae, several species have been investigated. Toyama *et al.* (1955a) reported as melting point of the sterols of *Viviparus histricus* 132–140°C and a content of $\Delta^{5,7}$ sterols as high as 19.3%.

Matsumoto and Tamura (1955a) studied *Viviparus japonicus*. The sterols in this experiment melted at 120–125°C and consisted of brassicasterol, poriferasterol, cholesterol and γ-sitosterol, while ultraviolet spectrometry showed the presence of $\Delta^{5,7}$ sterols, which, however, could not be isolated. Voogt (1969b, 1971b) analyzed the sterols of *Viviparus fasciatus* (Müller). The results are given in Table V. The sterols mentioned by Matsumoto and Tamura are also found in *V. fasciatus*, as it is nearly impossible to distinguish poriferasterol from its epimer stigmasterol and γ-sitosterol from β-sitosterol. In the reports on *Cipangopaludina japonica* and *Cipangopaludina malleata*, melting points and $\Delta^{5,7}$ content of the sterols of flesh and of viscera are given (Toyama *et al.*, 1955a).

Of the family Littorinidae, *Littorina littorea* was studied by Kind and Herman (1948). They found that the unsaponifiable matter contained 61% sterols, of which 85% consisted of cholesterol. Further clionasterol and 7-dehydroclionasterol were present. Tanaka and Toyama (1959b) found that cholesterol was the principal component of the sterols of *Littorina brevicula* and that $\Delta^{5,7}$ sterols made up only 2.2%. *L. littorea* was also studied by Voogt (1969b, 1971b). The results are given in Table V. As nine components were identified, these results are strongly different from those mentioned above.

Crepidula fornicata (family Calyptraeidae) was studied by Voogt

TABLE V

STEROL COMPOSITION OF SOME MESOGASTROPODS[a]

Sterol	*Viviparus fasciatus*	*Littorina littorea*	*Crepidula fornicata*	*Natica cataena*
C_{26} sterols	0.3	Trace	3.2	3.4
Δ^{22}-Cholestenol, $\Delta^{5,22}$-Cholestadienol	1	Trace	—	1
$\Delta^{5,7,22}$-Cholestatrienol	2	Trace	7	9.5
Cholesterol	72	67	58	52
$\Delta^{5,7}$-Cholestadienol	2	—	—	8
Desmosterol	—	15	—	—
Brassicasterol	3	4.5	4	12
Unknown (C_{28})	—	—	9.5	—
Campesterol	4.5	4	1.5	—
24-Methylenecholesterol	2	4	8	7.5
Stigmasterol	2.5	—	2.5	2
β-Sitosterol	10	3	3	4
Fucosterol	1	2	—	—
Unknown (C_{29})	—	—	2	—

[a] Percent of total sterols.

(1971c). The sterols of this species are summarized in Table V. Of the family Naticidae, *Neverita didyma* was studied twice by Toyama *et al.* (1955a). The unsaponifiable matter contained 32% sterols of which 3.7% possessed the $\Delta^{5,7}$ structure. The sterols of *Natica cataena* have been analyzed by Voogt (1971c) and are listed in Table V.

Tonna luteostoma (superfamily Tonnacea) has been the subject of several studies. Toyama *et al.* (1955b) found that the unsaponifiable matter contained 43% sterols, which consisted nearly exclusively of cholesterol. The $\Delta^{5,7}$ sterol (making up 8.6%) could not be identified. The melting point of the sterols of *Tonna tessellata* (Toyama and Tanaka, 1956b) was higher than that of the sterols of *T. luteostoma.* The content of $\Delta^{5,7}$ sterols was only 1%. In the latter study, similar results are given for *Ficus subintermedius* and *Apollon perca.* Tanaka and Toyama (1957a) have restudied the $\Delta^{5,7}$ sterol of *T. luteostoma,* which was tentatively identified as $\Delta^{5,7,22,25}$ cholestatetraenol. In 1958, the same authors (Tanaka and Toyama, 1958) mentioned that the sterols contained cholesterol and clionasterol besides the $\Delta^{5,7}$ sterol.

ii. Fatty acids. Little is known about the fatty acid distribution in mesogastropods. Tanaka and Toyama (1959b) have determined the overall composition of the unsaturated fatty acids of *Littorina brevicula* and calculated values for di- (11.75%), tri- (13.4%), tetra- (9.07%), penta- (6.43%), and hexaunsaturated fatty acids (2.05%). These values were also determined for *Tonna luteostoma* (Tanaka and Toyama, 1957) and were 5.6, 10.3, 9.0, 7.4, and 3.0% respectively. Tibaldi (1966) analyzed the fatty acids of *Natica josephina.* They are listed in Table VI. The fatty acid distribution resembles that of the archeogastropods in Table

TABLE VI
FATTY ACIDS OF *Natica josephina*

Fatty acid	Content (%)
14:0	0.8
14:1	0.4
16:0	14.9
16:1	1.6
18:0	1.3
18:1	16.4
20:1	16.2
20:2	16.9
20:5	11.4
22:2	7.3
22:6	6.2

III. Though higher in content than in the archeogastropods, 20:5 and 22:6 are not yet the predominant unsaturated fatty acids.

iii. Phospholipids. Phospholipids consisted of seven components in *Littorina brevicula* and *Littorina squalida* and made up 32.3% and 32.8%, respectively, of total lipids (Vaskovsky and Kostetsky, 1969). Among these components were phosphatidylcholine, phosphatidylethanolamine, and sphingomyelin.

Hori *et al.* (1965) discovered in *Heterogen longispira* a previously unknown phosphosphingoside which they identified as ceramideaminoethyl phosphate (*N*-acylsphingosylphosphorylethanolamine). In 1966 (Hori *et al.*, 1966a), they confirmed this structure and reported the presence of this phospholipid in *Sinotaia histrica.* Analysis of fatty acids present in this phospholipid showed that 16:0 made up 81%. Other fatty acids present were 18:0 (9.3%), 17:0 (7.3%) and 15:0? (1.2%). Further distribution studies of this phospholipid were undertaken, and in 1967 (Hori *et al.*, 1967b) they reported the presence of this substance in *Batillaria multiformis* and *Semisulcospira bensoni.* Thus, in Archeogastropoda investigated, only CAEP is found and ceramide-2-aminoethylphosphate is lacking; in Mesogastropoda investigated both substances are present.

In *Crepidula plana,* Hack *et al.* (1962) found phosphatidylcholine (lecithin), monophosphoinositide, and a small quantity of polyglycerolphosphatide (cardiolipin). Of the lecithin, only a small part occurred as plasmalogen. The most important plasmalogen was the ethanolamine analog. Interesting is the presence of a plasmalogen with serine. Another report about α-glyceryl ethers in mesogastropods is that of Tanaka and Toyama (1958) who obtained batyl alcohol and selachyl alcohol from *Tonna luteostoma.* Karnovsky *et al.* (1946) found 3.6% glyceryl ethers in the unsaponifiable matter of *Oxystele tigrina.*

c. Neogastropoda.

i. Sterols. Several representatives of the family Muricidae have been investigated for their sterol content. One of the oldest reports is concerned with *Rapana thomasiana* (Tsujimoto and Koyanagi, 1934b). The sterols of this animal should contain, besides cholesterol, conchasterol (24-methylenecholesterol). An additional datum about this animal is provided by Toyama *et al.* (1955b). Sterols made up 32.95% of the unsaponifiable matter and contained 1.6% provitamin D.

Of *Chicoreus sinensis,* it is given that the unsaponifiable matter contained 36.41% sterols, of which 3.4% was provitamin D (Toyama and Tanaka, 1956b). These values were 60.33 and 10.0%, respectively, for *Chicoreus asianus* (Kita and Toyama, 1959).

The sterols of *Murex brandaris* have been analyzed by Voogt (1971d). The results are given in Table VII. Toyama *et al.* (1955a) reported

TABLE VII
STEROLS OF *Murex brandaris*

Sterol	Content (%)
C_{26} sterol	0.9
$\Delta^{5,7,22}$-Cholestatrienol	1.0
Cholesterol	58.8
$\Delta^{5,7}$-Cholestadienol	2.2
24α-Methyl-Δ^{22}-Cholesterol	4.4
Unknown	3.3
Campesterol	3.8
24-Methylenecholesterol	2.4
Stigmasterol	2.8
β-Sitosterol	16.0
24α-Ethyl-$\Delta^{5,7}$-cholestadienol	2.0
C_{30} sterols	2.5

for *Thais bronni* and *Thais clavigera* $\Delta^{5,7}$ sterol contents of 1.2 and 2.3%, respectively. Studying fresh instead of dried animals, Kita and Toyama (1959) found that these values were 14.1 and 17.3%, respectively, while the unsaponifiable matter yielded 35.69 and 52.33% sterols. Kind and Goldberg (1953) reported that the sterols of *Urosalpinx cinereus* consisted of 90% cholesterol and 10% of a not identified $\Delta^{5,7}$ sterol.

The family Buccinidae has been little investigated. Toyama *et al.* (1955a) found that the sterols of *Babylonia japonica* obtained from the viscera contained 4% $\Delta^{5,7}$ sterols, but that $\Delta^{5,7}$ sterols were absent in the sterols of the flesh. Dorée (1909) reported the presence of cholesterol in *Buccinum undatum*, Bock and Wetter (1938) found in *Buccinum*, besides cholesterol, 27.5% 7-dehydrocholesterol. The sterols of *Buccinum perryi* (Toyama and Tanaka, 1956a) consisted chiefly of cholesterol.

The sterols of *Hemifusus ternatanus*, belonging to the family Busyconidae, made up 38.40% of the unsaponifiable matter and contained 7.5% $\Delta^{5,7}$ sterols (Toyama *et al.*, 1955b). Cholesterol was the main sterol, but a considerable amount of diunsaturated sterols was present.

The sterols of *Nassa obsoleta* (family Nassidae) were investigated by Kind *et al.* (1948). The principal sterol was cholesterol, but clionasterol might be present as second sterol. In 1953, Kind and Goldberg reported that the presence of clionasterol had been clearly shown.

Of *Fusinus perplexus* (Fasciolariidae), the only thing known is that its sterols (33.53% of unsaponifiable matter) contain 2.3% $\Delta^{5,7}$ sterols. Finally, two representatives of the Toxoglossa have been studied, but data are again very fragmentary. The unsaponifiable matter of *Turris unedo* and *Asprella sieboldi* contained, respectively, 55.92% and 43.63% sterols (Toyama and Tanaka, 1956b). The $\Delta^{5,7}$ sterol content was 0.56% and 0.85%, respectively.

In conclusion, one can say that the older investigations have yielded only few details about the occurrence and distribution of sterols. Only cholesterol was reported repeatedly, and in some cases clionasterol was mentioned. The $\Delta^{5,7}$ sterol content was determined many times, but the identity of $\Delta^{5,7}$ sterols was nearly always unknown. Comparison of the data raises the suspicion that several data after all are of little value, as the results depend strongly on the method of animal processing. Recent investigations have yielded more details about the sterols present in Prosobranchia and have shown that in general complex mixtures are encountered. Perhaps there are some tendencies going from Archeogastropoda via Mesogastropoda to Neogastropoda. Thus, it might be concluded that $\Delta^{5,7}$ sterol content in Archeogastropoda is very low, high in Mesogastropoda and again low in Neogastropoda (excepting the report on *Buccinum*). This is also suggested in the list given by Bergmann (1962). In an attempt to show such tendencies Voogt (1970) has determined the relative composition according to carbon content of the sterols present in some Prosobranchia. The results are summarized in Table VIII. From this table, it might be concluded that C_{27} sterols are predominant throughout the subclass Prosobranchia, especially in the order Archeogastropoda. In mesogastropods, there is an increase in sterols of all other types at the cost of C_{27} sterols. It should be

TABLE VIII

RELATIVE COMPOSITION ACCORDING TO CARBON CONTENT OF THE STEROLS IN SOME PROSOBRANCHIA

	Content (%)				
Species	C_{26}	C_{27}	C_{28}	C_{29}	C_{30}
Patella coerulea	0.2	92.6	2.6	4.6	—
Monodonta turbinata	0.4	94.6	2.8	2.1	—
Littorina littorea	1.0	83.7	10.6	5.2	—
Viviparus fasciatus	0.7	78.1	9.9	11.4	—
Crepidula fornicata	3.0	63.7	24.9	8.3	—
Natica cataena	3.4	71.9	19.5	5.3	—
Neptunea antiqua	4.4	85.1	9.2	1.1	0.1
Murex brandaris	0.4	61.6	15.4	20.3	2.4
Purpura lapillus	1.3	89.5	7.9	1.2	—

mentioned here that *Littorina* and *Viviparus* have retained the feeding habit of the Archeogastropoda but that *Crepidula* has adopted the feeding habit of Bivalvia and that *Natica* feeds on Bivalvia. This is reflected in the results listed. Thus, in the mesogastropods, *Littorina* and *Viviparus* can readily be compared with the archeogastropods. *Crepidula* and *Natica* present strong resemblance, perhaps reflecting dietary influences. With the exception of *Murex*, C_{27} sterols are higher again in neogastropods (*Neptunea* and *Purpura* feed on bivalves), but C_{29} sterols are significantly lower. Although conclusion may be premature, one gets the impression that each group possesses a characteristic pattern. To make certain, more representatives should be investigated and difference should be made between the sterols that are of endogenous or exogenous origin.

ii. Phospholipids. Zama (1963) studied the lipid composition in muscle and liver of *Neptunea arthritica.* Lipids were divided into neutral fats and conjugated lipids, and the latter were divided into total phospholipids and cerebrosides (glycolipids). The composition of the phospholipids was determined and is given in Table IX. Vaskovsky and Kostetsky (1969) have found that the phospholipids of *Nucella heuseana* and *Rapana thomasiana* make up 32.2% and 39.0%, respectively, of the total lipids and consist of six components, which include phosphatidylcholine, phosphatidylethanolamine, and sphingomyelin. Quin (1965) reported that *Busycon canaliculatum* contained a considerable amount of lipids with carbon-bonded phosphorus. Per 100 gm. of dry tissue 215 mg. phosphorus was present. These lipids were not further investigated, however.

TABLE IX
LIPID COMPOSITION OF *Neptunea arthritica*

Lipid	Content (%)	
	Muscle	Liver
Neutral fat	59.7	64.4
Total conjugated lipids	40.3	35.6
Cerebrosides	52.3	24.6
Total phospholipids	47.7	75.4
Phosphatidylcholine	43.7	31.3
Phosphatidylethanolamine	21.3	32.8
Phosphatidylserine	7.5	10.4
Acetal lipid	2.3	1.7
Sphingomyelin	25.7	23.9

In their paper on the distribution of ceramide-2-aminoethyl phosphonate, Hori *et al.* (1967a) report the presence of this substance in *Purpura bronni* and *Pugilina ternatana*. Sphingomyelin is present in the former but absent in the latter animal. In another paper (Hori *et al.*, 1967b), these authors mention the absence of ceramideaminoethyl phosphate in these two animals. A retrospect on the subclass Prosobranchia shows that CAEP is encountered in all orders, but that up till now ceramideaminoethyl phosphate has been found only in the order Mesogastropoda. This is an interesting fact, which deserves confirmation as in this case archeogastropods and neogastropods should show again the resemblance mentioned already for sterols.

iii. Lipids with ether bonds. Rapport and Alonzo (1960) have studied this type of lipids in *Busycon canaliculatum* by determining the aldehydogenic and α,β-unsaturated ether content. The aldehydes made up 147.0 μmole per gram of lipid, which equals 10%, while the α,β-unsaturated ethers made up 118.0 μmole per gram of lipid. That means that 80% of the aldehydes was α,β-unsaturated. The remaining 20% have not been investigated, but this percentage might be saturated glyceryl ethers, while the 80% represents the plasmalogens.

Lipids of *Thais lamellosa* consisted of 62.3% neutral fats and 37.7% phospholipids (Thompson and Lee, 1965). The glyceryl ether content in these fractions was 1.9 and 7.0%, respectively. The ether linked side chains have been determined, and in neutral fats they consisted of 14:0 (11%), 15:0 (4%), 16:0 (chimyl alcohol, 49%), 16:1 (5%), 18:0 (batyl alcohol, 21%), and 19: branched (10%); in phospholipids they consisted of 14:0 (13%), 15:branched (1%), 15:0 (5%), 16:0 (37%), 17:0 (2%), 18:0 (12%), 20:0 (13%), and 19:unsaturated (?) (16%).

2. *Opisthobranchia*

Little is known about lipid distribution in this subclass.

a. Pleurocoela.

i. Sterols. Toyama and Tanaka (1956b) observed that the sterols of *Philine japonica*, making up 29.20% of unsaponifiable matter, did not contain $\Delta^{5,7}$ sterols, but exhibited in ultraviolet spectrometry an absorption maximum at 277–278 mμ. Sterols could not be identified, however. *Aplysia kurodai*, belonging to the same order, was studied by Tanaka and Toyama (1959a). The nonconjugated sterols consisted mainly of cholesterol, and the conjugated sterol showed resemblance or was identical with the $\Delta^{5,7}$ sterol isolated from *Tonna luteostoma* (Tanaka and Toyama, 1957a) and which was identified by these authors as $\Delta^{5,7,22,25}$-cholestatetraenol.

ii. Fatty Acids. Fatty acids of *Aplysia kurodai* contained only 22.9%

saturated acids (Tanaka and Toyama, 1959a). The unsaturated fatty acids (77.1%) consisted of mono- (35.1%), di- (7.6%), tri- (0.74%), tetra- (11.7%), penta- (19.6%), and hexaunsaturated (2.4%) acids. The composition reported by Tibaldi (1966) for *Aplysia fasciata* is rather different. The following fatty acids were found: 14:0 (3.4%), 14:1 (1.5%), 16:0 (23.1%), 16:1 (2.4%), 18:0 (2.2%), 18:1 (15.6%), 20:1 (7.6%), 20:2 (15.3%), 20:5 (6.0%), 22:2 (15.3%), and 22:6 (4.1%).

iii. Phospholipids. Roots and Johnston (1965) analyzed the phospholipids in the ganglia of *Aplysia punctata.* Three types of phospholipids were observed—phosphatidylserine, phosphatidylethanolamine, and a small fraction behaving as phosphoinositide. Of the first two, which occurred mainly in plasmalogen form, phosphatidylserine was predominant.

b. Pteropoda. Our knowledge of this order is confined to one report. Baalsrud (1950) studied the sterols of *Limacina helicina* and *Clione limacina.* The unsaponifiable matter contained 62.1 and 40.8% sterols, respectively. In these sterols, no cholesterol could be traced. Attempts to fractionate the sterols were unsuccessful. This author proposed the name pteropodasterol for this sterol, which possessed twenty-nine carbon atoms and contained two double bonds (one in 5,6 position and one in the side chain). The exact structure is not yet known. Beside this pteropodasterol, some $\Delta^{5,7}$ sterol was present.

c. Acoela.

i. Sterols. Dendrodoris rubra possessed sterols that melted at 121–123°C and contained 14.6% $\Delta^{5,7}$ sterols (Tanaka and Toyama, 1959b). The sterols of *Actinocyclus japonicus* (34.44% of unsaponifiable matter) did not contain $\Delta^{5,7}$ sterols, but showed a single absorption maximum at 277–279 mμ. Sterols could not be characterized (Toyama and Tanaka, 1956b). Voogt (1970) determined the relative sterol composition according to carbon content in *Lamellidoris bilamellata, Archidoris tuberculata,* and *Dendronotus frondosus.* The results are given in Table X.

TABLE X
RELATIVE COMPOSITION ACCORDING TO CARBON CONTENT OF THE STEROLS IN SOME ACOELA

	Carbon content (%)				
Genus	C_{26}	C_{27}	C_{28}	C_{29}	C_{30}
Lamellidoris	2.0	72.6	25.4	—	
Archidoris	3.0	80.4	14.3	2.3	
Dendronotus	4.3	81.3	12.1	2.7	

This table shows that in Acoela, C_{27} sterols are predominant, followed by C_{28} sterols.

ii. Fatty acids. Tibaldi (1966) has studied the fatty acids of *Pleurobranchaea meckeli.* These acids consisted of 14:0 (1.7%), 14:1 (0.6%), 16:0 (25.0%), 16:1 (1.2%), 18:0 (1.4%), 18:1 (7.2%), 20:1 (8.3%), 20:2 (19.5%), 20:5 (6.3%), 22:2 (19.5%), and 22:6 (4.0%). This composition resembles that of *Aplysia fasciata* in that 16:0, 20:2, and 22:2 are the most important acids. In both cases, 20:5 and 22:6 content are rather low.

3. *Pulmonata*

a. Basommatophora. Little is known about lipid distribution in this order.

i. Sterols. Voogt (1970) has found that sterols of *Planorbarius corneus* consist of 89.4% C_{27}, 6.8% C_{28}, and 3.9% C_{29} sterols, but from this single datum no conclusions can be drawn of course.

ii. Phospholipids. Liang and Strickland (1969), in studying the phospholipids of *Lymnaea stagnalis,* observed the following types in the amounts indicated: phosphatidylserine 6%, phosphatidylethanolamine 30.7%, CAEP 7.5%, phosphatidylcholine 9.8%, phosphoinositide 3.6%, and two unidentified types X_1 and X_2 making up 2.2% and 0.3%, respectively. Thus, the phospholipids containing choline and ethanolamine made up about 80% of the total amount. These two types have been investigated further. The choline-containing type (49.8%) could be divided into diacylphosphatidylcholine (26.4%), glyceryl ethers (22.2%), and plasmalogens (1.2%). For the ethanolamine-containing type (30.7%), these values were 24.9, 4.7, and 1.0%, respectively. It should be mentioned that in the latter type 3.7% occurred as the phosphonic analog, of which 1% was diacyl-AEP.

b. Stylommatophora.

i. Sterols. Kita and Toyama (1959) studied the sterols of *Onchidium verruculatum* (family Oncidiidae) and found that they made up 44.53% of unsaponifiable matter and contained 1.8% $\Delta^{5,7}$ sterols. Interesting was the low melting point (104–109°C) combined with a specific rotation $[\alpha]_D° = +3.9$. These two data suggest an unusual composition, and specially the latter datum points to Δ^7-sterols. There were indications that 24-methylenecholesterol is present.

Sterols of *Succinea putris* (Voogt and Van der Horst, 1969) were composed of $\Delta^{5,22}$-cholestadienol (1.7%), cholesterol (83.8%), $\Delta^{5,7}$-cholestadienol (5.0%), brassicasterol (trace), campesterol (3.7%), ergosterol (trace), stigmasterol (0.6%), and β-sitosterol (4.4%).

The sterols of *Arion empiricorum* were studied by Bock and Wetter (1938). They found cholesterol to be the main sterol accompanied by 25% ergosterol. Voogt (1971a) studied the sterols of *Arion rufus* and did not find ergosterol. The results of this investigation were $\Delta^{5,22}$-cholestadienol (1%), cholesterol (78%), brassicasterol (6%), desmosterol (6.5%), campesterol (4%), stigmasterol (2%), and β-sitosterol (2%). Thus, sterols consisted of about 86% C_{27}, 9% C_{28}, and 4% C_{29} sterols.

Of the sterols of *Bradybaena similaris,* the only thing known is that they contain 9.4% $\Delta^{5,7}$ sterols (Toyama *et al.,* 1955a). Studies of *Euhadra herklotsi* (Takagi and Toyama, 1958) gave $\Delta^{5,7}$ content varying from 0 to 5.5%. Sterols consisted chiefly of cholesterol. Conversely, the sterols of *Incillaria confusa* (Tanaka and Toyama, 1957b) seem to be composed mainly of β-sitosterol. The sterols of *Arianta arbustorum* consisted of 0.3% $\Delta^{5,22}$-cholestadienol, 89.8% cholesterol, 4.9% $\Delta^{5,7}$-cholestadienol, 2.6% campesterol, 0.3% stigmasterol, and 1.9% β-sitosterol (Voogt and Van der Horst, 1969).

Bock and Wetter (1938), in studying the sterols of *Helix pomatia* with regard to provitamin D content, showed that sterols were composed mainly of cholesterol and ergosterol (9.6%). Thiele (1960) reported that in *Helix pomatia* cholesterol was the main sterol and that only traces of ergosterol were found. The $\Delta^{5,7}$-cholestadienol content varied from 5.8 to 12.3%.

ii. Fatty acids. The fatty acids of *Euhadra herklotsi* (Takagi and Toyama, 1958) contained about 20% saturated acids consisting chiefly of stearic acid. The composition of the unsaturated acids was about 44% mono-, 25% di-, 5% tri-, 3,5% tetra-, and 2% pentaunsaturated fatty acids. Thiele and Kröber (1963), have analyzed the fatty acids of *Helix pomatia,* which are given in Table XI. Van der Horst and Voogt (1969a,b) determined the fatty acid composition of *Arianta arbustorum* and *Succinea putris.* The combined results are given in Table XII. These fatty acids can be divided into several groups, which is shown in Table XIII.

iii. Phospholipids and glyceryl ethers. Thiele (1959) reported that lipids of *Helix pomatia* contained about 45% phospholipids and that the plasmalogen content was rather low, being maximal in March and minimal in June, which is the breeding time. Thompson and Hanahan (1963) studied the phospholipids in two terrestrial slugs, *Arion ater* and *Ariolimax columbianus,* with special reference to α-glyceryl ethers. Lipids of *Arion* contained about 40% phospholipids, which in turn consisted of almost 26 mole percent glyceryl ethers. The distribution of alkyl side chains was remarkable, as chimyl alcohol (16:0) made up 86–89% of the ethers. The other components were 18:0, 15:0, 17:br, and 13:0 (?). The absence of unsaturated side chains is striking. Analysis of the phospholipids of *Ariolimax* yielded 8.6% phosphatidic acid, 31.3% phosphatidylethanolamine, 14.1% phosphatidylinositol and phosphatidylserine, and 46.0% phosphatidylcholine (all percentages are mole percent). The content of glyceryl ethers in these fractions was 22, 13.3, 8.6, and 49 mole percent, respectively. The alkyl side chain of these glyceryl ethers and those in neutral lipids were analyzed and results are given in Table XIV. As in *Arion*, no unsaturated side chains are

TABLE XI
COMPOSITION OF FATTY ACIDS OF *Helix pomatia*[a]

Fatty acid	Free fatty acids (%)	Diglycerides (%)	Triglycerides (%)	Steryl esters (%)
12:0	0.21	0.55	0.90	0.40
13:0	0.04	1.78	0.29	1.11
Z[b]	0	0.90	1.06	3.31
14:0	0.38	1.21	1.63	0.44
15:0	0.26	0.77	0.82	0.40
Y[b]	0.42	1.58	3.68	7.25
16:0	4.65	16.57	21.89	8.11
16:1	0.66	1.06	2.04	0.32
17:0 + 16:2	0.66	1.69	2.37	1.20
X[b]	0.42	4.81	6.42	26.52
18:0	8.62	20.59	15.50	13.54
18:1	8.91	15.08	16.70	8.70
18:2	10.39	4.15	2.49	3.56
18:3	2.46	0.64	0	0.65
18:4	1.79	0	1.94	0.44
20:0	0.36	2.94	1.32	4.52
20:1	5.19	5.55	0.90	3.75
20:2	11.37	1.32	0.80	0.39
20:3	0.42	0	0	0.16
20:4	18.69	1.34	2.21	2.05
20:5	1.00	0	1.83	0.81
22:0	0.84	1.56	0.12	0.44
22:1	0.94	1.82	1.49	1.68
22:2	2.61	0	0	0.45
22:3	3.34	1.15	0	0
22:4	6.59	0	2.05	0
22:5	3.27	3.59	0	1.70
24:0	1.72	0.34	0.93	0.51
Other before 18:0	1.77	3.56	6.47	5.88
Other after 18:0	2.00	5.44	4.14	1.87

[a] Composition is given in percents.
[b] Unidentified

TABLE XII

Fatty Acids of *Arianta arbustorum* and *Succinea putris*

Fatty acid	Content (%)		Fatty acid	Content (%)	
	Arianta	*Succinea*		*Arianta*	*Succinea*
Before 14:0 iso	2.9	—	20:1	8.22	1.81
14:0 iso	0.02	0.05	20:2	2.30	2.65
14:0 anteiso	0.05	0.10	20:3	0.89	2.18
14:0	0.62	1.12	20:4	Trace	Trace
14:1	0.14	0.06	21:0 iso	0.53	0.31
14:2	0.15	0.04	21:0 anteiso	0.38	0.22
14:3	0.57	0.20	21:0	0.29	0.10
15:0 iso	0.54	1.15	21:1	0.62	—
15:0 anteiso	0.18	—	21:2	1.41	—
15:1	0.54	0.26	21:3	1.86	Trace
15:2	0.48	0.16	21:4	Trace	—
15:3	1.18	Trace	22:0 iso	0.14	—
15:4	0.13	—	22:0 anteiso	—	0.12
16:0 iso	0.65	1.24	22:0	0.26	0.13
16:0	10.90	17.96	22:2	0.38	0.22
16:1	1.79	1.96	22:3	3.35	0.59
16:2	0.14	0.12	22:4	Trace	Trace
16:3	0.29	Trace	23:0 iso	0.07	Trace
16:4	Trace	—	23:0 anteiso	—	Trace
17:0 iso	1.38	2.99	23:0	0.13	Trace
17:0 anteiso	1.52	2.33	23:2	1.94	—
17:0	2.07	2.11	23:3	0.39	—
17:1	0.44	0.88	24:0 anteiso	Trace	0.12
17:2	0.35	0.15	24:0	0.28	0.04
17:3	Trace	Trace	24:2	0.26	—
18:0 iso	0.68	0.91	24:3	1.16	Trace
18:0	18.23	16.78	24:4	0.11	—
18:1	15.83	18.01	25:0	Trace	—
18:2	2.61	7.40	25:2	0.09	—
18:3	Trace	2.91	25:3	0.11	—
18:4	—	Trace	26:0 anteiso	0.07	—
19:0 iso	1.79	2.92	26:0	0.04	—
19:0 anteiso	1.39	1.24	26:2	0.26	—
19:0	1.01	0.88	26:3 + 25:4	0.15	—
19:1	0.60	0.53	27:0 anteiso	Trace	—
19:2	Trace	Trace	27:2	0.39	—
19:3	2.85	0.81	27:3	0.39	—
19:4	Trace	Trace	28:0 iso	Trace	—
20:0 iso	0.22	0.36	28:0 anteiso	Trace	—
20:0 anteiso	—	0.03	28:0	0.19	—
20:0	0.77	0.14	28:3	0.28	—

TABLE XIII
CLASSIFICATION OF FATTY ACIDS OF *Arianta arbustorum* AND *Succinea putris*

Type of fatty acid	Content (%)	
	Arianta	*Succinea*
Saturated	45.3	55.5
Unsaturated	54.7	43.8
Straight chain saturated	78.15	73.2
Branched chain saturated	21.50	26.8
Isobranched saturated	13.50	18.3
Anteisobranched saturated	8.00	7.8
Monounsaturated	52.64	58.1
Diunsaturated	19.38	24.6
Triunsaturated	25.36	15.7
Tetraunsaturated	1.70	Trace

TABLE XIV
DISTRIBUTION[a] OF ALKYL SIDE CHAINS IN GLYCERYL ETHERS OF *Ariolimax columbianus*

Side chain	Triglycerides	Phosphatidic acid	Phosphatidyl-ethanolamine	Phospho-inositide; phosphatidyl-serine	Phosphatidyl-choline
15:0	4	<1	1	<1	6
16:0	72	38	32	37	94
17:0 br	12	8	14	5	0
17:0	3	4	8	4	0
18:0	9	38	28	42	0
Unknown	0	0	6	0	0
19:0 br	0	11	11	12	0

[a] Mole percent of total.

present, and again chimyl alcohol is the main glyceryl ether, making up 94% of the total lecithin fraction.

C. Bivalvia

1. *Sterols*

Sterols of numerous representatives of this class have been investigated. However, several of the data obtained are rather incomplete.

Data have been summarized by Toyama (1958) and Bergmann (1962). Papers dealt with here are those which contain more data than are given in any available reviews of them, and papers on investigations that are too recent to be reviewed.

a. Toxodonta.

Family Archidae. Data on *Anadara inflata* are given by Toyama *et al.* (1955a). Kita and Toyama (1959) showed in the sterols of *Barbatia obtusoides* 9.8% $\Delta^{5,7}$ sterols and the presence of 24-methylenecholesterol.

b. Anisomyaria.

i. Family Mytilidae. Bergmann and Ottke (1949) demonstrated the presence of brassicasterol in *Modiolus demissus.* In their study of the provitamin D in *Modiolus,* Rosenberg and Waddell (1951) concluded that it was not identical with 7-dehydrocholesterol, while Petering and Waddell (1951) found that it was different from all provitamins known and called it provitamin D_m. Fagerlund and Idler (1956) found 24-methylenecholesterol in *Modiolus.*

With regard to *Brachidontes senhousia,* it should be remarked that the conjugated sterols (>30% of sterols), besides the indicated $\Delta^{5,7,22}$ or $\Delta^{5,7,23}$ sterol, possibly contained small amounts of a $\Delta^{5,7,9(11)}$ sterol (Takagi and Toyama, 1959).

The sterols of *Mytilus planulatus* were studied by Fantl (1942) and contained greatly varying amounts of cholesterol. In one case, 25% cholesterol was found. Van der Vliet (1946) reported that the sterols of *Mytilus edulis* consisted probably of cholesterol, brassicasterol, $\Delta^{5,22}$-cholestadienol, and perhaps stigmasterol as well as the provitamins D, 7-dehydrocholesterol, ergosterol, and $\Delta^{5,7,22}$-cholestatrienol. Later (Van der Vliet, 1948b), a provitamin designated D_x, was added to this series. Toyama and Tanaka (1956a) reported the occurrence of poriferasterol, clionasterol, β-sitosterol, and cholesterol in *Mytilus edulis.* Voogt (1969a) found that the sterols of *Mytilus edulis* consisted of 9% C_{26}, 44.5% C_{27}, 37% C_{28}, 8.5% C_{29}, and about 0.5% C_{30} sterols. Further analysis yielded 1.2% $\Delta^{5,22}$-cholestadienol, 34% cholesterol, 8% 7-dehydrocholesterol, 4.5% campesterol, 24.5% brassicasterol, 6.8% 24-methylenecholesterol, 3.1% stigmasterol, and 3.8% β-sitosterol.

ii. Family Pinnidae. Pinna pectinata japonica has been studied frequently by Toyama and co-workers. In 1955 (Toyama *et al.*, 1955a, V, VII), data are reported about *P. pectinata,* but these are rather incomplete. Takagi *et al.* (1956) isolated from *P. pectinata japonica* clionasterol and a $\Delta^{5,7}$ sterol, which could not be identified but resembled a $\Delta^{5,7}$ sterol of *Modiolus demissus.* Tanaka and Toyama (1959b) noted that poriferasterol, cholesterol, and clionasterol were found in the same species.

iii. Family Pectinidae. Fagerlund and Idler (1956) reported the presence of 24-methylenecholesterol in *Pecten caurinus.* This sterol and clionasterol were also isolated from *Pecten yessoensis* by Kita and Toyama (1959). 22-Dehydrocholesterol was isolated from the sterols of *Placopecten magellanicus* by Tamura *et al.* (1964a), while application of gas chromatography enabled this working group (Wainai *et al.*, 1964) to identify the other sterols occurring in *Placopecten.* These were composed of 42.4% C_{27}, 41.9% C_{28}, and 9.6% C_{29} sterols, with 6.1% C_{26} and $<1\%$ C_{30} sterols. 22-Dehydrocholesterol made up 14.1% of total sterols. Other sterols identified were cholestanol, cholesterol, 24-methylenecholesterol, and brassicasterol, while Δ^5-ergostenol, β-sitosterol, and fucosterol were tentatively identified. Kritchevsky *et al.* (1967) determined the sterols in an unspecified scallop. They found 4.4% C_{26} sterols, 13.4% 22-dehydrocholesterol, 25.7% cholesterol, 14.1% brassicasterol, 19.5% 24-methylenecholesterol, and some C_{29} sterols.

iv. Family Anomiidae. Sterols of *Anomia simplex* have been analyzed (Kind and Goldberg, 1953) as this animal is not attacked by *Urosalpinx.* Its sterols contained 1% $\Delta^{5,7}$ sterols, a trace of cholesterol, and further a mixture of C_{28} sterols [brassicasterol(?) and chalinasterol(?)].

v. Family Ostreidae. The discovery by Idler and Fagerlund (1955), of 24-methylenecholesterol in *Ostrea gigas,* in which this substance made up 36% of total sterols is important. This sterol should be identical with conchasterol from the same animal (Tsujimoto and Koyanagi, 1934c) and with ostreasterol from *Ostrea virginica* (Bergmann, 1934). Stigmasterol was also observed in oysters (Bergmann, 1937). Kritchevsky *et al.* (1967) gave the following composition of sterols in oyster (not specified): 5.6% C_{26} sterol, 2.9% 22-dehydrocholesterol, 41.4% cholesterol, 16.0% brassicasterol, 25.9% 24-methylenecholesterol, and 8.2% C_{29} sterols.

c. Eulamellibranchia.

i. Family Unionidae. In addition to data given in the recommended reviews, it should be mentioned here that Matsumoto and Tamura (1955b) identified the $\Delta^{5,7}$ sterol of *Cristaria spatiosa* as 7-dehydrostigmasterol.

ii. Family Corbiculidae. Tsujimoto and Koyanagi (1934a) observed in *Corbicula leana* a sterol which might be identical with clionasterol. Then Matsumoto and Toyama (1943) isolated a highly unsaturated sterol from *C. leana* which they called corbisterol. The identity of corbisterol was elucidated by Tamura *et al.* (1956) as 7-dehydrostigmasterol. Matsumoto and Toyama (1944) added to this series cholesterol and brassicasterol. In 1954, Toyama and Tanaka reported the presence of poriferasterol, clionasterol, cholesterol, brassicasterol, and corbisterol.

iii. Family Cardiidae. Fagerlund and Idler (1956) demonstrated 53.3% 24-methylenecholesterol in *Cardium cordis.* Voogt and Van der Horst

(1969) studied *Cardium edule* and found C_{26} sterol, 22-dehydrocholesterol, cholesterol (25%), brassicasterol (24%), campesterol, 24-methylenecholesterol (11.5%), stigmasterol, β-sitosterol, and fucosterol.

iv. Family Tridacnidae. In *Tridacna gigas,* a sterol provisionally named shakosterol was observed by Tsujimoto and Koyanagi (1935). Shakosterol seems to be identical with or to consist substantially of brassicasterol (Bergmann and Low, 1947).

v. Family Veneridae. Saxidomus giganteus was one of the two first species from which 24-methylenecholesterol was isolated (Idler and Fagerlund, 1955). Tsujimoto and Koyanagi (1934a) reported the presence of a sterol in *Meretrix meretrix* that might be identical with clionasterol. Toyama and Yajima (1943) reported in this animal the sterol meretristerol, which, however, turned out to consist of at least four components—brassicasterol, poriferasterol, clionasterol, and cholesterol (Toyama *et al.*, 1953). Tsujimoto and Koyanagi (1934a) mentioned the presence of a sterol, possibly identical with clionasterol, in *Tapes philippinarum.* According to Yasuda (1966) 7-dehydrostigmasterol, accompanied by a second sterol, was the main sterol in *Tapes japonica.* Kritchevsky *et al.* (1967) determined the sterols in a clam (not specified) and found 4.8% C_{26} sterol, 7.9% 22-dehydrocholesterol, 36.7% cholesterol, 14.1% brassicasterol, 20.2% 24-methylenecholesterol, and 16.2% C_{29} sterols.

To get a first impression of the distribution of sterols according to carbon content in bivalves, Voogt (1970) analyzed the sterol mixtures of *Mytilus edulis, Ostrea edulis, Cyprina islandica,* and *Anodonta cygnea.* The results are given in Table XV. This table suggests that unlike in Gastropoda the C_{27} sterols are not the most important sterols in Bivalvia, but that this place is shared with the C_{28} sterols. Comparison of Table XV with Tables VIII and X and with the values given for *Planorbarius* and *Arion* shows that sterol composition in Bivalvia differs strongly from that in Gastropoda. Finally it should be emphasized once more that

TABLE XV
STEROL COMPOSITION[a] ACCORDING TO CARBON CONTENT IN SOME BIVALVES

Species	C_{26}	C_{27}	C_{28}	C_{29}	C_{30}
Mytilus edulis	8.9	44.5	36.9	8.4	—
Ostrea edulis	5.1	30.8	50.7	12.7	1.0
Cyprina islandica	5.0	40.3	40.9	13.5	—
Anodonta cygnea	1.0	24.8	35.9	38.3	—

[a] Percent of total sterols.

for obtaining full information about distribution of sterols in bivalves the reviews mentioned already may be consulted.

2. *Fatty Acids*

Investigations of fatty acids in Bivalvia have been concentrated on a few representatives, in general economically important species. Many of the older data are of little value and will be omitted if they are not informative. Of course. the mussel, *Mytilus,* has been studied extensively. The results of the studies on *Mytilus californianus* (Rodegker and Nevenzel, 1964) and *Mytilus canaliculus* (Bannatyne and Thomas, 1969) are summarized in Table XVI. This table shows rather great differences in the fatty acid composition of these two species, and it would be of interest to know whether these differences are due to different places in systematics or to different localities of capture.

TABLE XVI
FATTY ACIDS IN TWO SPECIES OF THE GENUS *Mytilus*

Fatty acid (short hand designation)	Content (%)			
		Mytilus californianus		
	Mytilus canaliculus	Male gonads	Female gonads	Remaining tissue
14:0	5.0	1.3	1.0	3.5
14:1	Trace	—	—	—
15:0	0.5	0.4	0.2	0.2
16:0	17.3	35.5	25.1	24.5
16:1	11.2	8.2	3.5	1.1
16:2	0.9	2.6	2.6	3.0
17:0	0.6	—	—	—
18:0	4.8	2.4	3.2	1.7
18:1	4.7	4.5	4.1	3.2
18:2	0.9	2.7	2.2	3.2
18:3	0.9	—	—	—
18:4	—	1.4	1.3	1.6
20:1	6.5	2.7	4.4	2.6
20:2	—	4.5	2.8	7.5
20:4	2.2	0.6	0.9	1.0
20:5	25.2	12.3	15.2	14.0
22:1	Trace	—	—	—
22:2	—	1.3	1.5	2.2
22:3	1.7	—	—	—
22:5	2.0	1.1	1.2	1.1
22:6	15.6	18.2	29.7	27.7

A second genus well studied is *Ostrea*. Results of these studies are listed in Table XVII. From this table it is clear that the fatty acid composition in *Ostrea edulis* differs strongly from those of the other oysters and can hardly be compared with them. Bannatyne has thoroughly discussed the differences between the other three oysters listed, and readers are referred to his paper. However, it should be remarked that only Gruger reports that the figures given by him are weight percentages, which is not known from the other authors, and which is essential to know for comparison. Bannatyne suggests that

TABLE XVII
FATTY ACIDS IN SOME OYSTERS

Fatty acid	Content (%)			
	Pacific oyster[a]	*Ostrea edulis*[b]	*Crassostrea gigas*[c]	*Ostrea lutaria*[d]
14:0	2.7	0.7	5.7	4.1
14:1	—	0.5	—	0.4
15:0	0.9	—	2.2	0.8
16:0	21.4	20.5	14.5	18.5
16:1	4.6	1.5	7.0	3.5
16:2	1.6	—	—	0.3
17:0	1.4	—	—	1.7
18:0	4.0	1.6	3.2	6.6
18:1	8.5	15.2	10.9	6.4
18:2	1.2	—	2.1	1.3
18:3	1.6	—	—	1.4
18:4	4.3	—	4.2	3.5
20:1	—	13.0	5.1[e]	6.7
20:2	—	18.8	—	—
20:4	1.9	—	3.9[f]	1.8
20:5	21.5	6.2	22.8	19.6
22:1	2.6	—	—	6.0
22:2	—	6.1	3.4(?)	—
22:3	—	—	—	0.8
22:4	0.5(?)	—	1.1(?)	—
22:5	1.0	—	—	0.8
22:6	20.2	7.3	10.7	15.8

[a] Gruger *et al.* (1964).
[b] Tibaldi (1966).
[c] Shimma and Taguchi (1964).
[d] Bannatyne and Thomas (1969).
[e] Inclusive 18:3.
[f] Inclusive 22:1.

the differences observed may reflect the different geographic origins of the animals. But first one should know whether the same notation has been used, while further differences may be due to different animal processing in which polyunsaturated fatty acids may have been partly lost.

From the genus *Tapes*, two species have been studied. Table XVIII shows the results of these studies, which are in good agreement. In Table XIX, fatty acid compositions of some other pelecypods have been gathered, and are given without comment. Finally, one can say that, apart from differences between the species, the bivalves show a typical marine distribution of their fatty acids, being 20:5 and 22:6 the most

TABLE XVIII
FATTY ACIDS IN TWO SPECIES OF THE GENUS *Tapes*

Fatty acid	Content (%)	
	Little neck clam[a]	*Tapes japonica*[b]
14:0	3.2	2.7
15:0	0.8	2.6
16:0	23.8	18.8
16:1	9.6	9.4
16:2	0.8	—
17:0	1.3	Trace
18:0	5.4	8.9
18:1	10.8	8.9
18:2	1.4	Trace
18:3	1.6	—[c]
18:4	3.0	3.1
20:1	3.5	7.7[c]
20:2(?)	1.2	1.8
20:3(?)	0.4	—
20:4	1.7	3.0[d]
20:5	10.0	10.9
22:1	2.6	—[d]
22:2(?)	—	7.8
22:4(?)	1.2	Trace
22:5	1.7	Trace
22:6	14.5	14.3

[a] Gruger *et al.* (1964).
[b] Shimma and Taguchi (1964).
[c] 18:3 is combined with 20:1.
[d] 20:4 is combined with 22:1.

TABLE XIX
FATTY ACIDS IN SOME BIVALVES

Fatty acid	Content (%)				
	Amphidesma ventricosum[a]	*Meretrix meretrix lusoria*[b]	*Mactra sulcataria*[b]	*Fulvia mutica*[b]	*Corbicula japonica*[b]
14:0	2.5	2.1	4.0	3.4	4.7
14:1	0.5	—	—	—	—
15:0	1.0	2.3	3.2	4.0	3.1
16:0	17.5	18.2	29.8	18.9	20.3
16:1	4.9	4.9	10.5	10.3	10.7
16:2	4.0	—	—	—	—
17:0	1.5	4.5	—	—	Trace
18:0	9.8	4.9	9.7	10.0	5.1
18:1	5.2	6.7	8.2	5.5	8.3
18:2	0.6	1.0	Trace	0.8	1.9
18:3	0.5	—[c]	—[c]	—[c]	—[c]
18:4	—	4.4	2.0	—	3.4
20:0	0.6	—	—	—	—
20:1	4.7	6.8[c]	6.7[c]	5.1[c]	9.0[c]
20:2	0.6	1.3(?)	Trace(?)	1.6(?)	Trace(?)
20:4	3.2	5.1[d]	1.7[d]	2.9[d]	4.8[d]
20:5	20.2	10.0	9.5	11.6	8.6
22:1	3.0	—[d]	—[d]	—[d]	—[d]
22:2	—	5.4(?)	1.9(?)	1.2(?)	2.9(?)
22:3	1.7	—	—	—	—
22:4	3.0	Trace(?)	3.8(?)	2.8(?)	3.0(?)
22:5	4.6	3.2	—	Trace	3.9
22:6	10.4	16.5	4.6	16.7	10.3

[a] Bannatyne and Thomas (1969).
[b] Shimma and Taguchi (1964).
[c] 18:4 is combined with 20:1.
[d] 20:4 is combined with 22:1.

important unsaturated acids. This is in contrast with what has been found for the gastropods investigated.

3. *Phospholipids*

These lipids have been investigated most extensively in *Corbicula* species. Hori and Itasaka (1961) observed six phospholipids in *Corbicula sandai,* and calculated that lipids containing ethanolamine made up 40%, those with serine and choline 30 and 25%, respectively. Small amounts of lipids containing inositol and amino sugars were observed as well as sphingolipids. In the same year (Itasaka and Hashimoto, 1961), part of the cephalins were identified as phosphatidylpeptides. Three types of amino components should be present, which might be amino sugars.

Hori and collaborators (1964b) isolated from *C. sandai* a sphingolipid containing ethanolamine, but failed to demonstrate beyond doubt phosphorylethanolamine after hydrolization. The fatty acids in sphingolipid were analyzed. Palmitic acid turned out to be the main one (42.5%). The other ones were 15:0 (0.4%), 16:1 (5.6%), 17:0 (?) (8.0%), unknown (0.5%), 18:0 (29.2%), 19:0 (2.5%), 20:0 (5.5%), 21:0 (1.1%), 22:0 (3.8%), 23:0 (0.5%), and 24:0 (0.3%). This is a normal fatty acid pattern with, however, a strong preference for saturated acids, especially for 16:0 and 18:0. In the same year, these authors reported that the ethanolamine-containing part was not phosphorylethanolamine but aminoethylphosphonic acid (Hori *et al.*, 1964c). Further, they mentioned the presence of this lipid in *Corbicula japonica,* in which it made up 66% of pyridine-insoluble lipids. This amount was 65% for *C. sandai.* The lipid was also found in *Unio biwae* (78%), *Inversidens hirasei* (67%), *Anodonta lauta rostrata* (53%), and *Cristaria plicata* (65%). Confirmation of the structure of ceramide-2-aminoethylphosphonate (CAEP) from *Corbicula sandai* was given in 1966 (Hori *et al.,* 1966b). In a series of papers, Hori *et al.* (1966b, 1967a,b, 1968c) studied the occurrence and distribution of CAEP and ceramideaminoethyl phosphate. In bivalves, only the former substance was present, and its presence was reported for *Hyriopsis schlegelii* and *Pinctada martensi.*

Arakawa *et al.* (1968) reported three other phosphosphingolipids in *Corbicula sandai.* These could not be identified, however. In 1969, Hori and Arakawa mentioned that there were strong indications for the presence of ceramide-*N,N*-acylmethylaminoethyl phosphonate and ceramide-*N*-acylaminoethyl phosphonate in *C. sandai.*

Pinctada martensii possessed two kinds of CAEP. The first contained palmitic acid (78%), margaric acid (18%), and stearic acid (4%), while the second one contained α-hydroxypalmitic acid (82%) and α-hydroxymargaric acid (16%) (Sugita *et al.,* 1968). From the work of Zama *et al.* (1960; Zama, 1963), data are given in Table XX about the phospholipids in *Pecten yessoensis, Chlamys nipponensis,* and *Mactra sachalinensis.* Yasuda (1967) reported that phospholipids of *Tapes japonica* were rich in ethanolamine. They consisted of 23.1% phosphatidylethanolamine, 9.2% ethanolamineplasmalogens, and 17.8% sphingoethanolamine, while the other phospholipids were composed of 19.8% phosphatidylcholine, 10.8% phosphatidylserine, and 6.0% phosphatidylinositol. This is in contrast with values given by Zama, because in that case the serine-containing phospholipids were the principal ones.

Shieh (1968) observed six phospholipids in *Placopecten magellanicus* of which phosphatidylcholine, phosphatidylethanolamine, and phosphatidylserine were the most important. The other three phospholipids have not been identified. He also determined the fatty acid distribution in total and the three main phospholipids. His results are given in Table XXI. Quin (1965) mentions the occurrence of aminoethylphosphonic

TABLE XX
PHOSPHOLIPIDS IN SOME BIVALVES

Lipid	Content[a]		
	Pecten yessoensis	*Chlamys nipponensis*	*Mactra sacchalinensis*
Neutral fat	15,657.7	1,070.6	893.6
Total conjugated lipids	1,271.3	525.4	552.4
Total phospholipids	808.2	508.4	532.1
Cerebroside	471.1	17.0	20.3
Phosphatidylcholine	260.7	151.4	216.8
Phosphatidylethanolamine	154.4	81.6	15.5
Phosphatidylserine	182.9	75.0	96.4
Acetal lipid	11.2	67.3	74.6
Sphingomyelin	199.0	133.1	128.8

[a] Values in milligrams percent of fresh weight.

TABLE XXI
FATTY ACID DISTRIBUTION IN PHOSPHOLIPIDS OF *Placopecten magellanicus*

Fatty acid	Total phospholipids (%)	Phosphatidylcholine (%)	Phosphatidylethanolamine (%)	Phosphatidylserine (%)
14:0	6.1	7.0	4.2	3.6
14:1	Trace	0	Trace	0
16:0	26.3	37.1	15.8	17.5
16:1	2.5	3.2	1.4	3.1
16:2	0.6	0.4	0.7	0.3
18:0	6.0	5.8	4.6	9.1
18:1	7.8	9.7	5.3	11.0
20:1	6.2	4.9	7.9	6.8
20:4	3.1	0.3	4.1	1.7
20:5	14.3	9.6	23.0	14.0
22:1	2.4	3.0	1.2	1.6
22:5	Trace	0	0.3	0.4
22:6	23.9	19.1	32.0	29.9
Others	Trace	Trace	Trace	Trace

acid in *Mytilus edulis* and *Venus mercenaria*. The number and content of phospholipids in several bivalves are given by Vaskovsky and Kostetsky (1969) (Table XXII).

Finally, Higashi *et al.* (1967) and Higashi and Hori (1968) studied the sphingolipids in spermatozoa of *Hyriopsis schlegelii*. There were five different types of which the first one is not identified. The second type contained ceramide galactoside and a small amount of ceramide glucoside. Fatty acids were mainly 16:0 and 18:0. The third type was CAEP. Stearic acid made up 96% of fatty acids and palmitic acid 4%. The fourth type was that of a mucolipid containing 18:0 and 16:0 as main fatty acids. Glucosamine was present. The neutral sugars were composed of fucose, xylose, mannose, and glucose (1:1:2:3). The fifth type was not identified, but contained perhaps an acylated hexosamine of the fourth type.

4. *Lipids with Ether Bonds*

Karnovsky *et al.* (1946) reported that the unsaponifiable matter of *Mytilus meridionalis* contained 4.1% glyceryl ethers. In *Corbicula leana*, Toyama and Tanaka (1954) encountered batyl and chimyl alcohol, with possibly selachyl alcohol. Thompson and Lee (1965) studied the α-glyceryl ethers in *Protothaca staminea* and found that they were present both in neutral lipids (1.85%) and phospholipids (3.64%). These amounts are considerably lower than in the chiton and gastropod

TABLE XXII
NUMBER AND CONTENT OF PHOSPHOLIPIDS IN SOME BIVALVES

Species	Number	Content[a]
Modiolus difficilis	7	30.0
Crenomytilus grayanus	7	51.4
Ostrea gigas	5	26.0
Chlamys nipponensis	6	25.0
Chlamys swifti	5	34.0
Patinopecten yessoensis	6	36.2
Mya arenaria	7	34.4
Spisula sacchalinensis	7	35.9
Arca boucardi	7	28.2
Anadara broughtoni	6	35.8
Glycimeris albolineatus	9	40.0
Mactra sulcataria	7	45.0
Pitaria pacifica	7	49.2

[a] Percent phospholipids in lipid extracts.

studied. It is remarkable that plasmalogens occurred exclusively in phospholipids. The side chains of the glyceryl ethers and plasmalogens have been analyzed and the results are shown in Table XXIII. Alkyl side chain composition in plasmalogens appears to be quite different from that in phospholipids. However, this difference might disappear when phospholipids are separated into classes. Rapport and Alonzo (1960) showed the presence of plasmalogens to rather high values (>10%) in several marine invertebrates, among which *Mya arenaria, Mytilus edulis, Pecten irradians* and *Venus mercenaria.* They also showed that only a small portion of total plasmalogens occurs in the choline containing lipids.

5. *Glycolipids*

Glycolipids have been studied well in *Corbicula sandai.* According to Hori *et al.* (1964a), four groups of glycolipids were found, with phosphorus in all groups. Glucosamine 6-phosphate was observed as well as the presence of sphingosine. The first group contained glucose and amino sugars, while the fatty acids were composed of C_{14} (0.4%), C_{15} (0.6%), C_{16} (60.9%), C_{17} (4.4%), C_{18} (15.2%), C_{19} (1.6%), C_{20} (4.9%), C_{21}

TABLE XXIII
ALKYL SIDE CHAINS IN GLYCERYL ETHERS AND PLASMALOGENS OF *Protothaca staminea*[a]

Side chain	Glyceryl ethers in: Neutral lipids	Glyceryl ethers in: Phospholipids	Plasmalogens
14:0	3	5	—
14:1	—	—	5
15:br	1	3	—
15:0	1	3	—
15:1	1	—	—
16:br	—	1	—
16:0 (chimyl alcohol)	23	35	4
16:1	5	6	—
17:br	8	11	5
17:0	3	3	3
18:br	4	3	10
18:0 (batyl alcohol)	36	21	58
18:1 (selachyl alcohol)	6	4	4
20:0	7	—	—
20:1	—	4	6
22:br	5	—	—

[a] Values are in mole percent.

(1.2%), C_{22} (6.3%), C_{23} (3.0%), and C_{24} (1.7%). The mean value was 17.4 carbon atoms. The second and third group contained D-xylose, L-fucose, D-glucose (1:1:1) and amino sugars (D-glucosamine and D-galactosamine 1:1). In the second group, the fatty acids were composed of C_{16} (12.2%), C_{17} (0.6%), C_{18} (16.3%), C_{19} (3.7%), C_{20} (18.5%), C_{21} (4.9%), C_{22} (21.0%), C_{23} (11.6%), and C_{24} (8.9%). The mean value was 19.8 carbon atoms.

Yamakawa and Ueta (1964) added to the known sugars of the third group mannose and an unknown one. Itasaka (1966a) identified this unknown component as 4-*O*-methylgalactose. They occurred in equimolar amounts. In the same year (Itasaka, 1966b), this author reported that in the second group, besides D-xylose, D-mannose and D-glucose, also 3-*O*-methylfucose was found. In 1968 (Itasaka, 1968), the same investigator mentioned that the only phosphorylated sugar present in the glycolipid was mannose 6-phosphate. The first group was restudied by Hori *et al.* (1968a). Ceramidemono- and dihexosides were observed in this fraction. The ceramidemonohexosides consisted of glucosylceramide (77%) and galactosylceramide (23%). The dihexoside was proposed to be mannosylglycosylceramide. Fatty acid distribution in each of this ceramides was determined.

Glycolipids of oyster have also been studied extensively. Akiya and Nakazawa (1955a) reported the isolation of a new glycolipid from the soft parts of the oyster. They proposed the formula $C_{44}H_{82}O_{21}N_2S$. In the same year (Akiya and Nakazawa, 1955b), they showed the presence of trisaccharide D-glucopyranosyl⟨4.1⟩D-glucopyranosyl⟨4.1⟩-L-fucopyranose. The terminal reducing group was attached to the remainder of the lipid. This remaining part turned out to be 14-methyl-4-pentadecenoic acid (Akiya and Nakazawa, 1956a,b). Nakazawa (1959) reports the presence of lactic acid, choline, and taurine and gives as total structure: *O*-(14-methyl-4-pentadecenoyl)choline-*N*-[*O*-(L-fucopyranosyl⟨1.4⟩-D-glucopyranosyl⟨1.4⟩-D-glucopyranosyl)-lactyl]taurate.

Hayashi and Matsubara (1966) found that glycolipid from oyster gills was a mucolipid consisting of fatty acid, sphingosine, neutral sugars (glucose, galactose, fucose, and an unidentified sugar), and amino sugars (glucosamine and galactosamine). The major fatty acids were 16:0, 17:0, 18:0, 20:0, and 22:0. This type of glycolipid resembles that of *Corbicula* described by Hori *et al.*, but differs strongly from that of oyster described by Nakazawa. Hayashi and Matsubara (1969) found that the glycolipid from the oyster mantle consisted of sphingosine, a sugar moiety, and fatty acid. Besides the C_{18} sphingosine, C_{16} sphingosine and an unidentified long chain base were present. The sugar moiety was composed of glucose, galactose, fucose, *O*-alkylfucose, and glucosamine. Further, an amino sugar phosphate was found. The fatty acids consisted of 16:0, 17:0, and 18:0.

D. Cephalopoda

1. *Sterols*

Deffner (1943), in studying *Sepia officinalis,* concluded on the ground of melting points of the sterols and sterol derivatives that only cholesterol was present. Zandee (1967) also concluded that the sterols consisted mainly of cholesterol. Voogt and Van Gennip (1969) found that the sterols consisted of 0.3% C_{26} sterols, 95.3% cholesterol, 3.1% desmosterol, and 1.3% 24-methylenecholesterol. Our knowledge about sterols in the order Teuthoidea is confined to the data given in the review by Bergmann. These data will not be dealt with here.

The data given by Bergmann (1962) about the sterols occurring in *Octopus vulgaris* (Order Octopoda) may be supplemented by the results of Hatano (1961) about *Octopus dofleini.* He found that unsaponifiable matter contained 50.7% sterol, which consisted mainly of cholesterol. Voogt and Van Gennip (1969) studied the sterols of *Octopus vulgaris* and found that they were composed of 0.8% C_{26}, 95.4% C_{27}, 3.8% C_{28}, and only trace amounts of C_{29} sterols. The C_{27} sterols consisted of cholesterol (about 87%) and desmosterol (about 7%). 24-Methylenecholesterol was the main C_{28} sterol. Stigmasterol and β-sitosterol may be present in the C_{29} fraction. These authors also investigated *Eledone aldrovandi.* Sterols contained 0.7% C_{26}, 95.9% C_{27}, 2.5% C_{28}, and 0.9% C_{29} sterols. Cholesterol made up about 90% of total sterols. Desmosterol, 24-methylenecholesterol, and β-sitosterol were present and possibly fucosterol. Though but few details are known at present, one gets the impression that sterol composition of cephalopods strongly resembles that of the neogastropods.

2. *Fatty Acids*

Toyama and Takagi (1962) demonstrated the presence of the polyunsaturated fatty acids 20:5(5,8,11,14,17) and 22:6(4,7,10,13,16,19) in cuttle fish oils. In a recent investigation of the fatty acids of *Sepia officinalis* by Van Gennip and Voogt (1969), at least ninety acids were observed with up to twenty-eight carbon atoms. The saturated fatty acids consisted of straight chain and iso- and anteiso branched fatty acids. In the unsaturated fatty acids, the presence of branched chain acids was shown. Tibaldi (1966) analyzed the fatty acids of *Loligo vulgaris,* Ito and Fukuzumi (1963) those of *Ommastrephes sloani pacificus,* while Jangaard and Ackman (1965) studied those of *Illex illecebrosus.* These results are summarized in Table XXIV.

Hatano (1958) studied the fatty acids in the acetone-soluble lipids from the liver of *Octopus dofleini.* He isolated and identified the acids 14:0, 16:0, 18:0, 20:0, 16:1, 18:1, 20:1, 16:3, 18:3, 20:4, 20:5, 22:5 and 24:5. Tibaldi (1966) reported the fatty acid composition for *Octopus vul-*

TABLE XXIV
FATTY ACIDS IN SOME CEPHALOPODA

Fatty acid	Content				
	Loligo vulgaris	*Ommastrephes sloani pacificus*	*Illex illecebrosus*[a]		
			Liver	Flesh	Viscera
14:0	0.1	3.9	4.4	2.2	3.3
14:1	1.0	—	0.1	—	—
15:0	—	—	0.2	0.3	Trace
15:1	—	—	Trace	—	—
16:0	28.4	15.4	13.7	27.6	20.3
16:1	0.6	5.9	9.0	0.4	6.4
16:2	—	—	0.5	0.2	Trace
16:3	—	—	0.1	0.1	—
17:0	—	—	0.4	0.3	Trace
17:1	—	—	0.1	0.1	Trace
18:0	0.3	2.3	1.8	4.4	4.7
18:1	0.7	16.4	16.3	4.9	8.1
18:2	—	1.5	0.8	0.3	Trace
18:3 (9,12,15)	—	0.9	0.4	0.1	—
18:3 (6,9,12)	—	—	0.3	—	—
18:4	—	0.7	0.8	0.1	0.4
19:1	—	—	0.1	Trace	—
20:1	5.2	11.8	12.4	4.9	9.4
20:2	7.6	—	0.3	0.3	Trace
20:3	—	—	0.1	—	—
20:4 (5,8,11,14)	—	1.3	0.4	0.8	0.5
20:4 (8,11,14,17)	—	0.6	0.4	Trace	Trace
20:5	15.6	12.1	13.9	15.8	18.9
22:1	—	7.7	8.2	0.5	2.6
22:2	2.1	—	—	—	—
22:4	—	—	0.3	0.1	0.5
22:5 (4,7,10,13,16)	—	—	0.1	Trace	Trace
22:5 (7,10,13,16,19)	—	1.3	1.3	0.3	Trace
22:6	27.2	15.8	16.9	37.1	24.6
24:1	—	2.5	0.3	—	0.4

[a] Values in weight percents.

garis to be 14:0 (0.4%), 14:1 (0.9%), 16:0 (16.7%), 16:1 (0.7%), 18:0 (2.7%), 18:1 (4.3%), 20:1 (15.1%), 20:2 (1.4%), 20:5 (23.2%), 22:2 (9.7%), and 22:6 (20.7%). Van Gennip and Voogt (1969) studied the fatty acid of *Octopus vulgaris* and *Eledone aldrovandi.* Results were comparable with those given above for *Sepia.* Details are not yet available.

3. *Phospholipids*

Zama (1963) studied the phospholipids in muscle and liver of *Ommastrephes sloani pacificus* and in the liver of *Octopus dofleini.* The results of this investigation are given in Table XXV. This table clearly shows that lipid distributions in the same organ of two cephalopods are quite different. However, within the same species there are also differences related to the sex of the animals, and even in one animal there are differences if different tissues are compared. Kreps *et al.* (1967, 1968) in studying the phospholipids in the nervous system in mollusks had similar results. Vaskovsky and Kostetsky (1969) found six different phospholipids in *Rossia* (Sepiolidae), making up 23% of total lipids.

Hori *et al.* (1967a) reported the presence of ceramide-2-aminoethyl phosphonate and sphingomyelin in *Polypus vulgaris.* The same authors (1967b, 1968c) reported the presence of CAEP and the absence of ceramide-2-aminoethyl phosphate in *Loligo edulis* and *Polypus vulgaris.*

TABLE XXV
LIPID CONTENT OF SOME CEPHALOPODA

Lipid	Content[a]				
	Ommastrephes			*Octopus*	
	Muscle	Liver (♂)	Liver (♀)	Liver (♂)	Liver (♀)
Total lipid[b]	1.68	55.45	53.84	14.42	13.33
Total conjugated lipids	1.508	1.862	2.123	0.502	0.643
Cerebrosides	0.023	0.142	0.077	0.031	0.086
Total phospholipids	1.485	1.720	2.046	0.471	0.557
Total sphingolipids	0.165	0.438	0.257	0.127	0.161
Sphingomyelin	0.142	0.296	0.180	0.096	0.075
Lecithin	0.999	1.026	1.494	0.294	0.295
Phosphatidylethanolamine	0.160	0.232	0.187	0.058	0.065
Phosphatidylserine	0.109	0.116	0.096	0.029	0.048
Acetal lipid	0.085	0.050	0.089	0.004	0.074

[a] Values in mmoles per 100 gm. of fresh tissue.
[b] Weight percents.

4. *Lipids with Ether Bonds*

Reports on this type of lipid in cephalopods seem to be confined to the two genera *Octopus* and *Loligo*. Karnovsky *et al.* (1946) found that glyceryl ethers made up 30.1% of the unsaponifiable matter from the hepatopancreas and 11.4% of this matter from the other viscera of *Octopus rugosus*. Hatano (1961) mentioned the occurrence of batyl alcohol and selachyl alcohol in the liver of *Octopus dofleini*.

Thompson and Lee (1965) also studied the phospholipids of *Octopus dofleini*. Glyceryl ethers made up 2.1 weight percent of neutral lipids and 6.5% of the phospholipids from the hepatopancreas. Remarkably, only trace amounts of glyceryl ethers were present in the neutral lipids of the tentacles, while the phospholipids of these organs contained 9.0% glyceryl ethers. Side chains of the glyceryl ethers were analyzed; and results are given in Table XXVI. Plasmalogens occurred both in neutral lipids and in phospholipids, but in the latter to a much higher degree. Side chains of the plasmalogens in phospholipids of octopus tentacle were analyzed. Their patterns, which differed rather strongly from that of the glyceryl ethers, were 15:0 (18%), 16:0 (14%), 17:br (2%), 17:0 (5%), 18:br (4%), 18:0 (41%), 19:br (8%), and 21:0 (8%).

TABLE XXVI
DISTRIBUTION OF ALKYL SIDE CHAINS OF GLYCERYL ETHERS IN *Octopus dofleini*

	Content[a]		
	Hepatopancreas		Tentacles
Side chain	Neutral lipids	Phospholipids	Phospholipids
14:0	—	1	2
15:br	—	2	1
15:0	—	1	2
16:br	—	—	1
16:0 (chimyl alcohol)	34	30	38
16:1	—	—	7
17:br	5	6	3
17:0	4	6	7
18:br(?)	4	4	4
18:0 (batyl alcohol)	34	41	20
18:1 (selachyl alcohol)	9	6	7
19:br	—	4	—
20:1	—	—	8

[a] Values are in mole percent.

Rapport and Alonzo (1960) found that in *Loligo pealeii* glyceryl ethers, containing phosphorus, were nearly exclusively α,β-unsaturated. They made up about 4% of total lipid. Lewis (1966) studied the glycerylether composition in *Loligo* sp. Results of this study are summarized in Table XXVII. The high content of chimyl alcohol is striking.

E. Conclusions

Recent investigations have shown that sterol mixtures in mollusks generally are more complicated than was suspected. Only Archeogastropoda, Neogastropoda, and Cephalopoda possess rather simple mixtures, and it might be very interesting to know whether this resemblance in the latter two groups is due to their carnivorous feeding habit, to phylogenetic relationship, or to a parallel evolution which has led to simplification of the mixture and preference for cholesterol. The complex mixture in Mesogastropoda, which resembles in some cases that of Bivalvia, is remarkable. Difference should be made between sterols of exogenous and endogenous origin to allow drawing conclusions. Fatty acid composition in Prosobranchia is somewhat different from the expected marine type. Unfortunately, no data are known on the Neogastropoda. Bivalvia and Cephalopoda show the marine type of fatty acid distribution. It is as yet impossible to give characteristics for the different classes within the phylum. Only those fatty acids that are synthesized by the animal itself should be determined, thereby eliminating dietary influences.

Ceramide-2-aminoethyl phosphonate is present in all classes of the Mollusca investigated. Ceramide-2-aminoethyl phosphate is absent throughout the phylum, excepted in Mesogastropoda (Hori *et al.*, 1968b).

TABLE XXVII
Glyceryl Ethers of *Loligo*

Side chain	Content (%)	Side chain	Content (%)
12:0	0.4	16:0	59.0
12:1	0.5	16:1	4.8
13:1	0.7	17:br	4.1
14:br	0.5	17:0	3.1
14:0	7.2	17:1	0.3
14:1	1.3	18:br	1.0
15:br	1.0	18:0	10.6
15:0	2.6	18:1	4.6
16:br	0.5	19:0	Trace

III. Metabolism of Lipids

In comparison with what we know about the distribution of lipids in the phylum Mollusca, our knowledge about metabolism of these lipids is very limited. But few investigators have worked in this field and for this reason our insight into the metabolic pathways is still limited.

A. Amphineura

In an experiment with *Lepidochiton* sp., the author (Voogt, 1970), has found that this animal is able to synthesize fatty acids and sterols from mevalonate. There was an intensive biosynthesis of these lipids, which is reflected in the observed specific radioactivity, 5900 dpm/mg. and 89500 dpm/mg., respectively.

B. Gastropoda

1. Prosobranchia

a. Archeogastropoda. Patella coerulea and *Monodonta turbinata* are able to synthesize 3β-sterols and fatty acids from acetate. After injection of sodium-1-^{14}C-acetate into the animals just mentioned, sterols and fatty acids showed radioactivity (Voogt, 1968b). These activities were for sterols 644 dmp/mg. and 730 dpm/mg., respectively. For fatty acids these values were 966 dpm/mg. and 975 dpm/mg. It is not known whether all or only one of the sterols are synthesized. In the latter case, the other sterols may originate directly from the diet or may have undergone transformations. The same holds for the fatty acids. Thus, no biosynthetic pathways have been established. It is only demonstrated that, as in vertebrates, acetate is a suitable precursor for the biosynthesis of sterols and fatty acids, and of course, that these animals possess the capacity of synthesizing these compounds.

b. Mesogastropoda. To check whether mesogastropods are able to synthesize fatty acids and sterols from a simple compound like acetate, the radioactive form of this substance was injected into *Littorina littorea* and *Viviparus fasciatus* (Voogt, 1969b). These animals have similar feeding habits as Archeogastropoda in that they are herbivorous. Results were comparable with those for Archeogastropoda, and it was concluded that the two mesogastropods investigated were able to synthesize fatty acids as well as sterols without specifying, however, which of the fatty acids and sterols were synthesized.

Crepidula fornicata, a filter feeder, thus exhibiting the feeding habit of Bivalvia, was also investigated by the author (Voogt, 1971c). In an experiment carried out in 1967, fatty acids showed 2919 dmp/mg. and sterols 285 dpm/mg. When the experiment was repeated in 1968, these values were 9000 dpm/mg. and 340 dpm/mg., respectively. Thus, *Crepidula,* though not exclusively herbivorous, also possesses the capacity of synthesizing sterols (and fatty acids).

Natica cataena was also studied (Voogt, 1971c), being an example of a carnivorous mesogastropod. In two experiments, carried out in 1967 and 1968, incorporation of radioactivity from acetate into lipids was extremely low. Fatty acids were distinctly labeled (570 dpm/mg.), however, but sterols showed activities of 8 dmp/mg. and 7 dmp/mg. Apparently, the speed of lipid metabolism is very low in this animal. For this reason, the experiment was repeated, using mevalonate-2-^{14}C. Fatty acids showed an activity of 160 dmp/mg. and sterols of 350 dpm/mg. Thus, it may be concluded that *Natica* can synthesize sterols and fatty acids, but that the biosynthesis proceeds at a low rate. Although only few data are known, it may provisionally be concluded that the capacities of synthesizing sterols and fatty acids appear as general properties throughout the order of Mesogastropoda.

c. Neogastropoda. Kind and Goldberg (1953), studying the sterols of *Urosalpinx cinereus,* found indirect evidence that sterols were synthesized by these animals. Their conclusion is based on the difference in sterol composition in *Urosalpinx* and its prey, which are oysters and other Bivalvia. However, modification (dealkylation) of the side chain of dietary sterols by *Urosalpinx* would suffice to explain the observed differences in sterol distribution. In an experiment with *Buccinum undatum* (Voogt, 1967b) using acetate-1-^{14}C, he found incorporation of radioactivity in fatty acids (660 dpm/mg.) but not in 3β-sterols. When the experiment was repeated, 3β-sterols showed 6 dpm/mg., which is not significant. In an experiment with mevalonate-2-^{14}C, sterols were radioactive (70 dmp/mg.). *Neptunea antiqua, Murex brandaris,* and *Purpura lapillus* showed the same picture as *Buccinum;* sterols were only labeled if mevalonate had been injected (Voogt, 1971d).

In all cases, fatty acids were radioactive. Thus, the Neogastropoda investigated are all able to synthesize fatty acids. They also can synthesize sterols if mevalonate is provided, but under the conditions in which the experiments were carried out, they failed to synthesize sterols if acetate was administered as a precursor. The same was observed for *Natica,* and the question arises whether this behavior is characteristic for carnivorous mollusks.

2. *Opisthobranchia*

From his studies on *Limacina helicina* and *Clione limacina,* Baalsruud (1950) concluded that these representatives of the Pteropoda could synthesize their sterols on the same ground as Kind and Goldberg (1953) did for *Urosalpinx cinereus*. The author studied *Dendronotus frondosus* and *Archidoris tuberculata*, which were injected with acetate-1-^{14}C. Fatty acids were highly labeled, 18000 dpm/mg. and 11800 dpm/mg., respectively. The unsaponifiable matter showed 18000 dpm/mg. and 6500 dpm/mg., respectively. Sterols, compared with the values of unsaponifiable matter, possessed only little radioactivity, 90 dpm/mg. and 64 dpm/mg., which did not decrease on further recrystallization. If this activity is not caused by impurities which could not be removed by recrystallization, sterols must have been synthesized, though at a very low speed. As many of Opisthobranchia are carnivorous, this is perhaps comparable with the situation in Neogastropoda.

3. *Pulmonata*

a. Basommatophora. The author studied the capacity of synthesizing sterols of three representatives of this order (Voogt, 1968a). *Limnaea stagnalis* and *Planorbarius corneus* were injected with sodium acetate-1-^{14}C, while *Limnaea peregra* was injected with sodium acetate-2-^{14}C. Fatty acids showed high radioactivities 31958 dpm/mg., 13320 dpm/mg., and 9400 dpm/mg., respectively. Radioactivities in sterols were 6559 dpm/mg., 2358 dpm/mg., and 1527 dpm/mg., respectively. Apparently these representatives of the Basommatophora are able to synthesize fatty acids as well as sterols from acetate.

b. Stylommatophora.

i. Sterols and Fatty Acids. From their investigations on *Arion empiricorum*, Bock and Wetter (1938) concluded that *Arion* was able to synthesize ergosterol because its presence was independent of the diet. Using sodium acetate-1-^{14}C, the author observed a rapid incorporation of radioactivity into the fatty acids and sterols of *Arion rufus* (Voogt, 1967a).

Vonk *et al.* (1950) studied the metabolism of *Helix pomatia* by means of heavy water. From this experiment, it was concluded that *Helix* could synthesize fatty acids. In another experiment with deuterium-labeled acetate, Vonk *et al.* (1956) observed biosynthesis of unsaponifiable lipids, and they suggested that sterols might be synthesized. This animal was restudied by Addink and Ververgaert (1963) by means of acetate-1-^{14}C. Results of Vonk *et al.* were confirmed in that fatty acids and sterols were synthesized. Voogt and Van der Horst (1969) studied *Succinea putris* and *Arianta arbustorum* by means of ^{14}C-labeled acetate. They observed biosynthesis of fatty acids and sterols in both animals.

Summarizing, one can say that all pulmonates investigated till now possess the capacity of synthesizing both fatty acids and sterols.

ii. Lipids Containing Ether Bonds. Thompson (1965) studied the biosynthesis of ether-containing phospholipids in *Arion ater*. After injection of glucose-6-^{14}C, glyceryl ethers were distinctly labeled. The incorporation of palmitate-1-^{14}C was measured after several intervals. Glyceryl ether phospholipids and diacyl phospholipids showed similar specific radioactivity, namely initially a steady increase, but finally a sharp drop in radioactivity. Plasmalogens were lower in specific activity in this experiment. This author (Thompson, 1966) demonstrated that plasmalogens were labeled lower only in experiments of short durations. The side chain of plasmalogens becomes labeled after periods of 2 days or longer if radioactive palmitic acid or glyceryl ethers are fed. From these results, it was supposed by Thompson that the side chain of the saturated glyceryl ether should be the direct precursor for the side chain of the alkenylglycerylether. A possible pathway of biosynthesis was presented. Evidence for this hypothesis was obtained from an experiment with chimyl alcohol, in which the side chain was ^{14}C-labeled and the glycerol moiety ^{3}H-labeled (Thompson, 1968).

C. Bivalvia

1. Sterols

It was reported (Fagerlund and Idler, 1960) that acetate-2-^{14}C was incorporated in sterols by *Mytilus californianus* and *Saxidomus giganteus*. It was observed that squalene-11,14-^{14}C was converted into sterols by *Saxidomus*. These authors (Fagerlund and Idler, 1961b) found that most of the radioactivity was present in a monounsaturated Δ^5 sterol and that only traces of radioactivity were present in 24-methylenecholesterol. It was also shown that *Saxidomus* was able to convert cholesterol-4-^{14}C into radioactive 24-methylenecholesterol. These authors also observed that *Saxidomus* was able to introduce unsaturation into the side chain of injected cholesterol-26-^{14}C at C_{22} and C_{25} (Fagerlund and Idler, 1961a). From studies on the seasonal variations in sterol components in muscle tissue of *Placopecten magellanicus,* Idler *et al.* (1964) obtained indications that metabolism and biosynthesis of cholesterol and 24-methylenecholesterol were related.

Tamura *et al.* (1964b) found indirect evidence that *Crassostrea virginica* was able to synthesize cholesterol and to maintain 24-methylenecholesterol level, perhaps by conversion of cholesterol, if animals were placed on a sterol-free diet. There were indications that a C_{29} sterol, tentatively identified as 20-isofucosterol, might be an essential sterol.

Voogt (1966) studied *Ostrea edulis* and found that no radioactivity was incorporated into sterols after injection of acetate-1-^{14}C in an experiment lasting 6.5 hours. In a second experiment lasting 7 days, oysters were injected every second day with acetate-1-^{14}C. The 3β-sterols possessed a specific radioactivity of 265 dpm/mg. After four recrystallizations, activity had increased to 348 dpm/mg. Salaque *et al.* (1966) found no incorporation of radioactivity into sterols of *Ostrea gryphea* from mevalonate-2-^{14}C or methionine-CH_3-^{14}C after incubation of the animals during 70 and 66 hours, respectively. They concluded that under the conditions of their experiments *Ostrea* did not synthesize sterols.

In a series of experiments the author found that *Mya arenaria* and *Cyprina islandica* had not incorporated radioactivity into sterols from acetate-1-^{14}C after 6 hours and that sterols of *Mytilus edulis* did not show radioactivity even after 120 hours. Specimens of *Anodonta cygnea* were injected every second day with acetate-1-^{14}C and killed after 6 days. Sterols were not or hardly labeled. When *Mytilus edulis* was injected with acetate-2-^{14}C, sterols did not show radioactivity after 7 days. Sterols of *Cyprina islandica* were not labeled 30 hours after injection of mevalonate-2-^{14}C. Thus, the picture is not clear; Fagerlund and Idler observed biosynthesis of sterols in *Mytilus californianus* and *Saxidomus giganteus,* starting from acetate-2-^{14}C and squalene. The author did not observe sterol biosynthesis after injection of acetate-2-^{14}C in *Mytilus edulis* and after injection of acetate-1-^{14}C in *Mya arenaria, Cyprina islandica, Mytilus edulis,* and *Ostrea edulis.* Sterols of *Anodonta cygnea* were hardly labeled, and in one experiment with *Ostrea edulis,* sterols were labeled distinctly, starting from acetate-1-^{14}C. Salaque *et al.,* found no incorporation of radioactivity from mevalonate-2-^{14}C into sterols of *Ostrea gryphea.* The same was observed by the author for *Cyprina islandica.*

Voogt and Oudejans (1968) studied the capacity of synthesizing sterols in *Cardium edule.* After injection of acetate-2-^{14}C, sterols were labeled, whereas after injection of acetate-1-^{14}C they were not. This was repeated by Voogt and Van der Horst (1968) who again found that sterols were labeled after injection of acetate-2-^{14}C. Thus, it seems as yet impossible to generalize for this class.

2. *Fatty Acids*

In all experiments of the author (and co-workers), mentioned above, fatty acids were distinctly labeled. Fatty acids of *Ostrea edulis* were separated into saturated and unsaturated. The radioactivities were 597 dpm/mg. and 477 dpm/mg., respectively. Thus, it may be concluded that Bivalvia are able to synthesize fatty acids, and in the case of *Ostrea* that this holds for both unsaturated and saturated fatty acids.

The observation of Shieh (1968), who studied the phospholipids in *Placopecten magellanicus,* is remarkable. After injection of acetate-1-^{14}C, fatty acids in phospholipids were not radioactive. From this result, he concludes that it seems likely that fatty acids in this animal may be of dietary origin.

3. *Phospholipids*

Shieh (1968) found that choline, ethanolamine, and serine were incorporated into the corresponding phospholipids by *Placopecten magellanicus*. Methionine could not be utilized in this experiment. Itasaka *et al.* (1969) observed that in *Hyriopsis schlegelii* $^{32}P_i$ was incorporated into all types of phospholipid present in this animal. Radioactivities were determined 24 and 48 hours after injection. The liver turned out to be the place in which the rate of ceramide-2-aminoethyl phosphonate biosynthesis was maximal. Liver homogenates also catalyzed incorporation of $^{32}P_i$ into CAEP.

D. Cephalopoda

1. *Sterols*

The absence of cholesterol biosynthesis in *Sepia officinalis* was reported by Zandee (1967), since after injection of acetate-1-^{14}C the animals failed to incorporate this substance into sterols. Voogt and Van Gennip (1969) studying *Eledone aldrovandi* and *Octopus vulgaris* also observed that no incorporation of acetate-1-^{14}C into sterols took place. However, when *Sepia officinalis* was injected with mevalonate-2-^{14}C sterols were distinctly labeled (1427 dpm/mg.). This situation is similar to that in Neogastropoda and *Natica* and thus seems to be characteristic for all (or most of) the carnivorous mollusks. Unfortunately, no explanation of this phenomenon is available as yet.

2. *Fatty acids*

In the experiments of Zandee and of Voogt and Van Gennip mentioned above, fatty acids were radioactive in all cases. From this it may be concluded that cephalopods are able to synthesize fatty acids.

REFERENCES

Ackman, R. G. (1963a). *J. Amer. Oil Chem. Soc.* **40**, 558.
Ackman, R. G. (1963b). *J. Amer. Oil Chem. Soc.* **40**, 564.
Ackman, R. G. (1964). *J. Fish. Res. Bd. Can.* **21**, 247.
Ackman, R. G. (1967). *Comp. Biochem. Physiol.* **22**, 907.

Ackman, R. G., and Burgher, R. D. (1963). *J. Chromatogr.* **11**, 185.
Addink, A. D. F., and Ververgaert, P. H. J. T. (1963). *Arch. Int. Physiol. Biochim.* **71**, 797.
Akiya, S., and Nakazawa, Y. (1955a). *J. Pharm. Soc. Jap.* **75**, 1322.
Akiya, S., and Nakazawa, Y. (1955b). *J. Pharm. Soc. Jap.* **75**, 1335.
Akiya, S., and Nakazawa, Y. (1965a). *J. Pharm. Soc. Jap.* **76**, 1401.
Akiya, S., and Nakazawa, Y. (1965b). *J. Pharm. Soc. Jap.* **76**, 1403.
Arakawa, I., Sugita, M., Itasaka, O., and Hori, T. (1968). *Shiga Daigaku Kyoiku Gakubu Kiyo, Shizenkagaku* **18**, 41; *Chem. Abstr.* **70**, 111704x (1969).
Baalsrud, K. (1950). *Acta Chem. Scand.* **4**, 512.
Bannatyne, W. R., and Thomas, J. (1969). *N. Z. J. Sci.* **12**, 207.
Beerthuis, R. K., and Recourt, J. H. (1960). *Nature (London)* **186**, 372.
Bergmann, W. (1934). *J. Biol. Chem.* **104**, 317.
Bergmann, W. (1937). *J. Biol. Chem.* **118**, 499.
Bergmann, W. (1949). *J. Mar. Res.* **8**, 137.
Bergmann, W. (1962). *Comp. Biochem.* **3**, 103.
Bergmann, W., and Low, E. M. (1947). *J. Org. Chem.* **12**, 67.
Bergmann, W., and Ottke, R. C. (1949). *J. Org. Chem.* **14**, 1085.
Bock, F., and Wetter, F. (1938). *Hoppe-Seyler's Z. Physiol. Chem.* **256**, 33.
Brooks, C. J. W., and Hanaineh, L. (1963). *Biochem. J.* **87**, 151.
Brooks, C. J. W., Horning, E. C., and Young, J. S. (1968). *Lipids* **3**, 391.
Carroll, K. K. (1961). *J. Lipid Res.* **2**, 135.
Cerbulis, J., and Wight Taylor, M. (1969). *Lipids* **4**, 363.
Clayton, R. B. (1962). *Biochemistry* **1**, 357.
Corey, E. J., Russey, W. E., and Ortiz de Montellano, P. R. (1966). *J. Amer. Chem. Soc.* **88**, 4750.
Danielsson, H., and Tchen, T. T. (1968). *Metab. Pathways* **2**, 117.
Dawson, R. M. C. (1962). *Comp. Biochem.* **3**, 265.
Deffner, M. (1943). *Hoppe-Seyler's Z. Physiol. Chem.* **278**, 165.
De Koning, A. J. (1966a). *Nature (London)* **210**, 113.
De Koning, A. J. (1966b). *J. Sci. Food Agr.* **17**, 460.
Dittmer, J. C. (1962). *Comp. Biochem.* **3**, 231.
Dorée, C. (1909). *Biochem. J.* **4**, 72.
Elovson, J. (1965). *Biochim. Biophys. Acta* **106**, 291.
Etienne, J., and Kahane, E. (1958). *Bull. Soc. Chim. Biol.* **40**, 211.
Etienne-Petitfils, J. (1956). Thesis, Paris.
Fagerlund, U. H. M., and Idler, D. R. (1956). *J. Org. Chem.* **21**, 372.
Fagerlund, U. H. M., and Idler, D. R. (1960). *Can. J. Biochem. Physiol.* **38**, 997.
Fagerlund, U. H. M., and Idler, D. R. (1961a). *Can. J. Biochem. Physiol.* **39**, 505.
Fagerlund, U. H. M., and Idler, D. R. (1961b). *Can. J. Biochem. Physiol.* **39**, 1347.
Fantl, P. (1942). *Aust. J. Exp. Biol. Med. Sci.* **20**, 55.
Freeman, C. P., and West, D. (1966). *J. Lipid Res.* **7**, 324.
Giese, A. C. (1966). *Physiol. Rev.* **46**, 244.
Goldfine, H., and Bloch, K. (1961). *J. Biol. Chem.* **236**, 2596.
Green, D. E., and Allmann, D. W. (1968). *Metab. Pathways* **2**, 37.
Gruger, E. H., Nelson, R. W., and Stansby, M. E. (1964). *J. Amer. Oil Chem. Soc.* **41**, 662.
Hack, M. H., Gussin, A. E., and Lowe, M. E. (1962). *Comp. Biochem. Physiol.* **5**, 217.

Haken, J. K. (1966). *J. Chromatogr.* **23**, 375.
Haken, J. K. (1969a). *J. Chromatogr.* **39**, 245.
Haken, J. K. (1969b). *J. Chromatogr.* **43**, 487.
Hamilton, R. J., Vanden Heuvel, W. J. A., and Horning, E. C. (1963). *Biochim. Biophys. Acta* **70**, 679.
Hatano, M. (1958). *Bull. Fac. Fish. Hokkaido Univ.* **9**, 207.
Hatano, M. (1961). *Bull. Fac. Fish. Hokkaido Univ.* **11**, 218.
Hayashi, A., and Matsubara, T. (1966). *J. Fac. Sci. Tech., Kinki. Univ.* **1**, 25.
Hayashi, A., and Matsubara, T. (1969). *J. Biochem. (Tokyo)* **65**, 503.
Hayashi, A., Matsubara, T., and Matsuura, F. (1969a). *Yukagaku* **18**, 118; *Chem. Abstr.* **70**, 103052t (1969).
Hayashi, A., Matsuura, F., and Matsubara, T. (1969b). *Biochim. Biophys. Acta* **176**, 208.
Higashi, S., and Hori, T. (1968). *Biochim. Biophys. Acta* **152**, 568.
Higashi, S., Nakagawa, M., and Hori, T. (1967). *Seikagaku* **39**, 175; *Chem. Abstr.* **67**, 71383w (1967).
Higgins, M. J. P., and Kekwick, R. G. O. (1969). *Biochem. J.* **113**, 36P.
Hirsch, J., and Ahrens, E. H. (1958). *J. Biol. Chem.* **233**, 311.
Hori, T., and Arakawa, I. (1966). *Shiga Daigaku Kyoiku Gakubu Kiyo, Shizenkagaku* **16**, 25.
Hori, T., and Arakawa, I. (1969). *Biochim. Biophys. Acta* **176**, 898.
Hori, T., and Itasaka, O. (1961). *Seikagaku* **33**, 169; *Chem. Abstr.* **55**, 25060g (1961).
Hori, T., Itasaka, O., and Hashimoto, T. (1964a). *J. Biochem. (Tokyo)* **55**, 1.
Hori, T. Itasaka, O., Hashimoto, T., and Inoue, H. (1964b). *J. Biochem. (Tokyo)* **55**, 545.
Hori, T., Itasaka, O., Inoue, H., and Yamada, K. (1964c). *J. Biochem. (Tokyo)* **56**, 477.
Hori, T., Itasaka, O., Inoue, H., and Akai, M. (1965). *Jap. J. Exp. Med.* **35**, 81.
Hori, T., Itasaka, O., Inoue, H., Gamo, M., and Arakawa, I. (1966a). *Jap. J. Exp. Med.* **36**, 85.
Hori, T., Itasaka, O., and Inoue, H. (1966b). *J. Biochem. (Tokyo)* **59**, 570.
Hori, T., Itasaka, O., Sugita, M., and Arakawa, I. (1967a). *Mem. Fac. Educ., Shiga Univ.* **17**, 23.
Hori, T., Arakawa, I., and Sugita, M. (1967b). *J. Biochem. (Tokyo)* **62**, 67.
Hori, T., Itasaka, O., and Kamimura, M. (1968a). *J. Biochem. (Tokyo)* **64**, 125.
Hori, T., Sugita, M., and Arakawa, I. (1968b). *Biochim. Biophys. Acta* **152**, 211.
Hori, T., Itasaka, O., and Arakawa, I. (1968c). *Abstr. Int. Congr. Biochem., 7th, 1967* F341.
Horning, E. C., Moscatelli, E. A., and Sweeley, C. C. (1959). *Chem. Ind. (London)* p. 751.
Horning, E. C., Haahti, E. O. A., and Vanden Heuvel, W. J. A. (1961). *J. Amer. Oil Chem. Soc.* **38**, 625.
Idler, D. R., and Fagerlund, U. H. M. (1955). *J. Amer. Chem. Soc.* **77**, 4142.
Idler, D. R., Tamura, T., and Wainai, T. (1964). *J. Fish. Res. Bd. Can.* **21**, 1035.
Ikekawa, N., Watanuki, R., Tsuda, K., and Sakai, K. (1968). *Anal. Chem.* **40**, 1139.
Itasaka, O. (1966a). *J. Biochem. (Tokyo)* **60**, 52.
Itasaka, O. (1966b). *J. Biochem. (Tokyo)* **60**, 435.
Itasaka, O. (1968). *J. Biochem. (Tokyo)* **63**, 347.

Itasaka, O., and Hashimoto, T. (1961). *Shiga Daigaku Gakugeigakubu Kiyo, Shizenkagaku* **11**, 21; *Chem.* Abstr. **57**, 5003d (1962).
Itasaka, O., Hori, T., and Sugita, M. (1969). *Biochim. Biophys. Acta* **176**, 783.
Ito, S., and Fukuzumi, K. (1963). *Yukagaku* **12**, 278.
Iverson, J. L., Firestone, D., and Eisner, J. (1965). *J. Assoc. Offic. Anal. Chem.* **48**, 482.
Jangaard, P. M., and Ackman, R. G. (1965). *J. Fish. Res. Bd. Can.* **22**, 131.
Karnovsky, M. L., Rapson, W. S., and Black, M. (1946). *J. Soc. Chem. Ind., London, Trans. Commun.* **65**, 425.
Kind, C. A., and Goldberg, M. H. (1953). *J. Org. Chem.* **18**, 203.
Kind, C. A., and Herman, S. C. (1948). *J. Org. Chem.* **13**, 867.
Kind, C. A., and Meigs, R. A. (1955). *J. Org. Chem.* **20**, 1116.
Kind, C. A., Slater, S. G., and Vinci, A. (1948). *J. Org. Chem.* **13**, 538.
Kita, M., and Toyama, Y. (1959). *Mem. Fac. Eng., Nagoya Univ.* **11**, 216.
Knights, B. A. (1966). *J. Gas Chromatogr.* **4**, 329.
Knights, B. A. (1967). *J. Gas Chromatogr.* **5**, 273.
Kreps, E. M., Krasilnikova, V. I., Patrikeeva, M. V., Smirnov, A. A., and Chirkovskaya, E. V. (1967). *Zh. Evol. Biokhim. Fiziol.* 3, 101.
Kreps, E. M., Smirnov, A. A., Chirkovskaya, E. V., Patrikeeva, M. V., and Krasilnikova, V. I. (1968). *J. Neurochem.* **15**, 285.
Kritchevsky, D., Tepper, S. A., Ditullo, N. W., and Holmes, W. L. (1967). *J. Food Sci.* **32**, 64.
Lewis, R. W. (1962). *Comp. Biochem. Physiol.* **6**, 75.
Lewis, R. W. (1966). *Comp. Biochem. Physiol.* **19**, 363.
Liang, C. R., and Strickland, K. P. (1969). *Can. J. Biochem.* **47**, 85.
Lovern, J. A. (1964). *Annu. Rev. Oceanogr. Mar. Biol.* **2**, 169.
Mangold, H. K., and Kammereck, R. (1961). *Chem. Ind. (London)* p. 1032.
Matsumoto, T., and Tamura, T. (1955a). *Nippon Kagaku Zasshi* **76**, 951.
Matsumoto, T., and Tamura, T. (1955b). *Nippon Kagaku Zasshi* **76**, 1413.
Matsumoto, T., and Toyama, Y. (1943). *J. Chem. Soc. Jap.* **64**, 326.
Matsumoto, T., and Toyama, Y. (1944). *J. Chem. Soc. Jap.* **65**, 258.
Matsumoto, T., Tamura, T., and Ito, S. (1955). *Nippon Kagaku Zasshi* **76**, 953.
Misra, U. K. (1966). *J. Sci. Ind. Res.* **25**, 257.
Nakazawa, Y. (1959). *J. Biochem. (Tokyo)* **46**, 1579.
Olson, J. A. (1965). *Rev. Physiol., Biochem. Exp. Pharmacol.* **56**, 173.
Petering, H. G., and Waddell, J. (1951). *J. Biol. Chem.* **191**, 765.
Popják, G. (1954). *Arch. Biochem. Physiol.* **48**, 102.
Quin, L. D. (1965). *Biochemistry* **4**, 324.
Rapport, M. M., and Alonzo, N. F. (1960). *J. Biol. Chem.* **235**, 1953.
Rodegker, W., and Nevenzel, J. C. (1964). *Comp. Biochem. Physiol.* **11**, 53.
Roots, B. I., and Johnston, P. V. (1965). *Biochem. J.* **94**, 61.
Rosenberg, H. R., and Waddell, J. (1951). *J. Biol. Chem.* **191**, 757.
Rossiter, R. J. (1968). *Metab. Pathways* **2**, 69.
Rouser, G., Kritchevsky, G., Heller, D., and Lieber, E. (1963). *J. Amer. Oil Chem. Soc.* **40**, 425.
Rudney, H. (1969). *Biochem. J.* **113**, 21P.
Salaque, A., Barbier, M., and Lederer, E. (1966). *Comp. Biochem. Physiol.* **19**, 45.
Sandermann, W. (1962). *Comp. Biochem.* **3**, 503.
Shieh, H. S. (1968). *Comp. Biochem. Physiol.* **27**, 533.

Shimma, Y., and Taguchi, H. (1964). *Bull. Jap. Soc. Sci. Fish.* **30**, 153.
Shorland, F. B. (1962). *Comp. Biochem.* **3**, 1–102.
Sipos, J. C., and Ackman, R. G. (1968). *J. Fish. Res. Bd. Can.* **25**, 1561.
Snyder, F. (1969). *Progr. Chem. Fats Other Lipids* **10**, 287.
Snyder, F., Malone, B., and Wykle, R. L. (1969a). *Biochem. Biophys. Res. Commun.* **34**, 40.
Snyder, F., Wykle, R. L., and Malone, B. (1969b). *Biochem. Biophys. Res. Commun.* **34**, 315.
Snyder, F., Malone, B., and Blank, M. L. (1969c). *Biochim. Biophys. Acta* **187**, 302.
Sugita, M., Arakawa, I., Hori, T., and Sawada, Y. (1968). *Seikagaku* **40**, 158.
Takagi, T., and Toyama, Y. (1956). *Mem. Fac. Eng., Nagoya Univ.* **8**, 177.
Takagi, T., and Toyama, Y. (1958). *Mem. Fac. Eng., Nagoya Univ.* **10**, 84.
Takagi, T., and Toyama, Y. (1959). *Mem. Fac. Eng., Nagoya Univ.* **11**, 175.
Takagi, T., Maeda, T., and Toyama, Y. (1956). *Mem. Fac. Eng., Nagoya Univ.* **8**, 169.
Tamura, T., Kokuma, K., and Matsumoto, T. (1956). *Nippon Kagaku Zasshi* **77**, 987.
Tamura, T., Wainai, T., Truscott, B., and Idler, D. R. (1964a). *Can. J. Biochem.* **42**, 1331.
Tamura, T., Truscott, B., and Idler, D. R. (1964b). *J. Fish. Res. Bd. Can.* **21**, 1519.
Tanaka, T., and Toyama, Y. (1957a). *Mem. Fac. Eng., Nagoya Univ.* **9**, 116.
Tanaka, T., and Toyama, Y. (1957b). *Mem. Fac. Eng., Nagoya Univ.* **9**, 123.
Tanaka, T., and Toyama, Y. (1958). *Mem. Fac. Eng., Nagoya Univ.* **10**, 77.
Tanaka, T., and Toyama, Y. (1959a). *Mem. Fac. Eng., Nagoya Univ.* **11**, 181,
Tanaka, T., and Toyama, Y. (1959b). *Mem. Fac. Eng., Nagoya Univ.* **11**, 204.
Terpstra, G. K., and Voogt, P. A. (1969). Unpublished data.
Thiele, O. W. (1959). *Z. Vergl. Physiol.* **42**, 484.
Thiele, O. W. (1960). *Hoppe-Seyler's Z. Physiol. Chem.* **321**, 29.
Thiele, O. W., and Kröber, G. (1963). *Hoppe-Seyler's Z. Physiol. Chem.* **334**, 63.
Thompson, G. A., Jr. (1965). *J. Biol. Chem.* **240**, 1912.
Thompson, G. A., Jr. (1966). *Biochemistry* **5**, 1290.
Thompson, G. A., Jr. (1968). *Biochim. Biophys. Acta* **152**, 409.
Thompson, G. A., Jr., and Hanahan, D. J. (1963). *J. Biol. Chem.* **238**, 2628
Thompson, G. A., Jr., and Lee, P. (1965). *Biochim. Biophys. Acta* **98**, 151.
Thomson, R. H. (1962). *Comp. Biochem.* **3**, 631.
Tibaldi, E. (1966). *Atti Accad. Naz. Lincei, Cl. Sci. Fis., Mat. Natur., Rend.* [8] **40**, 921.
Toyama, Y. (1958). *Fette, Seifen, Anstrichm.* **60**, 909.
Toyama, Y., and Shibana, F. (1943). *J. Chem. Soc. Jap., Pure Chem. Sect.* **64**, 323.
Toyama, Y., and Takagi, T. (1954). *Nippon Kagaku Zasshi* **75**, 1241.
Toyama, Y., and Takagi, T. (1962). *Fette, Seifen, Anstrichm.* **64**, 134.
Toyama, Y., and Tanaka, T. (1953). *Bull. Chem. Soc. Jap.* **26**, 497.
Toyama, Y., and Tanaka, T. (1954). *Bull. Chem. Soc. Jap.* **27**, 264.
Toyama, Y., and Tanaka, T. (1956a). *Mem. Fac. Eng., Nagoya Univ.* **8**, 29.
Toyama, Y., and Tanaka, T. (1956b). *Mem. Fac. Eng., Nagoya Univ.* **8**, 40.
Toyama, Y., and Yajima, M. (1943). *J. Chem. Soc. Jap.* **64**, 878.

Toyama, Y., Takagi, T., and Tanaka, T. (1953). *Bull. Chem. Soc. Jap.* **26,** 154.

Toyama, Y., Takagi, T., and Tanaka, T. (1955a). *Mem. Fac. Eng., Nagoya Univ.* **7,** 1.

Toyama, Y., Tanaka, T., and Maeda, T. (1955b). *Mem. Fac. Eng., Nagoya Univ.* **7,** 145.

Tsujimoto, M., and Koyanagi, H. (1934a). *J. Soc. Chem. Ind., Jap.* **37,** 81B.

Tsujimoto, M., and Koyanagi, H. (1934b). *J. Soc. Chem. Ind., Jap.* **37,** 85B.

Tsujimoto, M., and Koyanagi, H. (1934c). *J. Soc. Chem. Ind., Jap.* **37,** 436B.

Tsujimoto, M., and Koyanagi, H. (1935). *J. Soc. Chem. Ind., Jap.* **38,** 118B.

Tsujimoto, M., and Toyama, Y. (1922). *Chem. Umschau Geb. Fette, Oele, Wachse Harze* **29,** 27.

Vanden Heuvel, F. A., and Court, A. S. (1968). *J. Chromatogr.* **38,** 439.

Vanden Heuvel, F. A., and Court, A. S. (1969). *J. Chromatogr.* **39,** 1.

Vanden Heuvel, W. J. A., and Horning, E. C. (1962). *Biochem. Biophys. Acta* **64,** 416.

Vanden Heuvel, W. J. A., Sweeley, C. C., and Horning, E. C. (1960). *J. Amer. Chem. Soc.* **82,** 3481.

Vanden Heuvel, W. J. A., Gardiner, W. L., and Horning, E. C. (1966). *J. Chromatogr.* **25,** 242.

Van der Horst, D. J., and Voogt, P. A. (1969a). *Comp. Biochem. Physiol.* **31,** 763.

Van der Horst, D. J., and Voogt, P. A. (1969b). *Arch. Int. Physiol. Biochim.* **77,** 507.

Van der Vliet, J. (1946). Ph.D. Thesis, University of Groningen.

Van der Vliet, J. (1948a). *Rec. Trav. Chim. Pays-Bas* **67,** 246.

Van der Vliet, J. (1948b). *Rec. Trav. Chim. Pays-Bas* **67,** 265.

Van Gennip, A. H., and Voogt, P. A. (1969). Unpublished data.

Vaskovsky, V. E., and Kostetsky, E. Y. (1969). *Chem. Phys. Lipids* **3,** 102.

Van Tamelen, E. E., Willet, J. D., Clayton, R. B., and Lord, K. E. (1966). *J. Amer. Chem. Soc.* **88,** 4752.

von Ardenne, M., Osske, G., Schreiber, K., Steinfelder, K., and Tümmler, R. (1965). *Kulturpflanze* **13,** 101.

Vonk, H. J., Mighorst, J. C. A., and De Groot, A. P. (1950). *Physiol. Comp. Oecol.* **2,** 51.

Vonk, H. J., Sedee, P., and De Ligny, W. (1956). *Int. Congr. Physiol. Sci., Lect. Symp., 20th, 1956* p. 934.

Voogt, P. A. (1966). Unpublished data.

Voogt, P. A. (1967a). *Arch. Int. Physiol. Biochim.* **75,** 492.

Voogt, P. A. (1967b). *Arch. Int. Physiol. Biochim* **75,** 809.

Voogt, P. A. (1968a). *Comp. Biochem. Physiol.* **25,** 943.

Voogt, P. A. (1968b). *Arch. Int. Physiol. Biochim.* **76,** 721.

Voogt, P. A. (1969a). Unpublished data.

Voogt, P. A. (1969b). *Comp. Biochem. Physiol.* **31,** 37.

Voogt, P. A. (1970). Ph.D. Thesis, University of Utrecht.

Voogt, P. A. (1971a). *In* "Experiments in Physiology and Biochemistry" (G. Kerkut, ed.). Academic Press, New York **4,** 1–33.

Voogt, P. A. (1971b). *Arch. Int. Physiol. Biochim.* **79,** 391.

Voogt, P. A. (1971c). *Comp. Biochem. Physiol* **39B,** 139.

Voogt, P. A. (1971d). To be published.

Voogt, P. A., and Oudejans, R. C. H. M. (1968). Unpublished data.

Voogt, P. A., and Van der Horst, D. J. (1968). Unpublished data.

Voogt, P. A., and Van der Horst, D. J. (1969). Unpublished data.
Voogt, P. A., and Van Gennip, A. H. (1969). Unpublished data.
Wainai, T., Tamura, T., Truscott, B., and Idler, D. R. (1964). *J. Fish. Res. Bd. Can.* **21**, 1543.
White, H. B. (1966). *J. Chromatogr.* **2**, 213.
White, H. B., and Williams, W. L. (1965). *Fed. Proc., Fed. Amer. Soc. Exp. Biol.* **24**, 662.
Wiss, O., and Gloor, U. (1969). *Biochem. J.* **113**, 20P.
Wykle, R. L., and Snyder, F. (1969). *Biochem. Biophys. Res. Commun.* **37**, 658.
Yamakawa, T., and Ueta, N. (1964). *Jap. J. Exp. Med.* **34**, 37.
Yasuda, S. (1966). *Yukagaku* **15**, 50.
Yasuda, S. (1967). *Yukagaku* **16**, 596.
Zama, K. (1963). *Mem. Fac. Fish., Hokkaido Univ.* **11**, 1.
Zama, K., Hatano, M., and Igarashi, H. (1960). *Bull. Jap. Soc. Sci. Fish.* **26**, 917.
Zandee, D. I. (1967). *Arch. Int. Physiol. Biochim.* **75**, 487.

CHAPTER 10

Nitrogen Metabolism in Mollusks

Marcel Florkin and S. Bricteux-Grégoire

I. Nitrogenous Constituents of Molluscan Organisms

A. Amino Acids

A large proportion of the dry matter of molluscan bodies is made up of amino acids, either free or combined to form macromolecules, among which the proteins are predominant. The pattern of the amino acid composition of the bulk of the proteins of molluscan tissues is similar to the pattern generally observed in animal tissues (Fig. 1 in

Duchâteau *et al.*, 1954). On the other hand, no common pattern is observed in the free amino acid pools of molluscan tissues (Tables I–V).

1. *Free Amino Acids*

Data have been gathered concerning the composition of the proteins of a given tissue and of the free amino acid pool of the same tissues (Tables I–III). These compositions are different, a conclusion which is not unexpected. Each cell contains a large number of proteins, enzymes or others, and they differ with respect to their composition and amino acid sequences. The concentrations of these proteins are different, and each has a different speed of turnover. In the muscle tissue of the rabbit, for instance, the turnover is twice as rapid for aldolase as it is for glyceraldehyde–3-phosphate dehydrogenase (M. V. Simpson and Velick, 1954; Heimberg and Velick, 1954). There is nothing astonishing, therefore, in the fact that the amino acid pattern of the proteins of a tissue differs from the free amino acid pool in the same tissue.

Chemical isolation and characterization of free amino acids from molluscan tissues goes back to the nineteenth century. Chittenden (1875) isolated glycine from the muscles of *Pecten irradians*, Ackermann (1922) obtained arginine from *Mytilus edulis*, and Ackermann *et al.* (1922) obtained arginine from *Eledone moschata*. Owing to analytical progress, and particularly to the advent of the microbiological method for the determination of amino acids, more complete analyses were made possible. Noland (1949) used this method for determinations of some amino acids in mollusks. In the laboratory of the present authors, the microbiological method was used in a number of analyses (Tables I–IV) and later replaced by the more reliable method of automatic analyses by column chromatography according to Spackman *et al.* (1958; Bricteux-Grégoire *et al.*, 1964a,b). While these analyses have been mostly concerned with the amino acids found in proteins, free amino acids also have been detected, but most studies have been directed toward the role of free amino acids as osmoregulators (see Chapter 13). Campbell and Speeg (1968a) have reported the concentration of whole body free amino acids in terrestrial gastropods. The levels of free amino acids in different tissues and in different species of mollusks have been reviewed recently by Campbell and Bishop (1970).

2. *Nonprotein Free Amino Acids*

Nonprotein free amino acids have also been detected. J. W. Simpson *et al.* (1959) found β-alanine in several species of mollusks. Awapara

TABLE I
DISTRIBUTION OF AMINO ACIDS IN THE PROTEIN AND NONPROTEIN COMPONENTS OF MUSCLES IN BIVALVIA[a]

Amino acid	*Mytilus edulis* adductor muscle[b]			*Ostrea edulis* total adductor muscles[b]			*Anodonta cygnea* total adductor muscles[b]		
	Total hydrolyzed	Nonprotein	Protein	Total hydrolyzed	Nonprotein	Protein	Total hydrolyzed	Nonprotein	Protein
Alanine	2300	163	2137	1350	364	986	720	3.2	716.8
Arginine	2770	199	2571	920	37.5	882.5	1040	13.3	1026.7
Aspartic acid	3160	45.4	3114.6	1110	14.7	1095.3	1290	1.6	1288.4
Glutamic acid	5020	152	4868	2230	149	2081	1730	10.7	1719.3
Glycine	1350	191	1159	1450	140	1310	700	4.8	695.2
Histidine	1420	5.8	1414.2	250	12.9	237.1	200	0.9	199.1
Isoleucine	1550	11.9	1538.1	860	10.8	849.2	620	2.3	617.7
Leucine	2380	7.4	2373.6	1380	7.3	1372.7	1040	1.3	1038.7
Lysine	2210	18.9	2191.1	1520	12.4	1507.6	1060	3	1057.0
Methionine	710	4.7	705.3	420	4.8	415.2	290	0.1	289.9
Phenylalanine	770	4.6	751.5	540	4.8	535.2	370	0.6	369.4
Proline	790	13.9	776.1	630	93.7	536.3	450	0.4	449.6
Threonine	1280	14.6	1265.4	590	5.5	584.5	510	1.3	508.7
Tyrosine	900	6.1	893.9	430	5.8	424.2	350	0.8	349.2
Valine	1210	6.9	1203.1	730	6.1	723.9	500	1.2	498.8
Total	27820	845.2	26964.9	14410	869.3	13540.7	10870	45.5	10824.5

[a] Determination by microbiological assay. Values stated in milligrams per 100 gm of fresh tissue.
[b] Duchâteau and Florkin (1951).

and Allen (1959) isolated β-aminoisobutyric acid from *M. edulis.* As β-alanine and β-aminoisobutyric acid are the products of pyrimidine metabolism, their study is of importance. Campbell and Speeg (1968a) found β-alanine and β-aminoisobutyric acid in *Helix aspersa* and *Otala lactea.* Negus (1968) found β-alanine plus an unknown amino acid in the estuarine snail *Hydrobia ulvae.*

α-aminobutyric acid has been detected in a number of mollusks including cephalopods (Deffner, 1961a) and freshwater and terrestrial gastropods (Friedl, 1961; Campbell and Speeg, 1968a).

Citrulline is an intermediate in arginine biosynthesis. Although its

TABLE II
DISTRIBUTION OF AMINO ACIDS IN THE PROTEIN AND NONPROTEIN

Amino acid	*Buccinum undatum* foot[b]			*Buccinum undatum* foot[b]			*Buccinum undatum* penis[c]		
	Total hydrolyzed	Nonprotein	Protein	Total hydrolyzed	Nonprotein	Protein	Total hydrolyzed	Nonprotein	Protein
Alanine	1170	153	1017	940	79.4	860.6	1150	52.2	1097.8
Arginine	1130	174	956	1310	184	1126	1800	199	1601
Aspartic acid	1230	9.9	1220.1	1360	14.4	1345.6	1830	21.5	1808.5
Glutamic acid	2020	72.2	1947.8	2370	88	2282	3050	74.5	2975.5
Glycine	1380	24.5	1355.5	1640	27	1613	1580	14.7	1565.3
Histidine	230	9.7	220.3	260	9.5	250.5	320	3.7	316.3
Isoleucine	590	16.9	573.1	770	5.9	764.1	1010	5.8	1004.2
Leucine	1090	15.3	1074.7	1360	7.3	1352.7	1680	5.1	1674.9
Lysine	910	13.6	896.4	1270	4.2	1265.8	1480	5.5	1474.5
Methionine	380	5.6	374.4	450	7.6	442.4	540	2.4	537.6
Phenylalanine	430	6.3	423.7	450	4.8	445.2	580	3.1	576.9
Proline	690	36	654	780	41	739	800	3.6	796.4
Threonine	590	24.1	565.9	650	16.2	633.8	890	7.4	882.6
Tyrosine	430	6.5	423.5	500	10.2	489.8	660	2.9	657.1
Valine	630	16.8	613.2	790	7.3	782.7	980	4.8	975.2
Total	12900	584.4	12315.6	14900	506.8	14393.2	18350	406.2	17943.8

[a] Determination by microbiological assay. Values stated in milligrams per 100 gm of
[b] Duchâteau *et al.* (1954).
[c] Duchâteau and Florkin (1951–1952).

presence in herbivorous gastropod mollusks (Campbell and Speeg, 1968a) might be related to their diet, owing to the high concentration of citrulline in some plants, it has also been found by Deffner (1961a) in carnivorous cephalopods such as *Loligo* and *Dosidicus* in about the same concentration (0.2–0.3 μmole/gm tissue).

Ornithine has been reported as a component of the free amino acid pool in certain tissues of the gastropods *O. lactea, H. aspersa, Limnaea stagnalis* (Campbell and Speeg, 1968a; Friedl, 1961). It has also been found in the lamellibranch *Crassostrea virginica* (Lynch and Wood, 1966) and in the cephalopods *Loligo pealii* and *Dosidicus gigas* (Deffner, 1961a).

COMPONENTS OF TISSUES IN GASTROPODS[a]

Haliotis tuberculata foot[c]			*Patella vulgata* foot[c]			*Limnaea stagnalis* foot[c]		
Total hydrolyzed	Nonprotein	Protein	Total hydrolyzed	Nonprotein	Protein	Total hydrolyzed	Nonprotein	Protein
560	19.9	540.1	1130	101	1029	1050	61.4	988.6
940	607.9	322.1	1410	320	1090	880	54.2	825.8
530	21.1	508.9	1150	53.5	1096.5	1860	84.2	1775.8
1030	51.9	978.1	2200	94.2	2105.8	1870	167	1703
1230	73.7	1156.3	1280	27.7	1252.3	1440	50.2	1389.8
94	7.4	86.6	170	3.1	166.9	210	7.8	202.2
330	16.8	313.2	—	13	—	810	20.7	789.3
510	8.8	501.2	1270	6.2	1263.8	1020	27.3	992.7
510	24.8	485.2	840	7	833	880	31.2	848.8
—	—	—	—	5.3	—	170	3.9	166.1
210	7.3	203.7	420	3.5	416.5	490	18.8	471.2
390	9.5	380.5	780	491	289	640	29.1	610.9
320	26.6	293.4	580	21.2	557.8	680	26	654
190	9.1	180.9	440	10.7	429.3	330	3.9	326.1
310	8.6	301.4	620	17.2	602.8	660	21.9	638.1
						12990	607.6	12382.4

fresh tissue.

3. *Proteins*

The conchiolin of molluscan shells has been extensively studied in Chapter 2 of this volume and respiratory proteins in Chapter 7.

Tropomyosin has been isolated from *Pinna nobilis* and *Octopus vulgaris* (Bailey, 1957), from *Aulacomya* (Milstein, 1967), from different muscles of *Mytilus* and *Holothuria* (Mognoni and Lanzavecchia, 1969), from *Pecten irradiens* (Bárány and Bárány, 1966), from *Pecten maximus* (Bailey and Rüegg, 1960), from *Venus mercenaria* (Riddiford and Scheraga, 1962). A rubberlike protein has been isolated by R. E. Kelly and Rice (1967) from the internal triangular hinge ligament of *Pecten.*

The lens protein of *O. vulgaris* has been studied by Bon *et al.* (1967),

TABLE III

Distribution of Amino Acids in the Protein and Nonprotein Components of Muscles in *Octopus vulgaris*[a]

Amino acid	Sucker[b]			Sucker[b]			Mantle[b]			Mantle[b]		
	Total hydrolyzed	Nonprotein	Protein	Total hydrolyzed	Nonprotein	Protein	Total hydrolyzed	Nonprotein	Protein	Total hydrolyzed	Nonprotein	Protein
Alanine	530	6.3	522.7	940	13.2	926.8	620	82	538	520	27.7	492.3
Arginine	760	179	581	1330	218	1112	760	305	455	740	136.3	603.7
Aspartic acid	780	5.7	774.3	1410	5.9	1404.1	800	66.5	733.5	680	13.5	666.5
Glutamic acid	1100	21.3	1078.7	2030	30.7	1999.3	1250	147	1103	1030	74.3	955.7
Glycine	1040	7.6	1032.4	1790	15.2	1774.8	1050	110.9	939.1	1350	34.8	1315.2
Histidine	150	0.5	149.5	280	1.3	278.7	170	9.6	161.4	140	1.4	138.6
Isoleucine	590	5.5	584.5	—	6	—	500	35.2	464.8	—	11.7	—
Leucine	610	3.5	606.5	1210	3.1	1206.9	680	38.1	641.9	580	4.7	575.3
Lysine	680	4.2	675.8	1290	4.1	1285.9	750	53.9	696.1	570	8.9	561.1
Methionine	—	1	—	—	—	—	—	—	—	—	—	—
Phenylalanine	290	1.1	288.9	550	1.8	548.2	330	16.8	313.2	270	2.5	267.5
Proline	460	2	458	710	0.5	709.5	490	58	432	490	11.8	478.2
Threonine	400	2.7	397.3	700	4	696	430	29.9	400.1	340	5.5	334.5
Tyrosine	270	1.2	268.8	500	3.4	496.6	310	15.9	294.1	250	3.1	246.9
Valine	350	2.3	347.7	670	2.5	667.5	400	21.5	378.5	310	4.9	305.1

[a] Determination by microbiological assay. Values stated in milligrams per 100 gm of fresh tissue.

[b] Duchâteau and Florkin (1951).

TABLE IV
FREE AMINO ACIDS FOR THREE GASTROPODS[a]

Amino acid	*Ampullaria glauca*		*Helix pomatia* hepatopancreas[c]		*Telescopium telescopium* L. crystalline style[d]
	Hepato-pancreas[b]	Shell gland[b]	Non-hydrolyzed	Hydro-lyzed	
Alanine	—	—	19.9	18.5	2.9
Arginine	15.9	3.7	—	—	—
Aspartic acid	27.6	2.0	11.5	29.9	4.2
Glutamic acid	67.7	12.9	110.1	113.6	2.6
Glycine	45.2	2.4	17.8	26.2	7.4
Histidine	4.6	0.2	2.9	2.1	—
Isoleucine	19.2	1.2	8.6	9.1	0.2
Leucine	26.0	1.9	6.8	8.6	—
Lysine	8.1	2.4	4.7	6.7	—
Methionine	14.7	0.0	2.6	2.6	—
Phenylalanine	9.3	1.1	2.8	4.4	—
Proline	14.9	1.2	9.8	11.5	—
Threonine	18.5	2.7	8.4	10.8	6.7
Tyrosine	4.4	0.6	—	—	6.3
Valine	19.0	1.5	6.3	7.7	5.5

[a] Determination by microbiological assay. Values stated as milligrams per 100 gm of fresh tissue.
[b] Duchâteau and Florkin, in Florkin (1954).
[c] Duchâteau and Florkin (1957).
[d] Swaminathan (1958).

while Smith (1969) presents data concerning the lens protein of the squid *Notodarus hawaïensis*. Some data exist on the collagen of *M. edulis* (Pikkarainen *et al.*, 1968). Hunt (1970a) has extensively studied the protein of the operculum of the gastropod mollusc *Buccinum undatum*. A very exhaustive study of mucopolysaccharide and glycoproteins of invertebrates has been published recently by Hunt (1970b). Many enzymes have been detected in mollusks but only a few have been isolated and purified. An oyster hemagglutinin which has no obvious similarities with mammalian immunoglobins has been isolated by Acton *et al.* (1969).

B. Taurine

This sulfonic amino acid

(2-aminoethanesulfonic acid; $NH_2CH_2CH_2SO_3H$)

has been reported for several classes of mollusks (Cephalopoda—Kruken-

TABLE V
FREE AMINO ACIDS IN THREE SPECIES OF CEPHALOPODS[a]

Amino acid	*Ommastrephes sloani pacificus* Non-hydrolyzed	*Ommastrephes sloani pacificus* Hydrolyzed	*Sepia esculenta*[c]	*Octopus ochellatus*[c]
Alanine	75	83	179	15
Arginine	99	98	280	146
Aspartic acid	Trace	12	29	22
Glutamic acid	26	61	44	29
Glycine	42	82	104	23
Histidine	140	139	12	1.8
Isoleucine	15	25	12	6.3
Leucine	18	28	23	6.3
Lysine	9	24	32	7.6
Methionine	12	11	32	3.6
Phenylalanine	6	8	9.3	4.3
Proline	479	553	379	8.5
Threonine	23	28	53	7.3
Tyrosine	4	3	8.4	1.7
Valine	20	31	26	9.2

[a] Determination by microbiological assay. Values stated as milligrams per 100 gm of fresh muscle.
[b] Konosu *et al.* (1958).
[c] Ito (1957).

berg and Wagner, 1885; Henze, 1905, 1913, 1914; Suzuki and Joshimura, 1909; Okuda, 1929; Ackermann *et al.*, 1924; Gastropoda—Mendel, 1904; Mendel and Bradley, 1906; Schmidt and Watson, 1918; *M. edulis*—A. Kelly, 1904). J. W. Simpson *et al.* (1959) have studied the distribution of taurine in mollusks and have found it in high concentration in all the marine species studied, but not in freshwater and terrestrial forms (or less than 0.1 μmole/gm). This fact should be correlated with the role of taurine as one of the effectors of intracellular osmolar concentration (Bricteux-Grégoire *et al.*, 1964a,b) which is decreased in marine bivalves transferred into brackish water. It may be postulated that the adaptation to life in fresh water is accompanied by an even more pronounced lowering of the level of the intracellular concentration of taurine.

Taurine is mainly an intracellular component. Its concentration in *M. edulis* amounts to 2.4 μmole/ml in the hemolymph, compared to 72 μmole/gm in the muscle. Isethionic acid (2-hydroxyethanesulfonic acid; $OHCH_2CH_2SO_3H$), which may be metabolically related to taurine, has been demonstrated in squid axoplasm by Koechlin (1954, 1955) where it accounts for 20% of the dialyzable anions, and in the blood of *Loligo* by Deffner (1961a). Enzymatic oxidation of cysteamine to taurine was observed in *M. edulis* (Yoneda, 1968).

C. Ammonia

Older data on blood ammonia in invertebrates, and in mollusks in particular, have been collected in tables by Delaunay (1931, 1934). In this research, no attention had been given to the eventual variations the ammonia could show *in vitro* during the period of time before analysis. Measuring the amount of ammonia in the blood of *Anodonta cygnea* collected under paraffin, Florkin and Houet (1938) detected the presence of 51–71 μg of ammonia per 100 ml. In the blood of the snail *Helix pomatia,* Florkin and Renwart (1939) found much larger values, between 0.7 and 2 mg of ammonia per 100 ml. Contrary to what occurs in the blood of mammals and of birds, there is no increase in the concentration of ammonia more than an hour after the collection of the blood. Ammonia has been reported present in the excreta of a number of mollusks (for references, see Albritton, 1955).

D. Amines

The decarboxylation of amino acids results in the formation of amines which can be considered as substitution products of ammonia and have the following structures: RNH_2, RR_1NH, and RR_1R_2N.

1. *5-Hydroxytryptamine (5-HT, Enteramine, Serotonin)*

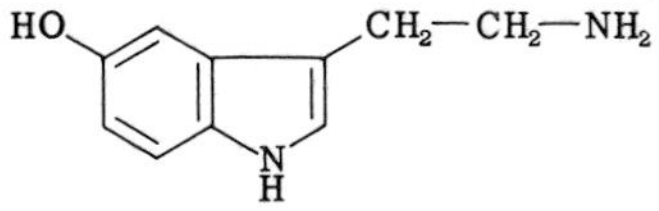

HO $CH_2—CH_2—NH_2$ N H

5-Hydroxytryptamine has been found in the posterior salivary glands of *Octopus vulgaris* and *Eledone moschata* (Vialli and Erspamer, 1940; Erspamer, 1940; Erspamer and Boretti, 1951), in brain, heart and mantle of *Helix aspersa* (Kerkut and Cottrell, 1963). Studying its repartition in different tissues of *Venus mercenaria* and *Busycon canaliculatum,* Welsh and Moorhead (1959) found it particularly abundant in nerve and heart tissue. Zs.-Nagy (1967) found it in the majority of nerve cells of the three ganglia of the lamellibranch *Anodonta cygnea,* but failed to find it in axons. Roseghini and Ramorino (1970) detected it in the optic ganglia, cerebral ganglia, and mantle nerve of the squid *Dosidicus gigas.*

Its synthesis from its immediate precursor, 5-hydroxytryptophan was found to occur in the presence of nerve extracts of *Mercenaria mercenaria* (Welsh and Moorhead, 1959), of nerve and heart extracts of *Helix pomatia* (Cardot, 1966a), of brain homogenates of *H. aspersa* (Kerkut and Cottrell, 1963), of ganglia of *Busycon canaliculatum* (Mirolli, 1968) and in *Octopus* (Hartman *et al.*, 1960).

According to Florey and Florey (1954), 5-hydroxytryptamine could represent the nerve transmitter in the heart of cephalopods. Kerkut *et al.* (1967) suggest that the cells at the junction of the visceral and right parietal ganglia of the brain of *H. aspersa* contain both dopamine and 5-hydroxytryptamine. The distribution of 5-hydroxytryptamine and dopamine in different ganglia of a fresh-water bivalve mollusk *Sphaerium sulcatum* has been studied by Sweeney (1968). The sequence postulated by S.-Rozsa and Zs.-Nagy (1967) in the heart of the snail *Limnaea stagnalis* is: cardiac nerve → natural excitatory transmitter → catecholamine → 5-hydroxytryptamine → heart muscle contracts.

2. *Tyramine (4-Hydroxyphenylethylamine, $HOC_6H_4CH_2CH_2NH_2$)*

Tyramine has been isolated by Henze (1913) in the form of a dibenzoyl derivative from extracts of the posterior salivary glands of *Octopus macropus*, and by Erspamer (1952) in the form of a dibenzoyl derivative and a picrate from the posterior salivary gland of *Octopus vulgaris*.

Tyramine has been considered by several authors to be the poison of cephalopod saliva (Henze, 1913; Baglioni, 1909). Fredericq (1947) has considered tyramine as playing a role in the humoral transmission of the impulses of the cardiomotor nerves in cephalopods. It appears, however, that tyramine is not necessarily present in the toxic secretions of octopods and even less in the toxic secretions of cephalopods (Erspamer and Boretti, 1951). Tyramine is found in large amounts in the posterior glands of *O. macropus* (1000–1500 μg per gram of fresh tissue) and of *O. vulgaris* (500–800 μg per gram of fresh tissue), but is lacking in the glands of *E. moschata* and *Sepia officinalis* (Erspamer, 1952). It has also been found in the urine of *Octopus hongkongiensis* (Emanuel, 1957).

3. *Octopamine (p-Hydroxyphenylethanolamine)*

OH
|
(benzene ring)
|
$CHOH—CH_2—NH_2$

Octopamine is found in large amounts in the salivary glands of *O. vulgaris* (700–1200 μg per gram fresh tissue). It is less abundant in the glands of *E. moschata,* and is present only in small amounts in the gland of *O. macropus* (Erspamer, 1952).

4. Histamine (4-imidazolethylamine)

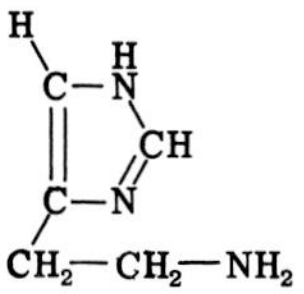

```
H
 \   H
  C—N
  ‖   \
  ‖    CH
  ‖   //
  C—N
 /
CH2—CH2—NH2
```

This compound has been found in the extracts of the posterior gland of *O. macropus* and of *E. moschata,* in varying and smaller amounts in *O. vulgaris* (Ungar *et al.*, 1937; Bacq and Ghiretti, 1951; Erspamer, 1952) and in the optic ganglia, cerebral ganglia, and mantle nerve of *Dosidicus gigas* (Roseghini and Ramorino, 1970). It has also been isolated from the salivary glands of the marine gastropod *Neptunea arthritica* (Asano and Ito, 1960). Its repartition in different tissues of the snail *H. aspersa* has been studied by Woodruff *et al.* (1969). The possible role of histamine as a neurotransmitter substance in *H. aspersa* was demonstrated by Kerkut *et al.* (1968).

5. Tetramine ($C_4H_{13}ON$ or $(CH_3)_4NOH$)

Tetramine has been isolated from the salivary glands of the marine gastropods *N. arthritica* Bernardi, *Neptunea intersculpta, Fusitriton oregonensis, Buccinum leucostoma* (Asano and Ito, 1959, 1960), and *Neptunea antiqua* (Fänge, 1958). It has also been identified in the nerve axoplasm of *D. gigas* (Deffner, 1961b).

E. UREA

Urea (H_2NCONH_2) has been found in small amounts in the tissues of *Helix pomatia* (Albrecht, 1923, Delaunay, 1931). It has been reported to be present in the excreta of a number of mollusks (see Albritton, 1955, Table 112, pp. 206 and 207).

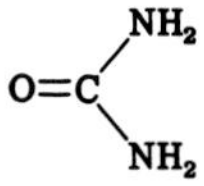

```
      NH2            NH2
     /              /
O=C          HN=C
     \              \
      NH2            NH2
```

Urea 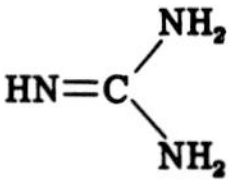 **Guanidine**

F. Guanidic Derivatives

Guanidine is an imino derivative of urea. A number of derivatives of guanidine have been found in mollusks:

1. An amino acid:

$$HN{=}C(NH_2){-}NH{-}(CH_2)_3{-}CH(NH_2){-}COOH$$

Arginine (α-amino-δ-guanidinovaleric acid)

2. ω-Guanidic acids:

$$HN{=}C(NH_2){-}NH{-}(CH_2)_3{-}COOH$$

γ-Guanidinobutyric acid

$$HN{=}C(NH_2){-}NH{-}(CH_2)_3{-}CO{-}COOH$$

α-Keto-δ-guanidinovaleric acid

3. Bases:

$$HN{=}C(NH_2){-}NH{-}(CH_2)_4{-}NH_2$$

Agmatine

$$HN{=}C(NH_2){-}NH{-}(CH_2)_4{-}NHCH_3$$

Methylagmatine

$$HN{=}C(NH_2){-}NH{-}(CH_2)_4{-}HN{-}C(NH_2){=}NH$$

Arcaine

4. Condensation products of arginine:

$$HN{=}C(NH_2){-}NH{-}(CH_2)_3{-}CH(COOH){-}NH{-}CH(CH_3){-}COOH$$

Octopine

In many animal species, arginine is the basis of the phosphagen phosphoarginine. For all mollusks so far examined, phosphoarginine appears to be the sole phosphagen utilized in the phylum (Meyerhof, 1928; Lohmann, 1936; Robin *et al.*, 1959a; Ennor and Morrison, 1958; Robin, 1960).

Phosphoarginine has been isolated as the barium salt from *Arca noae* (Roche *et al.*, 1960), from the cephalopods *Sepia officinalis, Loligo forbesi* and *Eledone cirrhosa* (Robertson, 1965), and from *Mytilus californianus* muscle (Seraydarian and Kalvaitis, 1964).

The enzyme arginine kinase catalyzing the reaction

$$\text{L-arginine} + \text{ATP} \rightleftharpoons \text{phosphoarginine} + \text{ADP}$$

has been demonstrated by Virden and Watts (1964) in muscle tissues from the bivalves *Pecten maximus* and *Chlamys opercularis.* The molluscan arginine kinase exists in two molecular weight forms, a 40,000 molecular weight enzyme, which seems to be associated with muscles of tonic function, and an 80,000 molecular weight enzyme, which is associated with muscles of more rapid movement (Moreland and Watts, 1967). The evolution of the phosphokinases in invertebrates is discussed by Watts and Watts (1968).

γ-Guanidinobutyric acid has been detected in a number of species, notably, *P. maximus, Mytilus edulis,* and *S. officinalis* (Thoai *et al.*, 1953). Agmatine is present in *Octopus vulgaris* (Irvin and Wilson, 1939) and in *Eledone moschata* (Ackermann and Mohr, 1937) as well as in the bivalve *A. noae* (Robin *et al.*, 1959b). There is no example, so far, of the presence of creatine, phosphocreatine, methylguanidine, glycocyamine, or taurocyamine in a mollusk (Roche *et al.*, 1957). Arcaine was found in *A. noae* (Ackermann, 1931; Kutscher *et al.*, 1931).

A number of other guanidine derivatives have been found in mollusks—methylagmatine in the muscle of *Octopus* (Iseki, 1931); octopine (Okuda, 1929; Ackermann and Mohr, 1937; Mayeda, 1936; Moore and Wilson, 1936, 1937; Irvin, 1938; Roche *et al.*, 1952b; Humoto, 1954); and α-keto-δ-guanidinovaleric acid, along with γ-guanidinobutyric acids in many species (Thoai *et al.*, 1952, 1953; Roche *et al.*, 1952b; Robin and Thoai, 1957).

G. Methylated Bases

1. N-Methylpyridiniumhydroxide (C_6H_9ON)

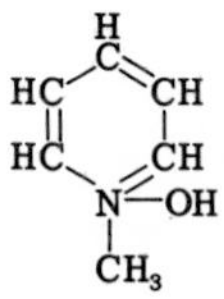

H
C
HC CH
HC CH
N—OH
CH$_3$

This compound has been isolated from *Mytilus edulis* by Ackermann (1922), from five species of the marine snail *Conus* by Kohn *et al.* (1960), and from cephalopod axoplasm (Deffner, 1961b).

2. *Trimethylamino Oxide* ($(CH_3)_3NO$)

Henze (1914), Kojima and Kusakabe (1956), Asano and Sato (1954), Konosu *et al.* (1958), Robertson (1965), and Norris and Benoit (1945) have reported the isolation of this compound from muscles of cephalopods or bivalves. Other authors failed to identify it, or trimethylamine, in the urine of *Octopus vulgaris* (Hoppe-Seyler and Linneweh, 1931), in muscles of cephalopods (Ackermann *et al.*, 1922, 1924), or in the adductor muscles of *M. edulis* (Bricteux-Grégoire *et al.*, 1964a) or of *Ostrea edulis* (Bricteux-Grégoire *et al.*, 1964b).

3. *Glycine Betaine* ($(CH_3)_3—N^+—CH_2—COOH$)

This substance has been found in high concentration in a number of mollusks—*Pecten irradians, Sycotypus canaliculatus, M. edulis, O. vulgaris, Eledone moschata, Arca noae* (see Ackermann, 1962), *O. edulis* (Bricteux-Grégoire *et al.*, 1964b), *Octopus honkongiensis* (urine) (Emanuel, 1957), *Patella* (Ackermann and Janka, 1954), *Arion empiricorum* (Ackermann and Menssen, 1960), *Polypus punctatus* (Takahashi, 1915), *Thais haemastomata* (Beers, 1967), and in three species of cephalopods (Robertson, 1965).

4. *Carnitine* (*β-Oxy-γ-butyrobetaine*) ($C_7H_{15}O_3N$ *or* $(CH_3)_3—N^+—CH_2—CHOH—CH_2COO^-$)

This betaine has been isolated from *Octopus octopodia* (Morizawa, 1927) and from *A. noae* (Kutscher and Ackermann, 1933). Its distribution in different tissues of *Ensis directus, Venus mercenaria, Busycon canaliculatum,* and *Loligo pealeii* has been studied by Fraenkel (1954). Carnitine, which plays a general role in the mitochondrial transport of fatty acids can be formed from γ-butyrobetaine, which has been detected in the marine snail *Conus* (Kohn *et al.*, 1960). The presence of an enzyme catalyzing the transfer of an acyl group from palmityl-CoA to carnitine has been demonstrated in the marine gastropod *Buccinum undatum* and the bivalves *M. edulis* and *Modiolus modiolus* (Norum and Bremer, 1966).

5. *Stachydrine* (*N-Methyl-*DL*-prolinemethylbetaine*) ($C_7H_{13}O_2N + H_2O$)

Stachydrine has been found in *A. noae* (Kutscher and Ackermann, 1933).

6. *Homarine (Methylbetaine of α-Pyridine-2-carboxylic Acid or Methylpicolinic Acid)*

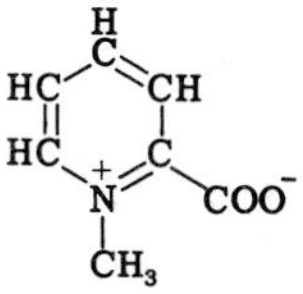

$$\text{HC}=\text{CH}-\text{CH}=\text{CH}-\text{C}(\text{COO}^-)=\text{N}^+(\text{CH}_3) \text{ (ring)}$$

Isolated from *A. noae* by Hoppe-Seyler (1933), homarine has since been found in the tissues of a number of marine mollusks belonging to the genera *Patella* (Ackermann and Janka, 1954), *Arion* (Ackermann and Menssen, 1960), *Pecten, Venus, Nassa, Loligo,* and *Busycon* (Gasteiger *et al.,* 1955) and in the muscle of the abalone (Nishita *et al.,* 1965). It has also been identified in the venom of the marine snail *Conus* (Kohn *et al.,* 1960). It does not appear to exist in freshwater invertebrates or in vertebrates (Nishita *et al.,* 1965).

The role of *N*-methylated bases in mollusks is not known. Some of them, like glycine betaine are certainly involved in osmotic regulation (Bricteux-Grégoire *et al.,* 1964a,b; see also Chapter 12 by Schoffeniels and Gilles). They may also act as methyl donors. Ericson (1960) has purified an enzyme (EC 2.1.1.5) from *Anodonta cygnea* hepatopancreas which catalyzes the reaction

$$\underset{\text{Glycine betaine}}{(CH_3)_3N^+{-}CH_2{-}COOH} + \underset{\text{Homocysteine}}{HS{-}CH_2{-}CH_2{-}\underset{\substack{|\\NH_2}}{CH}{-}COOH} \rightarrow$$

$$\underset{\text{Dimethylglycine}}{(CH_3)_2N^+H{-}CH_2{-}COOH} + \underset{\text{Methionine}}{CH_3{-}S{-}CH_2{-}CH_2{-}\underset{\substack{|\\NH_2}}{CH}{-}COOH}$$

Campbell and Speeg (1968a) have identified sarcosine, or monomethylglycine, which may be a product of dimethylglycine, as a free component in the hepatopancreas of *Helix aspersa.*

$$\underset{\text{Sarcosine}}{CH_3{-}NH{-}CH_2{-}COOH}$$

H. Purines and Uric Acid

Besides their presence as constituents of nucleic acids, the existence of purines in the free state has been demonstrated in the bodies of mollusks. Uric acid, which is major excretory product in terrestrial species, has been found in various amounts in extrarenal tissues of several mollusks by Needham (1935). Ishida (1956) has found uric acid in soft tissues and in the byssus of *Mytilus edulis* and *Ostrea gigas.* In addition to uric acid, other purines are also present in the free state. The presence of hypoxanthine in the urine of cephalopods was demonstrated by von Fürth (1900—confirmed by Hoppe-Seyler and Linneweh, 1931). Guanine and hypoxanthine have been detected in the urine of *Octopus hongkongiensis* (Emanuel, 1957) and in *Arion empiricorum* (Menssen, 1960). E. Farina *et al.* (1962) identified adenine, guanine, and hypoxanthine in the white bodies of *Octopus vulgaris.* Hypoxanthine was found in nerve axoplasm of *Loligo pealii* and *Dosidicus gigas* (Deffner, 1961a).

Free tissue purines are commonly found in gastropods. Thiele (1963) was able to extract 6.5 mg of a mixture of purines consisting mainly of xanthine and guanine from the body of one snail. Xanthine and guanine occur as free bases in the *Helix pomatia* albumin gland (Jezewska, 1968a) and xanthine and hypoxanthine in the hermaphrodite gland (Jezewska, 1968b). The accumulation of uric acid and other purines during estivation has been studied in the blood and in different tissues of the snail *Pila globosa* by Raghupathiramireddy and Swamy (1967), in *Otala lactea* by Lee and Campbell (1965), and in *H. pomatia* by Jezewska (1968a) and Jezewska and Sawicka, (1968).

Purines and pyrimidines also occur in combined state in mollusks, as nucleosides and ribonucleotides, as well as being constituents of nucleic acids (Seki *et al.*, 1967; Jezewska and Sawicka, 1968). The ribonucleotide content of molluscan tissues has been reviewed by Campbell and Bishop (1970).

II. Metabolism of Nitrogenous Compounds

A. Amino Acid Oxidases and Deamination

Blaschko and Hawkins (1951) have shown the presence of D-amino acid oxidase in the extracts of the hepatopancreas of *Octopus vulgaris* and *Sepia officinalis.* A specific D-amino acid oxidase for D-glutamic and D-aspartic acids has also been found in the hepatopancreas of *Eusepia officinalis, Loligo forbesi, O. vulgaris,* and *Eledone cirrosa* (Blaschko and Himms, 1955). Blaschko and Hawkins (1952) have confirmed the existence of D-amino acid oxidase in the hepatopancreas of *Octopus* and of *Sepia* and have shown its existence in the hepatopancreas of *Helix aspersa* (confirmed by Blaschko and Hope, 1956). Sarlet *et al.* (1950) have not found a D-amino acid oxidase in the hepatopancreas of *Anodonta cygnea* (confirmed by Blaschko and Hawkins, 1952). Furthermore, Villee (1947) has not found the enzyme in the hepatopancreas of *Buccinum* or in the adductor muscle of the scallop; nor have Blaschko and Hope (1956) observed this enzyme in the hepatopancreas of *Mytilus edulis*. Rocca and Ghiretti (1958) purified the D-glutamate oxidase from *O. vulgaris* and identified the prosthetic group of the octopus enzyme as flavine adenine dinucleotide (FAD). It is inhibited by barbiturates (Casola *et al.*, 1964).

The first demonstration of the presence of an L-amino acid oxidase in a mollusk is due to Roche *et al.* (1952a). These authors noticed that in the extracts of the organs of many invertebrates, arginine is present, as well as α-keto-δ-guanidinovaleric acid and γ-guanidinobutyric acid. This led them to the hypothesis that, as opposed to vertebrates which transfer arginine through the action of arginase, these invertebrates oxidize arginine in the presence of their own L-amino acid oxidase which acts on arginine, while the corresponding enzyme of vertebrates does not. In fact, they found that an L-amino acid oxidase is present in the hepatopancreas of a number of mollusks (*M. edulis, S. officinalis, Aplysia* sp.) which oxidizes L-arginine, L-ornithine, L-lysine, and L-citrulline, and which is without action on proline. (The mammalian L-amino acid oxidase acts on proline, but not on arginine, ornithine, and lysine).

Blaschko and Hope (1956) have confirmed the existence in the hepatopancreas of *M. edulis* of the enzyme discovered by Roche *et al.* (1952a). Glahn *et al.* (1955) have isolated from the hepatopancreas of *Cardium tuberculatum* an L-amino acid oxidase which oxidizes L-histidine. This enzyme was later purified by Roche *et al.* (1959) and found to require Mg^{2+} for activity. The hepatopancreas of neither *Pecten maximus* nor *Ostrea edulis* appears capable of effecting this oxidation, while in *Solen ensis* and *Mya arenaria* the hepatopancreas does it as well as that of *C. tuberculatum*. The two types of L-amino acid oxidases (type *Mytilus* and type *Cardium*) act on L-citrulline, L-ornithine, and L-histidine, whereas the amino acid oxidase of snake venoms is without action on these substrates. They differ in their optimum pH (*Cardium*, 9.2; *Mytilus*, 7.0), and the *Cardium* enzyme appears to be less active on basic amino acids. In contrast to the mammalian enzyme, which does not act on the basic amino acids, neither of the molluscan enzymes act on the aliphatic amino acids alanine, glycine, isoleucine, valine, aspartic acid, or glutamic acid.

Robin and Thoai (1957) have shown that the hepatopancreas and the muscles of *Limnaea stagnalis* contain an L-amino acid oxidase which acts on L-arginine with the formation of α-keto-δ-guanidinovaleric acid and γ-guanidinobutyric acid. This enzyme also acts on lysine and ornithine (Olomucki *et al.*, 1960). For a review, see Table VI in Campbell and Bishop (1970).

B. Metabolism of Arginine

1. *Conversion of Arginine to Guanidic Derivatives*

The conditions of biogenesis which explain the distribution of guanidic derivatives have been reviewed by Thoai (1960). Some of the derivatives are direct or indirect products of arginine degradation. Mollusks as well as other phyla of invertebrates possess, as stated above, an L-amino acid oxidase acting on arginine (contrary to the case in vertebrates). In many mollusks, the simultaneous presence of arginine and L-amino acid oxidase in muscles and in hepatopancreas, demonstrated in a number of cases, could explain the presence of α-keto-δ-guanidinovaleric acid and of γ-guanidinobutyric acid as a result of an oxidative deamination of arginine *in situ.*

It is assumed that agmatine results from the decarboxylation of arginine and that methylagmatine results from the methylation of agmatine. It is possible to consider arcaine as resulting from the amidination of agmatine. In favor of this scheme is the finding that arcaine accompanies agmatine in animals living in polluted water, e.g., in *Arca noae* found on rocky bottoms (Kutscher *et al.*, 1931; Robin *et al.*, 1959b). However, the reaction itself has not been observed thus far in any mollusk, and other hypotheses have been proposed to explain the origin of arcaine (Zervas and Bergmann, 1931; Kutscher and Ackermann, 1931).

Octopine was first isolated from the muscle of *Octopus* by Morizawa (1927). Moore and Wilson (1937), Irvin (1938), and Irvin and Wilson (1939) have demonstrated that arginine is the natural precursor of octopine. Knoop and Martius (1939), who chemically synthetized octopine by reductive condensation of arginine and pyruvic acid, proposed that the biosynthesis of octopine involves a condensation of arginine with pyruvic acid to form a Schiff's base which is then reduced to octopine as shown in Scheme I. That this scheme is the biosynthetic one acting *in vivo* has been proved by Thoai and Robin (1959a,b, 1960, 1961) using extracts of mantle of *Sepia officinalis,* adductor muscles of *Pecten maximus* and *Cardium edule,* and foot muscle of *Cardium edule* and of *Limnea stagnalis* to which NADH, arginine, and pyruvic acid (not lactic acid) were added. Octopine is thus derived from the arginine liberated by the hydrolysis of the phosphagen phosphoarginine and the pyruvic acid resulting from glycolysis which is not reduced to lactic acid.

The overall reaction of octopine synthesis is

$$\text{arginine} + \text{pyruvic acid} + \text{NADH} + \text{H}^+ \rightarrow \text{octopine} + \text{NAD}^+ + \text{H}_2\text{O}$$

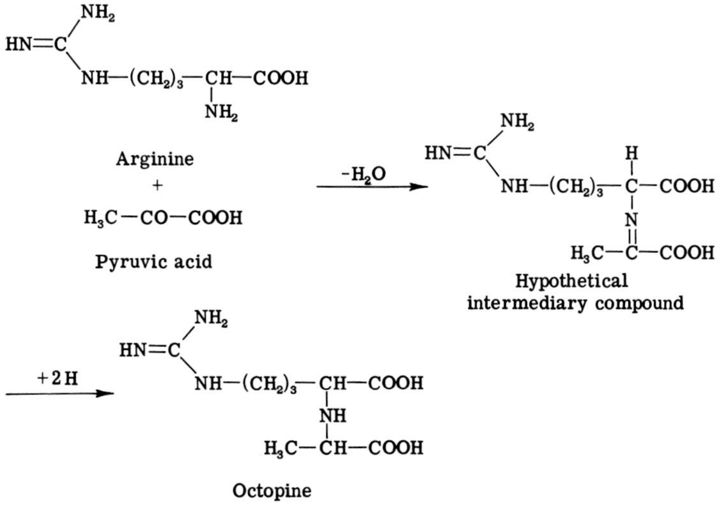

$$\text{HN}{=}\text{C}\begin{matrix}\diagup \text{NH}_2 \\ \diagdown \text{NH}{-}(\text{CH}_2)_3{-}\underset{\displaystyle |\atop \displaystyle \text{NH}_2}{\text{CH}}{-}\text{COOH}\end{matrix}$$

Arginine
+
$H_3C-CO-COOH$
Pyruvic acid

$$\xrightarrow{-H_2O} \text{HN}{=}\text{C}\begin{matrix}\diagup \text{NH}_2 \\ \diagdown \text{NH}{-}(\text{CH}_2)_3{-}\overset{\displaystyle \text{H}\atop \displaystyle |}{\underset{\displaystyle |\atop \displaystyle \text{N}}{\text{C}}}{-}\text{COOH} \\ \qquad\qquad \| \\ \qquad\qquad \text{H}_3\text{C}{-}\text{C}{-}\text{COOH}\end{matrix}$$

Hypothetical intermediary compound

$$\xrightarrow{+2H} \text{HN}{=}\text{C}\begin{matrix}\diagup \text{NH}_2 \\ \diagdown \text{NH}{-}(\text{CH}_2)_3{-}\underset{\displaystyle |\atop \displaystyle \text{NH}}{\text{CH}}{-}\text{COOH} \\ \qquad\qquad | \\ \qquad\qquad \text{H}_3\text{C}{-}\text{CH}{-}\text{COOH}\end{matrix}$$

Octopine

SCHEME I.

The enzyme system is not found in the hepatopancreas, but only in muscle. Some of the properties of octopine synthetase have been described by Thoai and Robin (1961).

In *Limnaea stagnalis,* as in other mollusks and other invertebrates, the tissues (and muscle more than the hepatopancreas) contain an L-amino acid oxidase acting on arginine with the formation of α-keto-δ-guanidinovaleric acid and γ-guanidinobutyric acid. In spite of the presence of an active arginase and a transformation in ornithine and urea, the metabolism of arginine follows, partially at least, the pathway of oxidative deamination (Robin and Thoai, 1957).

2. *Conversion of Arginine to* γ*-Aminobutyric Acid*

Another metabolic pathway in *Limnea* is an oxidative decarboxylation with formation of γ-guanidinobutyramide (Thoai *et al.,* 1957). The enzyme responsible for this was first observed by Thoai *et al.* (1955, 1956) in *Streptomyces griseus* Waksman. γ-Guanidinobutyramide is further hydrolyzed to γ-guanidinobutyrate. The presence of an enzyme, γ-guanidinobutyrate ureohydrolase, catalyzing the reaction

$$\gamma\text{-guanidinobutyrate} \xrightarrow{H_2O} \gamma\text{-aminobutyrate} + \text{urea}$$

has been demonstrated by Baret *et al.* (1967) and by Campbell and Speeg (1968a) in *Otala lactea* and by Porembska *et al.* (1968) in *Helix pomatia.* As this enzyme seems to be widespread in the mollusks, the arginine oxygenase pathway may play an important physiological role in these organisms as the source of γ-aminobutyrate in nerve tissues, where glutamate decarboxylase seems to be absent.

3. *Conversion of Arginine to Urea*

The enzyme arginase brings about the formation of ornithine and urea from arginine. In 1918 Clementi found this enzyme in a number of invertebrates, including *H. pomatia,* and Baldwin (1935a) confirmed the presence of arginase in *Helix.* At that time, it was commonly held that the existence of arginase in an organism was linked with the ureotelic nature of its nitrogen metabolism, and that the snail *H. pomatia* excreted a part of the nitrogen in the form of urea. Baldwin and Needham (1934) brought arguments supporting the interpretation according to which the small amounts of urea eventually produced in the metabolism of *Helix* may be derived from exogenous arginine. Bricteux-Grégoire and Florkin (1962a,b) have provided direct evidence in support of this notion. They injected guanido-^{14}C-arginine into the hepatopancreas of a snail, and 2 hours later recovered 5% of the activity in the form of urea.

A systematic study of the distribution of arginase activity in seventy species of mollusks has been made by Gaston and Campbell (1966). The results of this survey indicate that the levels of arginase activity in different species and in different tissues may be extremely variable. The activity is usually higher in the hepatopancreas of gastropods but is by no means restricted to this organ nor to this class of mollusks.

Molluscan arginase was first purified from the hepatopancreas of *Otala lactea* (Campbell, 1966). Some differences in molecular properties are found between the molluscan and the rat enzyme.

Baret *et al.* (1965, 1967) have studied the distribution of arginase and of γ-guanidinobutyrate ureohydrolase in twenty-five species of mollusks and proved that they are separate enzymes. Porembska *et al.* (1968) also isolated arginase from hepatopancreas of *H. pomatia,* which after purification, was free of γ-guanidinobutyrate ureohydrolase activity and a distinct γ-guanidinobutyrate ureohydrolase devoid of arginase activity.

The rapid degradation of arginine may pose a problem in snails. Speeg and Campbell (1969) suggest that only endogenously synthesized arginine is available for protein synthesis.

C. Metabolism of Sulfur Amino Acids

We have seen (Section I,B) that taurine has been detected in marine mollusks, but not in freshwater or terrestrial species. Allen and Awapara

(1960) have studied comparatively two bivalves, *Mytilus edulis,* a marine form, and *Rangia cuneata,* a freshwater species of the same group. They have shown that both convert methionine to cysteine and oxidize it to products that can give taurine by decarboxylation. The pathway of the sulfur-containing amino acids is not different from that

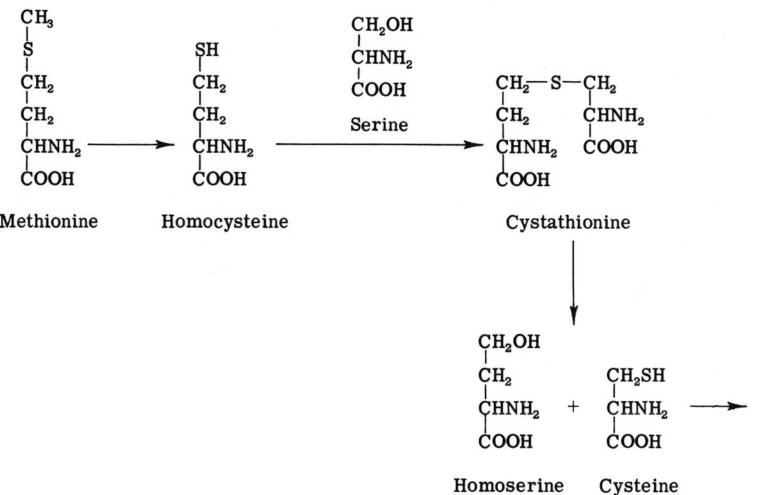

$$\underset{\text{Cysteine sulfinic acid}}{SO_2H-CH_2-CHNH_2-COOH} \xrightarrow{\textit{Mytilus edulis}} \underset{\text{Hypotaurine}}{SO_2H-CH_2-CH_2NH_2} + CO_2$$

Cysteine sulfinic acid $\xrightarrow{\textit{Rangia cuneata}}$ Cysteic acid; Hypotaurine $\longrightarrow$ Taurine

$$\underset{\text{Cysteic acid}}{SO_3H-CH_2-CHNH_2-COOH} \longrightarrow \underset{\text{Taurine}}{SO_3H-CH_2-CH_2NH_2} + CO_2$$

SCHEME II.

found in mammals. Cysteine sulfinic acid is oxidized to cysteic acid and later decarboxylated to taurine in *R. cuneata,* but in *M. edulis,* cysteic acid is not formed and hypotaurine, the product of the decarboxylation of cysteine sulfinic acid, is the intermediate. The various reactions are shown in Scheme II.

Hypotaurine (2-aminoethanesulfinic acid) has been found in tissues of mollusks by Shibuya and Ouchi (1957) and by Ouchi (1959). According to J. W. Simpson *et al.* (1959), the absence of taurine in freshwater and terrestrial forms is due to the rapid metabolizing of taurine in the tissues, whereas it is retained in the tissues of marine forms and plays a role in intracellular osmotic pressure.

D. Metabolism of Histidine

The general pathway of histidine degradation appears to follow the urocanic pathway (the histidine reaction) Urocanic acid is a constituent of urocanylcholine or murexine. Murexine was first isolated from the

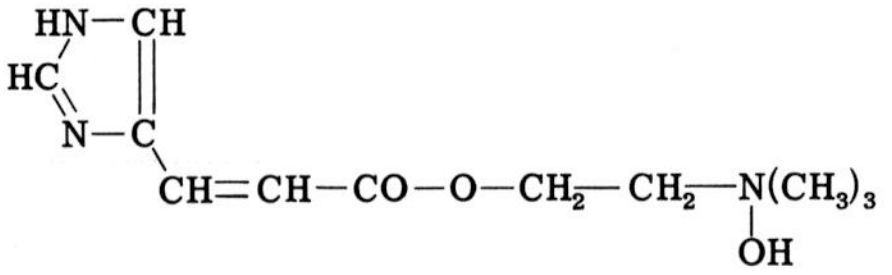

```
HN—CH
HC    ||
  \\  ||
   N—C
      \
       CH=CH—CO—O—CH2—CH2—N(CH3)3
                            |
                            OH
```

Murexine

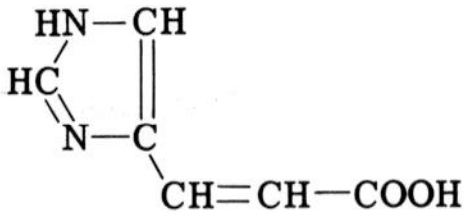

```
HN—CH
HC    ||
  \\  ||
   N—C
      \
       CH=CH—COOH
```

Urocanic acid

hypobranchial extracts of *Murex trunculus, Murex brandaris,* and *Tritonalia erinacea* (Erspamer and Dordoni, 1947; Erspamer, 1948). It appears as a characteristic of species producing indigoids and has not been found in the hypobranchial extracts of the following species: *Vulsocerithium vulgatum, Turitella communis, Euthria cornea, Dolium galea,* nor in extracts of different organs of *Helix pomatia, Mytilus galloprovincialis,* or *Ostrea edulis.* In the case of the indigoid-producing species, the metabolism of histidine is specialized, at least partly, toward the synthesis of a poisonous secretion.

Another pathway of histidine metabolism, which does not follow the urocanic pathway, has been found in other mollusks. It consists of a conversion into imidazole compounds and has been described by Thoai *et al.* (1954) in the hepatopancreas of *Mytilus edulis* L., an organ rich in L-amino acid oxidase. The same authors have shown that the action of the preparations of hepatopancreas on L-histidine leads to the formation of two groups of substances: (1) imidazolpyruvic acid and imidazolacetic acid, which are the products expected from the action of L-amino acid oxidase on histidine, and (2) 4-formylimidazole and 4 hydroxymethylimidazole.

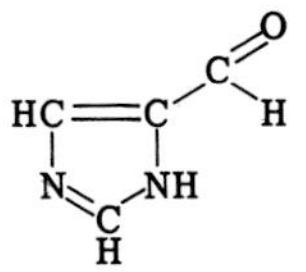

4-Formylimidazole

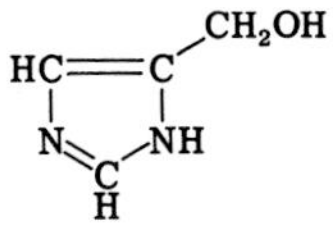

4-Hydroxymethylimidazole

The formation of these compounds cannot be explained by the same pathway, and they point to the existence of another pathway still unknown. The vertebrates appear to open the imidazole ring without shortening the side chain of histidine (through urocanic acid and α-formimino-L-glutamic acid). At least in some mollusks, the side chain is apparently shortened before the cycle is opened. Roche *et al.* (1955) have observed that there are several pathways of enzymatic degradation leaving the imidazole nucleus intact while liberating the α-amino group in the form of ammonia.

E. Decarboxylases and the Genesis of Amines

The amines found in the posterior salivary glands of some cephalopods (5-hydroxytryptamine, tyramine, octopamine, histamine, tryptamine) appear as the products of a decarboxylation of the corresponding amino acids, 5-hydroxytryptophan, tyrosine, *p*-hydroxyphenylserine, histidine, and tryptophan, respectively (see Table VI). The decarboxylases in the posterior glands of *Octopus apollyon* can be shown by adding various substrates, as Table VI illustrates. It will be seen that more decarboxylases are found than are required for the actual number of amines present in the gland.

Octopamine may result from the decarboxylation of *p*-hydroxyphenylserine (Table VI), but it is known that octopamine may also result

TABLE VI
AMINO ACIDS DECARBOXYLATED BY THE POSTERIOR SALIVARY GLANDS OF *Octopus apollyon*[a]

Amino acid		Product
Tryptophan	→	Tryptamine + CO_2
5-Hydroxytryptophan	→	5-Hydroxytryptamine (enteramine, serotonin) + CO_2
3,4-Dihydroxy-phenylalanine (dopa)	→	3,4-Dihydroxyphenyl-ethylamine (dopamine, hydroxy-tyramine) + CO_2
3,4-Dihydroxy-phenylserine	→	3,4-Dihydroxyphenyl-ethanolamine (Arterenol, noradrenaline, norepinephrine) + CO_2
m-Hydroxy-phenylserine	→	*m*-Hydroxyphenyl-ethanolamine + CO_2

TABLE VI (*Continued*)

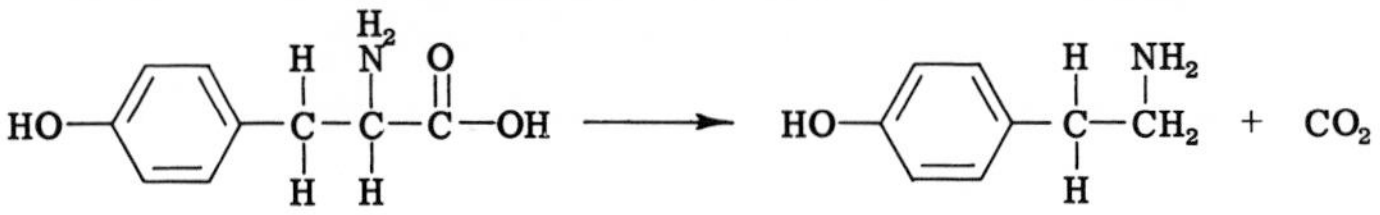

Substrate		Product
Tyrosine	→	Tyramine + CO_2
m-Tyrosine	→	*m*-Hydroxyphenylethylamine + CO_2
p-Hydroxyphenylserine	→	*p*-Hydroxyphenylethanolamine (octopamine) + CO_2
Phenylserine	→	Phenylethanolamine + CO_2
Histidine	→	Histamine + CO_2

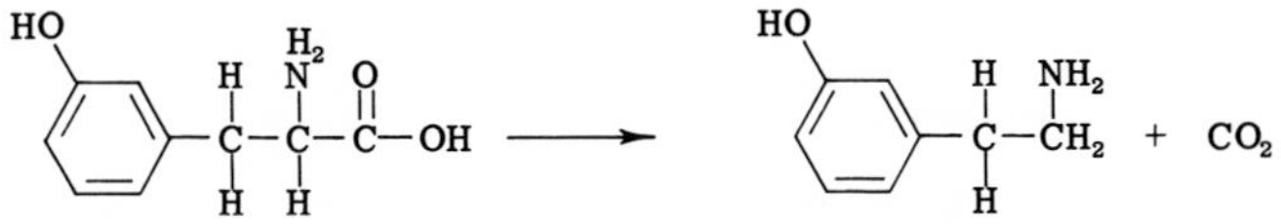

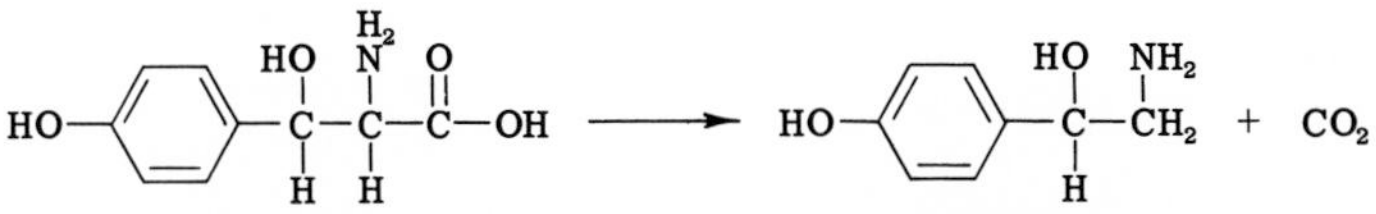

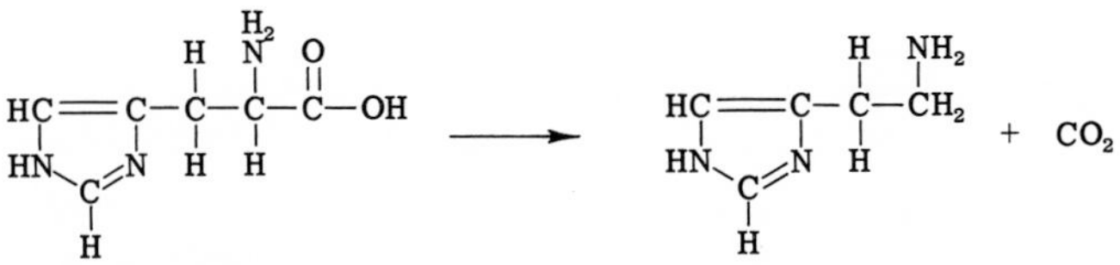

[a] After Hartman *et al.* (1960).

from the oxidation of tyramine. The reactions involved are shown in the following equation:

$$HO-C_6H_4-CH_2-CH_2(NH_2) \xrightarrow{+\frac{1}{2}O_2} HO-C_6H_4-CHOH-CH_2(NH_2)$$

Tyramine → Octopamine

$$HO-C_6H_4-CHOH-CH(NH_2)-COOH \xrightarrow{-CO_2} HO-C_6H_4-CHOH-CH_2(NH_2)$$

p-Hydroxyphenylserine → Octopamine

Octopamine itself can be oxidized into hydroxyoctopamine or noradrenaline (norepinephrine):

$$HO-C_6H_4-CHOH-CH_2(NH_2) \xrightarrow{+\frac{1}{2}O_2} (HO)_2C_6H_3-CHOH-CH_2(NH_2)$$

Octopamine → Noradrenaline

The decarboxylation of tyrosine has been demonstrated by Hartman *et al.* (1960) in the octopus and by Cardot (1966c) in *Helix pomatia.*

Dopa decarboxylase and 5-hydroxytryptophane decarboxylase appear to be a single enzyme acting on both dopa and 5-hydroxytryptophane. In *H. pomatia* nerve and heart preparations, the decarboxylation of each substrate is inhibited by the other and the decarboxylation of 5-hydroxytryptophane is inhibited by substrates and inhibitors of mammalian dopa decarboxylase (Cardot, 1966a).

The action of glutamate decarboxylase would give rise to γ-aminobutyrate which has been found in mollusks and is known to affect *H. aspersa* ganglia (Kerkut and Walker, 1961). This glutamate decarboxylase, however, has not been found in either muscle, nerve, or ganglia of *Mytilus edulis* or *H. aspersa* (Huggins *et al.*, 1967). It seems that the origin of γ-aminobutyrate may be from arginine (see Section II,B) rather than from glutamate (Campbell and Bishop, 1970).

F. Amine Oxidases

Biologically active amines occur in molluscan tissue (see Chapter 13 by Endean). Amine oxidases which inactivate these amines have a wide distribution in mollusks. Monoamine oxidase activity has first been reported in cephalopods, in which tyramine was thought to be the poison of the salivary glands, for instance in *Sepia officinalis* and *Octopus vulgaris* (Blaschko and Hawkins, 1951), and in *Loligo forbesi, Pecten maximus,* and *Pecten opercularis* (Blaschko and Philpot, 1953; Blaschko and Hope, 1957). Low levels of activity may also occur in tissues from several species of mollusks (for a review, see Campbell and Bishop, 1970). In *Helix,* there is a monoamine oxidase which acts on tyramine and tryptamine, but is only slightly active on 5-hydroxytryptamine (Cardot, 1965, 1966b). According to Kerkut and Cottrell (1963), *Helix pomatia* has a monoamine oxidase which converts 5-hydroxtryptamine to 5-hydroxy-3-indolylacetic acid. Blaschko and Milton (1960) and Blaschko and Levine (1960) found an hydroxyindole oxidase in *Mytilus edulis* which oxidizes 5-hydroxytryptamine but is not inactivated by iproniazid, the inhibitor of monoamine oxidase. This oxidase is active on a broad spectrum of compounds (see Table VIII in Campbell and Bishop, 1970).

G. Enzymes of the "System of Ureogenesis"

One may ask whether the arginine from which urea is formed by the action of arginase is derived from the system of urea synthesis such as that found in ureotelic vertebrates. The sequence of steps of ureogenesis in ureotelic vertebrates is formulated as follows:

$$CO_2 + NH_3 + 2\ \text{ATP}\ (+\ \text{glutamyl derivative}) \rightarrow \text{carbamyl phosphate} \quad (1)$$
$$\text{Carbamyl phosphate} + \text{ornithine} \rightarrow \text{citrulline} \quad (2)$$
$$\text{Citrulline} + \text{aspartate} + \text{ATP} \rightarrow \text{argininosuccinate} \quad (3)$$
$$\text{Argininosuccinate} \rightarrow \text{arginine} + \text{fumarate} \quad (4)$$
$$\text{Arginine} \rightarrow \text{ornithine} + \text{urea} \quad (5)$$

The enzymes catalyzing the individual steps are the following:

Carbamyl phosphate synthetase	EC 2.7.2.5	(1)
Ornithine transcarbamylase	EC 2.1.3.3	(2)
Argininosuccinate synthetase	EC 6.3.4.5	(3)
Argininosuccinate lyase	EC 4.3.2.1	(4)
Arginase	EC 3.5.3.1	(5)

Linton and Campbell (1962) and Campbell and Speeg (1968b) have studied the different organs of the land snails *Otala lactea* and *Helix aspersa* with respect to the enzymes of the system of ureogenesis and have demonstrated the presence of the complete set of enzymes. Porembska and Heller (1962) have found enzyme (2) in hepatopancreas, muscle, and kidney of *Helix pomatia* but failed to detect enzymes (3) and (4). Boonkoom and Horne (1968) found enzymes (2), (3), and (4) in hepatopancreas of *Bulimulus schiedeanus, Euglandina singleyana, Limax maximus, Mesodon roemeri, Rumina decolata,* and the aquatic mollusk *Heliosoma trivolvis.*

From the results of a recent study of the distribution of the ornithine cycle enzymes in twelve species of gastropods, Horne and Boonkoom (1970) conclude that only the prosobranch *Marisa cornaurietis* and *Sinotaia ingalesiana* are able to biosynthetize citrulline. In all snails studied they found ornithine transcarbamylase (enzyme 2) and arginase (enzyme 5).

In an experiment *in vitro,* incubating a homogenate of hepatopancreas with bicarbonate-^{14}C, Bricteux-Grégoire and Florkin (1964) did not detect any incorporation of activity into citrulline or arginine. However a small incorporation of bicarbonate-^{14}C into urea could be shown in the intact snail, after injection of bicarbonate-^{14}C into the hepatopancreas. Labeled urea was also isolated when a homogenate of hepatopancreas was incubated with ornithine-^{14}C and carbamyl phosphate, thus suggesting the presence of enzymes (2–5).

Although the presence of carbamyl phosphate synthetase in mollusks has been questioned, the results of Campbell and Speeg (1968a) showing the incorporation and the distribution of bicarbonate-^{14}C in arginine and in several intermediates clearly prove the presence of enzymes (1–5) in the land snails *O. lactea* and *H. aspersa.*

Recently, Tramell and Campbell (1970) have detected and partially characterized a carbamyl phosphate synthetase in hepatopancreas tissue of the land snail *Strophocheilus oblongus.* The enzyme is localized in mitochondria, utilizes glutamine, and shows an absolute requirement for *N*-acetyl-L-glutamate. ATP and Mg^{2+} are also required.

H. Biogenesis of the Purine Ring

Because of the apparent discrepancy between the high level of arginase activity in the snails and their uricotelism, several authors have concluded that uric acid, which is the main terminal nitrogenous product of amino acid metabolism, is synthetized in the hepatopancreas from urea (Wolf, 1933; Baldwin, 1935b; Grah, 1937). According to this view, which was hinted at again recently by Linton and Campbell (1962), the biosynthesis of uric acid should follow different pathways in mollusks and in birds. *Helix pomatia* would accomplish the biosynthesis of the purine ring along the pathway proposed by Wiener (1902), by which two molecules of urea are combined with a molecule of tartronic acid to form a molecule of uric acid. With respect to the birds, this scheme is obsolete, since Buchanan and his collaborators (1948) have demonstrated that in these animals uric acid biosynthesis is accomplished from simple molecules—carbon dioxide, acetate, formate, glycine, aspartic acid, and glutamic acid. Heller and Jezewska (1959) have shown that these precursors are also utilized by the moth *Antheraea pernyi* for the synthesis of purines.

According to Wiener's scheme for the synthesis of the purine ring, the labeled carbon of urea-^{14}C should be found in the C-2 and C-8 of uric acid, as shown in Scheme III.

However, when urea-^{14}C was injected into the snail *H. pomatia* (Bricteux-Grégoire and Florkin, 1962a), no activity appeared in C-2 and C-8. The greater part of the activity was found to be localized in the C-6 and C-4, i.e., in the positions also found to be chiefly labeled following the administration of bicarbonate-^{14}C to the pigeon (Buchanan *et al.*, 1948). The interpretation of these results is that the urea injected into the snail is decomposed by urease in the kidney (Baldwin and Needham, 1934; Heidermanns and Kirchner-Kühn, 1952) and that the carbon dioxide resulting from this action takes part in the biosynthesis of uric acid according to the pathway discovered by Buchanan *et al.* (1948).

According to this pathway, N-1 of the purine ring is derived from aspartate, C-2 and C-8 from the formyl group of tetrahydrofolic acid derivatives (the one-carbon pool), N-3 and N-9 from the amide N of

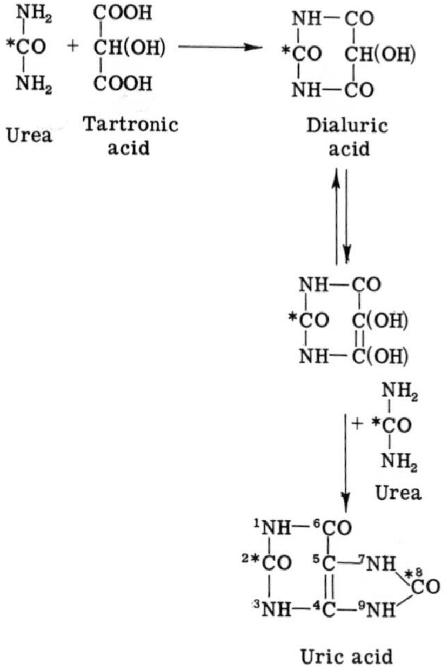

NH_2–*CO–NH_2 (Urea) + COOH–CH(OH)–COOH (Tartronic acid) → NH–CO / *CO CH(OH) / NH–CO (Dialuric acid)

⇅

NH–CO / *CO C(OH) / NH–C(OH)

+ NH_2–*CO–NH_2 (Urea)

↓

^{1}NH–^{6}CO / ^{2}*CO ^{5}C–^{7}NH / *^{8}CO / ^{3}NH–^{4}C–^{9}NH

Uric acid

SCHEME III.

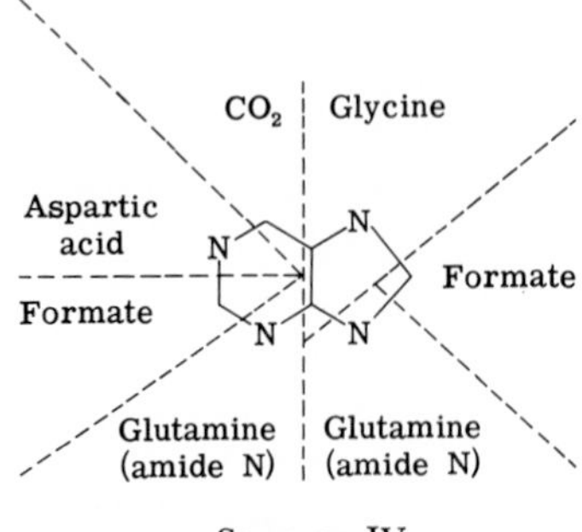

SCHEME IV.

glutamine, and C-4, C-5, and N-7 from glycine and C-6 from carbon dioxide. The origin of the different atoms of the purine ring is shown in Scheme IV.

Jezewska *et al.* (1964) have confirmed the utilization of C-1 of glycine for the synthesis of uric acid C-4 and C-5 in *H. pomatia.* Lee and Campbell (1965) have confirmed the origin of C-6 of uric acid from carbon dioxide in *Otala lactea,* of C-4 and C-5 from C-1 and C-2 of glycine, and of C-2 and C-8 from the one-carbon pool. These results have been confirmed in *H. pomatia* by Gorzkowski (1969). Lee and Campbell (1965) also showed that N-3 and N-9 may arise from glutamine and that 4-amino-5-imidazole carboxamide was a direct precursor of uric acid in *Otala,* probably via the ribosyl derivative. The results of their study of the inhibition of uric acid synthesis also indicate that the synthesis of the purine ring by snails involves ribosyl intermediates as it occurs in vertebrates.

I. Biogenesis of the Pyrimidine Ring

The sequence of steps leading to pyrimidine biosynthesis in vertebrates is as follows:

$$\text{Carbamylphosphate} + \text{L-aspartate} \rightarrow \text{carbamyl L-aspartate} + P_i \quad (1)$$

$$\text{Carbamyl L-aspartate} \rightarrow \text{L-dihydroorotic acid} \quad (2)$$

$$\text{L-Dihydroorotic acid} + \text{NAD} \rightarrow \text{orotic acid} + \text{NADH}_2 \quad (3)$$

$$\text{Orotic acid} + \text{5-phosphoribosyl pyrophosphate} \rightarrow \text{orotidine 5'-phosphate} + PP_i \quad (4)$$

The individual steps are catalyzed by:

Aspartate transcarbamylase	EC 2.1.3.2	(1)
Dihydroorotase	EC 3.5.2.3	(2)
Dihydroorotate dehydrogenase	EC 1.3.3.1	(3)
Orotate phosphoribosyltransferase	EC 2.4.2.10	(4)

Porembska *et al.* (1966) have demonstrated the presence of labeled CMP and UMP in RNA from *Helix pomatia* hepatopancreas, following the injection of orotic acid-6-^{14}C, thus demonstrating the existence of reaction (4). They also found incorporation of the label into all four RNA nucleotides after injection to the snails of bicarbonate-^{14}C and aspartate-U-^{14}C, thus showing the presence of carbamyl phosphate synthetase (see Section II,G) and the existence of reactions (1–3).

Pyrimidine biosynthesis has been very little studied in mollusks. It seems, however, that the experiments of Porembska *et al.* (1966) allow one to conclude that pyrimidine biosynthesis in mollusks is similar to that in other organisms.

J. Purinolytic Enzymes and Purinolysis (Including Urease)

The complete course of purinolysis leads from the purine nucleus to ammonia through the successive actions of uricase, allantoinase, allantoicase, and urease, as shown in Fig. 1. This series of enzymes is found, for instance, in *Mytilus edulis* (Przylecki, 1922; Brunel, 1938; Florkin and Duchâteau, 1943). That the successive steps of the purinolysis pathways can take place in different organs is shown in the case of *Meretrix meretrix*, in which guanine appears to be deaminated to xanthine in the gills, where the guanine deaminase is present. The xanthine could then be converted to uric acid in the foot and hepatopancreas where xanthine dehydrogenase predominates. The resulting uric acid is degraded mainly in the hepatopancreas, while the breakdown of urea to ammonia takes place in the gills and in the hepatopancreas (Ishida, 1955b).

The presence of enzymes effecting the interconversion of nucleotides such as it occurs in other organisms has been inferred from several studies reviewed by Campbell and Bishop (1970) (Fig. 2). Several phosphohydrolases have been found in mollusks. B. Farina (1964) has reported the presence of an enzyme liberating inorganic phosphate from 2′-, 3′-, and 5′-AMP in hepatopancreas of *Octopus vulgaris*. Umemori (1967) has shown the presence of phosphohydrolases acting on 5′-AMP, 3′-AMP, 5′-IMP, and GMP in the midgut gland of the clam *M. meretrix*. Ishida *et al.* (1969) have identified a 5′-AMP phosphohydrolase in the oyster *Crassostrea nippona*. The activity of ATP phosphohydrolase in hepatopancreas, mantle, and foot muscle of the snail *Pila globosa* has been studied by Raghupathiramireddy and Swamy (1967) in normal and estivating conditions. The decrease of 80% of the activity of this enzyme in estivating snails contributes to the increased level of ATP in these tissues.

Nucleotidase activity (ribo- and deoxyribohydrolase) has been re-

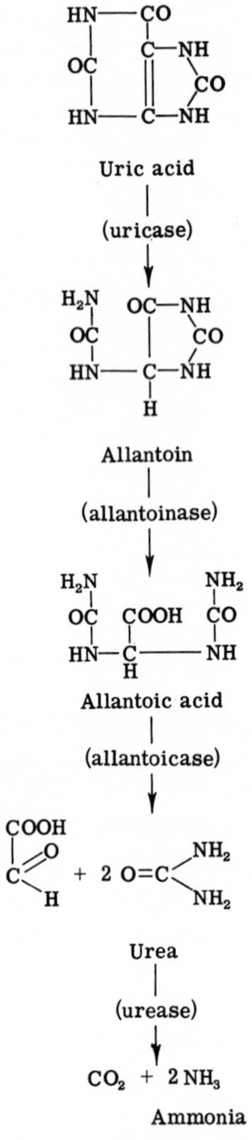

Uric acid

(uricase)

Allantoin

(allantoinase)

Allantoic acid

(allantoicase)

COOH–CHO + 2 $O{=}C(NH_2)_2$

Urea

(urease)

CO_2 + 2 NH_3

Ammonia

FIG. 1. The uricolytic enzyme system. The successive steps of this reaction can take place in different organs of an individual.

ported by B. Farina (1964) and B. Farina *et al.* (1966) in the hepatopancreas of *Octopus.*

Data on the distribution of the enzymes of purinolysis have been published by a number of authors. Adenine deaminase (adenase) has been reported by Mendel and Wells (1909) in *Sycotypus* and by Truszkowski (1928), as well as by Duchâteau *et al.* (1941), in *Anodonta.* Working with more up-to-date methods, Ishida (1954) and Jezewska (1968b) failed to detect adenase activity in *M. meretrix* and in *Helix pomatia,* respectively.

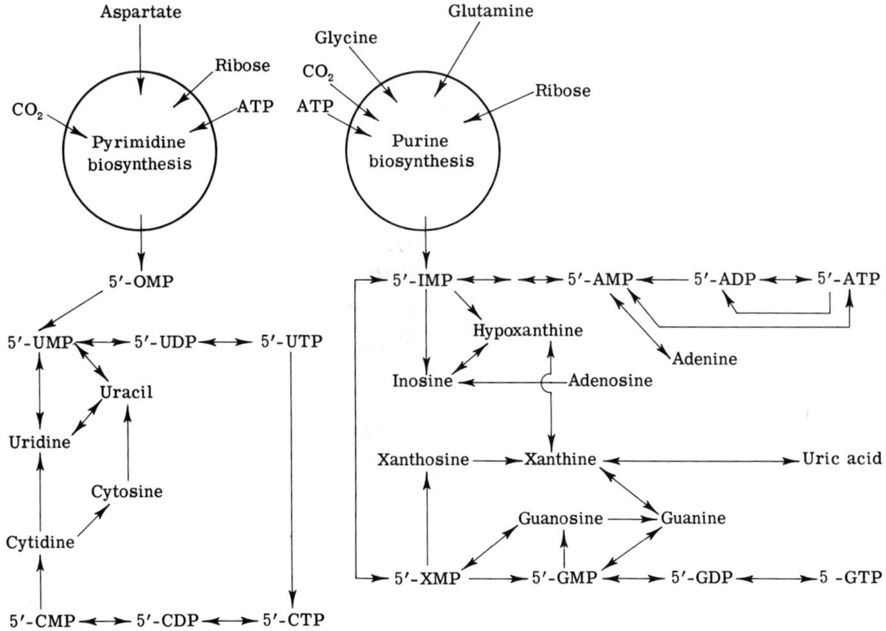

FIG. 2. Interconversion of the pyrimidine and purine ribonucleotides. (From Campbell and Bishop, 1970.)

Guanine deaminase (guanase) has been found in *Anodonta* (Truszkowski, 1928; Duchâteau *et al.*, 1941) in the gills of *Meretrix* (Ishida, 1954) and in *Octopus* hepatopancreas (B. Farina, 1964).

Duchâteau *et al.* (1941) were not able to demonstrate the presence of adenosine deaminase, or of guanosine deaminase in *Anodonta,* but Aikawa (1959) observed the presence of adenosine deaminase in *Meretrix.* Therefore, the view proposed by Duchâteau *et al.* (1941) according to which the system of the deamination of aminopurines acts not on nucleosides but on free purines in invertebrates must be abandoned. Adenosine deaminase has been isolated and purified from the midgut gland of *M. meretrix* by Aikawa (1966). The enzyme is active on adenosine and deoxyadenosine, but inactive on 5′-AMP, 5′-deoxyAMP, adenine, guanine, guanosine, GMP, cytosine, cytidine and CMP. Adenosine deaminase, together with 5′-AMP phosphohydolase have been demonstrated in the clam *C. nippona* by Ishida *et al.* (1969) and by Jezewska (1968b) in the hermaphrodite gland of *H. pomatia.*

Xanthine oxidase has been found in all mollusks examined: *H. pomatia* (Baldwin and Needham, 1934), *Anodonta* and *Planorbis* (Florkin and Duchâteau, 1941), *Anadara inflata, M. meretrix, Venerupis philip-*

pinarum, Crassostrea gigas, and *M. edulis* (Tsuzuki, 1957), and *P. globosa* (Lal and Saxena, 1952). Whereas the xanthine oxidase is present in the hepatopancreas of *Anodonta* (Florkin and Duchâteau, 1941), Ishida (1955b) found the main concentration in the foot of *Meretrix.*

Uricase has been detected in all bivalves studied—*M. edulis* (Brunel, 1938; Ishida *et al.*, 1956), *M. meretrix lusoria* and *Venerupis philippinarum* (Ishida and Tsuzuki, 1955), *Anodonta* (Przylecki, 1926; Truszkowski and Chajkinowna, 1935; Truszkowski and Goldmanowna, 1933; Florkin and Duchâteau, 1943), *Anadara inflata, Mactra sulcataria,* and *Crassostrea gigas* (mainly in the hepatopancreas) (Ishida *et al.*, 1956). In a survey on purinolytic enzymes, Razet and Dagobert (1968) demonstrated the presence of uricase in fifteen species of freshwater Mollusca and in thirty species of terrestrial Mollusca.

Allantoinase and allantoicase are found in *M. edulis* (Brunel, 1938) as well as in the hepatopancreas of *Anodonta cygnea* (Florkin and Duchâteau, 1943). The latter does not contain urease (Truszkowski and Chajkinowna, 1935). In marine bivalves studied so far, urease has been detected in *M. edulis* (Przylecki, 1922), *Mactra sulcataria* (mainly in the hepatopancreas) (Ishida, 1955a), *V. philippinarum* (Ishida, 1955a), and *A. inflata* (mainly in the gills) (Ishida, 1955a).

In gastropods, uricase, but not allantoinase or allantoicase, is found in *Planorbis* (Florkin and Duchâteau, 1943). On the other hand, no uricase could be detected in the hepatopancreas of *Sycotypus* (Mendel and Wells, 1909) or of *H. pomatia* (Truszkowski and Chajkinowna, 1935) in spite of indirect arguments to the contrary (Spitzer, 1937; Grah, 1937; Plum, 1935). The kidney of the snail contains urease (Baldwin and Needham, 1934; Heidermanns and Kirchner-Kühn, 1952; Speeg and Campbell, 1968a).

III. Terminal Products of Nitrogen Metabolism in Mollusks

A very extensive study of nitrogen excretion in mollusks has been published by Potts (1967). This question has also been reviewed by Campbell and Bishop (1970). It has long been known that in the frog and man 80–85% of the total nitrogen of urine is in the form of urea, whereas in birds and snakes, the uric acid of the excreta accounts for approximately the same proportion. On the other hand, in the urine of *Sepia officinalis,* the ammonia nitrogen represents 65% of the total nonprotein nitrogen. These are examples of clear cut cases of ureotelic, uricotelic, or ammoniotelic nitrogen metabolism, respectively. The term in each case means that a terminal product predominates. Such a conclusion is justified when it can safely be assumed that the nitrogen in the excretion analyzed represents the total nitrogen actually excreted. It would be quite misleading, for instance, to take the distribution of the nitrogenous substances present in the urine of the carp *Cyprinus carpio* as indicative of the relative proportions of the different nitrogenous excretion products that are partly eliminated through the gills (see Florkin, 1945).

Some years ago, one of us raised doubt about the presumed terminal products of nitrogen metabolism as a result of analyses of the water in which animals were kept and into which they eliminate not only the products of their kidneys but also their feces (Florkin, 1945). Some excretion products may be retained in the kidneys in an insoluble form. We must therefore be very cautious in the interpretation of such data as those collected in Table 112 of Albritton's "Standard Values in Nutrition and Metabolism" (1955), or in Table 22 of Prosser and Brown's "Comparative Animal Physiology" (1961).

A. Uricotelism of Terrestrial Gastropods

Let us consider the case of the snail *Helix pomatia.* According to Hesse (1910) *H. pomatia* excretes 3.85 mg of nitrogen per kilogram. When this observation was performed, it was known that the whitish content of the kidneys was largely uric acid, as Jacobson had established in 1820, and as many authors had confirmed. Marchal (1889) had isolated the uric acid contained in the kidneys of one hundred fifty snails, and after its purification had concluded that a kidney contains more than 7 mg uric acid. In fact, at the end of hibernation, as Baldwin and Needham (1934) have shown, a snail's kidney contains a mean weight of 32 mg of uric acid, i.e., about three-quarters of the dry weight of the organ. Uric acid is mostly excreted in its acid form, a small portion being in the form of urates (Heidermanns, 1953).

In the tables referred to above, the data concerning *H. pomatia* are those first published by Delaunay (1927) and made generally known later by the publications of Needham (1935) and by Baldwin (1947). Delaunay analyzed a "water extract" of the kidneys prepared in such a way that about 80% of the purine compounds were left unextracted, as Jezewska *et al.* (1963) have pointed out. Delaunay considered the snail to have two kinds of excreta—solids composed mainly of purines and liquid consisting mainly of ammonia and urea.

To collect the "liquid excreta," Delaunay kept the snails partially immersed in distilled water and determined the nitrogenous compounds given out in this water. This he called "liquid excretion," but in the tables referred to above they are considered as representing the excreta of *Helix.* Jezewska *et al.* (1963) have proceeded in a different way in their study of the nitrogen compounds in snail's excretion. The hibernating operculated snails were stored in a refrigerator at 4°C. In the spring, they were transferred to room temperature, and after breaking hibernation, were fed with lettuce and cauliflower leaves in glass beakers, the bottom of which was covered with a layer of 2–4 mm of water. The snails remained on the walls of the beakers, where they deposited their feces as well as their renal excreta. These were easily distinguished. The yellowish renal excreta were collected during April, May, and June and air dried.

During hibernation, as well as during the active period, snails were dissected and their kidneys were isolated. The content of each kidney was rinsed out with distilled water into a 100 ml flask. A saturated solution of lithium carbonate was added under conditions of shaking and moderate heating until all concretions were dissolved. The air-dried renal excreta were treated in the same way. The results of the analyses performed on the dissolved excreta are given in Table VII. They show that 90% of the total nitrogen in excreta and in the content of the kidneys consist of uric acid, xanthine, and guanine. Urea, ammonia, and allantoin were not found. During the feeding period, the total nitrogen content in the kidneys was only 50% of that in hibernating snails.

These data allow us to explain the nature of the erroneous data still often printed. Delaunay's "water extract" represented only 522 mg nitrogen per 100 gm of the kidneys of feeding snails, whereas the figures of Jezewska *et al.* (1963) recalculated on the same basis represent about 10,000 mg. Therefore the evaluation of urea, ammonia, and amino acids given by Delaunay would amount to only 5% of the total nitrogen of the excreta analyzed by Jezewska *et al.* (1963). Such small amounts are very likely overlooked in Table VII, the data of which were obtained on small samples. The results of Speeg and Campbell (1968b) obtained on *Otala lactea* confirm those of Jezewska *et al.* (1963). The purinotelic character of nitrogen metabolism in the terrestrial snail is obvious and the proportions in the excreta of end products other than uric acid, guanine, and xanthine amount to mere traces.

Slugs, like snails, were found to be purinotelic (Jezewska, 1969). During estivation terrestrial snails show a marked accumulation of uric acid in the kidney (Jezewska *et al.*, 1963; Speeg and Campbell, 1968b; De Jorge and Petersen, 1970). This accumulation has been used by Speeg and Campbell (1968b) to estimate the rate of synthesis of purines in *O. lactea* which has been found to be about 20 μg purine nitrogen per gram of tissue per 24 hours. Studies of incorporation of glycine-^{14}C into purines allowed the same authors to calculate a rate of synthesis of purines of 20–80 μg purine nitrogen per gram of tissue per 24 hours.

TABLE VII

NITROGEN COMPOUNDS IN KIDNEY AND EXCRETA FROM KIDNEY IN THE SNAIL *Helix pomatia*[a,b]

Material	Experiment no.	Total nitrogen (mg)	Nitrogen: Uric acid		Xanthine		Guanine		Total purines		Unknown compound		Nitrogen accounted for	
			mg	%	mg	%	mg	%	mg	%	mg	%	mg	%
Kidney during hibernation														
	1	43.9	25.0	56.9	8.6	19.5	5.8	13.2	39.5	90.0	0.0	0.0	39.4	90.0
	2	38.1	24.3	63.7	4.8	12.5	5.0	13.1	34.1	89.5	0.0	0.0	34.1	89.5
	3	59.4	44.6	75.0	4.8	8.0	5.9	9.9	55.3	93.0	0.0	0.0	55.3	93.0
	Average	47.1	31.3	66.4	6.0	12.7	5.6	11.8	42.9	91.0	0.0	0.0	42.9	91.0
Kidney during the feeding period														
	4	21.9	6.0	27.3	8.1	36.9	4.2	19.1	18.3	83.5	3.3	15.0	21.6	98.6
	5	22.1	5.0	22.6	8.7	39.3	5.3	24.0	19.0	85.9	3.2	14.0	22.2	100.4
	6	26.3	11.1	42.2	8.4	31.9	3.6	13.6	23.2	88.1	1.4	5.3	24.6	93.5
	7	15.1	7.8	51.6	3.6	23.8	3.6	23.8	15.0	99.3	0.0	0.0	15.0	99.3
	Average	21.3	7.5	35.2	7.2	33.9	4.2	19.7	18.9	88.7	1.9	8.6	20.8	97.6
First excreta after hibernation														
	8	33.6	21.7	64.6	12.6	37.5	0.0	0.0	34.3	102.1	0.0	0.0	34.3	102.1
Excreta during the feeding period														
	9	33.1	12.8	38.7	12.0	36.3	3.8	11.5	28.6	86.4	2.1	6.3	30.7	92.7
	10	34.7	15.8	45.5	8.8	25.3	7.4	21.3	32.0	92.2	2.1	6.1	34.1	98.3
	11	31.1	14.1	45.3	9.0	28.9	6.9	22.2	30.0	96.5	0.5	1.6	30.5	98.0
	Average	33.0	14.2	43.0	9.9	30.0	6.0	18.2	30.2	91.5	1.5	4.5	31.7	96.1

[a] The average values of two determinations are given, and the results are presented in milligrams nitrogen per kidney or per 100 mg of excreta, and as percentage of total nitrogen. The difference between the determinations did not exceed 0.5 mg per kidney or per 100 mg of excreta.

[b] Jezewska *et al.* (1963).

These values can be compared to the ammonia nitrogen excretion by bivalves and freshwater gastropods which are, respectively, 42 μg ammonia nitrogen per gram of tissue per 24 hours for *Modiolus demissus* (Lum and Hammen, 1964) and 38 μg ammonia nitrogen per gram of tissue per 24 hours for *Limnaea stagnalis* (Bayne and Friedl, 1968).

Speeg and Campbell (1968a) have demonstrated a form of extrarenal ammonia excretion in the snails *O. lactea* and *Helix aspersa.* The ammonia may occur, at least in part, through the action of urease. Injected urea increases the rate of ammonia production and $^{15}NH_3$ is formed from injected urea-^{15}N. The surface involved in the loss of ammonia appears to be the outer surface of the mantle underlying the shell. Observations made by Speeg and Campbell (1968a) indicate that the blood could become sufficiently alkaline by the loss of carbon dioxide at the body surfaces to bring about the dissociation of NH_4^+ ions into ammonia gas.

The small level of urea excretion mentioned in early reports and in the recent work of Haggag and Fouad (1968) on *Eremia desertorum* may be due to bacterial degradation of uric acid in the excreta (Campbell and Bishop, 1970).

B. Ammoniotelism of Cephalopods

In the case of the cephalopod *Sepia officinalis,* the data clearly indicate the nature of the main terminal product of nitrogen metabolism, in this instance ammonia. The results of Delaunay published in 1925 are shown in Table VIII. In this case, Delaunay used the urine taken from the urinary bladder. These results have been confirmed by Potts (1967)

TABLE VIII
Distribution of Nonprotein Nitrogen in the Urine of *Sepia officinalis*[a]

Parameter	Nonprotein nitrogen	Ammonia nitrogen	Urea nitrogen	Amino acid nitrogen	Purine nitrogen: Total	Purine nitrogen: Uric acid	Unidentified nitrogen
Mg per 100 ml	142	92	3	12	7	3.2	28
Percent of nonprotein nitrogen	—	64.4	2.1	8.4	4.9	2.2	20.2
Mg per 100 ml	125	87	1.8	9	4	2.6	23.2
Percent of nonprotein nitrogen	—	69.6	1.4	7.2	3.2	2.1	18.6

[a] Delaunay (1925, 1927).

who finds 65–70% of total nonprotein nitrogen in the form of ammonia in the urine of *Sepia.* Emanuel and Martin (1956) report an average value of 73% from *Octopus hongkongiensis* urine.

Potts (1965) demonstrated a glutaminase activity in renal appendage slices of *Octopus.* Thus, part of urinary ammonia may arise from this source; glutamine concentration, however, does not decrease greatly in the blood when it passes through the octopus kidney. On the basis that taurine concentration shows a 50% decrease during passage of the blood through the kidney, Potts (1965) suggests that taurine contributes to the ammonia excretion by being first metabolized to isethionic acid. In addition to urinary excretion, Potts also indicated the possibility of an extrarenal excretion of ammonia in the gills of *Octopus dofleini* which probably only involves blood NH_4^+. Potts found a decreased gill venous blood NH_4^+ content compared with gill arterial blood, together with a slight acidification of the blood after passage through the gills. Glutaminase activity could not be demonstrated in the octopus gill.

C. Marine and Freshwater Gastropods

The case of marine and freshwater gastropods is not so clear cut. Duerr (1968) has studied the excretion of ammonia by seven species of marine prosobranch snails held under bacteriostatic conditions in order to reduce bacterial degradation of end products and to avoid misleading conclusions regarding ammonia excretion. Unless such precautions are taken, results of ammonia determination on the medium or on the excreta are always questionable. Duerr found the rate of ammonia nitrogen excretion by the marine snails to be from 4 to 85 μg per gram total body weight per 24 hours. No urea was detected in the medium. In a previous study, Duerr (1967) had shown the presence of uric acid in all prosobranch snails studied. Duerr concludes that an individual snail can switch from excreting predominantly ammonia to excreting predominantly uric acid, but that this ability is not related to habitat or phylogeny. However, as the relative contribution of purine nitrogen and ammonia nitrogen to the total nitrogen excretion is not known, Duerr's conclusions must be taken as tentative.

In *Limnaea stagnalis,* Friedl and Bayne (1966) and Bayne and Friedl (1968) found some urea excretion, about seven times less important on a molar basis than ammonia excretion, which is 34 μg per gram total weight per 24 hours. As no rate of uric acid synthesis is known, it is difficult to correlate the relative contribution of ammonia nitrogen and purine nitrogen to the total nitrogen excretion.

The excretion of a freshwater snail *Lanistes baltemia* was followed during several months by Haggag and Fouad (1968). In relation to the period of the year, this snail excretes predominantly ammonia or urea, the content of uric acid in the kidney being very low during the whole year. The possibility of bacterial degradation of end products, however, does not seem to have been taken into account in this study. On the other hand, Duerr (1967) found a high level of uric acid in different species of prosobranch snails, especially in *Thais lamellosa.* The whole body uric acid content of freshwater pulmonate snails and marine prosobranch snails examined by Duerr (1967) is in the range reported by Needham (1935) for terrestrial snails. It is obvious that all gastropods accumulate uric acid and purines, and that the amounts of uric acid in gastropod kidney and tissues are far in excess of those expected if the uric acid simply reflected the general nucleic acid metabolism. Campbell and Bishop (1970) conclude that compared to other organisms, the gastropod mollusks, irrespective of their habitat, appear to be uniquely characterized by a "gouty metabolism."

D. Lamellibranchs

Nitrogen excretion in lamellibranch mollusks has been very little studied. Spitzer (1937) found some uric acid in the excreta of *Mytilus edulis* and *Unio,* but the results of the most complete analysis of nitrogen excretion of the oyster *Crassostrea virginica* made by Hammen *et al.* (1966) and of the mussel *Modiolus demissus* by Lum and Hammen (1964) allow one to conclude to the ammoniotelism of these two species. In *Crassostrea,* about 65% of the total nitrogen is excreted in the form of ammonia, 13% in the form of urea, and 17% is unidentified. In another experiment, they found 72% excreted in the form of ammonia and 28% as urea. Some nitrogen is also excreted in the form of α-amino nitrogen, especially in *Modiolus,* where it is about three times less important than ammonia (Lum and Hammen, 1964).

IV. Conclusions

It appears from all of our present knowledge of nitrogen metabolism in mollusks that the ammonia resulting from amino acid deamination is partly used in purine synthesis, through the same pathway as the one described in birds and partly excreted as such. When communication with the aquatic external medium is direct, the animal simply drains a large part of the ammonia into the medium, and the nitrogen used in purine synthesis also finally appears in the form of eliminated ammonia resulting from the action of a very extended purinolytic system. When the ecology of the animal involves a water shortage, as is the case in the gastropods that have become terrestrial, more ammonia is diverted toward purine synthesis, and the enzymatic purinolytic system is reduced in length and extension, the result being that the terminal products of amino acid metabolism, as well as of purine metabolism, are excreted in the form of purines and uric acid.

This does not mean that the Krebs–Henseleit cycle cannot function in mollusks. As Campbell and Bishop (1970) have pointed out, biochemists have generally looked upon the ornithine–urea cycle solely as a means of synthetizing the end product urea. However, as illustrated in Fig. 3, this cycle also provides for the synthesis of arginine, which is required on the one hand for protein biosynthesis, and on the other hand for the synthesis of phosphoarginine, the muscle phosphagen of mollusks.

Some questions have been raised as to the presence and action of

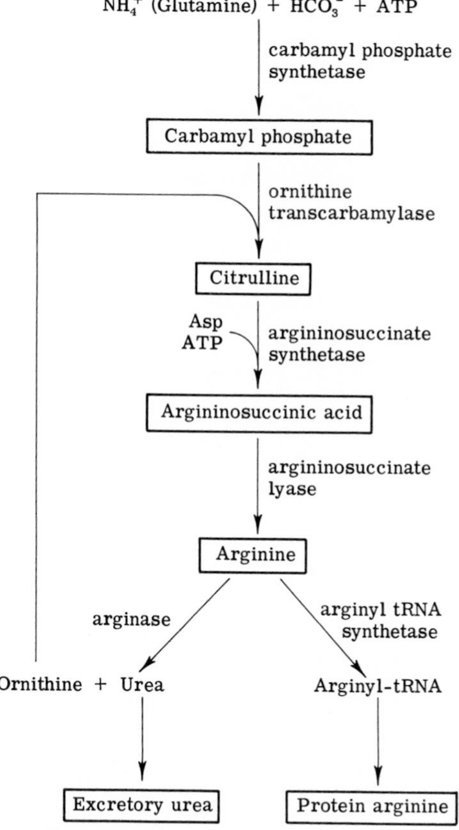

FIG. 3. Protein arginine and urea biosynthesis. (From Campbell and Bishop, 1970.)

arginase in mollusks. In *Neurospora,* both the biosynthetic arginine system and the degradative system are present, but Castañeda *et al.* (1967) suggest that the two systems are compartmentalized in such a way that the degradative system acts only on exogenous arginine. This conclusion has been ruled out in the case of *Otala lactea* and *Helix aspersa* by Speeg and Campbell (1969), who have shown that endogenously synthetized arginine, at least that formed from citrulline, is effectively hydrolyzed by arginase. The urea formed is then converted into ammonia and carbon dioxide by urease, which is known to be present in the kidney (Baldwin and Needham, 1934) and in the hepatopancreas (Campbell and Speeg, 1968a). This mechanism is in part responsible for the production of ammonia gas by the snails, since it has been shown that injected arginine stimulates the production of ammonia (Speeg and Campbell, 1968a).

Urea can also be partly derived from purinolysis. From a measurement of the turnover of urea-^{14}C injected into snails *Otala lactea,* Speeg and Campbell (1969) have suggested that the constant synthesis and turnover of urea might be directed toward maintaining the necessary alkaline conditions for the precipitation of shell carbonate.

Another important feature of molluscan nitrogen metabolism is the extended use of amino acids as substrates for decarboxylation, for example, in cephalopods, with production of a number of amines, the physiological uses of which are still conjectural.

REFERENCES

Ackermann, D. (1922). *Z. Biol.* **74,** 67.
Ackermann, D. (1931). *Verhandl. Phys-med. Ges. Würzburg* **56,** 48.
Ackermann, D. (1962). *Ber. Physik.-Med. Ges. Würzburg* **70,** 1.
Ackermann, D., and Janka, R. (1954). *Z. Physiol. Chem.* **298,** 65.
Ackermann, D., and Menssen, H. G. (1960). *Z. Physiol. Chem.* **318,** 212.
Ackermann, D., and Mohr, M. (1937). *Z. Physiol. Chem.* **250,** 249.
Ackermann, D., Holtz, F., and Kutscher, F. (1922). *Z. Biol.* **77,** 241.
Ackermann, D., Holtz, F., and Kutscher, F. (1924). *Z. Biol.* **80,** 155.
Acton, R. T., Bennett, J. C., Evans, E. E., and Schrohenloher, R. E. (1969). *J. Biol. Chem.* **244,** 4128.
Aikawa, T. (1959). *Sci. Rept. Tohoku Univ., Fourth Ser.* **25,** 73.
Aikawa, T. (1966). *Comp. Biochem. Physiol.* **17,** 271.
Albrecht, P. G. (1923). *J. Biol. Chem.* **56,** 483.
Albritton, E. C. (1955). "Standard Values in Nutrition and Metabolism". Saunders, Philadelphia.
Allen, K., and Awapara, J. (1960). *Biol. Bull.* **118,** 173.
Asano, M., and Ito, M. (1959). *Tohoku J. Agr. Res.* **10,** 209.
Asano, M., and Ito, M. (1960). *Ann. N.Y. Acad. Sci.* **90,** 674.
Asano, M., and Sato, H. (1954). *Tohoku J. Agr. Res.* **5,** 191.
Awapara, J., and Allen, K. (1959). *Science* **130,** 1250.

Bacq, Z. M., and Ghiretti, F. (1951). *Bull. Classe Sci., Acad. Roy. Belg.* [5] **37**, 79.
Baglioni, S. (1909). *Z. Biol.* **52**, 130.
Bailey, K. (1957). *Biochim. Biophys. Acta* **24**, 612.
Bailey, K., and Rüegg, J. C. (1960). *Biochim. Biophys. Acta* **38**, 239.
Baldwin, E. (1935a). *Biochem. J.* **29**, 252.
Baldwin, E. (1935b). *Biochem. J.* **29**, 1538.
Baldwin, E. (1947). "Dynamic Aspects of Biochemistry," p. 249. Cambridge Univ. Press, London and New York.
Baldwin, E., and Needham, J. (1934). *Biochem. J.* **28**, 1372.
Bárány, M., and Bárány, K. (1966). *Biochem. Z.* **345**, 37.
Baret, R., Mourgue, M., Broc, A., and Charmot, J. (1965). *Compt. Rend. Soc. Biol.* **159**, 2446.
Baret, R., Mourgue, M., and Broc, A. (1967). *Bull. Soc. Chim. Biol.* **49**, 25.
Bayne, R. A., and Friedl, F. E. (1968). *Comp. Biochem. Physiol.* **25**, 711.
Beers, J. R. (1967). *Comp. Biochem. Physiol.* **21**, 11.
Blaschko, H., and Hawkins, J. (1951). *Biochem. J.* **49**, xliv.
Blaschko, H., and Hawkins, J. (1952). *Biochem. J.* **52**, 306.
Blaschko, H., and Himms, J. M. (1955). *J. Physiol. (London)* **128**, 7P.
Blaschko, H., and Hope, D. B. (1956). *Biochem. J.* **62**, 335.
Blaschko, H., and Hope, D. B. (1957). *Arch. Biochem. Biophys.* **69**, 10.
Blaschko, H., and Levine, W. G. (1960). *Brit. J. Pharmacol.* **15**, 625.
Blaschko, H., and Milton, A. S. (1960). *Brit. J. Pharmacol.* **15**, 42.
Blaschko, H., and Philpot, F. J. (1953). *J. Physiol. (London)* **122**, 403.
Bon, W. F., Dohrn, A., and Batink, H. (1967). *Biochim. Biophys. Acta* **140**, 312.
Boonkoom, V., and Horne, F. R. (1968). *Am. Zool.* **8**, 817.
Bricteux-Grégoire, S., and Florkin, M. (1962a). *Arch. Intern. Physiol. Biochim.* **70**, 144.
Bricteux-Grégoire, S., and Florkin, M. (1962b). *Arch. Intern. Physiol. Biochim.* **70**, 496.
Bricteux-Grégoire, S., and Florkin, M. (1964). *Comp. Biochem. Physiol.* **12**, 55.
Bricteux-Grégoire, S., Duchâteau-Bosson, G., Jeuniaux, C., and Florkin, M. (1964a). *Arch. Intern. Physiol. Biochim.* **72**, 116.
Bricteux-Grégoire, S., Duchâteau-Bosson, G., Jeuniaux, C., and Florkin, M. (1964b). *Arch. Intern. Physiol. Biochim.* **72**, 267.
Brunel, A. (1938). *Compt. Rend* **206**, 858.
Buchanan, J. M., Sonne, J. C., and Delluva, A. M. (1948). *J. Biol. Chem.* **173**, 81.
Campbell, J. W. (1966). *Comp. Biochem. Physiol.* **18**, 179.
Campbell, J. W., and Bishop, S. H. (1970). *In* "Comparative Biochemistry of Nitrogen Metabolism" (J. W. Campbell, ed.), Vol. 1, p. 103. Academic Press, New York.
Campbell, J. W., and Speeg, K. V., Jr. (1968a). *Comp. Biochem. Physiol.* **25**, 3.
Campbell, J. W., and Speeg, K. V., Jr. (1968b). *Z. Vergleich Physiol.* **61**, 164.
Cardot, J. (1965). *Compt. Rend. Soc. Biol.* **159**, 1612.
Cardot, J. (1966a). *Compt. Rend. Soc. Biol.* **160**, 798.
Cardot, J. (1966b). *Compt. Rend. Soc. Biol.* **160**, 1264.
Cardot, J. (1966c). *Compt. Rend. Acad. Sci.* **262**, 2085.
Casola, L., Giuditta, A., and Rocca, E. (1964). *Arch. Biochem. Biophys.* **107**, 57.
Castañeda, M., Martuscelli, J., and Mora, J. (1967). *Biochim. Biophys. Acta* **141**, 276.

Chittenden, N. H. (1875). *Ann. Chem.* **178**, 266.
Clementi, A. (1918). *Atti Accad. Nazl. Lincei, Rend., Classe Sci. Fis. Mat. Nat.* **27**, 299.
Deffner, G. G. J. (1961a). *Biochim. Biophys. Acta* **47**, 378.
Deffner, G. G. J. (1961b). *Biochim. Biophys. Acta* **50**, 555.
De Jorge, F. B., and Petersen, J. A. (1970). *Comp. Biochem. Physiol.* **35**, 211.
Delaunay, H. (1925). *Compt. Rend. soc. biol.* **93**, 128.
Delaunay, H. (1927). Thesis, Sci. Nat., Paris.
Delaunay, H. (1931). *Biol. Rev.* **6**, 265.
Delaunay, H. (1934). *Ann. Physiol. Physicochim. Biol.* **10**, 695.
Duchâteau, G., and Florkin, M. (1951). Unpublished data.
Duchâteau, G., and Florkin, M. (1951–1952). Unpublished data.
Duchâteau, G., and Florkin, M. (1957). Unpublished data.
Duchâteau, G., Florkin, M., and Frappez, G. (1941). *Bull. Classe Sci., Acad. Roy. Belg.* [5] **27**, 169.
Duchâteau, G., Florkin, M., and Sarlet, H. (1954). *Arch. Intern. Physiol.* **62**, 512.
Duerr, F. G. (1967). *Comp. Biochem. Physiol.* **22**, 333.
Duerr, F. G. (1968). *Comp. Biochem. Physiol.* **26**, 1051.
Emanuel, C. F. (1957). *Z. Vergleich. Physiol.* **40**, 1.
Emanuel, C. F., and Martin, A. W. (1956). *Z. Vergleich. Physiol.* **39**, 226.
Ennor, A. H., and Morrison, J. F. (1958). *Physiol. Rev.* **38**, 631.
Ericson, L. E. (1960). *Nature* **185**, 465.
Erspamer, V. (1940). *Arch. Sci. Biol. (Bologna)* **26**, 295.
Erspamer, V. (1948). *Experientia* **4**, 226.
Erspamer, V. (1952). *Arzneimittel-Forsch.* **2**, 253.
Erspamer, V., and Boretti, G. (1951). *Arch. Intern. Pharmacodyn.* **88**, 296.
Erspamer, V., and Dordoni, F. (1947). *Arch. Intern. Pharmacodyn.* **74**, 263.
Fänge, R. (1958). *Acta Zool. (Stockholm)* **39**, 39.
Farina, B. (1964). *Ital. J. Biochem.* **13**, 75.
Farina, B., Mezzasoma, I., and Leone, E. (1966). *Boll. Soc. Ital. Biol. Sper.* **42**, 1440.
Farina, E., Farina, B., and Leone, E. (1962). *Ital. J. Biochem.* **11**, 141.
Florey, Ernest, and Florey, Elisabeth. (1954). *Z. Naturforsch.* **9b**, 58.
Florkin, M. (1945). "L'évolution du métabolisme des substances azotées chez les animaux," Masson, Paris. (*Actualites Bio. Ch. No. 3*).
Florkin, M. (1954). *Bull. Soc. Zool. France* **79**, 369.
Florkin, M., and Duchâteau, G. (1941). *Bull. Classe Sci., Acad. Roy. Belg.* [5] **27**, 174.
Florkin, M., and Duchâteau, G. (1943). *Arch. Intern. Physiol.* **53**, 267.
Florkin, M., and Houet, R. (1938). *Arch. Intern. Physiol.* **47**, 125.
Florkin, M., and Renwart, H. (1939). *Arch. Intern. Physiol.* **49**, 127.
Fraenkel, G. (1954). *Arch. Biochem. Biophys.* **50**, 486.
Fredericq, H. (1947). *Biol. Rev. Cambridge Phil. Soc.* **22**, 297.
Friedl, F. E. (1961). *J. Parasitol.* **47**, 773.
Friedl, F. E., and Bayne, R. A. (1966). *Comp. Biochem. Physiol.* **17**, 1167.
Gasteiger, E. L., Gergen, J. A., and Haake, P. C. (1955). *Biol. Bull.* **109**, 345.
Gaston, S., and Campbell, J. W. (1966). *Comp. Biochem. Physiol.* **17**, 259.
Glahn, P. E., Manchon, P., and Roche, J. (1955). *Compt. Rend. Soc. Biol.* **149**, 509.
Gorzkowski, B. (1969). *Acta Biochim. Polon.* **16**, 193.
Grah, H. (1937). *Zool. Jahrb., Abt. Allgem. Zool. Physiol. Tiere* **57**, 355.

Haggag, G., and Fouad, Y. (1968). *Z. Vergleich. Physiol.* **57**, 428.
Hammen, C. S., Miller, H. F., Jr., and Geer, W. H. (1966). *Comp. Biochem. Physiol.* **17**, 1199.
Hartman, W. J., Clark, W. G., Cyr, S. D., Jordon, A. L., and Leibhold, R. A. (1960). *Ann. N.Y. Acad. Sci.* **90**, 637.
Heidermanns, C. (1953). *Naturwissenschaften* **40**, 403.
Heidermanns, C., and Kirchner-Kühn, I. (1952). *Z. Vergleich. Physiol.* **34**, 166.
Heimberg, M., and Velick, S. F. (1954). *J. Biol. Chem.* **208**, 725.
Heller, J., and Jezewska, M. M. (1959). *Bull. Acad. Polon. Sci., Classe II* **7**, 1.
Henze, M. (1905). *Z. Physiol. Chem.* **43**, 477.
Henze, M. (1913). *Z. Physiol. Chem.* **87**, 51.
Henze, M. (1914). *Z. Physiol. Chem.* **91**, 230.
Hesse, O. (1910). *Z. Allgem. Physiol.* **10**, 273.
Hoppe-Seyler, F. A. (1933). *Z. Physiol. Chem.* **222**, 105.
Hoppe-Seyler, F. A., and Linneweh, W. (1931). *Z. Physiol. Chem.* **196**, 47.
Horne, F., and Boonkoom, V. (1970). *Comp. Biochem. Physiol.* **32**, 141.
Huggins, A. K., Rick, J. T., and Kerkut, G. A. (1967). *Comp. Biochem. Physiol.* **21**, 23.
Humoto, Y. (1954). *Z. Physiol. Chem.* **297**, 47.
Hunt, S. (1970a). *Biochim. Biophys. Acta* **207**, 347.
Hunt, S. (1970b). "Polysaccharide-Protein Complexes in Invertebrates." Academic press, New York.
Irvin, J. L. (1938). *J. Biol. Chem.* **123**, lxii.
Irvin, J. L., and Wilson, D. W. (1939). *J. Biol. Chem.* **127**, 565.
Iseki, T. (1931). *Z. Physiol. Chem.* **203**, 259.
Ishida, S. (1954). *Sci. Rept. Tohoku Univ., Fourth Ser.* **20**, 249.
Ishida, S. (1955a). *Bull. Marine Biol. Sta. Asamushi, Tohoku Univ.* **7**, 50.
Ishida, S. (1955b). *J. Coll. Arts Sci., Chiba Univ., Nat. Sci. Ser.* **1**, 270.
Ishida, S. (1956). *Bull. Marine Biol. Sta. Asamushi, Tohoku Univ.* **8**, 19.
Ishida, S., and Tsuzuki, K. (1955). *J. Coll. Arts Sci., Chiba Univ., Nat. Sci. Ser.* **1**, 267.
Ishida, S., Tsuzuki, K., and Nitahara, I. (1956). *J. Coll. Arts Sci., Chiba Univ., Nat. Sci. Ser.* **2**, 82.
Ishida, S., Umemori, Y., and Aikawa, T. (1969). *Comp. Biochem. Physiol.* **28**, 465.
Ito, K. (1957). *Bull. Japan. Soc. Sci. Fisheries* **23**, 497.
Jacobson, L. L. (1820). Cited after Wolf (1933).
Jezewska, M. M. (1968a). *Bull. Acad. Polon. Sci., Ser. Sci. Biol.* **16**, 73.
Jezewska, M. M. (1968b). *Bull. Acad. Polon. Sci., Ser. Sci. Biol.* **16**, 77.
Jezewska, M. M. (1969). *Acta Biochim. Polon.* **16**, 313.
Jezewska, M. M., and Sawicka, T. (1968). *Bull. Acad. Polon. Sci., Ser. Sci. Biol.* **16**, 197.
Jezewska, M. M., Gorzkowski, B., and Heller, J. (1963). *Acta Biochim. Polon.* **10**, 55.
Jezewska, M. M., Gorzkowski, B., and Heller, J. (1964). *Acta Biochim. Polon.* **11**, 135.
Kelly, A. (1904). *Beitr. Chem. Physiol. Pathol.* **5**, 377.
Kelly, R. E., and Rice, R. V. (1967). *Science* **155**, 208.
Kerkut, G. A., and Cottrell, G. A. (1963). *Comp. Biochem. Physiol.* **8**, 53.
Kerkut, G. A., and Walker, R. J. (1961). *Comp. Biochem. Physiol.* 3, 143.

Kerkut, G. A., Sedden, C. B., and Walker, R. J. (1967). *Comp. Biochem. Physiol.* **23**, 159.

Kerkut, G. A., Walker, R. J., and Woodruff, G. N. (1968). *Brit. J. Pharmacol. Chemother.* **32**, 241.

Knoop, F., and Martius, C. (1939). *Z. Physiol. Chem.* **258**, 238.

Koechlin, B. A. (1954). *Proc. Natl. Acad. Sci. U.S.* **40**, 60.

Koechlin, B. A. (1955). *J. Biophys. Biochem. Cytol.* **1**, 511.

Kohn, A. J., Saunders, P. R., and Wiener, S. (1960). *Ann. N.Y. Acad. Sci.* **90**, 706.

Kojima, Y., and Kusakabe, H. (1956). *J. Sci. Res. Inst. (Tokyo)* **50**, 193.

Konosu, S., Akiyama, T., and Mori, T. (1958). *Bull. Japan. Soc. Sci. Fisheries* **23**, 561.

Krukenberg, C. F. W., and Wagner, H. (1885). *Z. Biol.* **21**, 25.

Kutscher, F., and Ackermann, D. (1931). *Z. Physiol. Chem.* **199**, 266.

Kutscher, F., and Ackermann, D. (1933). *Z. Physiol. Chem.* **221**, 33.

Kutscher, F., Ackermann, D., and Flössner, O. (1931). *Z. Physiol. Chem.* **199**, 273.

Lal, M. B., and Saxena, B. B. (1952). *Nature* **170**, 1024.

Lee, T. W., and Campbell, J. W. (1965). *Comp. Biochem. Physiol.* **15**, 457.

Linton, S. N., and Campbell, J. W. (1962). *Arch. Biochem. Biophys.* **97**, 360.

Lohmann, K. (1936). *Biochem. Z.* **286**, 28.

Lum, S. C., and Hammen, C. S. (1964). *Comp. Biochem. Physiol.* **12**, 185.

Lynch, M. P., and Wood, L. (1966). *Comp. Biochem. Physiol.* **19**, 783.

Marchal, P. (1889). *Mem. Soc. Zool. France* **3**, 31.

Mayeda, H. (1936). *Acta Schol. Med., Univ. Imp. Kioto* **18**, 218.

Mendel, L. B. (1904). *Beitr. Chem. Physiol. Pathol.* **5**, 582.

Mendel, L. B., and Bradley, H. C. (1906). *Am. J. Physiol.* **17**, 167.

Mendel, L. B., and Wells, H. G. (1909). *Am. J. Physiol.* **24**, 170.

Menssen, H. G. (1960). *Z. Physiol. Chem.* **318**, 59.

Meyerhof, O. (1928). *Arch. Sci. Biol. (Bologna)* **12**, 536.

Milstein, C. P. (1967). *Biochem. J.* **103**, 634.

Mirolli, M. (1968). *Comp. Biochem. Physiol.* **24**, 847.

Mognoni, G. A., and Lanzavecchia, G. (1969). *Atti Accad. Nazl. Lincei, Rend. Classe Sci. Fis., Mat. Nat.* **46**, 610.

Moore, E., and Wilson, D. W. (1936). *J. Biol. Chem.* **114**, lxxi.

Moore, E., and Wilson, D. W. (1937). *J. Biol. Chem.* **119**, 573.

Moreland, B., and Watts, D. C. (1967). *Nature* **215**, 1092.

Morizawa, K. (1927). *Acta Schol. Med., Univ. Imp. Kioto* **9**, 285.

Needham, J. (1935). *Biochem. J.* **29**, 238.

Negus, M. R. S. (1968). *Comp. Biochem. Physiol.* **24**, 317.

Nishita, K., Arai, K. I., and Saito, T. (1965). *Bull. Fac. Fisheries, Hokkaido Univ.* **16**, 114.

Noland, J. L. (1949). *Biol. Bull.* **97**, 263.

Norris, E. R., and Benoit, G. J., Jr. (1945). *J. Biol. Chem.* **158**, 433.

Norum, K. R., and Bremer, J. (1966). *Comp. Biochem. Physiol.* **19**, 483.

Okuda, Y. (1929). *J. Coll. Agr., Imp. Univ. Tokyo* **10**, 281.

Olomucki, A., Thoai, N. V., and Roche, J. (1960). *Colloq. Intern. Centre Natl. Rech. Sci. (Paris)* **17**, 171–179.

Ouchi, S. (1959). *J. Biochem. (Tokyo)* **46**, 765.

Pikkarainen, J., Rantanen, J., Vastamäki, M., Lampiaho, K., Kari, A., and Kulonen, E. (1968). *European J. Biochem.* **4**, 555.

Plum, K. (1935). *Z. Vergleich. Physiol.* **22**, 155.

Porembska, Z., and Heller, J. (1962). *Acta Biochim. Polon.* **9**, 385.
Porembska, Z., Gorzkowski, B., and Jezewska, M. M. (1966). *Acta Biochim. Polon.* **13**, 107.
Porembska, Z., Gasiorowska, I., and Mochnacka, I. (1968). *Acta Biochim. Polon.* **15**, 171.
Potts, W. T. W. (1965). *Comp. Biochem. Physiol.* **14**, 339.
Potts, W. T. W. (1967). *Biol. Rev.* **42**, 1.
Prosser, C. L., and Brown, F. A., Jr. (1961). "Comparative Animal Physiology," 2nd ed. Saunders, Philadelphia, Pennsylvania.
Przylecki, S. J. (1922). *Arch. Intern. Physiol.* **20**, 103.
Przylecki, S. J. (1926). *Arch. Intern. Physiol.* **27**, 159.
Raghupathiramireddy, S., and Swami, K. S. (1967). *Can. J. Biochem.* **45**, 603.
Razet, P., and Dagobert, D. (1968). *Arch. Sci. Physiol.* **22**, 173.
Riddiford, L. M., and Scheraga, H. A. (1962). *Biochemistry* **1**, 95.
Robertson, J. D. (1965). *J. Exp. Biol.* **42**, 153.
Robin, Y. (1960). *Colloq. Intern. Centre Natl. Rech. Sci. (Paris)* **17**, 377.
Robin, Y., and Thoai, N. V. (1957). *Compt. Rend. Soc. Biol.* **151**, 2093.
Robin, Y., Pradel, L. A., and Thoai, N. V. (1959a). *Pubbl. Staz. Zool. Napoli* **31**, 153.
Robin, Y., Pradel, L. A., Thoai N. V., and Roche, J. (1959b). *Compt. Rend. Soc. Biol.* **153**, 21.
Rocca, E., and Ghiretti, F. (1958). *Arch. Biochem. Biophys.* **77**, 336.
Roche, J., Thoai, N. V., and Glahn, P. E. (1952a). *Experientia* **8**, 428.
Roche, J., Thoai, N. V., Robin, Y., Garcia, I., and Hatt, J. L. (1952b). *Compt. Rend. Soc. Biol.* **146**, 1899.
Roche, J., Glahn, P. E., Hedegaard, J., and Manchon, P. (1955). *Ric. Sci., Suppl.* **25**, 392.
Roche, J. Thoai, N. V., and Robin, Y. (1957). *Biochim. Biophys. Acta* **24**, 514.
Roche, J., Glahn, P. E., Manchon, P., and Thoai, N. V. (1959). *Biochim. Biophys. Acta* **35**, 111.
Roche, J., Robin, Y., Thoai, N. V. and Pradel, L. A. (1960). *Comp. Biochem. Physiol.* **1**, 44.
Roseghini, M., and Ramorino, L. M. (1970). *J. Neurochem.* **17**, 489.
Sarlet, H., Grivegnée, R., Faidherbe, J., and Frenck, G. (1950). *Arch. Intern. Physiol.* **57**, 286.
Schmidt, C. L. A., and Watson, T. (1918). *J. Biol. Chem.* **33**, 499.
Seki, N., Arai, K. I., and Saito, T. (1967). *Bull. Fac. Fisheries, Hokkaido Univ.* **17**, 184.
Seraydarian, M. W., and Kalvaitis, Z. (1964). *Comp. Biochem. Physiol.* **12**, 1.
Shibuya, S., and Ouchi, S. (1957). *Nature* **180**, 549.
Simpson, J. W., Allen, K., and Awapara, J. (1959). *Biol. Bull.* **117**, 371.
Simpson, M. V., and Velick, S. F. (1954). *J. Biol. Chem.* **208**, 61.
Smith, A. C. (1969). *Comp. Biochem. Physiol.* **30**, 551.
Spackman, D. H., Stein, W. H., and Moore, S. (1958). *Anal. Chem.* **30**, 1190.
Speeg, K. V., Jr., and Campbell, J. W. (1968a). *Am. J. Physiol.* **214**, 1392.
Speeg, K. V., Jr., and Campbell, J. W. (1968b). *Comp. Biochem. Physiol.* **26**, 579.
Speeg, K. V., Jr., and Campbell, J. W. (1969). *Am. J. Physiol.* **216**, 1003.
Spitzer, J. M. (1937). *Zool. Jahrb., Abt. Allgem. Zool. Physiol. Tiere* **57**, 457.
S.-Rózsa, K. and Zs.-Nagy, I. (1967). *Comp. Biochem. Physiol.* **23**, 373.
Suzuki, U., and Joshimura, K. (1909). *Z. Physiol. Chem.* **62**, 1.

Swaminathan, S. (1958). *J. Zool. Soc. India* **10**, 150.

Sweeney, D. (1968). *Comp. Biochem. Physiol.* **25**, 601.

Takahashi, E. (1915). *J. Coll. Agr., Tohoku Imp. Univ.* **6**, 289.

Thiele, O. W. (1963). *Z. Physiol. Chem.* **332**, 319.

Thoai, N. V. (1960). *Colloq. Intern. Centre Natl. Rech. Sci. (Paris)* **92**, 297–323.

Thoai, N. V. and Robin, Y. (1959a). *Biochim. Biophys. Acta* **35**, 446.

Thoai, N. V. and Robin, Y. (1959b). *Bull. Soc. Chim. Biol.* **41**, 735.

Thoai, N. V. and Robin, Y. (1960). *Colloq. Intern. Centre Natl. Rech. Sci. (Paris)* **92**, 353–375.

Thoai, N. V., and Robin, Y. (1961). *Biochim. Biophys. Acta* **52**, 221.

Thoai, N. V., Roche, J., and Robin, Y. (1952). *Compt. Rend.* **235**, 832.

Thoai, N. V., Roche, J., and Robin, Y. (1953). *Biochim. Biophys. Acta* **11**, 403.

Thoai, N. V., Glahn, P. E., Hedegaard, J., Manchon, P., and Roche, J. (1954). *Biochim. Biophys. Acta* **15**, 87.

Thoai, N. V., Hatt, J. L., and An, T. T. (1955). *Biochim. Biophys. Acta* **18**, 589.

Thoai, N. V., Hatt, J. L., and An, T. T. (1956). *Biochim. Biophys. Acta* **22**, 116.

Thoai, N. V., Robin, Y., and Pradel, L. A. (1957). *Compt. Rend. Soc. Biol.* **151**, 2097.

Tramel, P. R., and Campbell, J. W. (1970). *J. Biol. Chem.* **245**, 6634.

Truszkowski, R. (1928). *Biochem. J.* **22**, 1299.

Truszkowski, R., and Chajkinowna, S. (1935). *Biochem. J.* **29**, 2361.

Truszkowski, R., and Goldmanowna, C. (1933). *Biochem. J.* **27**, 612.

Tsuzuki, K. (1957). *J. Coll. Arts Sci., Chiba Univ., Nat. Sci. Ser.* **2**, 239.

Umemori, Y. (1967). *Comp. Biochem. Physiol.* **20**, 635.

Ungar, G., Ungar, A., and Parrot, J. L. (1937). *Compt. Rend. Soc. Biol.* **126**, 1156.

Vialli, M., and Erspamer, V. (1940). *Arch. Fisiol.* **40**, 293.

Villee, C. A. (1947). *Biol. Bull.* **93**, 220.

Virden, R., and Watts, D. C. (1964). *Comp. Biochem. Physiol.* **13**, 161.

von Fürth, O. (1900). *Z. Physiol. Chem.* **31**, 353.

Watts, R. L., and Watts, D. C. (1968). *Nature* **217**, 1125.

Welsh, J. H., and Moorhead, M. (1959). *Science* **129**, 1491.

Wiener, H. (1902). *Beitr. Chem. Physiol. Pathol.* **2**, 42.

Wolf, G. (1933). *Z. Vergleich. Physiol.* **19**, 1.

Woodruff, G. N., Oniwinde, A. B., and Kerkut, G. A. (1969). *Comp. Biochem. Physiol.* **31**, 599.

Yoneda, T. (1968). *Bull. Fac. Fisheries, Hokkaido Univ.* **19**, 140.

Zervas, L., and Bergmann, M. (1931). *Z. Physiol. Chem.* **201**, 208.

Zs.-Nagy, I. (1967). *Acta Biol. Acad. Sei. Hung.* **18**, 1.

CHAPTER 11

Endocrinology of Mollusca

Micheline Martoja

I. Introduction

Mollusca show a high degree of evolution, which has long led us to believe that they are probably endowed with endocrine regulations. Unfortunately, current methods of endocrinological research are not easily applied in the case of Mollusca. The study of neurosecretion presents unexpected difficulties. In many Mollusca, the classic methods* do not permit the identification of neurosecretory products, while the same methods stain structures not related to neurosecretion such as gliomes or waste pigments. The consequence is that true neurosecretory cells may remain unidentified while common nerve cells mimic neurosecretory aspects. On the other hand, secretion is only exceptionally seen outside perikarya; the absence of images of axonal transport and the impossibility of determining the fate of the secretion singularly complicate the search for endocrine glands and neurohemal organs.

In the course of a critical review of endocrinological data obtained

* In the immense majority of cases, the neurosecretory product, which is normally acidophilic, acquires a distinct basophilic character after oxidation; it is this character which is used in the majority of the methods of histological detection of the neurosecretory cells. The two best known methods use chromate–hematoxylin and paraldehyde fuchsin. They will be designated here as classic methods.

in gastropods, Simpson *et al.* (1966a) have correctly called attention to the many physiological and biochemical gaps in our information and the resulting lack of certainty; the reserves they formulated may be extended to all mollusks. Nevertheless, as demonstrated by recent reviews (E. Scharrer and B. Scharrer, 1954, 1963; Gersch, 1964; Gabe, 1965a, 1967; Durchon, 1967; Martoja, 1968), signs of the existence of neurosecretory phenomena have been recognized in all classes of mollusks, with the exception of Monoplacophora, which have not yet been studied in that respect. But the existence of endocrine glands or neurohemal organs has been suggested or demonstrated only in the case of gastropods and cephalopods. A glimpse of the role of neurosecretory activity in different physiological processes begins to be caught, but the functional role of the different endocrine glands has been made clear in only very few instances. As for the chemical nature of molluscan hormones, it remains unknown.

Finally, it is necessary to emphasize the fact that the unequivocal definition of certain endocrinological concepts (see, for example, Bern, 1966) remains difficult. For instance, according to the site of their production, the same substances may represent either neurotransmitters or hormones; and it is difficult to state precisely a distinction between neurosecretory cells and aminergic neurons, serotoninergic neurons, etc. It is therefore necessary to state that the terms hormone and neurosecretion are used here in their most traditional meaning.

II. Amphineura and Scaphopoda

From the endocrinological viewpoint, the classes Amphineura and Scaphopoda are the least well known of all the mollusks. Only "possible" neurosecretory cells according to Bern* have been identified in these classes, but no physiological data are at hand concerning this "possible" neurosecretion. It has never been suggested that endocrine glands are present in them. Neurosecretory cells have been described in three species of Polyplacophora belonging to three different families (*Lepidochiton cinereus, Ahanthochitona discrepans, Chiton olivaceus*) (Martoja, 1967). They are scattered in the cortex of the buccal ganglia.

* According to Bern (1963) neurosecretory cells are said to be "possible" when their cytoplasm contains a histologically detectable secretory product; they are said to be "probable" when the secretory product undergoes modifications which can be related to a known biological cycle; finally, they are said to be "certain" when the hormonal nature of the secretion product has been demonstrated experimentally. It is clear that in all cases, the nervous nature of the aforesaid cell must be demonstrated.

Their secretion is slightly acidophilic, but it becomes very basophilic after oxidation, a property which is the basis of their detection by classic methods.

Axonal transport is detectable in the neuropile, but the ultimate fate of the secretion product remains unknown. Furthermore, these cells appear to have a continuous activity and consequently no conclusion can be drawn concerning their relationship to a physiological process. Nevertheless, since the buccal ganglia of chitons are considered to be the equivalent of a neurovegetative system, it is possible that their secretions have a trophic function. Apparently no research has been carried out on neurosecretion in Aplacophora.

In *Dentalium entale*, the only species of Scaphopoda studied in that respect, the neurosecretory cells are located in the anterior buccal ganglia, in the cerebral ganglia, and in the pleural ganglia (Gabe, 1949, 1967). Each buccal ganglion contains a single voluminous neurosecretory cell. In the cerebral and pleural ganglia, the neurosecretory cells are of the same size as the surrounding neurons and do not differ from them by the presence of secretory products. There are thus two types of neurosecretory perikarya in *Dentalium*. The accumulation sites are unknown. In both cases the secretion is acidophilic and remains so after permanganate oxidation. In contrast to other mollusks, periodic acid–Schiff (PAS) reaction is negative.

III. Lamellibranchia

The neurosecretory cells, described simultaneously for the first time in a score of species (Gabe, 1955), have been carefully studied from the morphological and experimental points of view. In this class of mollusks, no endocrine gland and no neurohemal organ has been identified.

A. Morphological Data

All authors, and in particular Gabe (1955), Lubet (1955, 1956, 1957, 1959, 1965, 1966), Fährmann (1961) Antheunisse (1963), Nagabhushanam (1963b, 1964a,b,c) Baranyi (1963, 1964a,b, 1966) Baranyi and Salánki (1963), have recognized the presence of neurosecretory cells in the cerebropleural and in visceral ganglia. Their number and location differ from species to species. According to Gabe, they are more numerous and more scattered in the primitive species than in the more specialized ones where they show a tendency to form groups. For instance, in the cerebral ganglia as well as in the visceral ganglia of *Nucula nucleus* they form a dense cap (Gabe, 1955). They are scattered in the corresponding ganglia of *Teredo* (Gabe and Rancurel, 1958). In *Mytilus edulis* they aggregate at a laterodorsal point on the anterior face of the cerebral ganglion and are distributed in the dorsal cell layer of the visceral ganglion (Lubet, 1955). In *Dreissena polymorpha,* they are concentrated in four distinct masses in the cerebral ganglion and five masses in the visceral ganglion (Antheunisse, 1963).

On the other hand, the presence of neurosecretory cells in the pedal ganglia remains very controversial. Gabe and Lubet conclude that they are not present in the ganglia of the species they have studied. According to Antheunisse a slight neurosecretory activity takes place in the pedal ganglia of *Dreissena polymorpha.* Fährmann (1961) recognizes an intense neurosecretory activity in the pedal ganglia of *Unio tumidus,* and Baranyi and Salánki (1963) reach the same conclusion in the case of *Anodonta cygnea.* But some of these conclusions are, according to Gabe (1965a, 1967) and Durchon (1967), the result of a confusion between products of neurosecretion and chromolipoids.

Several types of neurosecretory cells probably exist in lamellibranchs, but the data are too scanty to serve as a basis for general conclusions. In *Mytilus edulis* and *Chlamys varia,* certain neurosecretory cells are pear shaped and of normal size, while others are multipolar and very small (Lubet). In *Crassostrea virginica,* two categories of cells are also identified—type I cells are pear shaped and stained by chromate–hematoxylin, while type II cells are oval and are not stained by this method (Nagabhushanam, 1963a, 1964a). Two kinds of inclusions have been described in *Unio tumidus* (Fährmann, 1961), but according to Gabe (1965a, 1967) and to Durchon (1967), the neurosecretory nature of type I granules has yet to be demonstrated. In the case of *Anodonta cygnea,* Baranyi and Salánki (1963) have been able to distinguish A neurosecretory cells representing 75% of the neurons, and B cells, large, oval, or pear shaped; moreover, neurosecretory granules have been identified in glial cells called C cells, apparently endowed with the function of the transport of the neurosecretory granules. According to Durchon (1967), only B cells are of a real neurosecretory nature. On the other hand, possible significant size differences have been noticed between the cells of cerebral ganglia and of the visceral ganglia of shipworms (Gabe and Rancurel, 1958) or between the cells of different masses described in *Dreissena polymorpha* (Antheunisse, 1963).

In conclusion, at the present time, the distinction can only be based on morphological or cytometrical criteria between the different categories of neurosecretory cell. This is a consequence of the uniformity of the tinctorial properties of the secretion in Pelecypods, which is normally acidophilic but becomes basophilic after permanganate oxidation in an acid medium. Type II cells of *Crassostrea* are not stained by chromate–hematoxylin and are the only known exceptions. Classic methods are therefore of use in the majority of cases, but without specificity, as remarked by Gabe and by Lubet. It is this lack of specificity which is the basis of the doubts raised, for instance, by the interpretations of Fährmann or of Baranyi and Salánki. Sterba's pseudo-isocyanine method (1964) renders the secretion product strongly fluorescent; this reaction is only carried out after permanganates oxidation as shown by Zs Nagy (1965) in the case of *Anodonta.*

Histochemical data are of a particularly rudimentary nature. Considering the PAS-negative character of the secreted product and the absence of lipids in it, Lubet concludes that this secretion could be of a proteinic nature in *Mytilus.* In *Crassostrea,* according to Nagabhushanam, it may contain polysaccharides. In *Anodonta,* the neurosecretory cells show an acid phosphomonoesterasic activity, the variations of which are parallel with those of the secretory activity. An alkaline phosphomonoesterasic activity has been identified in the C cells of *Anodonta,* but this could not be considered to be a characteristic of the neurosecretory cell, as these cells, of a recognized glial nature, are only the vectors of the secretion (Baranyi, 1964a, 1966).

The stages of the secretory cycles have repeatedly been studied with the light microscope. A parallel between the attenuation of the basophilic character resulting from the presence of cytoplasmic RNA and the accumulation of neurosecretory material has been observed in *Anodonta* (Salánki *et al.,* 1965). Very few studies of neurosecretion in lammelibranchs have been performed at the ultrastructural level (Fährmann, 1961; Zs Nagy, 1964).

The images of axonal transport rarely go beyond the neuropile of the ganglia in which the secretory perikarya are located. Consequently, we remain practically in complete ignorance of the transport and fate of the products of neurosecretion. This disappearance of the neurosecretory material remains unexplained. If it is the result of a change in the physical or chemical state, the ultrastructural study and the use of the pseudoisocyanic method, which is more sensitive than the classic methods, could solve this problem, as suggested by the results obtained by Fährmann and by Zs Nagy. But the disappearance of the neurosecretory material may be due to other causes, involving particular mechanisms. Several authors (Fährmann, Antheunisse, Baranyi, and Salánki) consider, for instance, that the glial cells play a role of relay or even of organs of accumulation; according to the same authors, this accumulation could have taken place in the perineurium, of the ganglia (Antheunisse, 1963). These authors also consider that the glial cells and epineuron could function as neurohemal organs.

B. Physiological Data

1. Stress

In lamellibranchs, stress may very rapidly modify the space distribution of the neurosecretion, whatever the degree of previous evolution of the secretory cycle may be (Lubet, 1966). Mussels are sensitive to thermal and osmotic stresses; a sudden rise of temperature (10°C) maintained for an hour or a sudden fall of salinity (20‰) result in an emptying of the neurosecretory cells of the cerebral ganglion, while a fall in temperature or an increase in salinity (45‰) are followed by a significant increase in the secretion product in the same cells. No variation takes place in the visceral ganglion (Lubet and Pujol, 1963, 1965). A similar phenomenon takes place in *Crassostrea,* but this time in the visceral ganglion. An electrical (20 volts for 30 minutes) or thermal (2 hours at 32°C) shock empties the secretion product (Nagabhushaham, 1964a). In addition to their physiological significance, these observations demonstrate with what care images attesting to the repletion or emptying of a neurosecretory cell of a lamellibranch, must be criticallv interpreted.

2. Reproduction

Reproductory cycle and neurosecretory cycle closely correspond. All authors agree on this point, but they sometimes differ in their interpretations. For instance, in two species of *Mytilus,* the accumulation of secretion in the pear-shaped cells of the cerebral ganglia is accomplished during the phase of gametogenesis, and its evacuation takes place when the gametes are ripe; fertilisins liberated by the gametes of the opposite sex are among the external stimuli able to produce the laying of eggs or ejaculation. The activity of the small multipolar neurones and of the cells of the visceral ganglia being almost permanent, it seems that the genital maturation depends only on the pear-shaped cells (Lubet, 1959). In the case of *Chlamys varia,* an even more definite synchronism is observed (Lubet, 1959). These facts have been confirmed in other species: in *Crassostrea virginica* and *Modiolus demissus,* type I cells appear to influence gonad maturation (Nagabhushanam, 1963a, 1964b), while in *Anodonta cygnaea* this effect appears to be due to A cells (Baranyi, 1964b). Nevertheless, Salánki and Baranyi (1965) observe that in *Anodonta* the period of inactivity of the neurosecretory cells is restricted to December and January, and they conclude that it is unjustified to link this aspect to the sexual cycle exclusively.

After a thorough study of the biology of *Dreissena polymorpha* and a recourse to surgical methods, Antheunisse goes as far as to decide in favor of pure coincidence, the sexual cycle being also superimposable on the seasonal cycle with all the thermal, trophic, and other variations that this implies. According to him, even an indirect influence is very improbable and he concludes that reproduction is independent of neurosecretion.

The experimental results are not comparable in all cases. In the fresh-water species *Dreissena,* the ablation of the cerebropleural ganglia does not stop the maturation of oocytes any more than the egg laying, which in spite of surgery, takes place a few weeks later (Antheunisse, 1963). On the contrary, in the marine species studied by Lubet, decerebration during the resting phase or at the beginning of gametogenesis distinctly delays sexual phenomena. Moreover, in decerebrated females, many ripe oocytes undergo lysis before egg laying and the others show an aspect which is the consequence of disturbances in the repartition of cytoplasmic RNA; carbohydrates and lipids show the same histological distribution as in normal females (Lubet, 1965). The same surgery, performed at the end of gametogenesis increases the number of emissions of genital products in mussel as well as in the oyster studied by Nagabhushanam (1964a). In this case, the role of the cerebral ganglion is undeniable, but Lubet remarks that it is impossible to decide if the mechanism is of hormonal or nervous nature. The cells of the visceral ganglia do not appear to be involved at any moment in the regulation of sexual life. The evacuation of the gametes is certainly disturbed after their ablation, but this is a consequence of the disorders introduced into the essential vegetative functions, as suggested by Lubet and Nagabhushanam.

3. *Annual Cycle*

In the case of *Dreissena,* the neurosecretory material begins to accumulate in the cerebral ganglion in autumn, and the maximum of activity takes place in winter. The emptying takes place in the spring. Summer corresponds to a period of inactivity. In the visceral ganglion, the cycle is almost exactly superimposable on that described above, but the period of emptying takes place during the summer (Antheunisse). The behavior of the B cells in *Anodonta* is very different; the stage of maximal repletion corresponds to June and their emptying takes place in August (Baranyi, 1964a,b, 1966). The *Unio* accumulation also corresponds to spring (Fährmann, 1961). In *Crassostrea,* only type I cells have an activity that varies during the year; the neurosecretory material begins to accumulate in January and this accumulation reaches a maximum in March. Between April and September, the number of cells containing a secretion diminishes; with rare exception, they are empty in October and December (Nagabhushanam, 1963a). In *Mytilus,* while the pear-shaped cells have a distinct annual cycle, the small star-shaped cells show a continuous activity (Lubet, 1959). The existence of a lunar cycle has also been recognized in *Mytilus* (Lubet, 1957).

4. *Growth and Metabolism*

Detailed experimental studies have been concerned with the relationships between the nervous system on one hand and metabolism or growth on the other hand in the case of *Mytilus* (Bourcart and Lubet, 1965; Lubet and Pujol, 1963, 1965; Lubet, 1965) and of *Crassostrea* (Nagabhushanam, 1964a).

The bilateral ablation of the cerebral ganglion of *Mytilus* results in growth delay, which is reversed more quickly, the younger the individual on which the surgery is performed. In older animals, the same ablation produces disorders of lipid metabolism. In addition, the byssus secretion is markedly reduced, an argument in favor of an influence of the neurohormonal products of cerebral origin on the synthesis of iodinated scleroproteins. Nevertheless, the periostracum, which is of the same composition, is reconstituted normally.

The ablation of the visceral ganglion has more severe consequences than the ablation of the cerebral ganglion; mussels, for instance, do not survive more than 3 or 4 months. During that period, a weakening of muscular tonus is observed, as well as a denutrition with a loss of reserve products. It must be emphasized that since the visceral ganglia contain rather large amounts of acetylcholine and 5-hydroxytryptamine, the results could be due to the loss of these chemical mediators. In *Crassostrea,* the ablation of the visceral ganglion slows down the heart rate and lowers the rate of water filtration.

5. *Osmoregulation*

The existence of osmotic stresses in lamellibranchs has been mentioned above, but in addition to these physiological shocks of short duration, the salinity of the medium is not without effect on the contents of neurosecretory cells. Nagabhushanam (1964a) has observed that, in the visceral ganglia of *Crassostrea,* the quantity of neurosecretory products varies as a function of the external salinity and that their ablation results in an increase in weight due to water absorption.

From the work of Lubet and Pujol (1963, 1965) and from data subsequently obtained by the same authors (mentioned in Lubet, 1966), it may be concluded that the cerebral ganglia of mussels control the penetration of water. Their ablation results in a certain inability of the animal to accomplish the type of isosmotic intracellular regulation defined by Florkin. In individuals decerebrated 2 months before being placed in diluted seawater, the increase in water content is much higher than in the nondecerebrated controls submitted to the same treatment. Compared to the controls, the decerebrated individuals show an increased content in amino nitrogen and in inorganic ions. Consequently, the cerebral ganglia appear to elaborate one or several hormonal factors acting on the ionic equilibria and on the regulation of the free amino acid concentration.

IV. Gastropoda

It appears that gastropods not only possess neurosecretory cells, but also neurohemal systems and endocrine glands. The endocrinological aspects are rather different in the three subclasses, and above all, the knowledge we have of them is very irregular.

Schematically, the nervous system of Gastropoda is composed of (1) a perioesophagal ring with three pairs of ganglia (cerebral ganglion, pleural ganglion, pedal ganglion) and a pair of buccal ganglia; and (2) a visceral loop composed of a succession of ganglia and connections interconnecting the pleural ganglia. The visceral loop bears a pair of parietal ganglia (in opistobranchs and pulmonates), a pair of intestinal ganglia (subintestinal and supraintestinal), and an unpaired visceral ganglion. In addition to these principal ganglia, there are rather numerous accessory ganglia, the presence of which varies from group to group (tentacular ganglion, osphradial ganglion, etc.). In the three subclasses, the evolutionary trend is manifested by a shortening of the connections and of the visceral loop (cephalization) and by a fusion of the ganglia, which have migrated towards the anterior end (cerebralization).

A. Morphological Data

1. *Neurosecretion*

a. Subclass Prosobranchia. The essential data concerning the neurosecretion of the prosobranchs are found in a small number of publications (Gabe, 1951, 1953a, 1954; Grzycki, 1951; Boddingius, 1960; Gorf, 1961, 1963; Nolte *et al.*, 1965; Simpson *et al.*, 1966a; Choquet and Lemaire, 1969).

The repartition of neurosecretory cells in prosobranchs varies slightly from species to species in a heterogeneous manner, as illustrated in the table compiled by Gabe (1953a). In most primitive Prosobranchia, the Archaeogastropoda, they are dispersed in the whole of the central nervous system, even reaching the initial part of the pedal nerve cords. In the Mesogastropoda and in the Neogastropoda, it seems that they never occur in the buccal and pedal ganglia. Heteropoda represent an extreme case of concentration, all the neurosecretory cells being grouped at the dorsal face of the cerebral ganglia.

Little information is available concerning possible differences between several categories of neurosecretory cells in the same species. Choquet and Lemaire (1969), it is true, have described two different types of neurosecretory cells in the cerebral ganglia of *Patella vulgata,* but they consider that these types might correspond to different stages of activity of a same category of cells.

Staining affinities vary according to the orders concerned. Gabe (1953a), in this respect, opposes Archaeogastropoda and Mesogastropoda to the Neogastropoda; in the former, the secretion is acidophilic and remains so, even after oxidation, while in the latter it may acquire a certain basophilia during certain stages of the secretory cycle; the chemical characteristics which could explain these variations have not been explored. Nevertheless, according to Boddingius (1960) or Choquet and Lemaire (1969), the neurosecretory cells of *Patella* (Archaeogastropoda) contain "Gomori-positive" granulations.

Histochemical data are scarce. A carbohydrate constituent is reported in Heteropoda (Gabe, 1951) and in *Vivipara vivipara* (Gort, 1961). In this species, Gorf also reports that some of the general protein reactions are negative (Millon reaction, xanthoproteic reaction) and that no protein is present which can be hydrolyzed by pepsine or by trypsin. As for the neurosecretory cells of Heteropoda, Gabe emphasizes that they differ from the common nerve cells of the same organisms by the presence of inorganic constituents, but that both categories show alkaline phosphatase activity.

Little is known of the ultrastructural aspect of the neurosecretion of prosobranchs. Two descriptions of the cerebral ganglia are available. In *Crepidula* Nolte *et al.* (1965) have observed elementary granules in the axons; and in *Calliostoma,* Simpson (cited after Simpson *et al.,* 1966a) has seen them in a few perikarya. It is interesting to note that these perikarya have no affinity for the classic methods.

b. Subclass Opistobranchia. Neurosecretion is not better known in the opistobranchs than in the prosobranchs. Since the discovery of neurosecretion by B. Scharrer (1935), a relatively small number of publications have been devoted to the subject (B. Scharrer, 1937; Gabe, 1953b;

Pavans de Ceccatty and von Planta, 1954; Lemche, 1955, 1956; Sanchez and Pavans de Ceccatty, 1957; Vicente, 1962, 1963, 1965a,b, 1966, 1969; Rosenbluth, 1963; Simpson *et al.*, 1963; Tiffon, 1964; Sakharov *et al.*, 1965; Minichev, 1965; Sakharov, 1966; Schmeckel and Weschler, 1968).

The different descriptions, and among them that of Gabe which includes thirty-five species and that of Vicente concerning twelve, show that the neurosecretory cells dispersed in the more primitive species become more and more grouped when ascending the evolutionary series. Gabe considers that the cerebral ganglia of the more primitive species, as well as the pedal ganglia of all species, are without them. On the other hand, Lemche (1955, 1956) considers that there are neurosecretory cells in all the nervous centers of *Cylichna cylindracea.* In spite of the restrictions formulated by Bern and Hagadorn (1965), Gabe (1965a), and Durchon (1967), who suspect that a confusion has arisen between secretion and chromolipoids, Minichev (1965) confirms the descriptions of Lemche in *Cylichna occulta* and attributes the dispersion of the neurosecretory cells to the lack of concentration of the nervous centers in this little evolved genus.

According to Vicente (1969), it is possible to recognize three types of neurons (A,B,C), and those belonging to category B, which are of medium size, are the most active from the neurosecretory viewpoint. For Schmekel and Weschler (1968), there are four types of neurons—giant, large, medium and small. Among them, the large ones are those performing a neurohumoral function. On the other hand, on the basis of morphological and physiological characteristics, two very different classes of neurosecretory cells have been recognized in the visceral ganglion of *Aplysia californica.* The first category (white cells) are included in the ganglion, while the second category (bag cells) surround the roots of the connectives of the visceral loop (Coggeshall *et al.*, 1966; Frazier *et al.*, 1967; Coggeshall, 1967).

Several authors have also discussed the existence of neurosecretory perikarya in the accessory ganglia, tentacular ganglion, or olfactory bulb, as well as their anatomical relationships with glandular formations such as the frontal organ (p. 369) or certain groups of subepithelial cells (Pavans de Ceccatty and von Planta, 1954; Tuzet *et al.*, 1957; Sanchez, 1962; Vicente, 1966, 1969).

The cytological data are very fragmentary. Images of nuclear extrusion have been observed in the course of the secretory cycle (Gabe, 1953b, Pavans de Ceccatty and von Planta, 1954; Vicente, 1962). The elementary neurosecretory granules and their relationship with the cellular organites have been described in several species (Rosenbluth, 1963; Sakharov *et al.*, 1965; Schmeckel and Weschler, 1968; Vicente, 1969).

As in the case of the Prosobranchia Neogastropoda, it is only during certain stages of the secretory cycle that the secretion is basophilic after permanganate oxidation (Gabe, 1953b), but here again the chemical interpretation is lacking. The classic methods are therefore seldom used. Furthermore, they show a manifest lack of specificity. This has been demonstrated in *Tritonia,* where the location of elementary granules does not coincide with that of the structures stained by the aforesaid methods. Sterba's method (1964) in which pseudoisocyanine is used appears more satisfactory in this species (Sakharov *et al.*, 1965). From the histochemical point of view, the secretion is rich in sulfur-containing proteins (Tiffon, 1964; Vicente, 1965b); an acid polysaccharide has also been detected in the secretion of the visceral ganglion of *Aplysia depilans.*

In *Aplysia californica,* a cytophysiological characteristic opposes the neurosecretory cells of the visceral ganglion to the other neurons of the same ganglion, the former being almost incapable of synthesizing acetylcholine (Giller and Schwartz, 1968).

c. Subclass Pulmonata. Since they were discovered by Grzycki (1951), the neurosecretory cells of Pulmonata have been more extensively studied than those of other gastropods. In the order Basommatophora, the neurosecretory cells are found in well defined regions of the cerebral, pleural, parietal, and visceral ganglia. After a first study of *Ferrissia* which led to the definition of five cellular types (A, B, C, D, E), among which certain are neurosecretory (Lever, 1957), the main works have been devoted to *Limnaea stagnalis* (Altmann and Kuhnen-Clausen, 1959; Lever *et al.*, 1961b; Joosse, 1964; Boer, 1965, 1967; Boer *et al.*, 1968a,b), *Planorbarius corneus* (Röhnisch 1964; Nolte, 1964), *Helisoma tenue* (Simpson *et al.*, 1964, 1966b), and *Australorbis glabratus* (Lever *et al.*, 1965).

The cerebral ganglia of Basommatophora are particularly rich in neurosecretory cells forming well individualized groups. In *L. stagnalis,* two of these groups, paired and symmetrical, are situated below the dorsal bodies (p. 370), their cells are said to be mediodorsal and laterodorsal. Two other clumps of cells, called caudodorsal, are situated at the dorsal face of the cerebral ganglia. A few neurosecretory cells are found dispersed in the intercerebral commissure. Finally, the lateral lobe—homolog of the procerebrum (see Van Mol, 1967)—contains, besides the bipolar neurons sending an axon toward the follicle gland (p. 372), very peculiar neurosecretory cells: one giant dorsal (canopy cell) and two others (droplet cells) running along the connective which binds the lobe to the main mass of the cerebral ganglion (Fig. 1). The other neurosecretory cells of *Limnaea* are distributed among the pleural ganglia, the right parietal ganglion and the visceral ganglion.

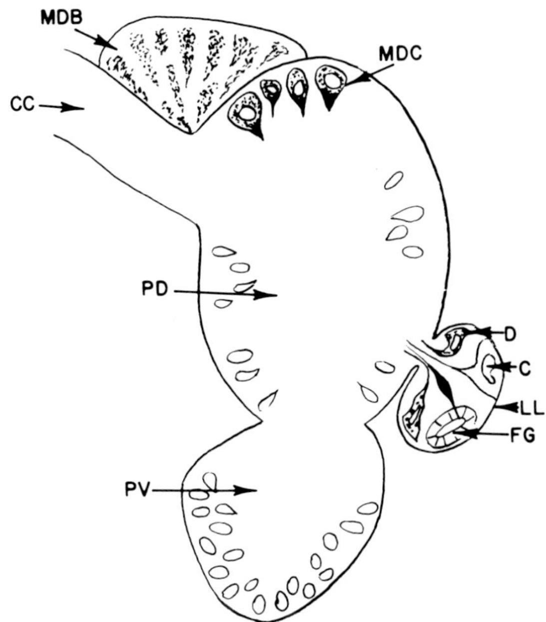

FIG. 1. Vertical section through the right cerebral ganglion of *Limnaea stagnalis*. CC = cerebral commissure, C = canopy cell, D = droplet cells, FG = follicle gland, LL = lateral lobe (= procerebrum), MDB = mediodorsal body, MDC = mediodorsal neurosecretory cells, PD = pars dorsalis, PV = pars ventralis. (After Lever and Joosse, 1961; Joosse, 1964.)

The laterodorsal cells of the cerebral ganglia and the three groups of the pleural ganglia are wanting in *Australorbis glabratus*. The cells of the parietal ganglion are less numerous, while those of the visceral ganglion are in a larger number. The mediodorsal and laterodorsal groups also exist in the cerebral ganglia of *Planorbarius corneus;* the cells of the lateral lobe have no connection with the follicle gland—a remark which applies to all primitive Basommatophora, as shown by Van Mol (1967). The other neurosecretory cells are concentrated in the visceral and left parietal ganglia. In *Helisoma tenue* (Fig. 2), the cells that can be stained by paraldehyde–fuchsin and containing elementary granules are found in the cerebral ganglia, including the lateral lobe, and in the visceral ganglion. Fuchsinophilic cells are also found in the parietal ganglion, but the infrastructural study shows that this characteristic is due to the presence of inclusions not related to neurosecretion; the same conclusion applies to certain neurons of the lateral lobe.

The description of Wauthier *et al.* (1961) concerning *Gundlachia*

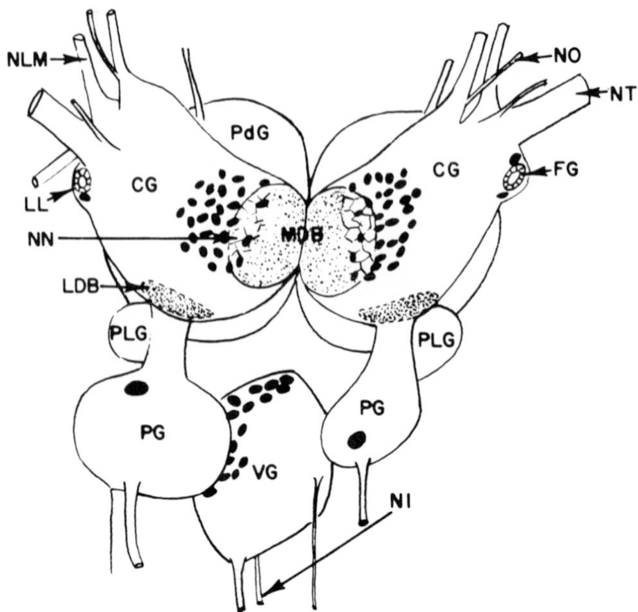

FIG. 2. The fuschsinophilic cells of the central ganglia of *Helisoma tenue*. CG = cerebal ganglion, FG = follicle gland, LDB = laterodorsal bodies, LL = lateral lobe, MDB = mediodorsal bodies, NI = nervus intestinalis, NLM = nervus labialis medius, NN = neurosecretory neuropile, NO = nervus opticus, NT = nervus tentacularis, PG = parietal ganglion, Pd G = pedal ganglion, Pl G = pleural ganglion, VG = visceral ganglion. (After Simpson *et al.*, 1966b).

differs markedly from those above. Adopting the classification of Lever, these authors show that B cells are located in the lateral lobes of the cerebral ganglia and in the pedal centers; A cells, loaded with inclusions, have a much more extensive distribution.

In the order Soleolifera, the neurosecretory cells of *Vaginulus* are located in the cerebral, parietal, and visceral ganglia (Quattrini, 1963, 1964; Nagabhushanam and Swarnamayye, 1963). Concerning the Succineidae, the systematic position of which remains a difficult problem (see Rigby, 1965; Cook, 1966; Van Mol, 1967), the distribution of neurosecretory cells is related to that in Basommatophora and Stylommatophora (Cook, 1966).

The neurosecretory cells are also well known in the order Stylommatophora. The first observation was accomplished in Helicidae (Sanchez and Pavans de Ceccatty, 1957). These cells are most frequently found in the cerebral, parietal, and visceral ganglia (Fig. 3) (Van Mol, 1960a, 1967; Krause, 1960; Jungstand, 1962; Quattrini, 1962; Kuhlmann,

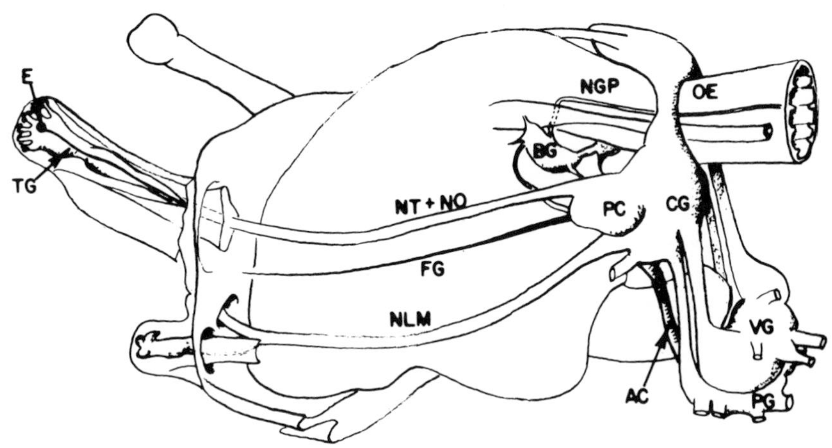

FIG. 3. The central nervous system of a pulmonate gastropod, *Arion rufus*. AC = arteria cerebralis with nervus arteriae cerebralis, BG = buccal ganglion, CG = cerebral ganglion, E = eye, FG = follicle gland, NGP = nervus gastricus posterior, NLM = nervus labialis medius, NO = nervus opticus, NT = nervus tentacularis, OE = esophagus, PC = procerebrum, PG = pedal ganglion, TG = tentacular ganglion, VG = visceral ganglion. (After Van Mol, 1967, simplified).

1963; Nolte, 1963; Nolte and Kuhlmann, 1964; Nagabhushanam and Swarnamayye, 1964; Menon, 1966; Smith, 1966, 1967). Their repartition seems rather homogeneous in this group where the most studied genera have been *Arion, Ariophanta, Cepaea, Helicigona, Helix, Milax, Onciidum,* and *Theba.*

The neurosecretory cells of the cerebral ganglion are always situated in the most peripheral layers of the medial lobe, the mesocerebrum. Contrary to what obtains in Basommatophora, the procerebrum does not contain any of these cells. The morphology of the neurosecretory cells of the visceral ganglion are sometimes very peculiar, and in *Helix* they have been called sacciform cells, or *sackzellen,* a denomination proposed by Krause (1960). In the slugs of the genus *Arion,* neurosecretory cells are also found in the buccal ganglia (Herlant-Meewis and Van Mol, 1959; Smith, 1967). Neurosecretory cells have also been described in the pedal centers of *Arion,* but they are present in very small numbers (Smith, 1967).

The presence of neurosecretory cells in the accessory ganglia of Pulmonata has also been suggested. In the osphradial ganglion of *Planorbarius corneus,* Benjamin and Peat (1968) have described lipidic inclusions, the variety of forms on which may correspond to the various phases of a secretory cycle. Nevertheless, it is mainly the tentacular ganglion of Stylommatophora which has been the main subject of attention. In this ganglion, Tuzet *et al.* (1957) recognized perikarya, the secretion of which spreads between the lobes of the piriform organ (p. 370). As for the collar cells surrounding the same ganglion (Fig. 4) and which have been known since Flemming (1870), they have been studied by several authors (Pelluet and Lane, 1961; Lane, 1962, 1963a,b, 1964a–d; Birbauer and Török, 1964, 1968; Birbauer *et al.*, 1965; Röhlich and Birbauer, 1966; Birbauer, 1969). The glandular nature clearly appears in electronic miscroscopy as well as in light microscopy, but it is more difficult to demonstrate their nervous nature. The authors accept nevertheless that they are very peculiar neurons, the axons of which penetrate in the mass of the tentacular ganglion, and they consider that the regression of certain typical characteristics of the nerve cells

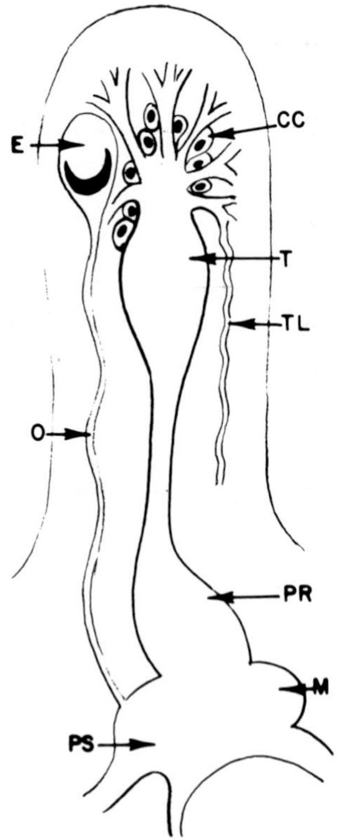

FIG. 4. Schematic illustration of the optic tentacle of Stylommatophora. CC = collar cells, E = eye, M = mesocerebrum, O = optic nerve, PR = procerebrum, PS = postcerebrum, T = tentacular ganglion, TL = fibrae tentaculares laterales. (After Birbauer and Török, 1968, simplified).

is due to an extreme specialization. But we may notice here that the so-called lateral cells, also found in the tentacles and to which a neurosecretory nature had been tentatively conferred, are only considered nowadays as commonplace gland cells (Birbauer and Török, 1968; Renzoni, 1968).

The appearance of neurosecretory cells during the course of postembryonic development has been confirmed in a few species. In *Limnaea stagnalis,* the secretion of the mediodorsal and of the laterodorsal cells becomes visible under the light microscope when the animal reaches a length of 75 mm., and in the case of caudodorsal cells, when it reaches 24 mm. (Joosse, 1964). The activity begins in a precocious way in *Helix pomatia* (Schloot, 1965) and in *Milax gagates,* in which the secretory processes are already initiated at the time of hatching (Quattrini, 1962). They manifest themselves in the buccal ganglion of *Arion rufus* when the animals are 2 weeks old, and in the cerebral ganglia at the time of puberty (Van Mol, 1962).

The diverse tentative classifications of cell types (Lever, Krause, Nagabhushanam, and Swarnamayye, already quoted above) attest to the existence of several categories of neurosecretory cells in Pulmonata. The criteria adopted in these classifications are mainly of a morphological nature, but differences in staining affinities also appear; certain cells always remain acidophilic, while others are stained by the classic methods and become fluorescent under the conditions of the pseudoisocyanine method.

No inquiry has so far been made concerning the nature of the secretion, except by histochemical methods. In *L. stagnalis,* these methods have led to the conclusion that there are three types of elaboration (Boer, 1965). That which is produced by the Gomori-positive cells (dorsal cells of the cerebral ganglia, for instance) is whitish blue *in vivo* and contains neither glycogen nor mucopolysaccharides. Its APS-positive characteristic might be due to lipids, or more precisely to lipoproteins, rich in cystein. The secretion of the acidophilic or Gomori-negative cells (caudodorsal cells, for instance) does not contain carbohydrates and the presence of lipids in it is dubious; it is mainly composed of proteins rich in tyrosine. These cells also contain important amounts of RNA, which is why they have been called Nissl's cells by Boer. Finally, the inclusions of the small cells scattered in all ganglia are of a lipidic nature. Their excellent preservation in paraffin sections suggests that they are associated with proteins. Analogous characteristics are found in *Succinea putris* (Cook, 1966), but in this species, the acidophilic cells themselves are not all identical. For instance, those of the right parietal ganglion and those of the left parietal ganglion present differences of tinctorial activity. The PAS-negative characteristic of the Gomori-positive secretion also exists in *Milax* (Quattrini, 1962). The existence of a phosphatasic activity in the cells of the circum-esophageal ring has been observed in *Helix* and *Planorbis* (Lane, 1966).

The collar cells of the tentacles of Stylommatophora contain phospholipids and probably cerebrosides (Lane, 1963b). They are, in addition, rich in cysteine (Birbauer *et al.*, 1967). A thiamine pyrophosphatasic activity (Lane, 1963b) as well as an intense acid phosphatasic activity (Birbauer, 1969) have been detected in them.

The infrastructural aspects of the neurons have been described by several authors (Gershenfeld, 1963; Baxter and Nisbet, 1963); Schlote, 1963; Schlote and Hanneforth, 1963; Amoroso *et al.*, 1964; Nolte, 1964; Quattrini, 1964, 1965; Simpson *et al.*, 1964, 1966b; Nolte *et al.*, 1965; Brink and Boer, 1967; Boer *et al.*, 1968a,b). At certain stages, the elementary granules are difficult to distinguish from other cell structures; the bulky inclusions composed of lipids and of pigments are more easy to differentiate (Thomas, 1951; Chou, 1957a,b; Chou and Meek, 1958).

2. *Transport of the Neurosecretion Product; Neurohemal Areas*

As the images of axonal transport are scarce, it is a difficult to understand the fate of the neurosecretion product, particularly in prosobranchs and opistobranchs. The scarcity of these products may be explained (1) by the rapidity of the flow, (2) by the intracytoplasmic dissolution before the liberation in the internal medium, (3) by chemical or physical modifications taking place during the transport, or (4) by the influence of elements foreign to neurons such as glial cells.

The neurohemal formations generally do not contribute to the formation of complex organs such as the sinus gland of Crustacea or the corpora cardiaca of the insects. Simple as they are, deprived of anatomical individuality and without glandular cells, these formations consist of an accumulation of nervous terminals in the neighborhood of blood capillaries. They are called neurohemal areas; they are particularly well-known in Pulmonata.

In the prosobranch *Crepidula fornicata*, the infrastructural study has revealed that the neurosecretory material follows a path leading to the anterior end of the ganglion and that there is another path in the posterior direction, reaching the point of emergence of the palleal nerves (Nolte *et al.*, 1965).

The lack of axonal transport so often noticed may, in the case of the opisotbranchs, be an "intracytoplasmic redissolution" observed by several authors (Gabe, 1953b; Tiffon, 1964; Vicente, 1965b). But elementary granules have been detected in the axons of the visceral ganglion of *Aplysia* and in the conjunctive capsule surrounding this ganglion. This capsule, bathed by hemolymph, could in this way play the role of a neurohemal organ (Rosenbluth, 1963).

The images of axonal transport and the neurohemal areas, are better defined in Pulmonata (see Fig. 3). Two of these areas are associated with the cerebral ganglia. One of them, laterally situated, first pointed out in *Arion rufus* (Van Mol, 1960a), corresponds to the territory of the cerebral artery nerve of the Stylommatophora (Kühlmann, 1963; Nolte, 1965) and to the territory of the median labial nerve of the Basommatophora (Joosse, 1964; Röhnisch, 1964; Nolte, 1964, 1967; Lever *et al.*, 1965; Boer *et al.*, 1968a); the other, dorsally situated, is represented by the intercerebral commissure and is in relation with the dorsal bodies (p. 370) in the Basommatophora as well as in the Stylommatophora (Kühlmann, 1963, Joosse, 1964; Röhnisch, 1964; Lever *et al.*, 1965; Nolte, 1964; 1965; Simpson *et al.*, 1966b; Van Mol, 1967). In the most evolved forms, where the dorsal body has become fragmented and situated in a position distant from the nervous tissue, specialized nerves, the commissural nerves, have been developed (Fig. 5).

Leaving the visceral complex, the neurosecretory product is drained by the intestinal nerve and by the palleal nerve. This has been proved in the case of the Basommatophora and Stylommatophora (Lever, Kühlmann, Röhnisch, Simpson *et al.*, Cook). The wall of the anterior aorta, innervated by the visceral ganglion, may be a site of accumulation in *Helisoma* (Simpson *et al.*, 1964). The secretion produced by the buccal ganglion flows through the posterior gastric nerve (Van Mol, 1962).

Another probable neurohemal system, without any relationship to those mentioned above, has been described in the heart of *Helix*. It appears in the form of a plexus of axonal terminals included in the auricular wall, at the junction with the ventricle; the secretion is directly liberated into the cardiac cavity. The authors (Cottrell and Osborne, 1969) compare this structure to the vena cava system of Cephalopoda (p. 379). On the other hand, fibers stainable by the classic methods have been detected in certain muscles (Foh and Schlote, 1965).

On account of the size of their constitutive elements, the neurohemal areas of the Gastropoda can only be studied by electronic microscopy. The research of Nolte (1965, 1967), of Simpson *et al.* (1966b) and of Boer *et al.* (1968a) on the Basommatophora show that the perineurium, with which the neurosecretory fibers are in contact, contains branches of tiny vessels. Inside a single fiber, the elementary granules appear in various ways; they are never found outside axons. It therefore seems that they undergo a fragmentation, and it is probably at the

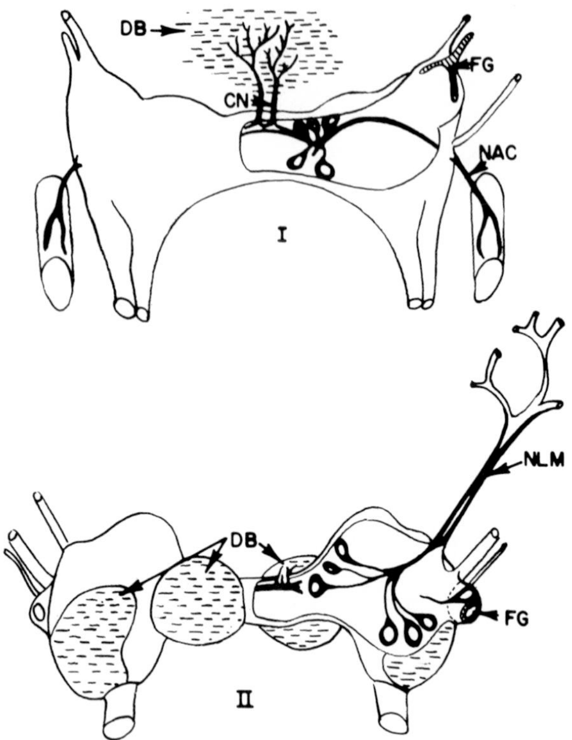

FIG. 5. The neurosecretory system in Stylommatophora (I) and in Basommatophora (II). CN = commissural nerve, DB = dorsal bodies, FG = follicle gland, NAC = nervus arteriae cerebralis, NLM = nervus labialis medius. (After Nolte, 1967).

molecular level that the neurohormonal principles cross the cell membranes. Nevertheless, in certain species, such as *Planorbarius,* the end of part of the axon is pierced with pores which may give passage to the neurosecretory material.

In the heart of *Helix,* the axons are stuffed with grains, the dimensions of which are between 1000 and 2500 Å and which contain neither 5-hydroxytryptamine nor primary catecholamine (Cottrell and Osborne, 1969).

3. *Endocrine Glands*

Certain organs of Gastropoda have been considered to be endocrine glands. For most of them, the presumptions rest on mere morphological bases. Only the dorsal bodies and the folliculous gland of Pulmonata have been the subject of detailed studies.

a. Prosobranchs and Opistobranchs. The lymphoid organ of the nudibranchs, related to the subesophagial commissure constituted by follicules grouped around arterial ramifications, is considered by Sanchez (1962) to be an endocrine gland. This organ is, after egg laying, the seat of an intense mitotic activity. Tuzet *et al.* (1957) compared the X organ of Crustacea to a frontal organ in nudibranchs, which is paired and unnerved and is near the olfactory bulbs, themselves located at the base of the rhinophores and containing neurosecretory cells. No hypothesis has been formulated concerning the function of this organ.

In the opistobranch and prosobranch archaeogastropods examined in this respect (*Aplysia, Hydromyles, Diodera, Patella, Trochochlea*), a particular tissue is coupled to the cerebral ganglia. The cytoplasm of this tissue elaborates a secretion, probably of a polypeptidic nature (Fig. 6). The formation has been called juxtaganglionar organ (Martoja, 1965a,b,c). Its presence has been confirmed in *Patella vulgata* (Choquet and Lemaire, 1969), as well as in several species of opistobranchs (Vicente, 1966, 1969). The juxtaganglionar organ seems to be present in all opistobranchs. On the other hand, it has been looked for in vain in the higher prosobranchs (Martoja, 1969). The exact relationship between this organ and the nervous system has not yet been made clear. Its physiological role remains unknown; nevertheless, it is known that this organ is little developed in the young animal and that it regresses after the phase of sexual activity. The atrophy it undergoes in the individuals in the incubatory phase in *Hydromyles globulosa* (a viviparous gymnosome) is particularly spectacular. Another apparently well established fact is its presence in the hermaphrodites,

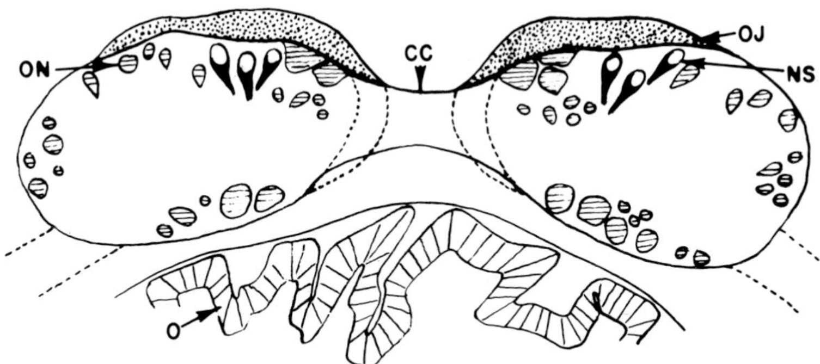

FIG. 6. Transversal view of cerebral ganglia of *Aplysia punctata*. Dotted line shows cerebral ganglia and juxtaganglionnar organ at a posterior level. CC = cerebral commissure, NS = neurosecretory cells, O = esophagus, OJ = juxtaganglionnar organ, ON = ordinary neurons.

either simultaneous (*Aplysia*) or successive (*Hydromyles, Patella,* for example), and its absence in the gonochoric species (Littorinidae, Muricacea, Buccinacea, Toxoglossa). Researches should therefore be carried out in order to elucidate the measure in which the juxtaganglionar organ is linked to hermaphroditism.

The hypothesis of an endocrinian role of the gonad has been suggested for prosobranchs. The elements responsible for this function could be the intertubular conjunctive for *Littorina* (Linke, 1934) or the fundamental syncytium of the testicle for *Pterotrachaea* (Franc and Gabe, 1951).

b. Pulmonata. The organ of Semper (Semper, 1856), coupled on one side to the buccal mass and on the other side to the inferior tentacle, could represent an endocrine organ in Stylommatophora according to Lane (1964d). In this organ, Quattrini (1956, 1957) has observed a cyclical secretory activity and a concommitant variation with the quantity of nuclear DNA. The observations of Van Mol (1967) and of Renzoni (1968) have shown nevertheless that the organ of Semper is an exocrine gland associated with the labial palp.

The chemoreceptor organ of the Basommatophora Ancylidae (André, 1893; Demal, 1955) has been considered to be a neuroendocrine organ by Wauthier *et al.* (1961, 1962), who gave it the name of tentacular paraganglion. A similar interpretation is proposed by Tuzet *et al.* (1957) for a tissue situated at the apex of the ocular tentacle of Stylommatophora Helicidae, a formation for which they proposed the name of piriform organ. In both cases, the cells are arranged in clusters or in lobules, and the organ is innervated by the tentacular nerve. According to Sanchez and Sablier (1962) this endocrine complex plays an important role in metabolic regulations in *Helix*.

The dorsal bodies of Gastropoda Pulmonata, described long ago by de Lacaze-Duthiers (1872) and Beddard (1882a,b) are now considered to be either endocrine glands or neurohemal organs, depending on the author (Lever, 1958b; Wauthier *et al.*, 1962; Joosse, 1964; Röhnisch, 1964; Nolte, 1965; Simpson *et al.*, 1966b; Cook, 1966; Kühlmann, 1966; Nolte and Machemer-Röhnisch, 1966; Van Mol, 1967; Boer *et al.*, 1968b). The dorsal bodies of Basommatophora are paired, compact, well individualized organs, each capping cerebral ganglia from which they are separated by a fibrous network corresponding to a conjunctive web according to Joosse, and to a neurosecretory neuropile according to Simpson *et al.* In certain species, they divide into two parts, forming a mediodorsal body and a laterodorsal body (Fig. 7). In Stylommatophora, they are diffuse organs represented only by cellular clusters, dispersed in the conjunctive surrounding the nervous centers. They are constituted

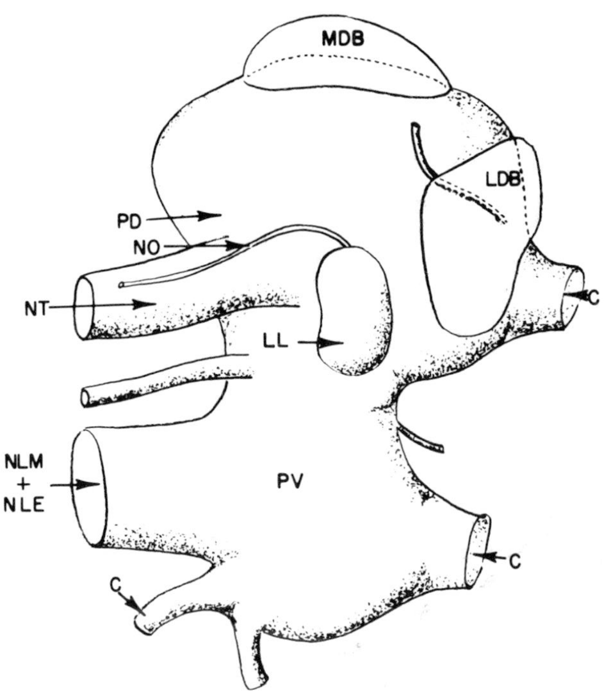

FIG. 7. Lateral view of the left cerebral ganglion of *Limnaea stagnalis*. C = connectives, LDB = laterodorsal body, LL = lateral lobe (procerebrum), MDB = mediodorsal body, NLE = nervus labialis externalis, NLM = nervus labialis medius, NO = nervus opticus, NT = nervus tentacularis, PD = pars dorsalis, PV = pars ventralis. (After Lever and Joosse, 1961; Van Mol, 1967).

by rather polymorphous cells; the mitochondria are numerous, often giant, and the ergastoplasm is well developed. The cytoplasms contain lipid inclusions visible with the light microscope and with the electron microscope (Joosse, 1964; Nolte, 1965; Simpson *et al.*, 1966b; Van Mol, 1967).

There is a similarity between the dorsal bodies and the juxtaganglionar organ of prosobranchs and opistobranchs. This similarity consists of cerebral ganglia and of the presence of a subjacent group of neurosecretory cells, but the histochemical nature of the secretion constitutes an important difference between the two structures, according to our present knowledge.

The dorsal bodies appear in the postembryonic state, and their growth persists during the whole life of the animal (Joosse, 1964; Van Mol, 1967). They are the seat of histological modifications in the course of the annual cycle. A hyperplasmic state corresponding probably to a recrudescence of activity is observed from April to October; winter, on the other hand, corresponds to a resting phase (Joosse, 1964).

Joosse (1964) and Joosse and Geraerts (1969), on the basis of successive ablations, consider that the dorsal bodies control vitellogenesis as well as the development and the functions of the glands annexed to the genital tract. On the basis of the ultrastructural characteristics and of the effect of certain physiological injections, Nolte and Massemer-Röhnisch (1966) are led to consider that the dorsal bodies may be involved in osmoregulation. The divergences referred to above show that the physiology of the dorsal bodies is far from being known, but that the histological characteristics of these organs and the changes they undergo in the course of the life of the organisms in which they are found renders the endocrine role highly probable.

The significance of the folliculous gland (cephalic gland, cerebral gland), first described by de Nabias (1899), is even more obscure than that of the dorsal bodies in spite of the fact that it has raised a similar degree of interest (Lever, 1958a; Sanchez and Bord, 1958; Lever *et al.*, 1959; Van Mol, 1960b, 1967; Nolte, 1963, 1966; Nishioka *et al.*, 1964; Nolte and Kühlmann, 1964; Röhnisch, 1964; Boer, 1965; Menon, 1965; Cook, 1966; Simpson *et al.*, 1966b; Brink and Boer, 1967). It is derived from an ectodermic invagination, the cerebral tube, from which is also derived the procerebrum, which is more often called the lateral lobe in Basommatophora. It persists during the whole life of the animal, except in *Amphibola* (Van Mol, 1967). Histologically, it presents a tubular or vesicular structure, the vesicles containing a colloid. The infrastructural methods show that the walls of the vesicles contain epithelial cells provided with microvilli belonging, according to the species, either to a single or to two categories. In the most evolved Basommatophora (*Limnaea, Chilina, Ancylidae*), the cells are intermixed with axonal terminals of the neurons situated in the procerebrum. According to the majority of authors, the colloid is a secretion product and it has been precisely stated that this substance resists alcoholic fixatives and is basophilic without oxidation, as opposed to the neurosecretion products; it contains acid mucopolysaccharides and cristalline inclusions (Boer, 1965; Nolte, 1965). According to Nishioka *et al.* (1964) and Simpson *et al.* (1966b), however, the colloid is only a felting of cilia and macrovilli and the cavity of vesicles contains no secretion.

The folliculous gland presents histological modifications chronologically superimposable on the phases of growth and on the periods of activity in hibernating species (Van Mol, Kühlmann, Nolte). At present, three hypotheses are formulated concerning the folliculous gland. Considering its relationships with the nervous cells in certain Basommatophora, Lever *et al.* (1959) suggest that the folliculous gland is a neuroendocrine organ. The much more frequent lack of relationships with the nervous system and the presence of microvilli characteristic of active exchanges between the cavity and the internal medium, lead Van Mol and Nolte to consider the folliculous gland as an endocrine gland. Nishioka, Simpson, and Bern, however, think that this gland is a vestigial sensory organ. Several authors consider that the gonad assumes an endocrine activity, but the essential data on this problem are of a physiological nature.

B. Physiological Data

1. *Reproduction*

The elaboration of gonadotropic hormones at the level of central ganglia appears almost certain in all Gastropoda. Tentacular hormones are probably also secreted. As for a possible endocrine function of the gonad, the data at hand remain contradictory.

In Prosobranchia Monotocardia, and especially the Heteropoda, the secretion accumulates in the neurosecretory cells of the cerebral ganglia during the juvenile period, while the depletion takes place at the time of the emission of the gametes (Gabe, 1951, 1953a, 1965b). In the case of *Crepidula fornicata*, a hermaphrodite protandrous species (see Coe, 1942), the quantity of elementary neurosecretory granules accumulated in certain axons of the central ganglia reaches its maximum at the time of sex inversion (Nolte *et al.*, 1965). Finally, when the females of *Vivipara vivipara* are submitted to the action of red monochromatic radiations, the neurosecretory activity increases in the cerebral ganglia and the number of eggs increases (Gorf, 1961).

The methods of organotypic cultures have revealed other important facts which tend to show that the cerebral ganglia simultaneously control the multiplication of gonies and their differentiation in one or the other sex. In *Calyptraea sinensis*, for instance, where the oogonial differentiation constitutes the normal mode of the evolution of gonies, spermatogenesis takes place only in the presence of a masculinizing hormone produced by the central nervous system. As for vitellogenesis, it can only take place if the phenomena of previtellogenesis have been unlocked by a hormone which is also secreted in the nervous system (Streiff, 1966, 1967b). Cerebral ganglia of the female *Vivipara vivipara* exert *in vitro* a distinct action on the testicle, where large oocyte-like cells appear (Griffond, 1969) and they lead to the feminization of the ovotestis in the Pulmonata *Helix aspersa* (Guyard, 1967). The influence of the central ganglia on the multiplication of the gonies has also been demonstrated by surgical methods in *Crepidula fornicata*. Their ablation during the male stage produces a degeneration of the gonad and the sex inversion does not take place (Streiff and Lubet, 1969).

The concept of tentacular hormones is derived from *in vitro* observations. The genital activity of *Patella vulgata* during the male stade, is controlled by two antagonist factors exerting a ganglionar stimulation and a tentacular inhibition, respectively. During the period of gametogenesis, the action of the cerebral ganglia dominates that of the tentacle and the reverse obtains during sexual rest (Choquet, 1965, 1967).

The biochemical nature of the hormones involved in the gonadic development remains unknown. It is only known that an injection of vertebrate hormones to *Murex trunculus* remains inoperative (Rose and Hamon, 1939a).

Thus, in the subclass Prosobranchia, the histophysiological and experimental data demonstrate that the central nervous system elaborates hormones conditioning sexual activity. Moreover, the experimental results lead to the consideration of the possibility of a hormone production by the tentacles. The elaboration of hormones by the gonad remains more controversial. The hypothesis was formulated on the basis of morphological observations on *Littorina littorea* by Linke (1934), who remarked that the gonad and the genital ducts go through the same involutive and evolutive phases in a completely synchronous way; his attempts at experimental control, however, remain fruitless. Certain later data seem to support Linke's views, at least partly. The content of the ovary of *Littorina litorrea* is, on the average, equivalent to 3 mg. of estradiol benzoate per kilo of fresh tissue, and a parallelism is observed between the quantity of estrogens contained in the ovary and the development of the genital tract. Estrogens are not found in the other tissues of the female, nor in the male; no androgen has been detected (Rohlack, 1959). On the contrary, the methods of organotypic culture applied to *Calyptraea sinensis* have led Streiff (1966, 1967a) to the conclusion that the edification of the male genital tract, the sex reversal, and the edification of the female genital tract are directly under the command of the cerebral and tentacular hormones, without any gonadal influence. Associations *in vitro* of organs from *Crepidula* and *Littorina* show that the hormones of the cerebropleural complex are not specific (Silberzahn *et al.*, 1969).

In Opistobranchia as well as in Prosbranchia, the neurosecretory cells of the cerebral ganglia exert an influence on the gonadal activity. Most authors observe a maximum of repletion when the animal is sexually mature, the emptying taking place at the time of gamete emission (Gabe, 1953b; Sanchez, 1962; Vicente, 1962, 1969). Nevertheless, according to

Minichev (1965), in the genus *Tritonia,* the neurosecretory activity of cerebropleural complex cells is very variable between the different individuals observed at the same stage of the sexual cycle.

The hypothesis of the production, by the cerebral ganglia, of a hormone controlling the genital activity is supported by experimental results such as the ablation of these ganglia abolishing mating and egg laying in *Aplysia rosea;* an antagonist hormone is secreted at the level of the tentacles, the ablation of which is accompanied moreover by a hypertropy of the juxtaganglionar organ (p. 369) (Vicente, 1966, 1969). On the other hand, the important role played by the "bag cells" (p. 359) in triggering off egg laying appears more and more clearly in *Aplysia* (Kupfermann, 1967; Strumwasser *et al.*, 1969; Toeva and Brackenbury, 1969). A possible endocrine function of the gonad has not been proposed in Opistobranchia.

The neurosecretory cells of the cerebral ganglia of Pulmonata also show a maximum of activity at the time of gametogenesis (Van Mol, 1960a, 1962, 1967; Joosse, 1964; Menon, 1966; Kühlmann and Nolte, 1967; Smith, 1967). The quantity of neurosecretory material accumulated in the neurohemal organs varies in the same manner, and the surgical removal of the mediodorsal cells of *Limnaea stagnalis* disturbs the genital function, and egg laying in particular, only if it is conjugated with ablation of the dorsal bodies (p. 370) (Joosse, 1964). The association in organotypic culture of the cerebral ganglia of adult *Helix* and of immature ovotestis immediately produces an outbreak of ovogenesis (Guyard, 1967). Contrary to the authors just referred to, Wauthier *et al.* (1962) consider that the essential in the control of sexuality is due to cells located in the visceral ganglion in *Gundlachia,* and Smith (1967) remarks that in *Arion ater* the variations of the neurosecretory activity as a function of the genital state are more important in the visceral ganglion than in the cerebral ganglion.

In Stylommatophora, the existence of a tentacular hormone controlling sexuality seems to be well demonstrated; this hormone may be secreted by the collar cells. Pelluet and Lane (1961), Pelluet (1964) considered that this tentacular hormone stimulated spermatogenesis in *Arion* and *Milax,* but more recently, Gottfried *et al.* (1967b) have shown that in *Ariolimax* it has an inhibitory influence on the functioning of the gonad. This conclusion corroborates the results obtained by Choquet (1965, 1967) in Prosobranchia. Sanchez and Sablier (1962), on the contrary, consider that the ablation of tentacles in *Helix* only results in a slight degeneration in the gonad attributable to denutrition, and Smith (1967) does not notice any variation in the collar cells that could be related to the genital state.

The injection of synthetic sexual hormones is without effect on the state of the gonad of the slugs *Milax* and *Arion* (Rose and Hamon, 1939b; Laviolette, 1954). Positive reactions have nevertheless been obtained in *Helix* and in *Limnaea*. Testosterone stimulates the male line and inhibits the female line, estradiol has the reverse effect, while progesterone stimulates both lines. Nevertheless, these substances have such a high degree of toxicity for gastropods that only experiments of short duration can be performed (Aubry, 1962). In the slug *Deroceras reticulatum*, the injection of pituitary hormones accelerates growth and maturation of the gonocytes (Bridgeford and Pelluet, 1952).

As in the case of Prosobranchia, the question of a possible endocrine function of the gonad remains confused. Parasitic castration, as observed in *Helix* (Garnault, 1889; Chalaux, 1935), *Succinea* (Wesenberg-Lund, 1939), or systematic experiments performed on *Limnaea* (Joosse, 1964) or the congenital absence of gonad (Boulenger, 1914; Geigy, 1935) are always accompanied by anomalies of the genital duct. Surgical castration, without effect if partial, leads to significant disturbances when it is total (Filhol, 1938; Abeloos, 1943; Goddard, 1960). When completed with grafts, receptor organ implantations, or injection of extracts, surgical castration has led Laviolette (1954) to conclude that in Arionidae and Limacidae, the gonad exerts a control on the development and function of the genital duct and that the substance elaborated is released in the hemolymph. It is still unknown in which cells this hormone may be elaborated, but an endogenous production of steroids *in vitro* by eggs or by spermatophores filled with spermatozoids has been detected in *Arion ater* (Gottfried and Lūsis, 1966; Gottfried *et al.*, 1967a).

But other data point to a parallelism of development of the genital ducts and of the gonad without a direct conditioning by the latter. In fact, there are exceptions to the role of synchronism formulated by Linke, and in particular, such exceptions have been observed in *Arion rufus* (Lūsis, 1961). A precocious surgical castration of *Bulinus contortus* leads to no modification in the edification of the female genital ducts, and in the adult, such surgery is not followed by an atrophy, but only by alterations in the dynamics of glandular function (Brisson, 1967). The growth of the albumin gland does not depend exclusively on the action of cerebral and tentacular hormones (Gottfried *et al.*, 1967b; Meenakshi and Scheer, 1969).

These contradictions could find an interpretation in the conclusions formulated by Gottfried *et al.* (1967b): "The tentacular hormone of the slug, ESH, is a gonadal inhibitor, whose release and/or biosynthesis is controlled directly by gonadal steroids. This introduces the concept of a negative–positive feedback mechanism, where an antigonadotrophin inhibits the secretion of gonadal sex steroids, the ESH itself being specifically controlled by the same steroids."

2. *Annual Cycle and Nycthemeral Cycle*

The existence of a correlation between the neurosecretory activity and the annual cycle has been observed in a few species. For *Limnaea stagnalis*, the phase of intense elaboration and transfer takes place in April, while the neurosecretory cells appear to be inactive in winter (Joosse, 1964). Likewise, in *Succinea putris*, the neurosecretory activity of the cerebral ganglia reaches a maximum at the end of hibernal sleep and decreases progressively afterwards until autumn; during hibernation, the neurosecretory cells are at rest (Van Mol, 1967).

The sacciform cells of the visceral ganglion of *Helix pomatia* have a low degree of activity at the beginning of winter, but the amount of secretion slowly increases during hibernation. At the end of this period, a fall in temperature results in a sudden lowering of the amount of neurosecretory product. These sacciform cells could therefore play a role in the maintenance of the hibernal sleep; this control could possibly be accomplished through an action on metabolism (Krause, 1960).

An agreement between the activity of the neurosecretory cells and the nycthemeral cycle has been recognized in *Vivipara vivipara* (Gorf, 1961) and *H. pomatia* (Jungstand, 1962). Such cycles are determined, at least partly, by light. A continuous or intensive illumination results in an increase of neurosecretory activity in *Planorbarius corneus* (Röhnisch, 1964) and in *V. vivipara* (Gorf, 1961). In this species, Gorf has shown, moreover, that the red radiations stimulate this activity, while green light has the opposite effect.

3. *Metabolism*

Neurosecretion may perhaps control the metabolism in Pulmonata Stylommatophora. This hypothesis has been formulated on the basis of a study of different kinds of cells, and different arguments have been presented. Considering their anatomical relationship with the posterior gastric nerve, Herlant-Meewis and Van Mol (1959) believe that the neurosecretory cells of the buccal ganglia of *Arion rufus* could exert a trophic function. In concluding a histophysiological study of hibernation, Krause (1960) attributes this role to the "*sackzellen*" of the visceral ganglion of *Helix pomatia*. Sanchez and Sablier (1962), from experiments in which they cut off the tentacles of *Helix*, conclude that the neurosecretory cells present in them control glucidic metabolism. Finally, Joosse and Geraerts (1969), after an ablation of the dorsal bodies and of the adjacent neurosecretory cells, suggest that this complex controls the general metabolism.

4. Osmoregulation

The conditions of the osmotic equilibrium are not well known in Prosobranchia, and in any case, their osmoregulatory properties appear to be rudimentary. In *Patella,* it may be observed nevertheless that when the animals are introduced into diluted seawater, an emptying of the neurosecretory cells is observed, and that when they are placed in a concentrated seawater there is an accumulation of granules stainable by paraldehyde-fuchsine (Boddingius, 1960). Changes in salinity or humidity also have an influence on the neurosecretory cells of *Vivipara* (Gorf, 1961).

Experiments involving ablations and implantations show that in Opistobranchia the pleural ganglia secrete an antiduretic hormone (Vicente, 1963, 1969). In the case of Pulmonata Basommatophora, the osmotic regulation appears to be of a complex nature (Hekstra and Lever, 1960; Lever *et al.*, 1961a; Lever and Joosse, 1961; Chaisemartin, 1968). From experimental data obtained in *Limnaea stagnalis,* it appears that a diuretic principle is elaborated in the left pleural ganglion, its action being equilibrated by an antidiuretic factor originating in the lateral lobes of the central ganglia. In addition, in these lateral lobes, it appears that the continuous regulation of the water equilibrium is due to the canopy cells, while the droplet cells only come into play under exceptional conditions.

V. Cephalopoda

The data on neurosecretion in Cephalopoda are less numerous than in Gastropoda and the physiological role of the phenomenon is totally unknown. Several glandular structures are believed to accomplish an endocrine function.

A. Morphological Data

1. Neurosecretion

a. Nautiloidea. In the cerebral cord of *Nautilus,* there are symmetrical clusters of neurons elaborating a secretion which can be stained by the classic methods. The images of axonal transport can be followed in the neuropile, but the final fate of the secretion product has not been elucidated (Bassot and Gabe, 1966). The functional significance of this secretion remains unknown.

b. Coleoidea. The most studied secretory system of Coleoidea depends on the visceral lobe. It is known in Octopoda (*Eledone cirrosa, Octopus vulgaris*) and in Decapoda (*Sepia officinalis*) under the name of "neurosecretory vena cava system" and it may be considered to be comparable to the pericardial organs of the Crustacea. It is constituted by unipolar neurons situated in the visceral lobe of the brain and in the median and lateral ganglions accompanying the visceral nerve and the infundibular posterior nerve. The axons join to form bundles crossing the wall of the vena cava, their terminals forming a dense network called neuropile under the endothelium. This neuropiliform layer includes, as well as the axonal terminals, conjunctive fibers and, in the octopodes, flattened cells provided with extensions. Inclusions in most cases stainable with paraldehyde-fuchsin can be detected in this network and by infrastructural methods; it may be recognized that they contain electron dense granules (Alexandrowicz, 1964, 1965, 1969). According to Alexandrowicz, no secretion can be detected in the perikarya nor in the nerves joining the vena cava, and consequently this author considers that it is at the level of the terminal fibers that this material is synthetized. Nevertheless, elementary granules have been detected in the piriform cells of the visceral lobe in several species (Martin, 1966; Martin *et al.*, 1968). Droplets stained by the classic methods have also been detected with the help of the light microscope in certain cells of the visceral lobe of *Octopus vulgaris;* their location and the fact that their axons loaded with secretion join the visceral nerve lead to the belief that they represent the perikarya from which originates the neurosecretory system of the vena cava discovered by Alexandrowicz. These cells are sensitive to stress, and this may explain the difficulty associated with a search for perikarya. On the other hand, they have no activity in the embryo nor in the larva, but their functioning appears as being permanent in the adult of both sexes (Bonichon, 1967). No hypothesis has been proposed concerning the physiological role of the vena cava system.

In the superior buccal lobe of *Octopus* there are possible neurosecretory cells as defined by Bern. They show no sign of activity in the embryo, but they function continuously in the adult. As the buccal lobe, located in the most front region of the supraesophageal mass, innervates the posterior salivary glands (p. 382) through the mandibular nerve, it is possible that these cells are implicated in the chromatic adaptation (Bonichon, 1968). In the same species, a similar interpretation has been proposed in the case of certain cells of the inferior buccal ganglion integrated in the somatogastric system (Bogoraze and Cazal, 1944). The neurosecretory nature of the cells of the stomatogastric system have been confirmed by infrastructural study (Barber, 1967).

The medium lobe of the ganglionary mass of the optic connective, commonly called the olfactory lobe also contain possible neurosecretory cells in *Octopus vulgaris.* In spite of the lack of observation of the fate of the secretion, the hypothesis of a functional relation between these cells and the optic glands (p. 380) deserves attention. Taking into account the role of these glands in the genital activity, such as interpretation could explain in particular why the secretion of the olfactory lobe cells is reduced in the young individuals (Bonichon, 1968).

There is no chemical information, not even of a histochemical nature, concerning neurosecretion in cephalopods, except for the tinctorial affinities, which consist of a stainability of the secretion by the classic methods.

2. *Endocrine Glands*

No endocrine gland has been recognized in Nautiloidea. The nearest approach to it is the recognition by Bassot and Gabe (1966) of a lobulous glandular tissue near the cerebral ganglia similar to the juxtaganglionar organ of the Gastropoda Prosobranchia and Opistobranchia (p. 369). No such tissue has been described in Coleoidea, but several other organs are considered with more or less certainty to be endocrine organs.

a. Optic Glands. The optic glands—peduncular glands of Cazal and Bogoraze, sometimes erroneously confused with the *corpus subpedunculatum* described by S. Thore (1936)— are small spherical masses, paired, orange colored, located in the orbit, and 0.5–1.5 mm. in length in *Octopus vulgaris.* They are connected to the optic tract through the olfactory lobe (Fig. 8). Discovered by Delle Chiaje (1828), they have been considered to be endocrine glands by Cazal and Bogoraze (1943, 1949) and they have since been the subject of important morphological, histophysiological, and experimental researches (Haefelfinger, 1954; Boycott and Young, 1956; Wells and Wells, 1959, 1969; Wells, 1960; Björkmann, 1963; Bonichon, 1967; Defretin and Richard, 1967). They are found in all Coleoidea.

The optic glands are composed of a homogeneous glandular parenchymatous tissue surrounded by a rather thick conjunctive capsule. The important vascular network which irrigates them is related to the orbital sinus. The innervation comes from the supraesophageal mass, probably from the vertical lobe. The parenchyme is composed of principal cells and of supporting cells.

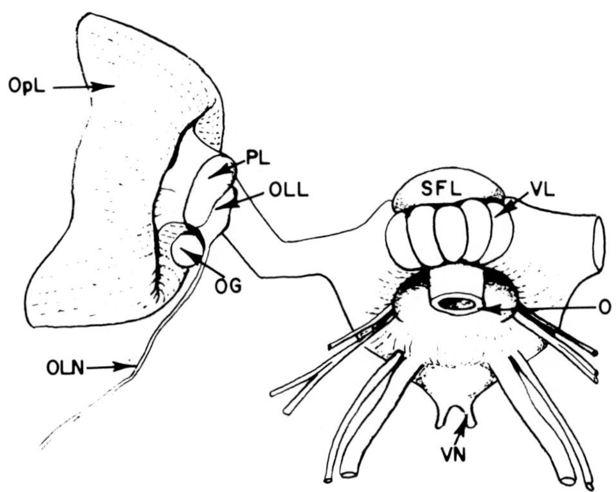

FIG. 8. The brain of a cephalopod, *Argonauta*. O = esophagus, OG = optic gland, OLL = olfactory lobe, OLN = olfactory nerve, OPL = optic lobe, PL = peduncle lobe, SFL = superior frontal lobe, VL = vertical lobe, VN = visceral nerve. (After Boycott and Young, 1956).

Cazal and Bogoraze were the first to call attention to the large number of mitochondria present in the principal cells. This is confirmed by the infrastructural observations of Björkmann (1963), who also emphasizes that the abundance of ribosomes indicates important biosynthetic activities. As for the supporting cells, the same authors compare them to glial cells. These new data bring support to the hypothesis of a neural origin as suggested by Boycott and Young.

The principal cells are of a natural orange color. In the octopus, they contain lipoids stained by Nile blue (Cazal and Bogoraze, 1949) and two types of granulations, among which some appear to belong to carboxylated mucopolysaccharides (Bonichon, 1967). In *Sepia,* the infrastructural study has revealed three types of secretory granules (Defretin and Richard, 1967).

b. Epistellar Bodies and Parolfactory Vesicles. The epistellar bodies of the octopods (Young, 1936; Cazal and Bogoraze, 1944, 1949; Nishioka *et al.*, 1962) and the parolfactory vesicles of the decapods (S. H. Thore, 1939; Haefelfinger, 1954; Boycott and Young, 1956; Nishioka *et al.*, 1966a,b), generally considered to be neuroendocrine organs, present great analogies in spite of their different anatomical locations. The epistellar bodies hang against the stellar ganglia in the palleal cavity, while the parolfactory vesicles are located at the top of the olfactory lobes in the orbital cavity. In both cases, we are confronted with vesiculous formations whose content, or colloid, has sometimes been interpreted as being a secretion. The relationship with the nervous system is very close and certain of the cells are modified neurons. By the problems they raise, these formations recall the folliculous glands of the Gasteropoda Pulmonata (p. 372). As a matter of fact, the infrastructural study performed by Nishioka *et al.* (1962) in *Octopus* has shown that the wall of the epistellar bodies contains neurons, glial cells, pigment cells, and epithelial cells, the arrangement of which is more similar to that of a photoreceptor organ than of an endocrine gland. Nevertheless, Young (1936) had observed that their ablation in *Eledone* resulted in a general lowering of the muscular tone. On the other hand they undergo histological modifications in the course of the annual cycle in *Octopus*. According to the infrastructural data obtained by Nishioka *et al.* (1966a,b), the parolfactory vesicles could also be considered to be rudimentary sensorial organs.

c. Branchial Glands. The branchial glands are considered to be endocrine glands on the basis of their histological structure (Hutchinson, 1928; Sereni, 1932; Cazal and Bogoraze, 1949). They are paired voluminous organs hanging to the base of the gills. The glandular parenchyma contains one single type of cell, each in contact with a capillary vessel. No excretory duct can be detected. A bilateral ablation results in ponderal growth inhibition. A unilateral ablation results in a compensatory hypertrophy of the remaining opposite gland.

d. Pericardial Glands. These appendices of the branchial hearts are generally considered to be excretory organs. Nevertheless, Kestner (1931) has proposed considering them as endocrine glands in *Sepia officinalis.* Death follows their bilateral ablation after a few days.

e. Posterior Salivary Glands. It is established that the posterior salivary glands secrete serotonin which regulates the chromatic adaptation by an action of the muscles of chromatophores (the chromatophores of cephalopods have a very peculiar structure; they are composed of a central pigmentary cell surrounded by plain muscles). Nevertheless, these substances, the action of which is also recognizable *in vitro,* cannot be considered hormones in the proper sense.

B. Physiological Data

The only physiological data regarding the endocrine correlations of cephalopods concern reproduction in which the optic glands are implicated as well as the gonad in its function as an endocrine gland. The majority of these researches have been performed on *Octopus.* The influence of the optic glands on genital maturation has been experimentally demonstrated in *Octopus vulgaris* (Boycott and Young, 1956; Wells and Wells, 1959, 1969; Wells, 1960). It has been shown that a bilateral section of the optic tract or a lesion of the subpedunculated lobe (lobe of the subesophageal mass (performed on miniature octopus results in a hypertrophy of the optic gland and of the gonad. If the optic glands have been previously excised, the same surgery does not result in gonad hypertrophy. Unilateral lesions of the nervous centers or of the optic tract result in a hypertrophy of the optic gland on the same side, a demonstration of the existence of a nervous control of this gland, but the effect suffered by the gonads is the same as when the bilateral ablation has been performed. Thus, gonad maturation appears to be controlled by a secretion of the optic glands. The subpedunculated lobe exerts an inhibitory role, and experiments in which the animals have been blinded show that the removal of this inhibition depends on light stimuli (Fig. 9).

The role of light in the functioning of the optic glands has also been demonstrated in *Sepia officinalis.* Their activity is the more intense the shorter the time light has been applied during the nycthemeral cycle (Defretin and Richard, 1967). The optic glands of *Octopus* exert, as well, a direct control on the development of the genital ducts at least in females. In the case of very young individuals, the ovary does not yet answer the solicitation of the optic glands when experimentally unlocked. Nevertheless, in such experiments the gonoducts and the annex glands undergo a precocious maturation.

The histological aspect of the ovary and of the genital ducts is the same, whether the development has been normal or experimentally induced. The hypertrophy of the optic gland is the consequence of an increased size of the principal cells and not of a cell multiplication. It is accompanied by a marked increase in the quantity of RNA in the whole gland and of an accumulation of the secretion in the principal cells (Wells, 1960).

The histophysiological study in the course of a normal cycle confirms that the secretory activity of the optic glands is increased at the time of sexual maturation. It is translated, as in experimental conditions, by an increase in cell volume and by the elaboration of a secretion which does not seem to be placed in reserve (Bonichon, 1967).

A possible endocrine function of the gonad has naturally been envisaged though no part of its structure appears to correspond to the production of any hormone. The hypothesis has been put to the experimental test and partly confirmed in males, but not in females. In the case of *Octopus*, ovariectomy has no effect on the genital ducts (Taki, 1944), the development of which is under the direct influence of the

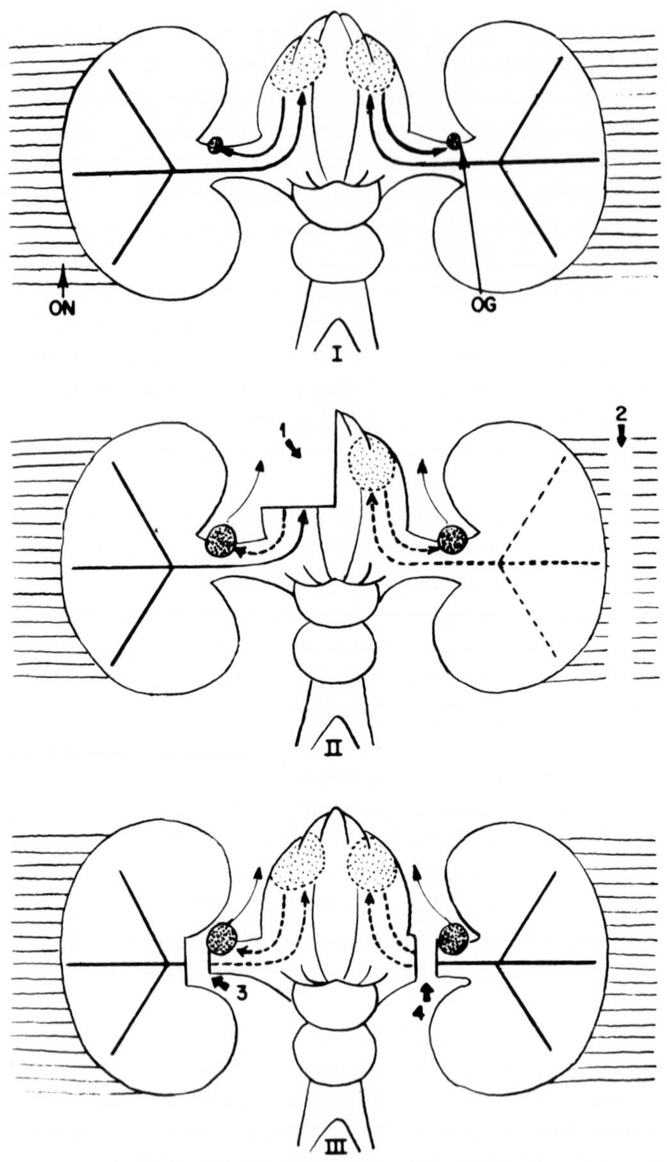

Fig. 9. The mechanism of hormonal control of sexual maturity in *Octopus*. (I) in immature, unoperated animal; (II and III) operations that allow the optic gland to secrete a product causing enlargement of the gonad and its ducts. (1) removal of the source of the inhibitory nerve supply, (2) optic nerves section, (3) removal of optic lobe, (4) optic tract section; OG = optical gland, ON = optic nerve. (After Wells and Wells, 1959).

optic gland (Wells, 1960). On the other hand, testicular ablation is followed by a degeneration of the deferent duct and of the spermatophore pouch (Taki, 1944). Moreover, it is followed by a hypertrophy of the optic glands similar to that which would result from an ablation of the subpedunculated lobe. The testicle thus acts on the state of the optic glands, and consequently, Wells and Wells (1969) compare the control of reproduction, in male Octopods, to the correlation hypothalamus–pituitary–gonad in mammals.

The determining factor of the differentiation of hectocotylus, which acts as copulative organ in the male, remains unknown. The eventual role played by a testicular hormone is still unsubstantiated. Sereni (1929) and Callan (1940) reached contradictory conclusions after a study of the regeneration of this organ in castrated males of *Octopus*. Sereni considers that the testicle exerts an influence on hectocotylyzation, while Callan observes the regeneration of a normal hectocotylus after castration.

Very few data are at hand concerning possible other endocrine glands. It is only known that an ablation of the epistellar bodies results in a general lowering of muscular tone in *Eledone,* that an ablation of the branchial glands stops ponderal growth, and that death follows the excision of the pericardial glands.

VI. Conclusions

Complete neurosecretory pathways are only known in very few species of Mollusca. Most frequently, only regions of accumulation of the neurosecretion product, facilitating its identification, have been recognized either in perikarya or in axonal terminals. The intermediate regions of the neurosecretory pathways appear much less distinctly, and the existence of transfers by mechanisms different from the axonal transport cannot be excluded.

In only one case—the optic gland of cephalopods—has the endocrine nature of an organ been irrefutably demonstrated. Serious presumptions are encouraged by morphological and experimental data in the case of other glandular formations, but uncertainties still persist. The exact anatomical and functional relationships between the neurosecretory system and the endocrine glands distinct from the nervous system, remains to be stated precisely in the great majority of cases.

The physiological aspect of neuroendocrine or endocrine correlations is not better known in mollusks. Even the analysis of secretory cycles sometimes gives equivocal results. It must be admitted that there are a wealth of indications that reproduction, metabolism, and osmoregulation, for example, are under neuroendocrine control, but the whole of the stages going from the stimulus to the answer has seldom been clarified.

Finally, no research work has yet been performed on the chemical constitution of molluscan hormones. In this respect, we must stress that no histochemical study is able to reveal the chemical constitution of hormones, as the stainable material inside the cell only represents their vector, their precursor, or the end products of their metabolism.

These uncertainties preclude any attempt toward a comparative endocrinology. Considering the degree of molluscan evolution, particularly in the case of cephalopods, it is principally with the vertebrates and the arthropods that fruitful comparisons can be made. Nevertheless, at the present time, it is not possible to undertake such an endeavour. At a time when research work has reached a very advanced stage in the case of vertebrates and certain arthropods, work on mollusks at the level of infrastructural studies, on the mechanism of hormone action at the cellular level, on the biosynthesis of hormones and their spatial structure, and on endocrinology has as yet rarely passed beyond the stage of anatomical data.

This situation may appear to be surprising, but it may be accounted for by the difficulties encountered in endocrinological research owing to certain anatomical and biological peculiarities of mollusks. Surgery is difficult in the phylum owing to the fact that the individuals are frequently enclosed in shells and very sensitive to trauma. The immense majority of mollusks are marine, and many of them are pelagic during a part of their existence. Consequently, they escape observation during that period, or raise difficult culture problems. There is no standard "laboratory mollusk," whose anatomy, physiology and biology are as well known as those of the albino rat, for instance, and it is certainly one of the obstacles to a progression in knowledge.

Finally, apart from metamorphosis and occasional sex inversion, the life of a mollusk does not present such spectacular phenomena as molting in arthropods or puberty in vertebrates, and the role played by these natural changes in the historical development of endocrinology is well known. Nevertheless, as the analysis of recent contributions reveals a clear renewal of technical approaches, it may be permitted to believe that endocrinological research is taking new directions in mollusks and that in a near future it will benefit from the experience acquired in other groups.

REFERENCES

Abeloos, M. (1943). *C.R. Acad. Sci.* **216**, 90–92.
Alexandrowicz, J. S. (1964). *J. Mar. Biol. Ass. U.K.* **44**, 111–132.

Alexandrowicz, J. S. (1965). *J. Mar. Biol. Ass. U.K.* **45**, 209–228.
Alexandrowicz, J. S. (1969). *J. Mar. Biol. Ass. U.K., Rep. Counc., 1968–1969.* **49**, 1118.
Altmann, G., and Kuhnen-Clausen, D. (1959). *Ann. Univ. Sarav., Sci.* **8**, 135–140.
Amoroso, E. C., Baxter, M. I., Chiquoine, A. D., and Nisbet, R. H. (1964). *Proc. Roy. Soc., Ser. B* **160**, 167–180.
André, E. (1893). *Rev. Suisse Zool.* **1**, 427–461.
Antheunisse, L. J. (1963). *Arch. Neer. Zool.* **15**, 237–314.
Aubry, R. (1962). *Arch. Anat. Microsc. Morphol. Exp.* **50**, 521–601.
Baranyi, I. B. (1963). *Acta Biol. (Budapest)* **13**, 63–64.
Baranyi, I. B. (1964a). *Acta Biol. (Budapest)* **15**, 35.
Baranyi, I. B. (1964b). *Biol. Kozlem.* **11**, 125–130.
Baranyi, I. B. (1966). *Acta Biol. (Budapest)* **16**, 255–260.
Baranyi, I. B., and Salánki, J. (1963). *Acta Biol. (Budapest)* **13**, 371–378.
Barber, V. C. (1967). *Nature (London)* **213**, 1042–1043.
Bassot, J. M., and Gabe, M. (1966). *C.R. Acad. Sci.* **263**, 1248–1251.
Baxter, M. L., and Nisbet, R. H. (1963). *Proc. Malacol. Soc. London* **35**, 167–177.
Beddard, F. E. (1882a). *Proc. Roy. Soc. Edinburgh* **82**, 576–581.
Beddard, F. E. (1882b). *Zool. Anz.* **48**, 281–292.
Benjamin, P. R., and Peat, A. (1968). *J. Comp. Neurol.* **132**, 617–630.
Bern, H. A. (1963). *In* "The General Physiology of Cell Specialisation." pp. 349–366. McGraw-Hill, New York.
Bern, H. A. (1966). *Symp. Soc. Exp. Biol.* **20**, 325–344.
Bern, H. A., and Hagadorn, I. R. (1965). *In* "The Structure and Function of the Nervous System of Invertebrates" (T. H. Bullock and G. A. Harridge, eds.), Vol. 2, p. 354–429. Freeman, San Francisco, California.
Birbauer, J. (1969). *5th Conf. Eur. Comp. Endocrinol.*, Utzecht, 1969 p. 16.
Birbauer, J., and Török, L. J. (1964). *Acta Biol. (Budapest)* **15**, 39.
Birbauer, J., and Török, L. J. (1968). *Acta Biol. (Budapest)* **19**, 133–143.
Birbauer, J., Török, L. J., and Teichmann, I. (1965). *Zool. Jahrb. Abt. Allg. Zool. Physiol. Tiere* **71**, 545–551.
Birbauer, J., Kiss, J., and Vigh, B. (1967). *Symp. Neurobiol. Inv. Acad. Sci., 1967* pp. 135–142.
Björkmann, N. (1963). *J. Ultrastruct. Res.* **8**, 195.
Boddingius, J. (1960). *Kon. Ned. Akad. Wetensch., Versl. Gewone Vergad. Afd. Natuurk.* **69**, 97–101.
Boer, H. H. (1965). *Arch. Neer. Zool.* **16**, 313–386.
Boer, H. H. (1967). *Gen. Comp. Endocrinol.* **9**, 507–508.
Boer, H. H., Douma, E., and Koksma, J. M. A. (1968a). *Symp. Zool. Soc. London* **22**, 237–256.
Boer, H. H., Slot, J. W., and van Andel, J. (1968b). *Z. Zellforsch. Mikrosk. Anat.* **87**, 435–450.
Bogoraze, D., and Cazal, P. (1944). *Arch. Zool. Exp. Gen.* **84**, 115–131.
Bonichon, A. (1967). *Vie Milieu* **18**, 227–263.
Bonichon, A. (1968). *C.R. Acad. Sci.* **266**, 1764–1766.
Boulenger, H. (1914). *Feuille Jeunes Natur.* **44**, 165.
Bourcart, C., and Lubet, P. (1965). *Rapp. Comm. Int. Explor. Scient. Mer. Mediterr.* **18**, 155–158.
Boycott, B. B., and Young, J. Z. (1956). *B. Hanström, Zool. Pap. in Honour sixty-fifth Birthday* Lund, pp. 76–105.

Bridgeford, H. B., and Pelluet, D. (1952). *Can. J. Zool.* **30**, 323–337.
Brink, M., and Boer, H. H. (1967). *Z. Zellforsch. Mikroski Anat.* **79**, 243–247.
Brisson, P. (1967). *C.R. Acad. Sci.* **264**, 131–133.
Callan, H. G. (1940). *Pubbl. Staz. Zool. Napoli* **18**, 15–19.
Cazal, P., and Bogoraze, D. (1943). *Bull Inst. Océanogr.* **847**, 1–9.
Cazal, P., and Bogoraze, D. (1944). *Arch. Zool. Exp. Gén.* **84**, 10–22.
Cazal, P., and Bogoraze, D. (1949). *Année Biol.* **25**, 225–238.
Chaisemartin, C. (1968). *C.R. Soc. Biol.* **162**, 1994–1998.
Chalaux, J. (1935). *Bull. Soc. Sci. Bretagne* **12**, 53–57.
Choquet, M. (1965). *C.R. Acad. Sci.* **261**, 4521–4524.
Choquet, M. (1967). *C.R. Acad. Sci.* **265**, 333–335.
Choquet, M., and Lemaire, J. (1969). *Bull. Soc. Zool. Fr.* **94**, 39–53.
Chou, J. T. Y. (1957a). *Quart. J. Microsc. Sci.* **98**, 47–58.
Chou, J. T. Y. (1957b). *Quart. J. Microsc. Sci.* **98**, 59–64.
Chou, J. T. Y., & Meek, G. A. (1958). *Quart. J. Microsc. Sci.* **99**, 279–284.
Coe, W. R. (1942). *J. Morphol.* **70**, 501–512.
Coggeshall, R. E. (1967). *J. Neurophysiol.* **30**, 1263–1287.
Coggeshall, R. E., Kandel, E. R., Kupfermann, I., and Waziri, R. (1966). *J. Cell Biol.* **31**, 363–368.
Cook, H. (1966). *Arch. Neer. Zool.* **17**, 1–72.
Cottrell, G. A., and Osborne, N. (1969). *Comp. Biochem. Physiol.* **28**, 1455–1459.
Defretin, R., and Richard, A. (1967). *C.R. Acad. Sci.* **265**, 1415–1418.
de Lacaze-Duthiers, H. (1872). *Arch. Zool. Exp. Gen.* **1**, 437–500.
Delle Chiaje, S. (1828). *Napoli* **4**, 38–116.
Demal, J. (1955). *Acad. Roy. Belg. Cl. Sci., Mem. 8th* **29**, No. 1 5–82.
de Nabias, B. (1889). *Trav. Sta. Arcachon* pp. 1–28.
Durchon, M. (1967). "L'endocrinologie des Vers et des Mollusques." Masson, Paris.
Fährmann, W. (1961). *Z. Zellforsch. Mikrosk. Anat.* **54**, 689–716.
Filhol, J. (1938). *Arch. Anat. Microsc. Morphol. Exp.* **34**, 155–439.
Flemming, W. (1870). *Arch. Mikrosk. Anat.* **6**, 439.
Foh, E., and Schlote, F. W. (1965). *Kurznachr. Akad. Wiss. Goettingen,* No. 7, pp. 33–38.
Franc, A., and Gabe, M. (1951). *Bull. Biol. Fr. Belg.* **85**, 340–346.
Frazier, W. T., Kandel, E. R., Kupfermann, I., Waziri, R., and Coggeshall, R. E. (1967). *J. Neurophysiol.* **30**, 1288–1351.
Gabe, M. (1949). *C.R. Acad. Sci.* **229**, 1172–1173.
Gabe, M. (1951). *Rev. Can. Biol.* **10**, 391–410.
Gabe, M. (1953a). *C.R. Acad. Sci.* **236**, 323–325.
Gabe, M. (1953b). *C.R. Acad. Sci.* **236**, 2166–2168.
Gabe, M. (1954). *Année Biol.* **30**, 6–22.
Gabe, M. (1955). *C.R. Acad. Sci.* **240**, 1810–1812.
Gabe, M. (1965a). "Neurosecretion." Pergamon Press, Oxford.
Gabe, M. (1965b). *Z. Morphol. Oekol. Tiere* **55**, 1024–1079.
Gabe, M. (1967). "Neurosécrétion." Gauthier-Villars, Paris.
Gabe, M., and Rancurel, P. (1958). *Bull. Inst. Fr. Afr. Noire, Sér. A* **20**, 73–79.
Garnault, P. (1889). *Bull. Sci. Fr. Belg.* **20**, 137–141.
Geigy, R. (1935). *Rev. Suisse Zool.* **32**, 207–213.
Gersch, M. (1964). "Vergleichende Endokrinologie des wirbellosen Tieren." Geest & Portig, Leipzig.
Gershenfeld, H. M. (1963). *Z. Zellforsch. Mikrosk. Anat.* **60**, 258–275.
Giller, J. E., and Schwartz, J. H. (1968). *Science* **161**, 908–911.

Goddard, C. K. (1960). *Aust. J. Biol. Sci.* **13**, 378–386.
Gorf, A. (1961). *Zool. Jahrb., Abt. Allg. Zool. Physiol. Tiere* **69**, 379–404.
Gorf, A. (1963). *Zool. Jahrb., Abt. Allg. Zool. Physiol. Tiere* **70**, 266–277.
Gottfried, H., and Lūsis, O. (1966). *Nature (London)* **212**, 1488.
Gottfried, H., Dorfman, R. I., and Wall, P. E. (1967a). *Nature (London)* **215**, 409–410.
Gottfried, H., Dorfman, R. I., Forchielli, E., and Wall, P. E. (1967b). *Gen. Comp. Endocrinol.* **9**, 454.
Griffond, B. (1969). *C.R. Acad. Sci.* **268**, 963–965.
Grzycki, M. S. (1951). *C. R. Sci. Math. Nat. Acad. Pol.* **1–2**, 5.
Guyard, A. (1967). *C.R. Acad. Sci.* **265**, 147–149.
Haefelfinger, H. R. (1954). *Rev. Suisse Zool.* **61**, 153–162.
Hekstra, G. P., and Lever, J. (1960). *Proc., Kon. Ned. Akad. Wetensch.* **63**, 271–282.
Herlant-Meewiss, H., and Van Mol, J. J. (1959). *C.R. Acad. Sci* **249**, 321–322.
Hutchinson, G. E. (1928). *Nature (London)* **121**, 674–675.
Joosse, J. (1964). *Arch. Neer. Zool.* **16**, 1–103.
Joosse, J., and Geraerts, W. J. (1969). *5th Conf. Eur. Comp. Endocrinol., (Utzecht) 1969* p. **75**.
Jungstand, W. (1962). *Zool. Jahrb., Abt. Allg. Zool. Physiol. Tiere* **70**, 1–23.
Kestner, A. (1931). Cited in Cazal, P., and Bogoraze, D. (1949). *Année Biol.* **25**, 225–238.
Krause, E. (1960). *Z. Zellforsch. Mikrosk. Anat.* **51**, 748–776.
Kühlmann, D. (1963). *Z. Zellforsch. Mikrosk. Anat.* **60**, 909–932.
Kühlmann, D. (1966). *Z. Wiss. Zool.* **173**, 218–231.
Kühlmann, D., and Nolte, A. (1967). *Z. Wiss. Zool.* **176**, 271–286.
Kupfermann, I. (1967). *Nature (London)* **216**, 814–815.
Lane, N. J. (1962). *Quart. J. Microsc. Sci.* **103**, 211–226.
Lane, N. J. (1963a). *J. Endocrinol.* **26**, 19–20.
Lane, N. J. (1963b). *Quart J. Microsc. Sci.* **104**, 401–412.
Lane, N. J. (1964a). *Comp. Neurochem., Proc. Int. Neurochem. Symp., 5th, 1962* pp. 281–301.
Lane, N. J. (1964b). *Quart. J. Microsc. Sci.* **105**, 31–34.
Lane, N. J. (1964c). *Quart. J. Microsc. Sci.* **105**, 61–71.
Lane, N. J. (1964d). *Quart. J. Microsc. Sci.* **105**, 331–342.
Lane, N. J. (1966). *Amer. Zool.* **6**, 139–157.
Laviolette, P. (1954). *Bull. Biol. Fr. Belg.* **88**, 310–332.
Lemche, H. (1955). *Experientia* **11**, 320–322.
Lemche, H. (1956). *Spolia Zool. Mus. Hauniensis* **16**, 1–278.
Lever, J. (1957). *Proc., Kon. Ned. Akad. Wetensch.* **60**, 510–522.
Lever, J. (1958a). *Proc., Kon. Ned. Akad. Wetensch.* **61**, 235–242.
Lever, J. (1958b). *Arch. Neer. Zool.* **13**, 194–201.
Lever, J., and Joosse, J. (1961). *Proc., Kon. Ned. Akad. Wetensch.* **64**, 630–639.
Lever, J., Boer, H. H., Duiven, R. J. T., Lammens, J. J., and Wattel, J. (1959). *Proc., Kon. Ned. Akad. Wetensch.* **62**, 139–144.
Lever, J., Jansen, J., and de Vlieger, T. A. (1961a). *Proc., Kon. Ned. Akad. Wetensch.* **64**, 531–542.
Lever, J., Kok, M., Meuleman, E. A., and Joosse, J. (1961b). *Proc., Kon. Ned. Akad. Wetensch.* **64**, 640–647.
Lever, J., de Vries, C. M., & Jaeger, J. C. (1965). *Malacologia* **2**, 219–230.
Linke, O. (1934). *Ver. Deut. Ges. Pathol.* **34**, 164–175.
Lubet, P. (1955). *C.R. Acad. Sci.* **241**, 119–121.

Lubet, P. (1956). *Ann. Sci. Nat. Zool. Biol. Anim.* **18**, 175–183.
Lubet, P. (1957). *Annee. Biol.* 33, 19–29.
Lubet, P. (1959). *Mem. Inst. Sci. Tech. Peches Mar., Paris* **23**, 175–183.
Lubet, P. (1965). *C.R. Soc. Biol.* **159**, 397–399.
Lubet, P. (1966). *Ann. Endocrinol.* **27**, 353–365.
Lubet, P., and Pujol, J. P. (1963). *C.R. Acad. Sci.* **257**, 4032–4034.
Lubet, P., and Pujol, J. P. (1965). *Rapp. Comm. Int. Explor. Scient. Mer. Mediterr.* **18**, 149–154.
Lūsis, O. (1961). *Proc. Zool. Soc. London* **137**, 433–468.
Martin, R. (1966). *Z. Zellforsch. Mikrosk. Anat.* **73**, 326–334.
Martin, R., Barber, V. C., and Boyde, A. (1968). *Proc. Reg. Conf. (Eur.) Electron Microsc., 4th, 1968* pp. 361–362.
Martoja, M. (1965a). *C.R. Acad. Sci.* **260**, 2907–2909.
Martoja, M. (1965b). *C.R. Acad. Sci.* **260**, 4615–4617.
Martoja, M. (1965c). *C.R. Acad. Sci.* **261**, 3195–3196.
Martoja, M. (1967). *C.R. Acad. Sci.* **264**, 1461–1463.
Martoja, M. (1968). *In* "Traité de Zoologie" (P.-P. Grassé, ed.), Vol. 5, No. 3, p. 926–986. Paris.
Martoja, M. (1969). Unpublished data.
Meenakshi, V. R., and Scheer, B. T. (1969). *Comp. Biochem. Physiol.* **29**, 841–845.
Menon, K. R. (1965). *J. Anim. Morphol. Physiol.* **12**, 107–108.
Menon, K. R. (1966). *Gen. Comp. Endocrinol.* **7**, 186–190.
Minichev, Y. (1965). *Vestn. Leningrad. Univ., Biol.* No. 9, pp. 13–25.
Nagabhushanam, R. (1963a). *Indian J. Exp. Biol.* **1**, 161.
Nagabhushanam, R. (1963b). *Nucleus* **6**, 99–102.
Nagabhushanam, R. (1964a). *Indian J. Exp. Biol.* **2**, 1–4.
Nagabhushanam, R. (1964b). *Curr. Sci.* **33**, 20–21.
Nagabhushanam, R. (1964c). *J. Mar. Biol. Ass. India* **5**, 316–317.
Nagabhushanam, R., and Swarnamayye, T. (1963). *J. Anim. Morphol. Physiol.* **10**, 171–173.
Nagabhushanam, R., and Swarnamayye, T. (1964). *Nucleus* **7**, 67–70.
Nishioka, R. S., Hagadorn, I. R., and Bern, H. A. (1962). *Z. Zellforsch. Mikrosk. Anat.* **57**, 406–421.
Nishioka, R. S., Simpson, L., and Bern, H. A. (1964). *Veliger* **7**, 1–3.
Nishioka, R. S., Yasumasu, I., and Bern, H. A. (1966a). *Nature (London)* **211**, 1181.
Nishioka, R. S., Yasumasu, I., Packard, V., Bern, H. A., and Young, J. Z. (1966b). *Z. Zellforsch. Mikrosk. Anat.* **75**, 301–316.
Nolte, A. (1963). *Gen. Comp. Endocrinol.* **3**, 721–722.
Nolte, A. (1964). *Naturwissenschaften* **51**, 1–3.
Nolte, A. (1965). *Zool. Jahrb., Abt. Anat. Ontog. Tiere* **82**, 365–380.
Nolte, A. (1966). *Z. Zellforsch. Mikrosk. Anat.* **75**, 120–128.
Nolte, A. (1967). *Symp. Neurobiol. Inv. Akad. Sci.,* 1967 pp. 123–133.
Nolte, A., and Kuhlmann, D. (1964). *Z. Zellforsch. Mikrosk. Anat.* **63**, 550–567.
Nolte, A., and Machemer-Röhnisch, S. (1966). *Z. Wiss. Zool.* **173**, 232–244.
Nolte, A., Breuker, H., and Kuhlmann, D. (1965). *Z. Zellforsch. Mikrosk. Anat.* **68**, 1–27.
Pavans de Ceccatty, M., and von Planta, O. (1954). *Bull. Soc. Zool. Fr.* **79**, 152–158.
Pelluet, D. (1964). *Can. J. Zool.* **42**, 195–199.

Pelluet, D., and Lane, N. J. (1961). *Can. J. Zool.* **39**, 789–805.
Quattrini, D. (1956). *Boll. Zool. Ital.* **23**, 679–684.
Quattrini, D. (1957). *Boll. Zool. Ital.* **24**, 243–251.
Quattrini. D. (1962). *Mon. Zool. Ital.* **70–71**, 56–96.
Quattrini, D. (1963). *Boll. Zool. Ital.* **29**, 357–369.
Quattrini, D. (1964). *Mon. Zool. Ital.* **72**, 1–12.
Quattrini, D. (1965). *Boll. Soc. Ital. Biol. Sper.* **41**, 475–477.
Renzoni, A. (1968). *Z. Zellforsch. Mikrosk. Anat.* **87**, 350–376.
Rigby, J. E. (1965). *Proc. Zool. Soc. London* **144**, 445–486.
Rohlack, S. (1959). *Z. Vergl. Physiol.* **42**, 164–180.
Röhlich, P., and Bierbauer, J. (1966). *Acta Biol.* (*Budapest*) **17**, 359–373.
Röhnisch, S. (1964). *Z. Zellforsch. Mikrosk. Anat.* **63**, 767–798.
Rose, M., and Hamon, M. (1939a). *C. R. Soc. Biol.* **131**, 106–108.
Rose, M., and Hamon, M. (1939b). *C.R. Soc. Biol.* **131**, 937–939.
Rosenbluth, J. (1963). *Z. Zellforsch. Mikrosk. Anat.* **60**, 213–236.
Sakharov, D. A. (1966). *Comp. Biochem. Physiol.* **18**, 957–959.
Sakharov, D. A., Borovyagin, V. L., and Zs.-Nagy, I. (1965). *Z. Zellforsch. Mikrosk. Anat.* **68**, 660–673.
Salánki, J., and Baranyi, I. B. (1965). *Ann Biol. Tihany* **32**, 77–82.
Salánki, J., Zs.-Nagy, I., and Vas, E. H. (1965). *Ann. Biol. Tihany* **32**, 77–82.
Salánki, J., Zs.-Nagy, I., and Vas, E. H. (1965). *Ann. Biol. Tihany* **32**, 111–116.
Sanchez, S. (1962). *Bull. Soc. Zool. Fr.* **87**, 309–319.
Sanchez, S., and Bord, C. (1958). *C.R. Acad. Sci.* **246**, 845–847.
Sanchez, S., and Sablier, H. (1962). *Bull. Soc. Zool. Fr.* **87**, 319–330.
Sanchez, S., and Pavans de Ceccatty, M. (1957). *C.R. Soc. Biol.* **151**, 2172–2173.
Scharrer, B. (1935). *Pubbl. Staz. Zool. Napoli* **15**, 132–142.
Scharrer, B. (1937). *Naturwissenschaften* **25**, 131–138.
Scharrer, E., and Scharrer, B. (1954). *In* "Handbuch der mikroskopischen Anatomie des Menschen" (W. von Möllondorff, ed.), Vol. 6, Part 5, p. 953–1066. Springer, Berlin.
Scharrer, E., and Scharrer, B. (1963). "Neuroendocrinology." Columbia, Univ. Press, New York.
Schloot, W. (1965). *Z. Zellforsch. Mikrosk. Anat.* **67**, 406–426.
Schlote, F. W. (1963). *Z. Zellforsch. Mikrosk. Anat.* **60**, 325–347.
Schlote, F. W., and Hanneforth, W. (1963). *Z. Zellforsch. Mikrosk. Anat.* **60**, 872–892.
Schmeckel, L., and Wechsler, W. (1968). *Z. Zellforsch. Mikrosk. Anat.* **89**, 112–132.
Semper, C. (1856). *Z. Wiss. Zool.* **8**, 366.
Sereni, E. (1929). *Amer. J. Physiol.* **90**, 512.
Sereni, E. (1932). *Arch. Zool. Ital.* **16**, 941–947.
Silberzahn, L. N., Streiff, W., Le Breton, J., and Lubet, P. (1969). *5th Conf. Eur. Comp. Endocrinol.*, (*Utrecht*) 1969 p. 150.
Simpson, L., Bern, H. A., and Nishioka, R. S. (1963). *J. Comp. Neurol.* **121**, 237–257.
Simpson, L., Bern, H. A., and Nishioka, R. S. (1964). *Amer. Zool.* **4**, 407–408.
Simpson, L., Bern, H. A., and Nishioka, R. S. (1966a). *Amer. Zool.* **6**, 123–138.
Simpson, L., Bern, H. A., and Nishioka, R. S. (1966b). *Gen. Comp. Endocrinol.* **7**, 525–548.
Smith, B. J. (1966). *J. Comp. Neurol.* **126**, 437–452.
Smith, B. J. (1967). *Malacologia* **5**, 285–298.

Sterba, G. (1964). *Acta Histochem.* **17**, 268–292.
Streiff, W. (1966). *Ann. Endocrinol.* **27**, 385–400.
Streiff, W. (1967a). *Ann. Endocrinol.* **28**, 461–472.
Streiff, W. (1967b). *Ann. Endocrinol.* **28**, 641–656.
Streiff, W., and Lubet, P. (1969). *5th Conf. Eur. Endocrinol.,* (*Utzecht*) *1969* p. 156.
Strumwasser, F., Jacklet, J. W., and Alvarez, R. B. (1969). *Comp. Biochem. Physiol.* **29**, 197–206.
Taki, J. (1944). *Jap. J. Malacol.* **13**, 267.
Thomas, O. L. (1951). *J. Comp. Neurol.* **95**, 73–100.
Thore, S. (1936). *Kgl. Fysiogr. Saellsk. Lund, Foerh.* **6**, 1–20.
Thore, S. (1939). *Pubbl. Staz. Zool. Napoli* **17**, 313–506.
Tiffon, Y. (1964). *C.R. Soc. Biol.* **158**, 754–756.
Toeva, L. A., and Brackenbury, R. W. (1969). *Comp. Biochem. Physiol.* **29**, 207–216.
Tuzet, O., Sanchez, S., and Pavans de Ceccatty, M. (1957). *C.R. Acad. Sci.* **244**, 2962–2964.
Van Mol, J. J. (1960a). *C. R. Acad. Sci.* **250**, 2280–2281.
Van Mol, J. J. (1960b). *Ann. Soc. Zool. Belg.* **91**, 45–55.
Van Mol, J. J. (1962). *2nd Conf. Eur. Endocrinol., 1962* p. 23.
Van Mol, J. J. (1967). *Acad. Roy. Belg. Cl. Sci., Mem., Collect.* **37**,(5) **7**–168.
Vicente, N. (1962). *Rec. Trav. St. Mar. Endoume* **25**, 293–304.
Vicente, N. (1963). *C.R. Acad. Sci.* **256**, 2928–2930.
Vicente, N. (1965a). *C.R. Acad. Sci.* **261**, 3193–3194.
Vicente, N. (1965b). *Rec. Trav. St. Mar. Endoume* **39**, 295–301.
Vicente, N. (1966). *C.R. Acad. Sci.* **263**, 382–385.
Vicente, N. (1969). *Rec. Trav. St. Mar. Endoume* **46**, 13–121.
Wauthier, J., Pavans de Ceccatty, M., Richardot, M., Buisson, B., and Hernandez, M. L. (1961). *Bull. Soc. Linn. Lyon* **30**, 70–87.
Wauthier, J., Pavans de Ceccatty, M., Richardot, M., Buisson, B., and Hernandez, M. L. (1962). *Bull. Soc. Linn. Lyon* **31**, 84–92.
Wells, M. J. (1960). *Symp. Zool. Soc. London* No. 2, 87–107.
Wells, M. J., and Wells, J. (1959). *J. Exp. Biol.* **36**, 1–33.
Wells, M. J., and Wells, J. (1969). *Nature* (*London*) **222**, 293–294.
Wesenberg-Lund, C. (1939). "Biologie der Süsswassertiere: Wirbellose Tiere." Vienna.
Young, J. Z. (1936). *Quart. J. Microsc. Sci.* **78**, 311–367.
Zs.-Nagy, I. (1964). *Ann. Biol. Tihany* **31**, 147–152.
Zs.-Nagy, I. (1965). *Ann. Biol. Tihany* **32**, 123–127.

CHAPTER 12

Ionoregulation and Osmoregulation in Mollusca

E. Schoffeniels and R. Gilles

I. Introduction

Mollusks are found in all types of aquatic biotopes, and many species are able to withstand changes in the salinity of their natural environment. This ability to succeed in various media depends for a large part on their capability to regulate the ionic and osmotic concentrations of both their blood and cells. This chapter deals, therefore, with the influence of the surroundings on the ionic and osmotic concentrations of body fluids and tissues and also with the mechanisms that mollusks have evolved to control these concentrations. Under a large range of external concentrations, mollusks are osmoconformers. Except in freshwater species where the blood osmotic concentration is maintained at a higher level than in the external medium, the blood remains normally isosmotic with the environmental medium. In these forms which are poecilosmotic, the body cells can bear concentration changes of the body fluid by means of active regulatory processes. When considering the osmotic regulation in mollusks, we are thus dealing with two types of mechanisms. We shall now investigate in more details these two regulatory systems. We shall first consider the so-called anisosmotic extracellular regulation (Florkin, 1962), which is involved in the regulation of the blood osmotic pressure. Then, we will discuss the mechanisms implicated in the regulation of the cell osmotic pressure termed by Florkin (1962) isosmotic intracellular regulation.

II. Anisosmotic Extracellular Regulation

A. Survey

As first noticed by Léon Fredericq (1901), the blood of marine invertebrates is normally isosmotic with the surrounding medium. As early as 1908, Botazzi showed by freezing point measurements the isosmoticity of the blood with seawater in a series of mollusks, including some Mesogastropoda, Opisthobranchia, and Cephalopoda. Later, many workers demonstrated the isosmoticity of the blood with the environmental medium for representatives of all the major classes of mollusks (Yazaki, 1929; Robertson, 1949, 1953; Todd, 1962).

The isosmoticity of the blood with the surrounding medium remains true even for marine euryhaline species, which can tolerate a wide range of external salinities. *Mytilus edulis* lives in the gulf of Finland at salinities down to 4–5‰ without any evidence of an osmotic regulation of the blood (Segerstråle, 1957). However, Milne (1940) records that the salinity inside the mantle cavity of *Mytilus* at low tide can be 24‰ when the salinity outside is only 7‰. As we shall show below, this apparent osmotic regulation is due to a mechanical phenomenon. Bivalves often close their valves tightly when placed in diluted media, thus prolonging their period of disequilibrium (see Section B,2). There is little information on the blood osmotic concentration of mollusks which can tolerate low salinities. Moreover, the few species used in experiments have only been put in media of salinities down to 50% seawater. In these experiments, the blood always remains isosmotic with the surrounding medium. However, there is some evidence that at lower salinities a regulatory process takes place. This is in fact the situation encountered when working with most of the euryhaline marine crustacea, where the blood remains isosmotic with seawater down to 50% seawater. At greater medium dilution, the osmotic pressure is regulated at a value which is approximately that measured in 50% seawater (Schoffeniels and Gilles, 1971). The same type of process seems to take place in some mollusks, however, with a lower regulatory power. For instance, the blood of the periwinkle *Littorina* is isosmotic in salinities down to 17‰ but becomes hyperosmotic in more diluted media, the mean blood Δ being 1.06°C against 0.48°C for the medium (Todd, 1962, 1964). In the same way, the estuarine bivalve *Scrobicularia plana* does not maintain an osmotic difference between blood and external medium in dilution down to Δ 1.05°C. However, a mean blood Δ of 0.74°C can be recorded for specimens adapted to a medium of Δ 0.59°C (Freeman and Rigler, 1957). This hyperosmotic state is probably due partially to a shell-closing mechanism (Section B,2), but it seems also that an anisosmotic regulation is at work when these marine euryhaline species are submitted to a hypoosmotic stress. In order to study the changes in blood ionic concentration during the adaptation of euryhaline bivalves to diluted media, it is necessary to consider the changes occurring in the perivisceral fluid. This fluid acts indeed as a balancing system between the blood and the environmental medium. As shown in Figs. 1, 2, and 3, the ionic composition of the perivisceral fluid of *M. edulis* follows that of the surrounding medium. The fact that isotoni-

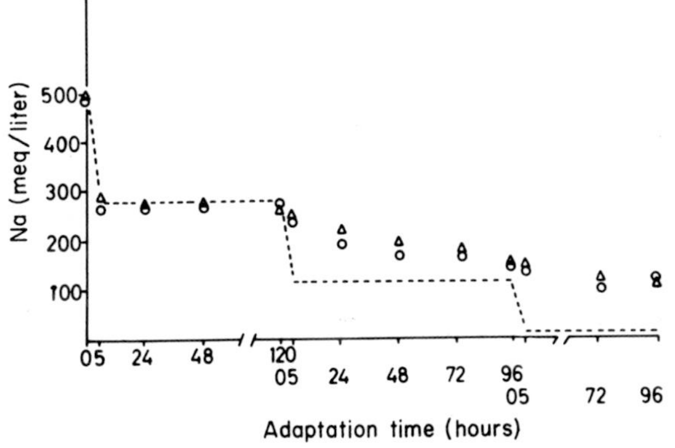

FIG. 1. Modifications of Na^+ concentration in the blood (○) and the perivisceral fluid (△) of *Mytilus edulis* during adaptation to diluted media. Broken line indicates osmotic pressure of the external medium. (From Gilles, 1971).

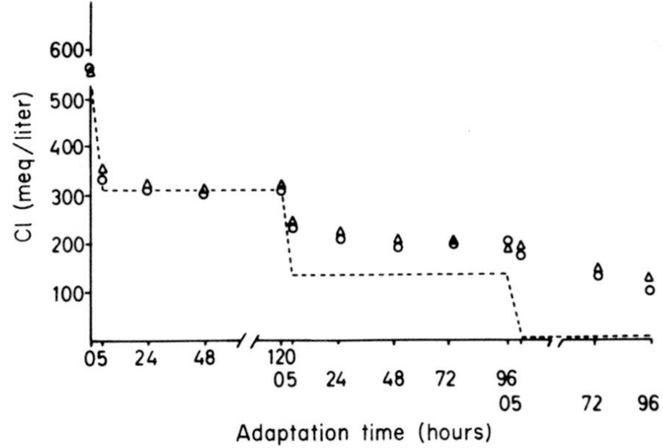

FIG. 2. Modifications of Cl^- concentration in the blood (○) and the perivisceral fluid (△) of *Mytilus edulis* during adaptation to diluted media. Broken line indicates osmotic pressure of the external medium. (From Gilles, 1971).

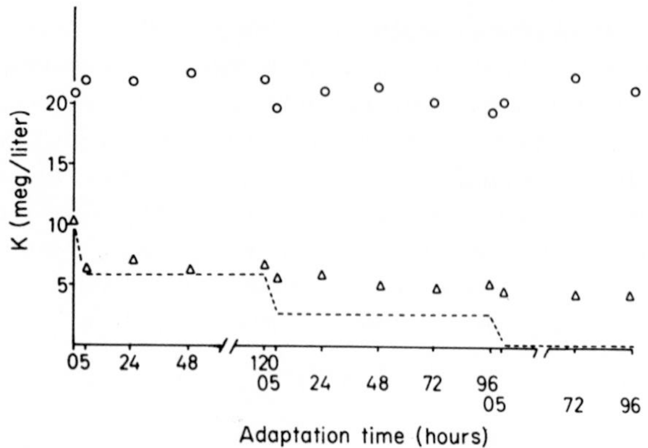

FIG. 3. Modifications of K^+ concentration in the blood (○) and the perivisceral fluid (△) of *Mytilus edulis* during adaptation to diluted media. Broken line indicates osmotic pressure of the external medium. (From Gilles, 1971).

city between the perivisceral fluid and the medium is not reached immediately after the beginning of the adaptation illustrates the buffering role that the perivisceral fluid plays in euryhaline species which can temporarily isolate themselves from the external medium. During the adaptation of *Mytilus* to diluted media, the blood Na^+ and Cl^- concentrations approximately follow those of the perivisceral fluid (Figs. 1 and 2). On the other hand, the blood K^+ concentration appears to be regulated at the value measured in the blood of animals living in seawater. The same phenomenon is also observed during the adaptation of *Glycymeris glycymeris* to dilute media (Gilles, 1971). Such a regulation of the K^+ blood concentration has also been demonstrated by Hoyaux and Jeuniaux (in preparation) when working on the salinity tolerance of the marine gastropods *Littorina littorea*, *Purpura lapillus*, and *Patella vulgata*. Such an ionic regulatory mechanism should involve an active process. We shall discuss this point later on.

Thus, in their normal environment as well as in media diluted down to about 50% seawater, marine mollusks are in osmotic equilibrium with the medium. However, as we have already shown for euryhaline species adapted to diluted media, a passive equilibrium between the environment and the blood plasma cannot account for the isosmoticity observed in marine species. Many workers have given analyses of the blood ionic content of marine mollusks (Bethe and Berger, 1931; Hayes and Pelluet, 1947; Cole, 1940). However, these surveys reveal many discrepancies. We shall thus mainly refer in this study to the work of Robertson (1949, 1953), who has undertaken detailed experiments on the blood ionic composition of marine invertebrates, using accurate chemical methods on animals which were in equilibrium with the surrounding medium. The obtained results demonstrate that most of the marine mollusks so far studied are capable of ionic regulation. The results given in Table I show that if Na^+ and Cl^- concentrations are generally close to those of the surrounding seawater, the blood K^+ level is regulated at a higher value, and often the SO_4^{2-} content is lower. This ionic regulation of the blood appears to be particularly important in the Cephalopoda *Sepia* and *Loligo*.

This regulation process, already shown in marine species, becomes more important in freshwater animals and in terrestrial species. Indeed, the blood of freshwater mollusks is much more concentrated in Na^+ and Cl^- than the environmental medium (Table II). In most cases, freshwater mollusks have also the ability to concentrate Ca^{2+}. This high calcium content probably helps the maintenance of a state of saturation or even supersaturation of the blood with respect to calcium carbonate. This should be related, at least partially to the shell formation or regeneration process (Wilbur, 1964). However, the body fluid of freshwater mollusks show many similarities with seawater in spite of the lower osmotic concentration. Sodium and chloride are generally the most abundant ions, while the concentrations of potassium and magnesium are very low. Mollusks thus have the faculty of anisosmotic extracellular regulation. We shall now investigate the processes implicated in this regulation.

B. Mechanisms Involved

The ionic regulation observed in both marine and freshwater animals, as well as the hyperosmotic state observed in freshwater forms, and in some euryhaline species adapted to diluted media implicates both passive and active mechanisms. Let us now consider some of these mechanisms.

1. Donnan Equilibrium

When dialyzing body fluids samples against seawater, Robertson (1949, 1953) found that a small Donnan effect occurs in some mollusks. However, the mean Donnan ratio for univalent and divalent ions (except Ca^{2+}) is only 1.01–1.02 in prosobranchs and cephalopods. When referring to Tables I and II, such values demonstrate that some or all the blood ions are regulated by other means than the Donnan equilibrium. In marine mollusks, the regulation consists mainly of raising K^+ and Ca^{2+} concentrations, blood Na^+ and Cl^- remaining practically in equilibrium with the surrounding medium. However, in cephalopods as well as in

TABLE I
BLOOD IONIC COMPOSITION OF SEVERAL MARINE MOLLUSKS

Species	Ionic concentration[a]						References
	Na	K	Ca	Mg	Cl	SO_4	
Acanthochitona discrepans	504	9.8	—	—	546	—	Gilles, 1971
Mytilus edulis	472	13.5	10.3	53.8	556.4	27.8	Robertson, 1949, 1953
Mytilus galloprovincialis	476.7	12.1	11	52.2	545.4	34	Robertson, 1949, 1953[b]
Pecten maximus	472	13	10.6	52.2	550.9	27.5	Robertson, 1949, 1953[b]
Ostrea edulis	472	12.9	10.4	54.9	550.9	28.4	Robertson, 1949, 1953[b]
Ensis ensis	476.3	15.5	11.1	53.3	545.4	24.7	Robertson, 1949, 1953[b]
Mya arenaria	476.7	10.7	11.0	53.3	550.9	28.6	Robertson, 1949, 1953
Glycymeris glycymeris	474	12.0	12.8	—	602	—	Gilles, 1971
Buccinum undatum	457.8	14.2	10.7	55.4	550.9	25.5	Robertson, 1949, 1953[b]
Neptunea antiqua	476.7	11.4	10.5	54.3	556.4	27.8	Robertson, 1949, 1953[b]
Pleurobranchus membranaceus	472.0	11.7	11.5	53.3	550.9	28.9	Robertson, 1949, 1953[b]
Archidoris pseudoargus	476.3	12.8	13.6	57.6	550.9	27.2	Robertson, 1949, 1953[b]
Littorina littorea	451	17.3	13.5	—	430	—	Hoyaux and Jeuniaux, 1971
Purpura lapillus	478	18.6	13.9	—	467.6	—	Hoyaux and Jeuniaux, 1971
Patella vulgata	505	19.6	14.8	—	505	—	Hoyaux and Jeuniaux, 1971
Sepia officinalis	439	20.6	9.4	52.7	578.4	6.2	Robertson, 1949, 1953[b]
Loligo forbesi	448.4	22.0	10.5	54.9	571.9	8.2	Robertson, 1949, 1953[b]
Eledone cirrhosa	457.8	15.2	11.0	55.4	501.9	24.8	Robertson, 1949, 1953[b]

[a] In millimoles per liter blood plasma.

[b] Data taken from Robertson (1949, 1953) are recalculated using a seawater of the following composition (in millimoles per liter): Na, 472; K, 10.0; Ca, 10.3; Mg, 53.8; Cl, 550.9; SO_4, 28.4.

TABLE II
BLOOD IONIC COMPOSITION OF SEVERAL FRESHWATER AND TERRESTRIAL MOLLUSKS

Species	Ionic concentration[a]					References
	Na	K	Ca	Mg	Cl	
Freshwater species						
Piba globosa	54.7	4.9	7.8	1.4	—	Saxena, 1957
Planorbarius corneus	85.9	2.3	3.0	0.55	21.0	Florkin, 1943
	50.0	1.9	5.9	1.6	—	Burton, 1968b
Limnaea stagnalis	47.4	2.8	1.5	2.4	42.6	Huf, 1934
	31.2	1.2	7.1	2.0	27.2	Florkin and Duchâteau, 1950
	49.1	2.0	4.0	1.5	—	Burton, 1968b
Anodonta cygnea	15.6	0.49	4.2	0.1	11.7	Potts, 1954
	—	0.40	7.7	0.30	—	Florkin and Duchâteau, 1949, 1950
Terrestrial species						
Helix pomatia	63	4.2	10.3	13.2	72	Florkin and Duchâteau, 1950
	51	3.7	10.0	12.6	—	Burton, 1968b
Helix aperta	75	3.5	7.5	4.9	—	Burton, 1966
Helix aspersa	68	2.9	6.2	3.6	—	Burton, 1968a
Otala lactea	71	3.4	7.1	5.7	—	Burton, 1966
Cepaea nemoralis	66.7	3.1	6.7	5.2	—	Burton, 1968b
Cepaea hortensis	59	2.7	7.0	5.8	—	Burton, 1968b
Strophocheilus oblongus	38	2.4	12.3	19.6	—	De Jorge *et al.*, 1965
Archachatina marginata	57	2.7	11.2	10.0	—	Burton, 1968b
	67	2.8	3.3	—	64	Michon and Alaphilippe, 1958
Achatina achatina	59	9.7	4.8	1	40	Drilhon and Florence, 1942
Arianta arbustorum	47	3.7	8.5	9.8	—	Burton, 1968b

[a] In millimoles per liter of blood plasma.

freshwater species, the concentration of all the ions, including Na^+ and Cl^-, appears to be regulated.

2. *Shell Closing Mechanism*

Many aquatic mollusks can withstand large changes in the salinity of their medium or even resist dessication for relatively long periods merely by closing their shell tightly when placed in abnormal conditions. *Hyridella australis* can survive at least 3 months in dried up mud and keep its blood chlorinity between 27 and 34 mg. ions per liter (Hiscock,

1953a). As already stated, the salinity inside the mantle cavity of *Mytilus* at low tide can be 24‰ (Milne, 1940).

The effect of the valve closing process on the osmotic pressure of the internal fluids of mollusks is well demonstrated by experiments carried out on bivalves and on the Polyplacophora *Acanthochitona discrepans* (Gilles, 1971). *Acanthochitona* is not able to isolate itself from the external medium when withstanding an osmotic stress. As shown in Fig. 4, the blood of this species remains isosmotic with the medium at each dilution tested. On the other hand, in the bivalve *Mytilus edulis*, the perivisceral fluid comes to osmotic equilibrium with the medium only 24 hours after the beginning of the adaptation when the salinity of the medium is suddenly decreased from 100% to 50% seawater. Moreover, in more diluted media (25% seawater or fresh water), the perivisceral fluid remains hyperosmotic to the medium for at least 96 hours. In experiments run on a longer period of time, the perivisceral fluid of *Mytilus* was still hyperosmotic after 10 days of adaptation to 25% seawater (osmotic pressure of the medium 297 mosm/liter and of the perivisceral fluid 551 mosm/liter). It has to be noted that in these experiments, the blood always remains in osmotic equilibrium with the perivisceral fluid. The same type of results have been obtained with the bivalve *Glycymeris glycymeris* (Gilles, 1971) and the gastropods *Littorina littorea, Purpura lapillus,* and *Patella vulgata* (Hoyaux and Jeuniaux, 1971). These last authors, moreover, demonstrate that the disequilibrium between the perivisceral fluid and the surrounding medium disappears if the operculum of *Littorina* is removed. This

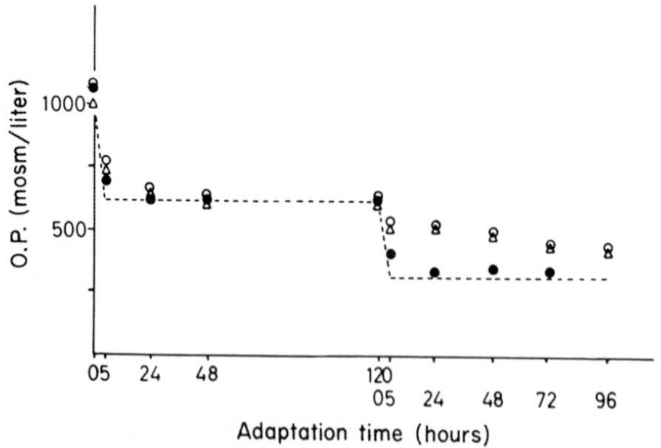

FIG. 4. Changes in the osmotic pressure of the blood (○) and perivisceral fluid (△) of *Mytilus edulis* and of the blood (●) of *Acanthochitona discrepans* during adaptation to diluted media. Broken line indicates osmotic pressure of the external medium. (From Gilles, 1971).

temporary hyperosmotic state is not observed if the animals are progressively adapted to a medium of low salinity as shown by Avens and Sleigh (1965) when working with *Littorina saxatilis*.

It appears thus that some mollusks may be helped in withstanding a sudden osmotic stress by isolating themselves from the external medium. The temporary hyperosmotic state so obtained, which has been sometimes interpreted as the reflection of an osmotic regulation in diluted media, is achieved by various processes. Bivalves such as *Mytilus* or *Glycymeris* close their valves tightly, gastropods with an operculum such as *Littorina* or *Purpura* retract themselves strongly into their shells, and *Patella* adheres firmly to the rock. In such a process, the perivisceral fluid acts as a buffering system between the blood and the external medium. At any rate, this shell-closing mechanism can only help the animal to wait for better conditions during a relatively short period of time. It cannot contribute to the ionic regulation observed in mollusks.

3. *Permeability to Salts and Water*

Data concerning the permeability of molluscan body wall to salts and water are scarce. In marine species, the permeability has been estimated only by following the volume changes of some soft-bodied gastropods. In fact, only an animal permeable to water would behave as an osmometer. On the other hand, if the animal is also permeable to salts, the new equilibrium reached will be less than that expected from a perfect osmometer. The volume change behavior of soft-bodied mollusks deviate significantly from that of a perfect osmometer. For instance, when the opistobranch *Aplysia* is transferred to a dilute medium, the initial swelling is much less than that of an ideal semipermeable animal. Moreover, within a few hours, the volume is regulated and returns to normal. If *Aplysia* is returned during the experiment to normal seawater, it shrinks down to a volume which is slightly smaller than the original one. This can be explained by the fact that the equilibrium in the dilute medium has been reached at least partially by loss of solutes (Bethe, 1934). Such experiments have also been performed on *Onchidium* (Dakin and Edmonds, 1931; Bethe, 1934; Van Weel, 1957) and on *Mytilus* in which the valves have been kept open (Maloeuf, 1937). However these experiments leave unanswered the question of whether the solutes are diffusing through the body wall or if some specific organ plays a part in the process. Bethe (1934), by ligaturing the gut of *Aplysia,* showed that this organ is not involved in the volume regulation.

In some freshwater mollusks, the permeability to water has been estimated by measurements of the rate of urine production. The given data just show that freshwater mollusks are permeable to water, but it is impossible to give absolute permeability values. For instance, the urine production, reduced to 1 gm. weight is 4350% of the body weight in *Anodonta* (Potts, 1954, cited in Potts and Parry, 1964). This hypothetical urine production is much larger than that found in any other animal. It may perhaps be related to the very large surface/volume ratio, but it may also imply a large absolute permeability. The comparison of such figures with those given for other phyla or even with figures which can be calculated for marine mollusks (Robertson, 1953; Martin, 1957), is therefore speculative. When considering the data obtained for other phyla, it can be seen that the permeability of freshwater species is lower than that of seawater animals (Schoffeniels and Gilles, 1971). Such a difference may be considered a step on the way to adaptive evolution to fresh water. However, it seems doubtful that such a mechanism, if it exists in mollusks, can play a part in the regulation of the blood ionic composition.

4. *Active Uptake of Ions*

In marine mollusks, the fluid excreted by renal tubules is isosmotic with the blood as already shown in 1907 in *Octopus* by Mayer and Rathery. However, more recent data have shown that the urine, although isosmotic with the blood, is far from having the same ionic composition. Data given by Robertson (1949, 1953) show that in the cephalopod *Eledone cirrhosa*, a reabsorption of potassium, calcium, and magnesium takes place at the level of the excretory organ. On the other hand, sulfate appears to be actively excreted from the blood, while sodium and chloride concentration are about the same in both blood and urine (Table III). From these data, it can be concluded that potassium, calcium, and magnesium should be more concentrated in the blood than in the environmental medium and that sulfate should be less concentrated. This is confirmed in Table III. Except for sodium, the results obtained with *Sepia officinalis* (Table III) give rise to the same conclusion. According to Robertson (1964), the fact that sodium is less concentrated in the urine than in the blood of *Sepia* may be related to the large amount of ammonium excreted in the urine. Requirements for ionic and osmotic balances are therefore achieved by lowering the sodium content of the urine.

Freshwater mollusks as well as terrestrial ones produce a urine hypotonic to the blood (Picken, 1937; Martin, 1957; Todd, 1962; Little, 1965). This reabsorption of ions inevitably decreases the amount of salt loss from the body. In the land snail *Achatina fulica*, the urine chloride concentration is about 30% lower than that of the blood, and the rate of urine formation does not appear to modify the reabsorption

TABLE III
IONIC COMPOSITION OF THE PLASMA AND RENAL SAC FLUID OF *Sepia officinalis* AND *Eledone cirrhosa*

Species and fluid	Ionic concentration[a]						References
	Na	K	Ca	Mg	Cl	SO_4	
Sepia officinalis							Robertson, 1953
Plasma	460	23.8	10.8	56.9	589.5	4.8	
Renal sac fluid	363	11.8	7.5	38.5	588.9	10.5	
Eledone cirrhosa							Robertson, 1949
Plasma	446.0	14.8	11.6	54.2	534.4	20.6	
Renal sac fluid	452.6	13.3	10.1	50.1	518.3	28.1	

[a] In millimoles per liter of blood plasma.

rate of this ion (Martin *et al.*, 1965). The reabsorption of ions from urine constitutes, together with the absorption from food, the main system involved in the anisosmotic regulation of terrestrial species. In freshwater species, the active reabsorption of salts greatly help the anisosmotic regulation. The loss of ions in a freshwater mollusk can indeed be decreased by about 50% of what it would be if the urine were isotonic to the blood (Picken, 1937; Potts, 1954). The loss of salts, however, remains considerable in view of the large amount of urine produced. Potts (1954) found that *Anodonta* produces in 24 hours a quantity of urine which amounts to about 24% of the body weight (including the shell). The same value is obtained for the freshwater prosobranch *Viviparus* (Little, 1965).

Thus, reabsorption of salts from urine appears to be the main system involved in the blood ionic regulation of marine and terrestrial species. However, in freshwater mollusks, this mechanism can only compensate part of the salt loss both through the urine and the body walls. The loss of salts must thus be balanced in these species by an active uptake from the surrounding medium. Unfortunately, information on absorption of ions such as those obtained in balance experiments carried out, for instance, with crustaceans (reviewed in Schoffeniels and Gilles, 1971) are lacking in the phylum Mollusca.

Picken (1937) already suggested that absorption of salts must take place in some part of the body surface. Proof of ion absorption was given by Krogh (1939) who found that some mollusks, such as *Anodonta, Unio, Limnea,* and *Paludina,* could take up salts from very dilute solutions. Most of these species can reduce the concentration of a 0.7–1.1 mM sodium chloride solution to 0.1–0.15 mM. The maximum rates of net uptake (i.e., the excess of uptake over loss) recorded by Krogh were 165 μM chlorine per kilogram per hour for *Limnaea,* 173 μM chlorine per kilogram per hour for *Viviparus,* and 300 μM chlorine per kilogram per hour for *Unio.* More recently, Schoffeniels (1951a,b) has demonstrated an absorption of both calcium and phosphate ions by isolated tissues of *Anodonta cygnea.* The rate of calcium exchange is the highest in the foot then decreases in the following tissues: blood, adductor muscle, mantle, and gills. Kirschner *et al.* (1960) show that the potential difference existing across the clam mantle tissue is dependent on the calcium ion content of the medium. This study suggests the participation of calcium in the genesis of the potential difference across the clam mantle tissue. It has also been shown that the oyster mantle transports calcium from the body side to the shell side (see Wilbur and Jodrey, 1955).

To sum up, we have seen that marine mollusks are in osmotic equilibrium with the surrounding medium. However, they regulate their blood ionic concentration in the same way as the freshwater and terrestrial mollusks. The hyperosmotic state observed in these latter species, like the ionic regulation of the blood in the marine animals, is the result of charge and discharge phenomena. The active transport of salts by the excretory organs, and particularly in freshwater species, at the body walls level. is certainly of prime importance in the maintenance of the blood ionic steady state. Euryhaline marine mollusks stay in osmotic equilibrium with the surrounding medium even in very diluted media. However, in these diluted media, a regulation of the blood ionic concentration takes place. This regulation only concerns the blood potassium level, and therefore, the cells must support important changes in the blood osmotic pressure. We shall now investigate the mechanisms implicated in the regulation of the cell osmotic pressure.

III. Isosmotic Intracellular Regulation

When considering the general problem of the adaptation of mollusks to fresh water or to the terrestrial habitat, it is clear that their cells must have adapted to considerable changes in the composition of the surrounding fluid (see the first part of this chapter). Moreover, mollusks certainly constitute the most representative phylum of the littoral area, in as much that they have often been chosen as landmarks in the definition of the shore levels. In these biotopes, animals have to withstand frequent and rapid changes in the osmotic pressure of their environmental medium. Since it is generally assumed that cells are in osmotic equilibrium with the blood, and as we have seen that marine mollusks are poecilosmotic animals, it can be expected that the changes in the blood concentration which occur during adaptation to dilute media must induce concomitant modification of the cells osmotic pressure. This process of cell osmotic pressure adaptation has been termed by Florkin (1962) the isosmotic intracellular regulation. Thus, in mollusks, which have practically no anisosmotic regulation power, the penetration in media of different salinities mainly depends on the power of the isosmotic regulation mechanism. This mechanism is of course related to the ability to modify the level of the intracellular components which act as osmotic effectors. Before attempting to study the mechanism itself, a survey of the compounds used by mollusks as effectors of the cellular osmotic pressure should first be made.

A. Effectors of Cell Osmotic Pressure

An osmotic steady state appears to exist between molluscan cells and blood. The freezing point of isolated fibers of *Mytilus* muscle is within 1.5% that of the blood (Potts, 1952). However, in contrast to body fluids, molluscan tissues are rich in potassium and poor in sodium and chloride. This is the situation found in most of the cells studied so far. Moreover, whereas about 99% of the osmotic concentration of the blood is made up of inorganic ions, these ions represents only about 40–50% of the cell osmotic concentration (Henze, 1905; Bialaszewicz and Kupfer, 1936; Potts, 1958; Bricteux-Grégoire *et al.*, 1964a,b,c; Robertson, 1965). Various organic compounds account for the rest of the osmotic pressure. As shown in Table IV, amino acids, together with some other amino compounds such as taurine or glycine–betaine are of general occurrence, and they account for about half of the total osmotic pressure. The presence of such compounds in mollusk tissues has been known for a long time. As early as 1904, Kelly showed the existence of glycine and taurine in *Pecten* muscle. In 1905, Henze found about 52 μM per gram of water of taurine in *Octopus*. A few years later, the same author demonstrated the presence of betaine and trimethylamine oxide in *Octopus* muscle (Henze, 1911, 1914).

Some of these compounds, which are anionic, undoubtedly play a great part in the cellular ionic balance. The organic molecules involved in this cation–anion balance vary from species to species and even in the same group from tissue to tissue. As shown in Table V, isethionic acid and dicarboxylic amino acids are of prime importance for the ionic balance of *Loligo* nerve. On the other hand, this balance mainly implicates inorganic phosphate in the muscle of another species of squid *Sepia*. The concentration of dicarboxylic amino acids appears to be negligible in cephalopod muscle—0.4–2.3 μM/gm. in *Sepia esculenta*

TABLE IV
OSMOTIC EFFECTORS IN MUSCLE OF SOME MOLLUSKS

Effector	Concentration (mM/kg. water)			
	Sepia officinalis mantle muscle[a]	*Eledone cirrhosa* mantle muscle[a]	*Mytilus edulis* fast adductor muscle[b]	*Anodonta cygnea* fast adductor muscle[b]
Inorganic compounds				
Sodium	30.8	33.3	158–191.4[c]	6.6
Potassium	189	167	125–192.1[c]	18.4
Calcium	1.88	3.08	8.5–9.6[c]	11.5
Magnesium	19.0	17.4	38	3.9
Ammonium	2.2	1.3	—	—
Chloride	45.0	54.8	187–307.3[c]	3.9
Sulfate	2.0	4.5	13	—
Inorganic phosphate	92.0	88.6	13.3	6.9
Organic compounds				
α-amino nitrogen	483	326	155[d]–389.4[c]	10.5[d]
Taurine	—	—	91–89.5[c]	0
Trimethylamine oxide	86.1	39.7	0[c]	—
Betaine	108.2	116.9	97.2[c]	—
Arginine phosphate	4.3	8.2	13.4	8.4
Adenosine triphosphate	8.2	5.9	4.9	1.7
Remaining acid-soluble phosphate	12.2	23.2	7.4	2.8
Reducing sugar	2.9	7.1	—	—
Glycerol	4.2	1.5	—	—
Osmotic pressure due to inorganic compounds (mosm./kg. water)[e]	588.76	572.16	560.50–851.90[c]	80.40
Osmotic pressure due to organic compounds (mosm./kg. water)[e]	787.88	587.22	421.88–569.25[c]	26.00
Total cellular osmotic pressure (mosm./kg. water)[e]	1376.64	1159.38	982.38–1411.15[c]	106.40

[a] Robertson (1965).
[b] Potts (1958).
[c] Bricteux-Grégoire *et al.* (1964a).
[d] Only carboxylic acids as determined by the method of Van Slyke *et al.* (1941).
[e] The osmotic pressure of the organic compounds have been calculated assuming that Δ of a molar solution of some organic compounds such as amino acids is −1.86°C (Prosser *et al.*, 1950).

TABLE V

Cation—Anion Balance in Cephalopod Tissues

Cation	Concentration (meq./kg. water)		Anion	Concentration (meq./kg. water)	
	Loligo pealii nerve[a]	*Sepia officinalis* muscle[b]		*Loligo pealii* nerve[a]	*Sepia officinalis* muscle[b]
Sodium	78.3	30.8	Chloride	168.6	45.0
Potassium	414.2	189	Phosphate	32.0	148.1
Calcium	8.4	1.9	Sulfate	—	4.0
Magnesium	24.1	38.0	Lactate	—	6.8
Ammonium	—	2.2	Isethionic acid	183.0	—
Arginine	3.9	64	Aspartic acid	87.9	—
Lysine	2.9	—	Glutamic acid	23.6	—
Ornithine	2.2	—	Arginine phosphate	—	4.5
			Adenosine triphosphate	—	31.1
Total	534.0	325.9	Hexose monophosphate	—	23.7
			Total	495.1	263.2

[a] Deffner and Hafter (1960).
[b] Robertson (1965).

(Endo *et al.*, 1962)—and to our knowledge, isethionic acid has not yet been identified.

It is also worth noting from Table IV that the amino acid content is much higher in the muscle of seawater species than in the muscle of the freshwater *Anodonta*. This is also the case for betaine and taurine; this last compounds, though found in large amounts in seawater species, is generally not present in freshwater ones (Simpson *et al.*, 1959). Such facts suggest that these molecules together with the inorganic ions play an important part in the isosmotic intracellular regulation process. This is well demonstrated by experiments where animals are adapted to media of various salinities. For instance, during the adaptation of *Gryphea angulata* from seawater to 50% seawater, there is an important decrease in the amino acid concentration in the muscle (Table VI). The concentration of the sodium, potassium, and chloride ions is also considerably decreased. The same type of result has been obtained by Potts (1958) on *Mytilus edulis* and also by Bricteux-Grégoire *et al.* (1964a,c) on *Ostrea edulis* and *M. edulis*. On the other hand, an increase in the ionic and amino acid concentrations can be recorded in the muscle of the freshwater mussel *Anodonta* when adapted to 18% seawater (Potts, 1958). During these adaptations, the slight changes in the water content cannot

TABLE VI
CONCENTRATION OF VARIOUS SUBSTANCES OF THE FAST ADDUCTOR MUSCLES OF *Gryphea angulata* ADAPTED TO SEAWATER OR TO 50% SEAWATER[a]

Substance	Concentration (mosm./kg. water)	
	Gryphea in seawater	*Gryphea* in 50% seawater
Inorganic ions		
Sodium	188.4	59.0
Potassium	157.5	105.3
Calcium	9.2	4.3
Chloride	162.3	44.7
Organic compounds		
Alanine	36.7	17.0
Arginine	9.7	4.8
Aspartic acid	23.3	19.8
Glutamic acid	11.9	9.8
Glycine	43.1	22.0
Histidine	0.5	—
Isoleucine	0.3	Trace
Leucine	0.5	Trace
Lysine	2.5	1.5
Phenylalanine	Trace	Trace
Proline	8.3	3.4
Serine	1.9	1.7
Threonine	2.7	1.7
Tyrosine	Trace	Trace
Valine	Trace	Trace
Taurine	125.0	71.0
Betaine	174.4	98.4
Trimethylamine oxide	0	0

[a] From Bricteux-Grégoire *et al.* (1964b).

account for the modifications of the concentration of the inorganic ions and the amino acids. One can thus consider these reversible concentration changes which operate an intracellular osmotic regulation as being the result of an active process. As far as the ionic concentration changes are concerned, the modifications of the potassium level are generally only slightly greater than that expected from the endosmosis and exosmosis of water (Potts, 1958). Moreover, the retention of potassium in invertebrates cells appears to be largely electrostatic (Robertson, 1953; Potts, 1958). Thus, if an active process is concerned in the modification of the potassium level, it can only account for a small part of it. As shown by Potts (1958), they are certainly the mechanisms by which sodium and chloride amounts are modified, which are the most important processes in the ionic regulation of molluscan cells. However, except for the recent work of Beadle (1969) and Beadle and Beadle (1969), which showed that egg and early embryo of *Biomphalaria* can regulate their salt and water content, nothing is known about these mechanisms. Most investigations have been hitherto concerned by the part played by organic molecules and particularly by amino acids and taurine in the isosmotic regulation process. Let us therefore briefly summarize the different results obtained in this field.

B. Amino Acids and Taurine as Osmotic Effectors

As first noticed by Florkin and co-workers (Camien *et al.*, 1951; Duchâteau *et al.*, 1952), the amount of free amino acids is higher in tissues of marine invertebrates than in tissues of freshwater ones. For instance, the concentration of th^ free amino acid pool in muscles of *Mytilus* and *Ostrea* is higher than in *Anodonta* (Duchâteau *et al.*, 1952), which suggests a participation of the free amino acids in the osmolar intracellular concentration under conditions met by species which can live in seawater or in diluted media. This led Florkin and co-workers to compare the amino acid pool of some euryhaline species when adapted to media of different salinities. This was first done with the Crustacea *Eriocheir sinensis* (Duchâteau and Florkin, 1955). The same type of study was undertaken a few years later by Potts (1958), who showed that the amino acid level is lower in the tissues of various euryhaline mollusks when adapted to diluted media than when adapted to concentrated ones. Many studies in recent years have dealt with the free amino acid content of euryhaline animals in relation to the salinity of the environmental medium. Most of these studies have been done on Crustacea (for review, see Florkin and Schoffeniels, 1965, 1969; Schoffeniels and Gilles, 1971). However, the modifications of the amino acid amount in response to a change in the surrounding salinity have been demonstrated in various invertebrate phyla such as echinoderms (Jeuniaux *et al.*, 1962; Lange, 1964; Stephens and Virkar, 1966), arachnomorphs (Bricteux-Grégoire *et al.*, 1966), annelids (Jeuniaux *et al.*, 1961; Duchâteau-Bosson *et al.*, 1961), and sipunculans (Virkar, 1966). This phenomenon, which was first thought to be restricted to invertebrates, seems also to be at work in some euryhaline vertebrates (Gordon, 1965; Lange and Fugelli, 1965; Fugelli, 1967; Baxter and Ortiz, 1966).

If all of these species possess an isosmotic regulation mechanism, only some of them have the function of blood osmotic pressure regulation. One may thus consider the isosmotic regulation a primitive mechanism to which, in several species, the anisosmotic extracellular regulation add a new range of possibilities (Gilles, 1970). This last mechanism, however, is not necessary to insure the survival of many species in diluted media. For instance, we have shown in the first part of this chapter that mollusks are poecilosmotic animals. However, an isosmotic regulation mechanism has been demonstrated in all the euryhaline mollusks studied so far. In this process, amino acids are greatly implicated in many species, as shown by the results given in Tables VII and VIII.

However, in the Polyplacophora *Acanthochitona discrepans,* as in the lamellibranch *Glycymeris,* the part played by amino acids in the isosmotic regulation appears to be restricted. In these two species, amino acids account only for 2.5% of the total osmotic pressure (Table VII). In the other species listed, amino acids play a more important part in the intracellular osmotic regulation. However, as far as individual amino acids are concerned, the results show many discrepancies. If glycine can be considered an osmotic effector in all the listed species, alanine does not participate in the cell osmotic regulation in *Tegula funebralis* (Table VIII). In the same way, arginine acts as an osmotic effector in most species except *Purpura lapillus,* and apart from a few cases, aspartic and glutamic acids are not involved in the osmotic regulation. As already stated, the concentration of the dicarboxylic amino acids is very low in cephalopod muscles, and the part they play in the osmotic regulation of these species can therefore be considered as being reduced. This is at variance with what can be observed in cephalopod nerves where these amino acids appear to be of prime importance (Table IV) (Lewis, 1952; Deffner and Hafter, 1960). It thus appears that, when dealing with the part taken by amino acids in the isosmotic intracellular regulation in mollusks, one has to be aware of these variations between species.

In any case, amino acids account only for a small part of the cell osmotic pressure (ranging from 2% in *Achanthochitona* to 15% in *Gryphea*). A great osmoregulatory role appears to be played by other organic molecules, such as taurine or glycine betaine (Tables VII and VIII). In molluscan tissues, the taurine concentration is so strikingly high that mollusks have been proposed as a commercial source of this compound. Taurine acts, moreover, as a cellular osmotic effector (Tables VII and VIII) (Awapara, 1962; Lange, 1963). For instance, this substance alone accounts for about 15% of the total osmotic pressure in *Patella* (Table VIII). Furthermore, as shown in Tables VII and VIII, the taurine concentration undergoes important changes in most species, except for *Tegula,* when the animals are adapted from seawater to twice diluted seawater. In *Tegula,* an unknown compound, thought to be phosphoethanolamine or phosphoserine, is present in high concentrations in the animals adapted to seawater; these concentrations decrease considerably during the adaptation to hypoosmotic media (Peterson and Duerr, 1969). This compound plays the same osmotic regulatory function in *Tegula* as taurine plays in the other species.

Most of the studies dealing with the isosmotic intracellular regulation have been performed by submitting the animals to a hypoosmotic medium. This is in fact the situation that a euryhaline marine mollusk normally encounters in nature. Very few experiments have been performed on the adaptation of euryhaline mollusk to hyperosmotic media. Peterson and Duerr (1969) show that during the adaptation of *Tegula* to 160% seawater, there is a decrease in the cellular osmotic pressure due to the free amino acids. However, an increase in ninhydrin-positive substances is recorded in 160% seawater. This suggests that some other organic compound acts in the isosmotic regulation in hyperosmotic media. Such a shift in the mechanism of cell osmotic regulation in concentrated media has already been demonstrated by Allen (1961) on *Rangia cuneata*.

Thus, in molluscan tissues, a great part of the contribution to the osmotic pressure of the intracellular components, as well as the contribution to the osmotic cellular adjustment when dealing with an euryhaline species withstanding an osmotic stress, is due to organic molecules. Among these molecules, taurine, betaine, and amino acids are of general occurrence. Some other compounds, such as trimethylamine oxide and isethionic acid, contribute to the isosmotic regulation in various tissues of some species. Since the concentration changes of these substances during adaptation of mollusks to various salinities cannot be explained by endosmosis and exosmosis phenomena, one must consider these modifications as being the result of an active regulation process. We shall now give the available information regarding the mechanisms responsible for these adaptive changes in concentration. Although such studies have been carried out in Crustacea down to the molecular level (Florkin and Schoffeniels, 1965, 1969; Schoffeniels, 1968; Gilles, 1969, 1970; Schoffeniels and Gilles, 1971), the situation is far from being the same when considering the phylum Mollusca.

C. Mechanisms Involved in the Regulation of Amino Acids and Taurine Concentration

The first question to arise concerns the origin of the organic compounds acting as osmotic effectors. There are two possibilities; either these substances are of extracellular origin, or they are formed within the cell. No attempt has been made so far to support the first hypothesis. As far as the second hypothesis is concerned, one can consider that

TABLE VII

ORGANIC COMPOUNDS AS OSMOTIC EFFECTORS IN MUSCLE OF VARIOUS MOLLUSKS ADAPTED TO SEAWATER OR TO 50% SEAWATER

Compound	Concentration[a]									
	Gryphea angulata[c]		*Ostrea edulis*[d]		*Mytilus edulis*[b]		*Glycymeris glycymeris*[e]		*Acanthochitona discrepans*[e]	
	Seawater	50% seawater	Seawater	50% seawater	Seawater	50% seawater	Seawater	50% seawater	Seawater	50% seawater
Alanine	24.47	13.02	33.56	25.36	18.40	13.02	7.60	7.10	1.80	1.70
Arginine	12.51	3.62	1.76	1.49	11.94	8.38	6.70	6.80	3.40	3.10
Aspartic acid	15.55	15.03	1.76	2.52	9.77	1.73	9.60	9.70	1.50	1.00
Glutamic acid	8.02	7.48	6.10	5.93	9.79	10.61	5.40	4.20	4.00	4.70
Glycine	28.80	16.80	32.67	24.93	61.87	18.40	2.10	1.30	0.60	0.70
Histidine	0.35	—	—	—	1.80	1.22	0.70	0.20	Trace	Trace
Isoleucine	0.19	Trace	Trace	Trace	0.31	0.39	0.20	0.10	Trace	0.10
Leucine	0.35	Trace	Trace	Trace	0.47	0.60	0.20	0.20	Trace	0.20
Lysine	1.64	1.16	2.56	1.70	2.81	1.71	0.60	0.70	0.30	0.40
Phenylalanine	Trace	Trace	Trace	Trace	0.51	0.28	0.80	0.05	Trace	0.10
Proline	11.19	5.24	10.24	7.03	6.29	2.80	Trace	Trace	0.40	0.40
Serine	1.27	1.29	0.77	0.53	5.99	3.05	3.40	1.40	0.70	0.90
Threonine	1.83	1.27	0.34	0.21	4.37	1.93	—	—	0.40	0.40
Tyrosine	Trace	Trace	Trace	Trace	0.77	0.48	Trace	Trace	0.10	0.10
Valine	Trace	Trace	Trace	Trace	0.94	0.50	—	—	0.20	0.20
Taurine	83.10	54.25	46.42	39.71	58.09	44.51	47.10	40.30	17.83	15.24
Betaine	116.24	75.15	63.33	52.91	62.99	40.66	—	—	—	—

Water content	73.7	81.2	76.3	82.3	69.8	80.2	73.4	81.2	75.1	82.6
Osmotic pressure due to amino acids[f]	160.06	88.82	130.71	94.10	216.54	90.19	56.46	43.45	19.83	18.83
Osmotic pressure due to taurine[f]	125.28	74.23	67.60	53.61	92.47	61.66	71.30	55.14	26.35	20.48
Osmotic pressure due to betaine[f]	175.24	102.83	92.22	71.43	100.27	56.33	—	—	—	—

[a] In micromoles per gram wet weight tissue.
[b] Bricteux-Grégoire *et al.* (1964a).
[c] Bricteux-Grégoire *et al.* (1964b).
[d] Bricteux-Grégoire *et al.* (1964c).
[e] Gilles (1971).
[f] The osmotic pressure due to the various compounds is given in microosmols per gram water. (Δ) of a molar solution of amino acids is −1.86°C) (Prosser *et al.*, 1950).

TABLE VIII

ORGANIC COMPOUNDS AS OSMOTIC EFFECTORS IN MUSCLES OF VARIOUS GASTROPODS ADAPTED TO SEAWATER OR TO 50% SEAWATER

Compound	Concentration[a]							
	Tegula funebralis[b]		*Patella vulgata*[c]		*Littorina littorea*[c]		*Purpura lapillus*[c]	
	Seawater	50% seawater	Seawater	50% seawater	Seawater	50% seawater	Seawater	50% seawater
Alanine	8.70	12.70	6.47	3.30	17.73	3.87	8.01	1.35
Arginine	12.00	3.40	10.03	3.38	6.30	5.31	5.31	18.36
Aspartic acid	5.90	4.90	2.20	0.89	1.89	0.90	2.25	0.90
Glutamic acid	4.40	4.00	4.18	2.27	3.96	2.43	4.41	2.70
Glycine	11.00	7.00	1.35	0.61	9.00	5.49	5.31	1.08
Histidine	0.60	0.30	0.05	0.09	Trace	0.27	0	0
Isoleucine	1.90	1.40	—	—	—	—	—	—
Leucine	0.80	1.00	0.45	0.70	0.81	0.99	0.81	0.09
Lysine	3.80	2.50	1.05	0.40	Trace	0.81	0.45	Trace
Phenylalanine	—	—	0.25	0.31	0.36	4.05	Trace	Trace
Proline	3.50	3.00	1.51	3.67	18.99	10.80	3.51	1.98
Serine	3.80	3.50	1.42	0.48	2.88	3.69	3.15	0.18
Threonine	3.30	2.90	0.86	0.31	0.99	2.07	0.36	0.18
Tyrosine	Trace	Trace	0.58	0.08	0.63	0.45	0.36	0.07
Valine	3.60	3.40	0.49	0.58	0.90	1.26	0.99	0.27
Taurine	11.50	12.50	106.50	28.88	40.14	22.23	38.34	21.87
Betaine	—	—	—	—	—	—	—	—
Water content	65	74	72.5	80.3	71.2	82	71	84.9
Osmotic pressure due to amino acids[d]	108.20	75.07	47.34	23.62	100.56	57.44	54.65	35.54
Osmotic pressure due to taurine[d]	19.66	18.32	163.22	39.96	62.64	30.12	60.00	28.62

[a] In micromoles per gram wet weight tissue.
[b] Peterson and Duerr (1969).
[c] Hoyaux and Jeuniaux (1971).
[d] The osmotic pressure due to the various compounds is expressed in microosmoles per gram water.

some of the organic compounds acting as cellular osmotic effectors, such as the amino acids, originate from a modification in the steady state between amino acids and proteins. In experiments on the osmotic adjustment of *Tegula funebralis*, Peterson and Duerr (1969) show that the protein content is about twice as high in the tissues of this species when adapted to 50% seawater than when adapted to seawater. However, in these conditions, the amino acid level remains unchanged. Moreover, when *Tegula* is adapted from seawater to 160% seawater, there is a decrease of 36.8 μM per gram of tissue dry weight in the amino acid quantity, while the protein level increases by 65.9 μg./mg. dry weight. A correlation between the modifications in the protein quantity and those in the amino acid level during the osmotic stress in *Tegula* is therefore hazardous. Moreover, as we have already pointed out, it is mainly the nonessential amino acids which act as osmotic effectors in euryhaline mollusks. On the other hand, organic compounds such as taurine, isethionic acid, or glycine betaine, which also play an important part in the isosmotic regulation in mollusks, cannot be considered constituents of proteins.

These considerations suggest that the mechanism involved in the regulation of the concentration of such compounds is probably other than a simple modification of the protein steady state. This mechanism could reside in a control of the metabolism of these compounds from carbohydrates. If such is the case, one might expect to observe changes in the oxidative metabolism in relation to the metabolic changes occurring during the isosmotic regulation process. One could therefore expect modifications in the oxygen consumption that could be related to salinity changes. A few cases are known in mollusks. *Hyridella* shows a progressive decrease in oxygen uptake when adapted to media of increased salinities (Hiscock, 1953b). In the same way, the oxygen consumption of *Hydrobia ulvae* is highter in 10% seawater than in full strength seawater (Negus, 1968), and the marine *Mytilus edulis* or *Mytilus galloprovincialis* shows decreases in respiration in both diluted and concentrated media (Maloeuf, 1937; Bouxin, 1931). This phenomenon has also been observed on isolated tissues such as gills (Pieh, 1936; Schlieper, 1957; Bielawski, 1961). On the other hand, Allen (1961) reports a decrease in muscle glycogen content when *Rangia cuneata* is placed into concentrated media. Such a result may also be interpreted as an indication of the energy spent in order to insure the isosmotic regulation.

Such findings favor the idea that the amount of the organic osmotic effectors is regulated by a mechanism controlling the degradation–synthesis balance. The fact that in many euryhaline mollusks osmotic adaptation is paralleled by a modification of the ammonia excretion is also in agreement with this hypothesis. As shown in Fig. 5, the ammonia excretion of *Macoma inconspicua* is higher when it is adapted to 50% seawater than when it is adapted to seawater. However, this is a transitory phenomena, since after 16 days of adaptation to the diluted medium, the ammonia excretion returns to the values recorded for animals adapted to seawater. This may indicate that a new metabolic steady state is reached after the stress period. In the same species, there is a significant decrease in the level of the free ninhydrin-positive substances during adaptation to 50% seawater (Emerson, 1969). Moreover, in the prosobranch species *Thais emarginata* and *Acmaea scutum,* where there is practically no change in the taurine and amino acid concentrations during the osmotic stress, there are no significant changes in the ammonia excretion during adaptation to various salinities (Emerson, 1969).

It thus appears that in mollusks using taurine and amino acids as osmotic effectors, a relationship between the regulation of the amount of these compounds and the nitrogen excretion can be demonstrated when these species withstand an osmotic stress. A decrease in the taurine and amino acid levels corresponds to an increased nitrogen excretion. This can be interpreted as indicating an increased degradation of these compounds. This is in agreement with the hypothesis according to which the regulation of the intracellular amino acid and taurine pool may

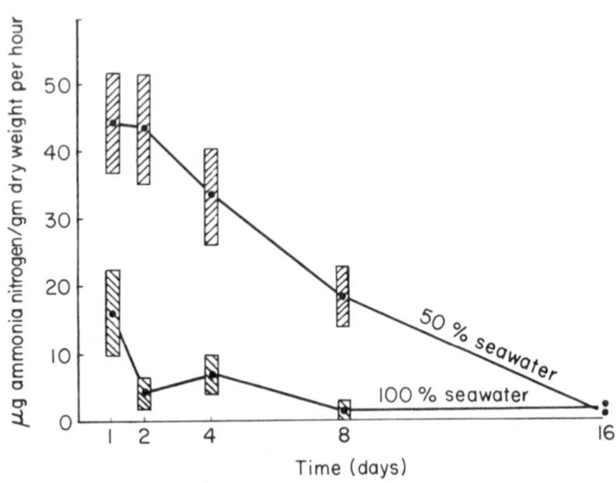

Fig. 5. Ammonia excretion of *Macoma inconspicua* as modified by salinity and time. Each mean value shown by a solid dot represents twelve determinations. The bars indicate the confidence limits at the 95% level. Values are based on dry weight of soft parts of animals. (From Emerson, 1969).

depend at least partly on a mechanism controlling the relative rate of anabolism and catabolism of these compounds. There is now evidence demonstrating the existence of such a control mechanism in crustacean cells (for review, see Florkin and Schoffeniels, 1969; Schoffeniels and Gilles, 1971).

In Crustacea, as in mollusks, amino acids play a prominent part in the cellular volume regulation. The concentration of these compounds appear to be under the control of the ionic composition of the cell. Inorganic ions have indeed an important effect on the activity of enzymes implicated in the amino acid metabolism (Schoffeniels, 1968; Gilles, 1969). One may suggest that such a control mechanism is also at play in the isosmotic intracellular regulation in mollusks. From the results obtained in this field with molluscan species, it is hazardous, however, to draw definite conclusions. A study of the intermediary metabolism in tissues of euryhaline mollusks subjected to an osmotic stress is still lacking, and further experimental results are needed to bring more evidence in favor of the interpretation given above.

IV. Summary

Marine mollusks are in osmotic equilibrium with the surrounding medium. However, they show some ability for ionic regulation. Freshwater and terrestrial species show an anisosmotic regulation mechanism. Despite their permeability to ions and water, these species maintain their blood ionic concentration at a higher value than that of the environmental medium. This hyperosmotic state, as well as the ionic regulation observed in marine species, is achieved by active processes. The active uptake of salts by the excretory system and by some other body part (the foot epithelium, the mantle, and the gills) is certainly of prime importance in this respect.

Euryhalinity in the phylum Mollusca, as in many other phylum, cannot be unequivocally defined as resulting from the existence of one single mechanism. As we have seen, it results from the association of a number of adaptations, some morphological, some physiological or biochemical. It is also clear that a mechanical device, such as the shell closing mechanism observed in intertidal bivalves and gastropods may help a given individual to withstand, at least temporarily, a dramatic change in the concentration of the outside medium. However, true euryhalinity can only exist if the animal possesses various mechanisms enabling it to cope with the osmotic stress. In many species, this implies the ability to regulate adequately the osmotic pressure of both blood and cells. Mollusks being poekilosmotic animals, euryhalinity in the phylum is mainly due to the ability of the animals to regulate their cellular osmotic pressure with respect to changes in the blood osmolarity.

In molluscan cells, ions contribute only to about 50% of the total osmoconcentration. The rest of the osmotic pressure is made up of several organic compounds, among which taurine, glycine betaine, and amino acids are of general occurrence. These substances together with sodium and chloride ions play an important part in the isosmotic regulation process which enables euryhaline mollusks to cope with the osmotic stress. It seems that a mechanism controlling the synthesis–degradation pathways of the organic osmotic effectors is at play in the regulation of their concentration. Nothing is known about the control mechanisms involved in the regulation of these metabolic pathways. A possible role of the inorganic composition of the intracellular medium on the activity of enzymes implicated in these pathways can be suggested. More information on the intracellular localization of the enzymes involved, on the ionic composition prevailing locally, and on the effect of ions on the enzymatic activity are needed before a more complete picture can be produced.

REFERENCES

Allen, K. (1961). *Biol. Bull.* **121**, 419.

Avens, A. C., and Sleigh, M. A. (1965). *Comp. Biochem. Physiol.* **16**, 121.

Awapara, J. (1962). *In* "Amino Acid Pools" (J. T. Holden, ed.), pp. 158–175. Elsevier, Amsterdam.

Baxter, C. F., and Ortiz, C. L. (1966). *Life Sci.* **5**, 2321.

Beadle, L. C. (1969). *J. Exp. Biol.* **50**, 473.

Beadle, L. C., and Beadle, S. F. (1969). *J. Exp. Biol.* **50**, 481.

Bethe, A. (1934). *Pfluegers Arch. Gesamte Physiol. Menschen Tiere* **234**, 629.

Bethe, A., and Berger, E. (1931). *Pfluegers Arch. Gesamte Physiol. Menschen Tiere* **227**, 571.

Bialaszewicz, K., and Kupfer, C. (1936). *Arch. Int. Physiol.* **42**, 398.

Bielawski, J. (1961). *Comp. Biochem. Physiol.* **3**, 250.

Botazzi, F. (1908). *Ergeb. Physiol.* **7**, 161.

Bouxin, H. (1931). *Bull. Inst. Oceanogr.* **569**, 1.

Bricteux-Grégoire, S., Duchâteau-Bosson, G., Jeuniaux, C., and Florkin, M. (1964a). *Arch. Int. Physiol. Biochim.* **72**, 116.

Bricteux-Grégoire, S., Duchâteau-Bosson, G., Jeuniaux, C., and Florkin, M. (1964b). *Arch. Int. Physiol. Biochim.* **72**, 835.

Bricteux-Grégoire, S., Duchâteau-Bosson, G., Jeuniaux, C., and Florkin, M. (1964c). *Arch. Int. Physiol. Biochim.* **72**, 267.

Bricteux-Grégoire, S., Duchâteau-Bosson, G., Jeuniaux, C., and Florkin, M. (1966). *Comp. Biochem. Physiol.* **19**, 729.

Burton, R. F. (1966). *Comp. Biochem. Physiol.* **17**, 1007.

Burton, R. F. (1968a). *Comp. Biochem. Physiol.* **25**, 501.

Burton, R. F. (1968b). *Comp. Biochem. Physiol.* **25**, 509.

Camien, M. N., Sarlet, H., Duchâteau, G., and Florkin, M. (1951). *J. Biol. Chem.* **193**, 881.
Cole, W. H. (1940). *J. Gen. Physiol.* **23**, 575.
Dakin, W. J., and Edmonds, E. (1931). *Aust. J. Exp. Biol. Med. Sci.* **8**, 169.
Deffner, G. G. J., and Hafter, R. E. (1960). *Biochim. Biophys.* Acta **42**, 189.
De Jorge, F. B., Ulhôa Cintra, A. B., Haeser, P. E., and Sawaya, P. (1965). *Comp. Biochem. Physiol.* **14**, 35.
Drilhon, A., and Florence, G. (1942). *Bull. Soc. Chim. Biol.* **24**, 96.
Duchâteau, G., and Florkin, M. (1955). *Arch. Int. Physiol. Biochim.* **63**, 249.
Duchâteau, G., Sarlet, H., Camien, M. N., and Florkin, M. (1952). *Arch. Int. Physiol. Biochim.* **60**, 124.
Duchâteau-Bosson, G., Jeuniaux, C., and Florkin, M. (1961). *Arch. Int. Physiol. Biochim.* **69**, 30.
Emerson, D. N. (1969). *Comp. Biochem. Physiol.* **29**, 1115.
Endo, K., Hujita, M., and Simidu, W. (1962). *Bull. Jap. Soc. Sci. Fish.* **28**, 1099.
Florkin, M. (1943). *Bull. Soc. Roy. Sci. Liege* **5**, 301.
Florkin, M. (1962). *Bull. Cl. Sci., Acad. Roy. Belg.* [5] **48**, 687.
Florkin, M., and Duchâteau, G. (1949). *Physiol. Comp. Oecol.* **1**, 29.
Florkin, M., and Duchâteau, G. (1950). *C.R. Soc. Biol.* **144**, 1132.
Florkin, M., and Schoffeniels, E. (1965). *In* "Studies in Comparative Biochemistry" (K. A. Munday, ed.), pp. 6–39. Pergamon Press, Oxford.
Florkin, M., and Schoffeniels, E. (1969). "Molecular Approaches to Ecology." Academic Press, New York.
Fredericq, L. (1901). *Bull. Cl. Sci., Acad. Roy. Belg.* [4] 428.
Freeman, R. F. H., and Rigler, F. H. (1957). *J. Mar. Biol. Ass. U.K.* **36**, 553.
Fugelli, K. (1967). *Comp. Biochem. Physiol.* **22**, 253.
Gilles, R. (1969). *Arch. Int. Physiol. Biochim.* **77**, 441.
Gilles, R. (1970). *Arch. Int. Physiol. Biochim.* **78**, 91.
Gilles, R. (1971). *Biol. Bull.* (in press).
Gordon, M. S. (1965). *Biol. Bull.* **128**, 218.
Hayes, F. R., and Pelluet, D. (1947). *J. Mar. Biol. Ass. U.K.* **26**, 580.
Henze, M. (1905). *Hoppe-Seyler's Z. Physiol. Chem.* **43**, **477.**
Henze, M. (1911). *Hoppe-Seyler's Z. Physiol. Chem.* **70**, 253.
Henze, M. (1914). *Hoppe-Seyler's Z. Physiol. Chem.* **91**, 230.
Hiscock, I. D. (1953a). *Aust. J. Mar. Freshwater Res.* **4**, 317.
Hiscock, I. D. (1953b). *Aust. J. Mar. Freshwater Res.* **4**, 330.
Hoyaux, T., and Jeuniaux, C. (1971). In preparation.
Huf, E. (1934). *Pfluegers Arch. Gesamte Physiol. Menschen. Tiere* **235**, 129.
Jeuniaux, C., Duchâteau-Bosson, G., and Florkin, M. (1961). *J. Biochem. (Tokyo)* **49**, 527.
Jeuniaux, C., Bricteux-Grégoire, S., and Florkin, M. (1962). *Cah. Biol. Mar.* **3**, 107.
Kelly, A. (1904). *Beitr. Chem. Physiol. Pathol.* **5**, 377.
Kirschner, L. B., Sorenson, A. L., and Kriebel, M. (1960). *Science* **131**, 735.
Krogh, A. (1939). "Osmotic Regulation in Aquatic Animals." Cambridge Univ. Press, London and New York.
Lange, R. (1963). *Comp. Biochem. Physiol.* **10**, 173.
Lange, R. (1964). *Comp. Biochem. Physiol.* **13**, 205.
Lange, R., and Fugelli, K. (1965). *Comp. Biochem. Physiol.* **15**, 283.
Lewis, P. R. (1952). *Biochem. J.* **52**, 330.

Little, C. (1965). *J. Exp. Biol.* **43**, 38.
Maloeuf, N. S. R. (1937). *Z. Vergl. Physiol.* **25**, 1.
Martin, A. W. (1957). *Recent Advan. Invertebr. Physiol., Symp., 1955* pp. 247–276.
Martin, A. W., Stewart, D. M., and Harrison, F. M. (1965). *J. Exp. Biol.* **42**, 99.
Mayer, A., and Rathery, F. (1907). *J. Anat., Paris* **43**, 25.
Michon, J., and Alaphilippe, F. (1958). *C. R. Soc. Biol.* **152**, 1349.
Milne, A. (1940). *Trans. Roy. Soc. Edinburgh* **50**, 107.
Negus, M. R. S. (1968). *Comp. Biochem. Physiol.* **24**, 317.
Peterson, M. B., and Duerr, F. G. (1969). *Comp. Biochem. Physiol.* **28**, 633.
Picken, L. E. R. (1937). *J. Exp. Biol.* **14**, 20.
Pieh, S. (1936). *Zool. Jahrb., Abt. Allgem. Zool. Physiol. Tiere* **56**, 129.
Potts, W. T. W. (1952). *Nature (London)* **169**, 834.
Potts, W. T. W. (1954). *J. Exp. Biol.* **31**, 164.
Potts, W. T. W. (1958). *J. Exp. Biol.* **53**, 749.
Potts, W. T. W., and Parry, G. (1964). "Osmotic and Ionic Regulation in Animals." Pergamon Press, Oxford.
Prosser, C. L., Bishop, D. W., Brown, F. A., Jahn, T. L., and Wulff, V. J. (1950). *In* "Comparative Animal Physiology" (C. L. Prosser, ed.), Saunders, Philadelphia, Pennsylvania.
Robertson, J. D. (1949). *J. Exp. Biol.* **26**, 182.
Robertson, J. D. (1953). *J. Exp. Biol.* **30**, 277.
Robertson, J. D. (1964). *In* "Physiology of Mollusca" (K. M. Wilbur and C. M. Yonge, eds.), Vol. 1, pp. 283–308. Academic Press, New York.
Robertson, J. D. (1965). *J. Exp. Biol.* **42**, 153.
Saxena, B. B. (1957). *Physiol. Zool.* **30**, 161.
Schlieper, C. (1957). *Anneé Biol.* **33**, 117.
Schoffeniels, E. (1951a). *Arch. Int. Physiol. Biochim.* **58**, 467.
Schoffeniels, E. (1951b). *Arch. Int. Physiol. Biochim.* **59**, 245.
Schoffeniels, E. (1968). *Arch. Int. Physiol. Biochim.* **76**, 319.
Schoffeniels, E., and Gilles, R. (1971). *In* "Chemical Zoology" (M. Florkin and B. T. Scheer, eds.). Vol. V, Part A, pp. 255–286.
Segerstråle, S. G. (1957). *In* "Treatise on Marine Ecology and Paleocology" (J. W. Hedgpeth, ed.), Vol. 1, pp. 751–800.
Simpson, J. W., Allen, K., and Awapara, J. (1959). *Biol. Bull.* **117**, 371.
Stephens, G. C., and Virkar, R. A. (1966). *Biol. Bull.* **131**, 172.
Todd, M. E. (1962). Ph.D. Thesis, University of Glasgow.
Todd, M. E. (1964). *J. Exp. Biol.* **41**, 665.
Van Slyke, D. D., McFadyen, D. A., and Hamilton, P. (1941). *J. Biol. Chem.* **141**, 671.
Van Weel, P. B. (1957). *Z. Vergl. Physiol.* **39**, 294.
Virkar, R. A. (1966). *Comp. Biochem. Physiol.* **18**, 617.
Wilbur, K. M. (1964). *In* "Physiology of Mollusca" (K. M. Wilbur and C. M. Yonge, eds.), Vol. 1, pp. 243–277. Academic Press, New York.
Wilbur, K. M., and Jodrey, L. H. (1955). *Biol. Bull.* **108**, 359.
Yazaki, M. (1929). *Sci. Rep. Tohoku Imp. Univ. Ser. 4* **4**, 285.

CHAPTER 13

Aspects of Molluscan Pharmacology

Robert Endean

I. Introduction

Two aspects of molluscan pharmacology have aroused considerable interest in recent years. One concerns chemical transmission within the nervous and neuromuscular systems of mollusks, and the other concerns toxic materials elaborated by mollusks. Despite its relative simplicity, the molluscan brain has basic structural and functional features which are similar to those found in vertebrate brain. However, studies of molluscan neuropharmacology have been facilitated by the presence in gastropods of giant neurons. These giant neurons can be readily identified because of their size (150–500 μ in diameter) and their positions with respect to other neurons. Also, their superficial location renders them amenable to investigations involving microelectrodes and the iontophoretic application of drugs. It is to be expected that detailed studies of the identifiable neurons present in mollusks will lead to an elucidation of the roles of several pharmacological compounds which are believed to function as neurotransmitters and will increase markedly our knowledge of the chemical heterogeneity of the brain. In this respect, it should be noted that many pharmacological agents which modify neuronal activity in vertebrates are reactive within molluscan nervous systems.

Then too, the ease with which the bivalve heart can be isolated and manipulated has led to some bivalve hearts being used as standard preparations for the bioassay of pharmacological agents such as acetylcholine (ACh) and 5-hydroxytryptamine (5-HT). It is possible that studies of chemical regulation in molluscan hearts will shed light on several aspects of chemical regulation in vertebrate hearts. Likewise, an understanding of the manner in which chemical transmitters initiate and release "catch," a phenomenon shown by certain muscles of mollusks, may elucidate ill understood aspects of persistent tonus in certain types of vertebrate muscle, such as vascular smooth muscle. Current research on ACh, 5-HT, dopamine, and other compounds associated with transmission in the neuromuscular systems of mollusks is in the forefront of studies designed to determine the biological roles of these compounds.

It is not inconceivable that some chemical transmitters may be concentrated in the potent biotoxins elaborated by some mollusks. However, compounds present in these toxic materials have been isolated and characterized chemically in only a few cases. Even so, pharmacologists have already been provided with a new biologically active peptide, eledoisin (Erspamer and Anastasi, 1962), and new choline derivatives (Erspamer and Dordoni, 1947; Whittaker, 1960). It can be expected that compounds with particularly interesting biological activity will be obtained from molluscan venoms such as those possessed by *Conus geographus* or *Hapalochlaena maculosa,* which have been responsible for human fatalities, or from material elaborated by *Mercenaria,* which exhibits oncolytic activity. Some of these biologically active compounds may be used as new therapeutic drugs and some may find a use as tools in the investigation of biological processes (Endean, 1966).

In the following account, morphological and physiological features of mollusks are discussed where they assist in understanding pharmacological aspects. In studies of toxic material produced by mollusks, cognizance should be taken of how the material is normally used by its possessor. Hence, relevant aspects of the general ecology and behavior of these animals are mentioned in the section dealing with molluscan biotoxins.

II. Chemical Transmission at Synapses within the Central Nervous System

In this account, attention is focused on recent work. References to the earlier work in this general field will be found in reviews by Florey (1965) and by Cottrell and Laverack (1968).

A. 5-HYDROXYTRYPTAMINE

There is no doubt that 5-HT occurs in molluscan ganglia (Florey and Florey, 1953; Welsh, 1957; Welsh and Moorehead, 1960). Indeed, Welsh and Moorehead (1960) noted that ganglia of bivalve mollusks contain the highest levels of 5-HT found in any nervous system. However, 5-HT may not be uniformly distributed within the molluscan nervous system. For example, Mirolli (1964) has shown that in *Busycon* the pedal ganglion contains 25 μg. per gram of tissue, the cerebral ganglion 14 μg. per gram of tissue, and the parietal ganglion 7–8 μg. per gram of tissue. Moreover, the ganglia of mollusks may contain less 5-HT than some other organs. For example, the kidney of *Busycon* contains much more 5-HT than the ganglia of this species (Mirolli, 1964).

Although the experiments of Gerschenfeld and Tauc (1961) and Kerkut and Walker (1962) with molluscan ganglia indicated that 5-HT exerted a strong depolarizing and excitatory action upon molluscan neurons, these experiments were not in themselves sufficient to establish 5-HT as a synaptic transmitter in the central nervous system of mollusks. Nor was the finding by Chase *et al.* (1968), who used tritiated 5-HT to establish that 5-HT could be accumulated by the visceral ganglia of *Aplysia* and released from them by electrical stimulation. Ascher *et al.* (1968) have shown that the perihepatic connective tissue of *Aplysia* in which no nerve endings have been reported and which normally contains 2.3 μg. per gram of 5-HT will accumulate tritiated 5-HT readily. Moreover, the release of accumulated 5-HT from this connective tissue can be triggered by electrical stimulation, perfusion with high K^+ seawater, or perfusion with ACh.

One type of neuron in the abdominovisceral ganglionic mass of the snail *Cryptomphallus aspersa* responds to iontophoretic applications of 5-HT (Gerschenfeld and Stefani, 1965, 1966). These neurons belong to a group known as CILDA (Cells with Inhibition of Long Duration) neurons and are characterized by the prolonged response which they show to presynaptic stimulation (Gerschenfeld and Tauc, 1964). Evidence supporting a transmitter role for 5-HT in snail CILDA neurons has recently been summarized by Gerschenfeld and Stefani (1968). Of particular importance is the fact that minute amounts of 5-HT are capable of depolarizing CILDA cells, and the depolarization frequently initiates action potentials. A complication is provided by the fact that CILDA cells can also be depolarized by ACh. Moreover, both ACh and 5-HT potentials in snail CILDA cells appear to be due to a change in membrane permeability to sodium ions. Atropine blocks the responses of the cells to both ACh and 5-HT. However, hexamethonium bromide (10^{-5} gm./ml.) blocks the ACh responses without affecting the 5-HT responses while lysergic acid diethylamide (LSD) (10^{-4} gm./ml.), tryptamine (10^{-5} gm./ml.), and 5-HT (10^{-5} gm./ml.) itself block 5-HT responses without affecting ACh responses. CILDA neurons show a complex excitatory input composed of slow and fast excitatory potentials (EPSP), and it was suggested that ACh was the transmitter for fast EPSP and 5-HT the transmitter for slow EPSP.

Glaizner (1968) found that twenty-four of forty-six neurons tested in the subesophageal ganglia of *Helix aspersa* responded to 5-HT. Eighteen gave a depolarizing response, but six were hyperpolarized, showing that 5-HT can have an inhibitory action on neurons.

Of particular relevance to the possible role of 5-HT as a neurotransmitter in the central nervous system of mollusks is its histological localization. Dahl *et al.* (1962, 1966) used fluorescence microscopy to localize 5-HT in molluscan neurons. The results obtained indicated that 5-HT-containing granules were present in nerve terminals, in the cell bodies, and in the proximal parts of the axons of certain neurons in the bivalves and gastropods studied. Zs.-Nagy *et al.* (1965) believed that 5-HT was associated with membranes of the endoplasmic reticulum in ganglion cells of the bivalve *Anodonta*. Although Cottrell (1966) showed that a large percentage of the total 5-HT in the nervous tissue of the bivalve *Mercenaria mercenaria* is particle bound, he was unable to identify these particles with disrupted portions of the endoplasmic reticulum. Indeed, Cottrell and Maser (1967) stated that the subcellular localization of 5-HT in the bivalve nervous system is still an open question. Sedden *et al.* (1968) have used fluorescence microscopy to reveal the presence of 5-HT-containing neurons in the pedal ganglion and a single large 5-HT-containing cell in the cerebral ganglion of the snail *H. aspersa*. The visceral and right parietal ganglia of this species contained cells in which both 5-HT and dopamine were detected.

The 5-HT receptors of CILDA cells appear to be localized at the axon hillock and possibly for some distance beyond the hillock (Gerschenfeld and Stefani, 1968). These authors believed that the sites of the 5-HT receptors on the neuronal membrane of the CILDA cell are different from those of the ACh receptors possessed by this cell type.

It is apparent from a survey of the literature that 5-HT satisfies several criteria for a role as a neurotransmitter in the central nervous systems of mollusks. It is present in molluscan ganglia in significant amounts and can be located histochemically in the terminals of certain neurons. It can be released by electrical stimulation of molluscan ganglia. Certain large neurons (CILDA neurons) can be depolarized by 5-HT, and these neurons possess highly sensitive 5-HT receptors which can be specifically blocked by LSD-25 and tryptamine.

On the other hand, the extent to which 5-HT normally functions as a synaptic transmitter in the central nervous systems of mollusks is questionable and requires clarification. Only relatively few neurons in molluscan ganglia appear to be sensitive to iontophoretically applied 5-HT. Many of these neurons are also sensitive to and depolarized by ACh. Histochemical tests have revealed that as well as being present in nerve terminals of certain molluscan neurons, 5-HT is present in the cell bodies and proximal parts of the axons of these neurons as well as in nonnervous tissues. Difficulty is encountered in accounting for the inactivation of 5-HT at synapses. Gerschenfeld and Stefani (1968) noted that monoamine oxidase inhibitors did not prolong the 5-HT response of CILDA cells and suggested a possible mechanism of inactivation by diffusion. In this respect, 5-HT behaves differently from the classic synaptic transmitter ACh which is rapidly inactivated after its release at vertebrate nerve terminals.

B. Acetylcholine and Cholinesterases

Acetylcholine has been found in high concentration in gastropod ganglia (see Tauc, 1966), in the ganglia of bivalves (Cottrell, 1966), and in cephalopod nervous systems (see Cottrell and Laverack, 1968). It was noted by Tauc and Gerschenfeld (1961, 1962) that some neurons in the brain of the tectibranch gastropod *Aplysia* were hyperpolarized while others were depolarized by ACh. These cells were termed H and D cells, respectively. Neurones which reacted to ACh in a similar manner were observed in snail ganglia (Kerkut and Thomas, 1964), and it was found that H cells could be converted to D cells by injecting chloride ions into the cells. Chloride ion appeared to be the principal ion responsible for changes in the membrane potential of the H cell, while sodium ions appeared to be responsible for ACh-induced depolarization of D cells (Kerkut and Meech, 1966).

Glaizner (1968) applied ACh iontophoretically to sixty-two cells in the subesophageal ganglia of *Helix aspersa* and found that thirty-four behaved as H cells, twenty-four as D cells, and four failed to respond. Different proportions of H and D cells were found in parietal and abdominal ganglia.

The suggestion by Tauc and Gerschenfeld (1961) that release of ACh at different synaptic endings of the same neuron in *Aplysia* ganglia might have an excitatory action on some neurons (D cells) and an inhibitory action on other neurons (H cells) was confirmed by Kandel *et al.* (1967).

Walker *et al.* (1968a) have made a detailed study of the pharmacology of the neurons of *H. aspersa.* They found that the response of H and D cells to ACh can be antagonized by both muscarinic and nicotinic blocking agents, and it appears that these neurons contain both muscarinic and nicotinic receptor sites. Also, some neurons were more sensitive to muscarinic agonists than they were to nicotinic agonists, while the reverse was true for other neurons. Still other neurons were equally sensitive to both muscarinic and nicotinic agonists. It seems that the relative proportions of muscarinic and nicotinic receptors varies from one neuron to another. However, it is likely that the response of snail neurons to cholinomimetics and cholinolytics shows some variation from species to species. Zeimal and Vulfius (1968) studied the activity of several cholinomimetics and cholinolytics on giant neurons of the snails *Planorbis corneus* and *Limnaea stagnalis.* Of the cholinomimetics studied, the most active were those known to excite the nicotinic cholinoreceptors of vertebrates. Of the cholinolytics tested, only compounds which are known to block the nicotinic receptors of skeletal muscles were effective.

ACh and cholinesterase have been demonstrated in the giant axon and in the distal synapses of the stellate ganglion of the squid *Loligo pealii* by Webb *et al.* (1966). Using ultracentrifugation techniques, Florey and Winesdorfer (1967) separated particles rich in ACh from *Octopus* brain tissues. Using density gradient centrifugation, Cottrell (1966) showed that a large proportion of the ACh present in nervous tissue of *Mercenaria mercenaria* was bound to subcellular particles and evidence was presented which suggested that the sedimentation properties of the ACh particles resembled those prepared in a similar manner from mammalian brain.

Cholinesterases have been located at a few synaptic junctions in the neuropile of the brain of the snail *H. aspersa* (Newman *et al.*, 1968). At least two different types of vesicle occur in presynaptic endings at these cholinergic junctions. One type consists of electron-dense vesicles ranging from 800 to 1000 Å in diameter, the other consists of clear vesicles ranging from 250 to 300 Å in diameter. However, Sakharov and Turpaev (1968) maintain that the central nervous systems of gastropods and bivalves differ markedly from that of cephalopods in that little or no cholinesterase is present in the gastropod or bivalve neuropile. On the other hand, a cholinesterase resembling acetylcholinesterase is well represented in the hemolymph of gastropods and bivalves, while blood cholinesterase of cephalopods has a lower activity. Sakharov and Turpaev (1968) postulated that removal of ACh from the junctional area occurs by diffusion in both gastropods and bivalves. However, they also postulated that an additional mechanism is used to remove ACh from receptors on the postsynaptic membrane in gastropods. This mechanism involves the competition for receptors which exists between the transmitter and its antagonist, the antagonist being released into the synaptic cleft under the influence of the transmitter itself. The possibility that the cholinergic antagonist is ATP, as in the heart of *Anodonta* (see p. 434) was raised.

C. CATECHOLAMINES

Although conflicting data on the amounts of noradrenaline and adrenaline in molluscan ganglia have been presented (Oestlund, 1954; von Euler, 1961; Puppi, 1964; Dahl *et al.*, 1966; Cottrell, 1967), there is little doubt that the major catecholamine in molluscan ganglia is dopamine (Sweeney, 1963; Cottrell and Laverack, 1968). The neuronal localization of dopamine in mollusks was demonstrated by Dahl *et al.* (1962, 1963, 1966) and by Kerkut *et al.* (1966a, 1967a).

In the neuropile region of the ganglia of the bivalve *Spisula solida*, Cottrell (1968) observed small granular vesicles ranging from about 300 to 1000 Å in diameter. The results of depletion experiments with reserpine and other studies indicated that dopamine was probably present within the granular vesicles. It seems likely that the granular vesicles noted by Cottrell (1968) are similar to the dense-core vesicles found in neurons of *Anodonta cygnea* by Zs.-Nagy (1964). Evidence has been presented that these vesicles contain dopamine (Zs.-Nagy, 1968).

Using intracellular recording techniques, Kerkut and Walker (1961, 1962) showed that some neurons of *Helix* are inhibited and others excited by low concentrations of dopamine. In most cases dopamine had an inhibitory effect, causing a decrease in the frequency of spontaneous action potentials. Dopamine was tested on forty-four neurons and noradrenaline on twenty-three neurons in the subesophageal ganglia of *Helix aspersa* by Glaizner (1968). It was found that twenty-one cells were hyperpolarized and one was depolarized by dopamine. Twelve cells were hyperpolarized by noradrenaline. Evidence was obtained indicating that separate receptors exist for noradrenaline and dopamine even where both react with the same cell. Similar results were obtained for some *Aplysia* neurons by Gerschenfeld and Tauc (1961, 1964) and for certain neurons in the brain of *Cryptomphallus aspersa* (Gerschenfeld, 1964).

One type of noncholinergic inhibitory potential found in molluscan neurons is known as Inhibition of Long Duration, or ILD. It was first described in certain *Aplysia* neurons following their stimulation and consisted of hyperpolarization of the membrane potential which was sustained for several seconds (Tauc, 1958, 1959). CILDA neurons exhibit ILD, and these cells were shown to be hyperpolarized and inhibited by catecholamines. They are especially sensitive to dopamine (Gerschenfeld and Tauc, 1964). Ascher *et al.* (1967) studied the effects of iontophoretic application of dopamine to certain *Aplysia* neurons and found that there was a similarity between the reversal potentials of the dopamine effect and ILD. This raised the possibility that dopamine was the transmitter for ILD. It was noted by Ascher (1968) that the reversal potential of the dopamine effect in *Aplysia* neurons varied with the external potassium concentration but was not affected if methyl sulfate replaced the external chloride. Kerkut *et al.* (1969a) have found that ILD could be produced in a known cell in the right parietal ganglion of *H. aspersa* by stimulation of the left pallial nerve, and they observed that ILD was caused by an increase in permeability to potassium ions and had a reversal potential of approximately —80 millivolts. They found that the ILD was blocked by ergometrine and concluded that dopamine was the most likely transmitter responsible for ILD in the cell studied.

Ascher (1968) found that in some *Aplysia* neurons the hyperpolarization induced by dopamine contained two components, one component being blocked by ouabain. The dopamine-induced hyperpolarization of a neuron in the snail brain can also be blocked by ouabain (Kerkut *et al.*, 1969b). This hyperpolarization is little affected by alteration in the external potassium concentration and is apparently different from that involved in ILD.

The effects of a number of compounds on the inhibition produced by dopamine in specific neurons of *H. aspersa* were investigated by Walker *et al.* (1968b). It was concluded that the dopamine receptor resembled the α- rather than the β-adrenaline receptor of vertebrates. However, they noted that ergometrine, which is not an α blocker, is a potent antagonist of dopamine. Woodruff and Walker (1969) obtained evidence for the presence in certain *H. aspersa* neurons of a specific dopamine receptor distinct from the action of dopamine on either a classic α- or β-adrenaline receptor and they noted two requirements for a compound to react with this dopamine receptor. One requirement was the presence of two hydroxyl groups on the 3 and 4 positions of the benzene ring, and the other was the presence of a terminal nitrogen either unsubstituted or with one methyl group.

It is possible that dopamine is a synaptic transmitter within the central nervous system of mollusks. Dopamine is present in significant amounts in molluscan ganglia. It has been localized in certain molluscan neurons. It has a marked effect (generally inhibitory) on certain molluscan neurons. However, the extent to which dopamine is normally involved as a synaptic transmitter in the central nervous systems of mollusks has still to be established.

D. Glutamate

During a study of the amino acid content of snail brain tissue, Kerkut and Cottrell (1962) found that glutamate was the most common amino acid present. This raised the possibility that glutamate is a transmitter in the snail central nervous system. Gerschenfeld and Lasansky (1964) detected in *Helix pomatia* and *Cryptomphallus aspersa* neurons which respond to glutamate at a threshold concentration of about 10^{-7} gm./ml. Some neurons were excited, others inhibited by glutamate. Oomura *et al.* (1965) have shown that one type of neuron in *Onchidium* was hyperpolarized by glutamate.

Kerkut *et al.* (1969a) found that low concentrations of glutamate (less than 1 μg./ml.) caused inhibition of a large neuron in the right parietal ganglion of *H. aspersa*. However, it was shown that dopamine rather than glutamate was the most probable transmitter normally causing inhibition in the cell studied. It was suggested by Miledi (1966) that L-glutamate may be a transmitter between certain neurons in the squid, but recent work (Miledi, 1969) does not support his original suggestion. The possibility that glutamate is a synaptic transmitter in the molluscan central nervous system warrants further attention.

E. γ-Aminobutyric Acid (GABA)

Tauc (1966) believed that GABA could not have any synaptic transmitter function in gastropod ganglion cells because the compound temporarily inhibits D cells and depolarizes H cells. However, the possibility that GABA acts as a transmitter at synapses between other neuronal types warrants consideration. Indeed, some ganglion cells of *Helix aspersa, Helix pomatia,* and *Cryptomphallus* are excited, while others are inhibited by GABA (Kerkut and Walker, 1961; Gerschenfeld and Lasansky, 1964). Further study of the possible role of GABA as a synaptic transmitter in the molluscan central nervous system is obviously warranted.

F. Histamine

Walker *et al.* (1968a) showed that certain neurons of *Helix aspersa* responded to histamine by being depolarized and excited, while the spontaneous activity of other neurons was inhibited by histamine. It was noted that both effects could be antagonized reversibly by mepyramine.

III. Chemical Transmission at Neuromuscular Junctions

Although there is no direct evidence that ACh is liberated at molluscan nerve–muscle junctions, there is much circumstantial evidence implicating ACh as a junctional transmitter. This evidence depends mainly on the effects of adding the compound, its antagonists, and its synergists to nerve–muscle preparations. Bacq (1947) has reviewed the early literature. The stimulating action of ACh on the radular muscle of *Buccinum undatum* is augmented by eserine (Fänge and Mattisson, 1958). A similar effect on the radular protractor muscle of *Busycon canaliculatum* was described by Hill (1958). Cambridge *et al.* (1959) showed that the stimulating effect of ACh on the anterior byssus retractor muscle (ABRM) of *Mytilus edulis* was potentiated slightly by eserine and inhibited by hexamethonium, atropine, and curarine. However, propantheline and methantheline were much more potent than other compounds in inhibiting the effect of ACh. Twarog (1960), using the ABRM of *M. edulis,* showed that ACh caused tonic contractile responses and that 5-HT acted as a relaxing agent which antagonized maintained tonic contractions induced by ACh and speeded relaxation, but did not prevent phasic contractions resulting from nerve stimulation. Jaeger (1962) demonstrated that 10^{-7} gm./ml. ACh stimulated the penis retractor muscle of the snail *Strophocheilos oblongus* to contract and that the contraction was blocked by mytolon and curarine and partially blocked by eserine. Further examples of studies dealing with the cholinerginicity of molluscan muscles have been listed by Hoyle (1964).

Kerkut and Leake (1966) showed that contractions of the pharyngeal retractor muscle (PRM) of *Helix aspersa* caused by ACh were augmented by curarine and eserine, whereas the contractions induced by brain stimulation were inhibited by these drugs. Because curarine is a standard drug used to block ACh in vertebrate nerve–muscle junctions, the augmentation of ACh-induced contractions by curarine in the PRM of *H. aspersa* suggests a complex situation.

Florey (1966) found that the actions of ACh and 5-HT on the chromatophore muscles of the squid *Loligo opalescens* were very similar to those described by Twarog (1960) for the ABRM of *Mytilus*. Since he could find no evidence for the presence of inhibitory fibers innervating the chromatophore muscles, Florey (1966) doubted that 5-HT could be involved as a junctional transmitter.

Elucidation of the catch mechanism of molluscan paramyosin muscles has been attempted repeatedly (see Hoyle, 1969). The ABRM of *Mytilus* has been the favored preparation. Like many other muscles of bivalves, it is doubly innervated, one set of nerve fibers causing contraction, the other relaxation. Catch has been described by Twarog (1966) as a prolonged state of resistance to stretch which persists in the absence of activation. Twarog (1967a,b) succeeded in penetrating individual muscle fibers with microelectrodes. She found that methantheline, which blocks cholinergic synapses, reduced and also prevented junctional potential formation. Since the drug could have been acting presynaptically, this evidence does not necessarily support the apparent cholinergicity of the ABRM. Catch can be prevented or reduced by 5-HT quantitatively over the range 10^{-8}–10^{-6} M, and it was considered likely that 5-HT is the agent released from relaxing nerves which permeate the fibers and that it acts directly (Hidaka *et al.*, 1967). However, Florey (1965) has produced strong arguments in support of the idea that 5-HT acts intracellularly.

When the radular muscles of *B. undatum* and *B. canaliculatum* were treated with ACh and 5-HT simultaneously, there was a strong tendency for rhythmical activity, while ACh alone provoked prolonged contraction (Fänge and Mattisson, 1958; Hill, 1958). Atropine blocked the effect of nerve stimulation and eserine prolonged the relaxation period after a tetanus. From this evidence, Fänge (1962) considered that a cholinergic system is probably involved in the neuromuscular excitation of the radular muscles. In these muscles, too, it is probable that the function of 5-HT is to release catch rather than inhibit excitation.

Kerkut *et al.* (1965) searched for transmitters by using the PRM–brain preparation of *H. aspersa*. They stimulated the brain electrically and collected the muscle perfusate at regular intervals. Glutamate was identified in the perfusate by thin-layer chromatography, and the amount detected in each perfusate was proportional to the number of stimuli applied to the brain. It was found also that both stimulated and unstimulated muscles contracted when glutamate was added to them. The threshold dose for the most sensitive preparation was 2×10^{-8} gm./ml. Further, the muscle was twenty to one hundred times more sensitive to L-glutamate than to D-glutamate. It was shown by Kerkut *et al.*

(1967b) that labeled glutamate appeared in the perfusate after the brain had been incubated in 1 μC. of glutamate-^{14}C for 3 hours and then stimulated. Further, incubation of the isolated brain in glucose-^{14}C and alanine-^{14}C also resulted in the appearance of labeled glutamate in the muscle perfusate. Kerkut and Leake (1966) found that L-glutamic acid stimulated the PRM and that GABA inhibited contraction. Mattisson and Arvidsson (1966) stimulated the radular muscle of *B. undatum* to contract for periods of 12 hours or more and then added glutamate, which caused marked contraction. Succinate caused relaxation and inhibition of contractions.

The search for transmitters by perfusion techniques is complicated by axoplasmic flow. Kerkut *et al.* (1967b) showed that glutamate traveled not only from brain to PRM of *H. aspersa* but also antidromically. Doubtless, other compounds which have no involvement in neuromuscular transmission are transferred normally from nerve to muscle and also in the reverse direction.

A few studies have been made on the ultrastructure of the molluscan nerve–muscle junction. Kerkut *et al.* (1966b) studied the neuromuscular junctions in the muscle surrounding the nerve trunk of the pharyngeal retractor nerve of *H. aspersa*. The junctions appeared as simple bandlike structures lying on the surface of the muscle fibers which often contained pinocytotic vesicles around their periphery. These vesicles differed in size and structure from the vesicles found in the nerve terminals which were 350–450Å in diameter. The gap between the nerve terminal and the muscle fiber in each case was 150–200 Å. Motor nerve endings in the suckers of the cephalopod *Octopus vulgaris* and in the lips of the cephalopod *Sepia officinalis* were described by Graziadei (1966). In both cases the presence of apposed membranes of the two elements (nerve and muscle) was clearly demonstrated. The basic features of a synaptic apparatus were present, i.e., increased membrane density, synaptic vesicles, and mitochondria in the presynaptic process. The overall thickness of the synaptic membrane complex was 250 Å with a gap of 100 Å between the pre- and postsynaptic membranes. Cloney and Florey (1968) have recently described neuromuscular junctions in the chromatophore muscles of the squid *L. opalescens*.

IV. Chemical Regulation of Molluscan Hearts

A. Anatomy and Innervation of Molluscan Hearts

The general structure of molluscan hearts and details of their innervation have been discussed recently by Hill and Welsh (1966). With the exception of cephalopods, which have closed circulatory systems, mollusks possess hearts which receive blood from and pump it to an open hemocoel. Usually, the molluscan heart is two-chambered, consisting of a receiving auricle and a pumping ventricle. However, other arrangements of auricles and ventricles occur. For example, the monoplacophoran *Neopilina galatheae* possesses two ventricles and two pairs of auricles, while most bivalves have two auricles and one ventricle. Accessory branchial hearts are possessed by cephalopods. Molluscan hearts generally are extrinsically innervated by the visceral nerve(s) which contains both inhibitory and excitatory fibers. Since Hill and Welsh (1966, p. 141) remarked that "the occurrence of nerve cells in molluscan hearts is still questionable," new data have been obtained on the intrinsic innervation of molluscan hearts.

Silver staining was employed by Phillis (1966) to reveal unipolar and bipolar nerve cells and a complex network of nerve fibers in the ventricular wall of the heart of the bivalve *Tapes watlingi.* It was thought possible that these cells may be involved in controlling cardiac functions. Using fluorescence techniques, S.-Rózsa and Zs.-Nagy (1967) demonstrated nerve cells in the auricle, auricle–ventricle junction, and ventricle of the heart of the snail *Limnaea stagnalis.* The neurons contained primary catecholamines which could be depleted by stimulation of the visceral nerve. The role of the nerve cells in the heart was considered to be synthesis and storage of substances involved in heart activity. These authors obtained no evidence that suggested to them that nerve cells of the heart constituted a neurosecretory system. On the other hand, Cottrell and Osborne (1969) found a dense network of swollen axons containing granular material in the atrioventricular junction region of the heart of the snail *Helix pomatia* and considered the network to be a neurosecretory system ideally situated for the release of an agent into the blood capable of causing a general stimulation of the animal such as might be necessary to arouse a snail from hibernation. The system appeared to them to be analogous to the nervous system present in the venae cavae of cephalopods which is assumed to have a neurosecretory function (Alexandrowicz, 1964, 1965; Martin, 1968).

B. Chemicals Isolated from Molluscan Hearts

1. *Acetylcholine*

ACh has been isolated from the hearts of gastropods, bivalves, and cephalopods by several authors (see Hill and Welsh, 1966). Phillis (1966) isolated an ACh-like substance from the heart of *Tapes watlingi* in amounts of the order of 0.19 μg. per gram wet weight and successfully antagonized its action on the heart with mytolon. S.-Rózsa and Zs.-Nagy

(1967) isolated a substance thought to be ACh from the heart of *Limnaea stagnalis.* Again, the activity of the substance on the heart was antagonized by mytolon. Carroll and Cobbin (1968) detected a compound indistinguishable from ACh in four different chromatographic solvent systems from aqueous extracts of the heart (and visceral ganglia) of *T. watlingi.*

2. *Cholinesterase*

From the branchial heart and systemic heart of the cephalopod *Octopus dofleini,* Loe and Florey (1966) measured 3.8 and 3.9 mg. ACh and chloride per 100 mg. tissue per hour, respectively as cholinesterase quotients. Cholinesterase activity in extracts of the heart of the snail *Helix aspersa* was measured by Korn (1969), who found a value of 317 μmoles ACh per gram per hour for the rate of enzymatic hydrolysis of ACh. He found further that the cholinesterase activity of auricle extracts was some fivefold richer per unit of weight than was the activity of ventricle extracts. The cholinesterase from the heart was more active on acetyl- than on butyryl- or propionylcholine.

On the other hand, Nistratova (1968) noted that many authors have found that eserine and other cholinesterase inhibitors had no effect or only weakly potentiated the effect of ACh on the molluscan heart and that cholinesterase was, in fact, present in low concentration or absent. He postulated that after ACh is liberated from the nerve endings in the heart of the freshwater bivalve *Anodonta* following visceral ganglion stimulation, ATP is liberated from either the nerve endings or the muscle cells. The presence of ACh inhibits ATPase activity leading to the accumulation of ATP in the perfusion fluid. The affinity of the receptor for ACh decreases in the presence of ATP, and finally ATP removes the ACh from its bond with the receptor by competition at high concentrations of ATP. This results in the cessation of ACh inhibition. A similar series of events was claimed for other mollusks, namely *Mytilus grayanus, Helix lucorum, Rapana bezoar, Neptunea eulimata, Ommatostrephus sloanei-pacificus,* and *Octopus* sp.

3. *5-Hydroxytryptamine*

Hill and Welsh (1966) listed values of 5-HT in the hearts of the bivalve *Mercenaria mercenaria* and the gastropods *Helix aspersa* and *Busycon canaliculatum.* Phillis (1966) isolated a 5-HT-like factor from extracts of the heart of *Tapes watlingi.* The excitatory effects of the extracts were largely abolished by methylsergide. S.-Rózsa and Perenyi (1966) separated 5-HT from perfusates of stimulated *Helix* hearts by chromatographic techniques. S.-Rózsa and Zs.-Nagy (1967) isolated a compound, thought to be 5-HT because it was antagonized by BOL, from the heart of *Limnaea stagnalis*. Chase *et al.* (1968) found 1.61 ± 0.31 μg. of 5-HT per gram of auricle of *Aplysia californica*. About half this amount of 5-HT was detected in the ventricle.

4. *Catecholamines*

Adrenaline and noradrenaline appear to be absent from molluscan hearts. Assays for dopamine have failed to reveal any appreciable dopamine content in the heart of *Mercenaria mercenaria* (Sweeney, 1963). Frontali *et al.* (1967) tested heart extracts of this species for the presence of dopamine by chromatographic and fluorescence comparison with pure dopamine. Again none was detected. Using fluorescence techniques, S.-Rózsa and Zs.-Nagy (1967) noted that greenish fluorescence was exhibited by nerve cells in sections of the heart of *Limnaea stagnalis*, suggesting that these cells contained some primary catecholamine.

5. *Histamine*

Woodruff *et al.* (1969) reported 0.25 μg./gm. ± 0.07 S.D. histamine by bioassay on guinea pig ileum and 0.30 μg./gm. ± 0.08 S.D. histamine by fluorimetric assay in extracts of *Helix aspersa* heart.

6. *Substance X*

Several authors have extracted cardioexcitatory material, distinguishable from 5-HT, from the nervous systems and hearts of many mollusks. Cottrell (1966) proposed that until this material has been chemically identified it should be known by the omnibus term substance X. This term was used by Hill and Welsh (1966) in their review of the subject. Substance X was reported in the brain of the amphineuran *Cryptochiton stelleri* (Greenberg, 1962), in the hearts and nervous systems of gastropods (Jullien *et al.*, 1956, 1961; Ripplinger, 1957; Gersch and Deuse, 1960; Meng, 1960; Kerkut and Laverack, 1960; Kerkut and Cottrell, 1963; Jaeger, 1966; Welsh and Frontali, 1966; Frontali *et al.*, 1967; S.-Rózsa and Zs.-Nagy, 1967), and in bivalve hearts and ganglia (Cottrell, 1966; Welsh and Frontali, 1966; Frontali *et al.*, 1967; Agarwal and Greenberg, 1969). There have been no reports of substance X in the Scaphopoda or Cephalopoda.

Attempts at elucidation of the chemical nature of substance X have produced conflicting results. Jaeger (1966) reported in the heart of the snail *Strophocheilus oblonga* a cardioexcitor substance which was nondialyzable, inactivated by pronase, and which did not migrate electrophoretically. He thought it was a large polypeptide. Cottrell (1966) used ultracentrifugation techniques in a study of the chemistry of particles obtained from macerated *Mercenaria mercenaria* ganglia. He reported that the greater part of substance X was particle bound. Later studies have indicated that substance X is a mixture of molluscan cardioexcitor compounds rather than a single compound. Frontali *et al.* (1967) thought that the substance X they extracted from the hearts of *M. mercenaria* and *Busycon canaliculatum* could be a mixture of peptides. That the principal cardioexcitatory compound of substance X is a peptide was challenged by Agarwal and Greenberg (1969), who found that extracts of the ganglia and hearts of the bivalves *Amblema neisleri* and *Elliptoides sloatianus* yielded as the most active cardioexcitor agent a compound of molecular weight <1500 with an absorption peak in the ultraviolet at 260 mμ. They postulated that their cardioexcitor agent could include a nucleotide and that it might be a long duration, long distance substitution for neurotransmitters involved in the long term maintenance of rhythmicity of active hearts.

C. Pharmacology of Molluscan Hearts

Many references to the pharmacology of molluscan hearts have been listed by Hill and Welsh (1966) and by Cottrell and Laverack (1968). It was shown by Greenberg and Windsor (1962) that many bivalve hearts are both inhibited and excited by ACh, depending on the concentration present. They found that 10^{-5} gm./ml. benzoquinonium chloride (mytolon) blocked both the depressor and excitor effects of ACh on isolated ventricles of thirty-nine species of twenty families of bivalves. However, the threshold and effectiveness of blockade were different for the two responses, and the relationship varied from species to species. Subsequently, Greenberg (1965) challenged the isolated ventricles of forty species of bivalve with a wide range of ACh doses. Two species groups emerged. Those species with low depressor threshold to ACh showed an equally low excitor threshold to ACh, and only amplitude of contraction was increased. Those species whose hearts showed a higher depressor threshold had a much higher excitor threshold, both tone and frequency being especially affected. The systematic distribution within the class Bivalvia of these types of response was discussed by Greenberg (1965).

Hill and Welsh (1966) listed the ACh antagonists which are effective in blocking the inhibitory action of ACh on some molluscan hearts. Carroll and Cobbin (1968) showed that the excitatory effect of ACh on the heart of the bivalve *Tapes watlingi* could be seen only when the inhibitory effect was blocked by benzoquinonium. Subsequently, they discovered (1969a) that the excitatory effect of ACh on this heart was a direct one, independent of any nerve supply and (1969b) that the most potent modifiers of the excitatory effect of ACh were structural analogs of tetraethylammonium.

Members of a series of choline esters had from 9.5 to 14,000 times less potency as inhibitors of *T. watlingi* heart than had ACh itself (Phillis, 1966). Thus, it is understandable that bivalve hearts have been used extensively for quantitative bioassay of ACh (Chong and Phillis, 1965; Loe and Florey, 1966; Florey, 1967). Amounts of ACh of the order of 1 ng. were estimated by Cottrell *et al.* (1968) using *Mya arenaria* hearts pretreated with ergometrine. Carroll *et al.* (1968) studied the heart of *T. watlingi in situ.* Addition of benzoquinonium (10^{-5} gm./ml.) resulted in marked cardiac excitation following visceral ganglion stimulation. The effect of ambenonium was essentially the same. However, methylsergide (10^{-5} gm./ml.) successfully abolished the cardiac excitation produced by visceral ganglion stimulation. These results on *in situ* preparations agree with those of Phillis (1966) on isolated heart–nerve preparations of *T. watlingi.*

Hill and Welsh (1966) have summarized the results of work carried out on the effects of ACh on gastropod hearts. In 1967 S.-Rózsa and Zs.-Nagy studied the effects of ACh on isolated hearts of the snail *Limnaea stagnalis.* The compound caused complete inhibition with a strong tonus decrease at 10^{-12} *M* and higher concentrations. Among the many compounds tested, only ACh produced negative inotropic and chronotropic effects. Hill and Thibault (1968) found that ACh would not restart isolated ventricles of the gastropod *Strombus gigas* made hypodynamic by long perfusion with seawater or by chilling, but it had the usual inhibitory effect on the beating of fresh ventricles.

Most molluscan hearts so far examined are excited by 5-HT, and indeed, bivalve hearts are used for the detection and estimation of 5-HT (Hill and Welsh, 1966). The excitatory effect is antagonized by bromolysergic acid diethylamide and methylsergide. In an extensive study of the effect of drugs on the heart of the snail *L. stagnalis,* S.-Rózsa and Zs.-Nagy (1967) showed that 10^{-10} *M* 5-HT produced positive inotropic and chronotropic effects. The effects disappeared after 10 minutes of pretreatment with BOL (10^{-4} *M*). Hill and Thibault (1968) found that although 5-HT would restart ventricles of the gastropod *S. gigas* made hypodynamic by long perfusion with sea water or by chilling, it did so only in concentrations which were high relative to threshold concentrations for positive inotropic effects on spontaneously beating ventricles. Chase *et al.* (1968) showed that the auricle of *Aplysia californica* had a marked capacity to accumulate tritiated 5-HT. This labeled 5-HT could be released by direct electrical stimulation or by electrical stimulation of the visceral nerves leading to the heart. The release of tritiated 5-HT upon nerve stimulation was not affected significantly by the addition of BOL (10^{-6} M) to the heart perfusate, but the drug blocked the excitatory effect of nerve stimulation upon the heart.

Substance X mimics the excitatory effect of 5-HT on molluscan hearts, but it is not antagonized by 5-HT blocking agents (Kerkut and Cottrell, 1963). Other unidentified substances which possess excitatory effects on molluscan hearts have been found by Kerkut and Laverack (1960) in extracts of several tissues of *Helix aspersa.* Frontali *et al.* (1967) discovered compounds with similar excitatory activity on molluscan hearts in extracts of hearts and ganglia of *Mercenaria mercenaria* and *Busycon canaliculatum.* S.-Rózsa and Zs.-Nagy (1967) found two unidentified heart excitatory substances in the hearts of *L. stagnalis.*

Some work has been done on the effects on the molluscan heart of substances such as GABA (γ-aminobutyric acid), dopamine, adrenaline, noradrenaline, and glutamine (S.-Rózsa and Zs.-Nagy, 1967; Cottrell and Laverack, 1968). GABA was without effect on the isolated heart of *L. stagnalis.* Glutamine, dopamine, adrenaline, and noradrenaline produced positive inotropic and chronotropic effects in this preparation. It is of interest that Cottrell (1967) used the heart of the cephalopod *Eledone cirrhosa* to assay noradrenaline at concentrations as low as 50 ng. Tetrodotoxin at a concentration of 10^{-5} gm./ml. caused no recognizable effect on the isolated heart of *Mytilus edulis* (Irisawa *et al.,* 1967).

V. General Discussion of Chemical Transmission in Mollusks

Neurons responsive to ACh, 5-HT, dopamine, noradrenaline, glutamate, GABA, and histamine have been detected in molluscan ganglia. Some giant neurons of gastropods are responsive to more than one of these agents. Obviously, a knowledge of the full range of sensitivity to these agents of each giant neuron in the gastropod brain is desirable and is now technically feasible in the brains of some species at least. Even so, the fact that a neuron responds to a pharmacological agent does not imply that the agent is necessarily a chemical transmitter. The question of what criteria a substance must possess before it can be regarded as a chemical transmitter has been raised by several authors (Paton, 1958; Curtis, 1961; McClennan, 1963; Florey, 1965; Gerschenfeld, 1966; Cottrell and Laverack, 1968). Cottrell and Laverack (1968) point out that there is no reason to exclude substances as potential transmitters because they do not comply with a great range of different criteria. In any case these criteria appear to have been derived from studies of a single transmitter substance (ACh) in vertebrates.

The available evidence indicates that 5-HT is probably a neurotransmitter. It is present in certain neurons and is particle bound. In the majority of cases it has an excitatory action on molluscan neurons. 5-HT receptors have been detected in CILDA neurons of the snail which are depolarized by 5-HT. The 5-HT receptors are located on the axon hillock and just beyond the hillock. There is, however, difficulty in accounting for the inactivation of 5-HT at synapses, and the possibility exists that 5-HT acts intracellularly (see Florey, 1965) rather than acting on the postsynaptic membrane. This possibility warrants further study. The possibility that the principal role of 5-HT in a mollusk is that of a hormonal substance present in many tissues might also be given attention. Release of 5-HT from these tissues could result in activation of 5-HT-sensitive neurons (e.g., CILDA cells), resulting in marked changes in the general physiology and behavior of the mollusk.

Many molluscan neurons respond to ACh. It has been shown that ACh released at different synaptic endings of the same neuron has an excitatory action on some neurons and an inhibitory action on others. Also, two different types of cholinergic receptors (nicotinic and muscarinic) have been found in some neurons. Difficulties have arisen concerning the inactivation of ACh at synapses and it has been postulated that removal of ACh by diffusion occurs in both gastropods and bivalves. In addition, it has been postulated that in gastropods the presence of ACh in the synaptic cleft triggers the release of an antagonist which competes for receptor sites. This possibility should be fully explored and studies might also be made to determine whether antagonistic substances are usually released from presynaptic cells along with transmitter substances or released from postsynaptic cells (or adjacent cells) under the influence of the transmitter. Such studies might be put into a wider frame of reference by determining whether there is an active chemical regulation of the sensitivities of receptor sites to various transmitter substances.

One of the interesting discoveries stemming from the use of giant gastropod neurons was the finding that dopamine is present in some neurons. Neurons can be excited or inhibited by dopamine, but usually the response is inhibitory. Dopamine appears to be the transmitter for ILD. It has been found that the dopamine receptor on neurons resembles the α- rather than the β-adrenaline receptor of vertebrates, and some of the requirements for a compound to react with a dopamine receptor have been established. It has been shown that cells possess separate receptors for noradrenaline and dopamine.

Some neurons have been found to respond to glutamate, some to GABA, and some to histamine, but further work is required before transmitter roles for these substances can be established. A search for other possible transmitter substances is obviously warranted. Further work is required to localize transmitters in specific neurons. Of particular interest in this respect was the discovery that two transmitters (5-HT and dopamine) can occur in the one neuron. The full significance of this finding requires elucidation and work is needed to establish whether each neuron normally carries more than one transmitter.

Studies of chemical transmission at the neuromuscular junctions of mollusks have been hampered by lack of knowledge regarding the details of innervation of the molluscan nerve–muscle preparations commonly employed. However, it would seem likely that ACh is an excitatory transmitter at some junctions at least. There is also a possibility that glutamate is an excitatory transmitter at some junctions. Cytochemical studies of the junctional vesicles recently demonstrated in molluscan neuromuscular junctions can be expected to throw light on the nature of chemical transmitters at these junctions. The factor responsible for releasing catch in muscles which display this phenomenon appears to be 5-HT. It is likely that 5-HT is released from inhibitory nerves, but the exact manner in which 5-HT exerts its effect on muscle cells requires clarification.

Although considerable pharmacological work has now been carried out on molluscan hearts and their innervation, knowledge of the chemical regulation of molluscan hearts is still rudimentary. There is good evidence that ACh and 5-HT are involved in the regulation of molluscan hearts, ACh having an inhibitory role and 5-HT an excitatory effect. It is generally believed that inhibitory extracardial nerves are cholinergic and excitatory ones serotoninergic. However, it is not known whether the extracardial nerve fibers make direct connection with cardiac muscle fibers. Indeed, recent work has confirmed earlier reports that neurons are present in the hearts of mollusks, and the possibility arises that these neurons are interposed between the extracardial nerve fibers and the cardiac muscle fibers. Further work is required to clarify this aspect.

The possibility that transmitters other than ACh and 5-HT are present in the neurons found in the molluscan heart and that they are involved in cardiac regulation warrants attention. S.-Rózsa and Zs.-Nagy (1967) have, in fact, produced evidence that neurons in the heart of *Limnaea stagnalis* contain primary catecholamines which could act as transmitters. These workers also noted that 5-HT is not localized in the cardiac neurons but occurs in the cardiac muscle cells. They postulated that excitation of the heart involves a sequence of processes. First, there is liberation of a transmitter from the presynpatic terminals of the extracardial nerve. This transmitter then induces the liberation of catecholamine stored in the cardiac neurons. The liberated catecholamine, in turn, causes the release of 5-HT localized in cardiac muscle cells. This 5-HT affects the metabolism of the heart muscles bringing about their excitation.

Subsequently S.-Rózsa (1968) postulated that the excitatory effects on *Helix* heart of a number of bioactive amines (5-HT, tryptamine, 5-methoxytryptamine, adrenaline, noradrenaline, dopamine, glutamine, histamine) are mediated through the activation of adenyl cyclase resulting in the accumulation of cyclic 3′,5′-AMP. Cyclic 3′,5′-AMP then produces an excitatory effect at the level of muscle cells. Further work is necessary to establish the validity of the scheme involving a system of several different messenger molecules for excitation of the snail heart proposed by the Hungarian workers, and to determine whether this or allied schemes operate in the hearts of other molluscan species. It should be noted that a number of cardioexcitatory substances have been isolated from molluscan tissues and one of these appears to be a nucleotide.

Further work is also required to ascertain whether ATP is normally released from nerve endings or heart muscle cells of bivalves generally following visceral ganglion stimulation (see Nistratova, 1968). Then, too, the exact mechanism whereby ATP acts as a competitive antagonist of ACh, if indeed it does, requires elucidation.

Molluscan hearts have proved valuable for assay of ACh, 5-HT, and noradrenaline. Part of their value as pharmacological tools lies in the ease with which they can be manipulated. Also, the use of isolated mollusk hearts does not demand the complexity of apparatus required to maintain optimal conditions for isolated mammalian hearts. For these and other reasons it is to be expected that molluscan hearts will be used increasingly in pharmacological studies and that our knowledge of the pharmacology of the hearts of other animal taxa will be augmented as a result of these studies. At the same time, it should be noted that isolated molluscan hearts might react differently from hearts *in situ*. Trueman (1967) has shown that heart activity in intertidal bivalves is affected by environmental factors such as the state of the tide, and Silvey (1968) has shown that stimulation of ganglia remote from the heart can result in a modification of the activity of the heart. Then, too, hearts from different species of mollusk often show marked differences in their behavior to drugs.

VI. Toxic Substances Elaborated by Mollusks

Numerous compounds which are toxic to other animals are known to be elaborated by many species of mollusk. Most of these biotoxins are produced by representatives of three classes of mollusks—Gastropoda, Bivalvia, and Cephalopoda. In the following account, attention is focused on the pharmacology of some of the biotoxins elaborated by mollusks and the uses to which these biotoxins are put by the animals possessing them. For a fuller account of aspects of the medical importance of many of these toxins, an account of their chemistry, and an introduction to the already extensive literature relating to these toxins, the reader is referred to the monographic work of Halstead (1965). Only toxins known or believed to be synthesized by mollusks are dealt with in the present account. Some toxins possessed by mollusks (e.g., saxitoxin) are derived from their food.

A distinction can be made between venomous and poisonous mollusks. Venomous mollusks elaborate toxic material (venom) in association with an apparatus, the venom apparatus, capable of actively injecting the material. Poisonous mollusks are those whose tissues generally, or tissues from specific regions of their bodies, contain material (poison), which when eaten by other animals, elicits toxic effects in these animals.

A. Gastropoda

1. *Buccinidae*

In 1952, Asano reported that consumption of the whelk *Neptunea arthritica* in Hokkaido, Japan, had resulted in many poisonings. He was able to locate the poison in the salivary glands of the gastropod. Severe headache, dizziness, vomiting, urticaria, visual disturbances and motor paralysis were among the symptoms displayed by the victims. Halstead (1965) referred to an unpublished report of Kanna and Hirai that similar poisonings had stemmed from ingestion of *Neptunea intersculpta* and that the toxic material was located in the salivary glands of this species.

Asano and Itoh (1959, 1960) showed that extracts of the salivary glands of *N. arthritica* contained material which produced salivation, lachrymation, miosis, motor paralysis, respiratory failure, and death in mice. These authors were able to show that the toxic principle in the extracts was tetramine, although histamine, choline, and choline ester were also present in the extracts, and these substances probably acted as synergists. The tetramine content of the salivary glands of *N. arthritica* ranged from 4.0 to 7.5 mg. per gram of gland and that of *N. intersculpta* from 5.5 to 9.0 mg. per gram of gland. Fänge (1960) found that approximately 1% of the salivary glands of the European whelk *Neptunea antiqua* consisted of tetramine and that this substance was responsible for the major part of the pharmacological activity exhibited by extracts of the glands.

The pharmacological activity of tetramine in mammals consists of curare-like effects and stimulation of the parasympathetic system. Such activity would account for the symptoms displayed by human victims of *N. arthritica* poisoning or the signs displayed by mice injected with extracts of the salivary glands of *N. arthritica*. Asano and Itoh (1960) have shown that these extracts are toxic to fish (*Cyprinus carpio*), and it is possible that the saliva of the various species of *Neptunea* is toxic to those marine organisms upon which these species feed. However, it is not known whether species of *Neptunea* use their saliva in offense or defense. The exact role of tetramine in their salivary glands awaits elucidation.

Extracts of the salivary glands of *Buccinum leucostoma* were shown by Asano and Itoh (1960) to be toxic to mice, but the active constituents in the extracts were not determined. Whittaker (1960) found that a substance with pharmacological properties resembling those of acetylcholine could be isolated from the hypobranchial gland of *Buccinum undatum*. The substance was identified as acrylylcholine, which has the following formula:

$$CH_2 = CH—CO_2—CH_2—CH_2—\overset{+}{N}—(CH_3)_3$$

Acrylylcholine causes contraction of smooth musculature and hypotension in mammals. It possesses a weak neuromuscular blocking action. While it is possible that the salivary toxin of *B. leucostoma* may be used in offense or defense, it is difficult to envisage the choline ester in the hypobranchial gland of *B. undatum* being used for such purposes.

2. *Conidae*

Members of the family Conidae possess a well developed venom apparatus which is used in the capture of prey. The structure of the venom apparatus has been studied by several authors (Bouvier, 1887; Bergh, 1895; Shaw, 1914; Alpers, 1931; Hermitte, 1946; Hinegardner, 1958; Kohn *et al.*, 1960; Martoja, 1960; Endean and Duchemin, 1967). Essentially, the apparatus consists of a hollow radular tooth held at the tip of a mobile and protrusible proboscis which encloses a tubular prepharynx which merges with the pharynx, a radular sac, a venom duct, and a venom bulb. The arrangement of these structures is depicted in Fig. 1. It has been shown by Endean and Duchemin (1967) that the epithelium lining the duct and bulb of *Conus magus* has a syncytial structure which first appears anteriorly in the duct near its junction with the pharynx. Near this point, small spherical bodies are elaborated in vacuoles in the syncytial epithelium. A gradual enlargement of the vacuoles

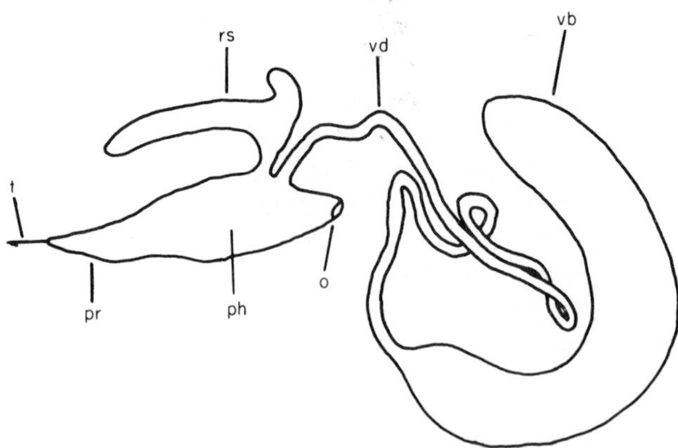

Fig. 1. Diagram of venom apparatus of *Conus striatus* showing a radular tooth (t) held at the tip of the proboscis (pr). A radular sac (rs) in which a number of radular teeth are stored opens into the pharynx (ph), which merges with the esophagus (o) which is cut. A venom duct (vd) opens anteriorly into the pharynx and posteriorly into the muscular venom bulb (vb).

and their contained spherical bodies occurs causing the duct epithelium to grow posteriorly toward the terminal venom bulb. Ellipsoidal venom bodies which possess obvious sheaths and which are derived from the spherical bodies occur in posterior regions of the duct (Fig. 3). These ellipsoidal bodies are the end products of biosynthetic activity in the duct epithelium. They are prominent in fluid found in the lumen of the most posterior regions of the venom duct and in the lumen of the venom bulb. These bodies and the fluid in which they occur constitute the venom of the cone.

Contraction of the thick musculature of the venom bulb results in venom being propelled along the lumen of the venom duct to the pharynx. From there it passes to an opening in the base of a hollow radular tooth held in the prepharynx when the cone is about to feed. Another opening is present near the barbed tip of the harpoonlike radular tooth, which is thrust into prospective prey and acts as a hypodermic needle for the passage of venom into the tissues of the prey. Additional teeth are formed and stored in the radular sac prior to use.

Kohn (1959) has shown that the various species of Conidae appear to be vermivorous, molluscivorous, or piscivorous. Occasionally, the venoms of some species must be used defensively, since there are records of numerous cases of human injury and of several fatalities stemming from envenomations involving these species (Clench and Kondo, 1943; Kohn, 1958, 1963; Rice and Halstead, 1968). However, bioassays of the venoms of thirty-seven species of Conidae have revealed that only piscivorous species possess venoms which are lethal to mice, and it was postulated that only piscivorous species of Conidae pose a real threat to man (Endean and Rudkin, 1963, 1965). Although the venoms of the molluscivorous species studied rapidly paralyzed the gastropods used for bioassay purposes, none of these venoms elicited any obvious activity in vertebrates. In particular, the venom of the molluscivorous *Conus textile* which has been held responsible for several human fatalities elicited no toxic signs in vertebrates. It would appear that *C. textile* has been confused with the piscivorous species *Conus geographus* in at least two reports of human fatalities stemming from cone envenomations. There are, in fact, no well authenticated records of serious human injury stemming from envenomations by *C. textile.* It is possible that the venoms of molluscivorous Conidae have a specific action on the neuromuscular system of mollusks.

Although the venoms of four vermivorous species studied (*Conus virgo, Conus tessulatus, Conus eburneus,* and *Conus rattus*) were toxic to fish, they were much less toxic to fish than the venoms of piscivorous species. The venoms of *Conus pulicarius, C. tessulatus, C. eburneus, Conus lividus,* and *Conus quercinus* produced tissue necrosis in mice and might be capable of causing local injury to humans. The venoms of the majority of vermivorous species studied effectively immobilized the polychete (*Phyllodoce malmgremi*) used for bioassay purposes. However, the polychete was rarely immobilized quickly by the venoms. Indeed, the venoms of some species of vermivorous Conidae had no obvious effects on the polychete. A study of the effects of such venoms on the normal prey of these species of Conidae might prove illuminating.

It was shown that the structure of the radular teeth of Conidae (Fig. 2) provides a means of distinguishing among piscivorous, vermivorous, and molluscivorous species (Endean and Rudkin, 1965). These bioassay studies also revealed that the potency of venom from posterior regions of the venom ducts of piscivorous and molluscivorous Conidae was usually much greater than venom from anterior regions of the ducts. However, the reverse was true for some species of vermivorous Conidae. Subsequently, Endean and Duchemin (1967) postulated that venom containing mainly mature venom granules from posterior regions of the venom duct of *C. magus* is injected into prospective prey. Possibly a similar situation prevails in the case of piscivorous and molluscivorous species of Conidae generally.

Kohn *et al.* (1960) found that aqueous or saline extracts of the venom of *Conus striatus* were lethal to mice and fish. The minimum lethal dose for mice by the intraperitoneal and intravenous routes ranged from

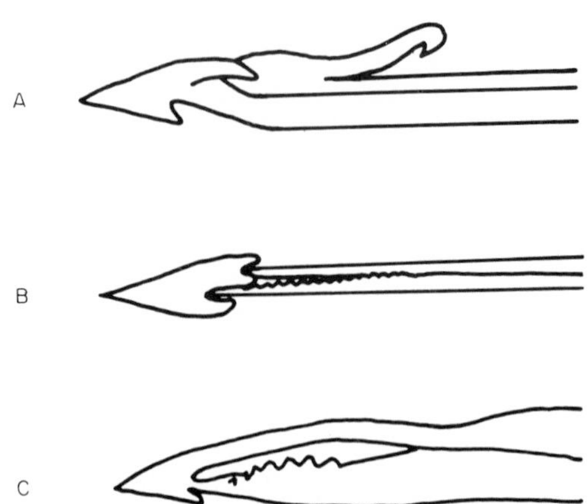

FIG. 2. (A) Anterior end of tooth of the piscivorous species *Conus magus* (×40). (B) Anterior end of tooth of the molluscivorous species *Conus ammiralis* (×35). (C) Anterior end of tooth of the vermivorous species *Conus tiaratus* (×130).

0.1 to 0.3 mg. of venom (freeze dried) per mouse. Mice injected with lethal doses generally showed ataxia, tonic spasms, hyperexcitability, partial paralysis, coma, and respiratory failure before death. Death was attributed to respiratory failure followed by cardiac arrest. It was suggested by Kohn *et al.* (1960) that the main action of the venom was interference with neuromuscular transmission, although they did not exclude the possibility that the venom had an action on the central nervous system.

In general, the results of toxicity studies on the venom of *C. striatus* made by Endean and Rudkin (1965) confirmed the results obtained by Kohn *et al.* However, Endean and Rudkin noted that toxic manifestations evinced in mice by the venom of *C. striatus* could be placed in two categories. Some specimens possessed venoms which caused a flaccid paralysis of the skeletal musculature of mice. The minimum lethal dose ranged from 11 to 57 mg. (wet weight) of venom per kilogram of mouse. Other specimens possessed venoms with a lower toxicity for mice but which caused hypothermia, convulsive spasms, spastic paralysis, and death when injected into mice in adequate amounts. Respiratory failure appeared to be the prime cause of death.

It was noted by Whyte and Endean (1962) and Endean and Rudkin (1963) that respiratory failure occurred in mice injected with large amounts of venom from *C. geographus* (Fig. 3) before cardiac failure occurred. Endean and Rudkin (1963) found that if lower but still lethal doses of venom were injected, mice became almost completely paralyzed

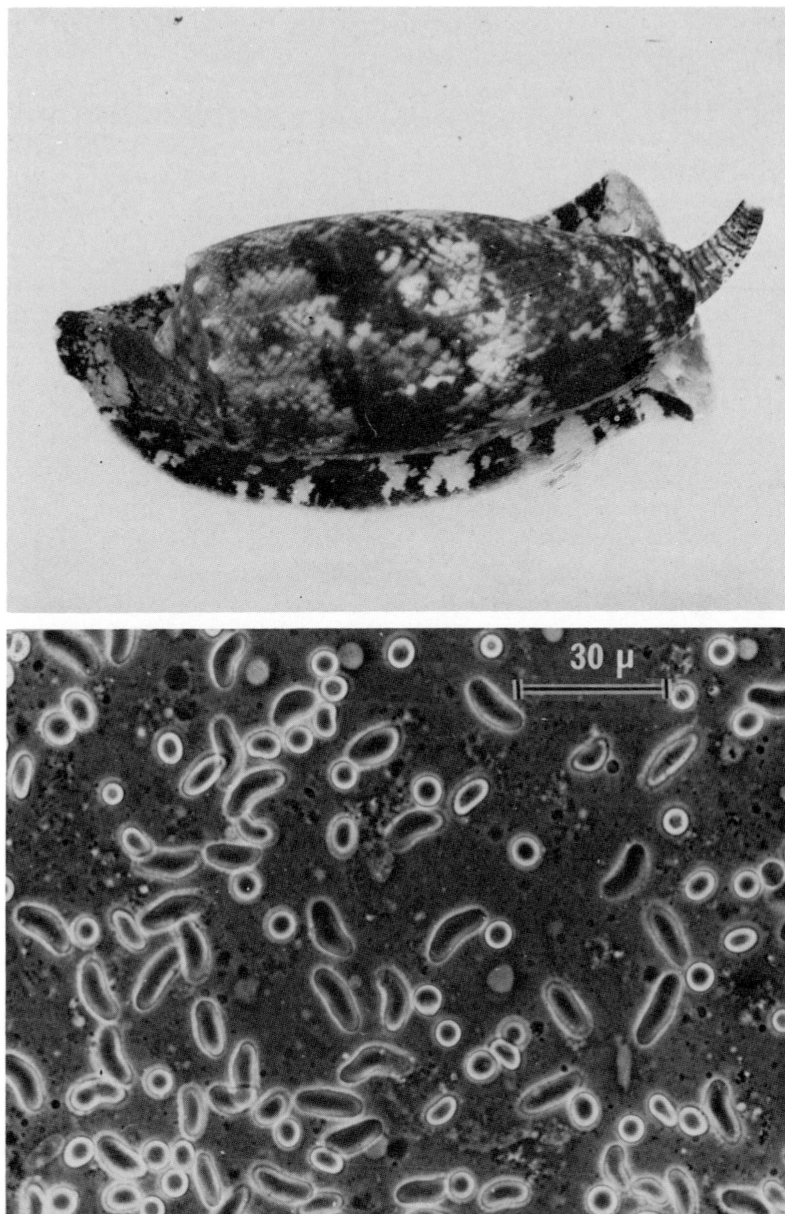

FIG. 3. (*Top*) The geographer cone (*Conus geographus*) moving on its extended foot. This species has been responsible for several human fatalities. (*Bottom*) Photomicrograph showing granules present in the venom of *Conus magus*.

before respiration failed. The paralysis was flaccid. They also found that *C. geographus* venom was toxic to representatives of all vertebrate classes and that the minimum lethal dose for mice ranged from 1.6 to 24.1 μg. (wet weight) of venom per mouse (mean weight of mice approximately 29 gm.).

Although the venom of *Conus catus* produced effects in mice similar to those produced by *C. geographus* venom, the venom of *Conus tulipa* produced comas in mice. A spastic paralysis of the skeletal musculature was produced in mice by venom from posterior regions of the venom ducts of specimens of *C. magus* (Endean and Rudkin, 1965), while venom from posterior regions of the venom ducts of *Conus stercusmuscarium* produced a flaccid paralysis of the skeletal musculature of mice.

It is noteworthy that deaths of mice injected with the venoms of *C. striatus, C. geographus, C. tulipa, C. magus, C. catus,* and *C. stercusmuscarium* appeared to be due to respiratory failure. No evidence was obtained from the bioassay studies which suggested that the venoms had direct actions on the mammalian cardiovascular system. However, extracts lethal to mammals were prepared from the venom ducts of *Conus californicus* by Whysner and Saunders (1963), and when small amounts of the extracts were administered to mammals, a fall in systemic arterial pressure, bradycardia, and an increase in respirations were observed. Marked changes in the cardiovascular system including a severe decline in blood pressure followed by respiratory arrest were caused by larger amounts of the extracts. It is possible that the chemistry and pharmacology of the venom of *C. californicus,* which preys on mollusks and polychetes (Saunders and Wolfsen, 1961), differ from those of Conidae that are exclusively piscivorous.

Venom from the posterior half of the venom duct of *C. geographus* will paralyze isolated mammalian skeletal muscle directly without prior neuromuscular blockade (Whyte and Endean, 1962). However, it was noted that the venom enhanced the effect of liminal doses of tubocurarine on the neuromuscular junction, and tubocurarine enhanced the muscle paralysis produced by small doses of venom. The action of the venom was not opposed by eserine. Whyte (1962) noted that smooth and cardiac musculature were not affected by the venom of *C. geographus.* Venom extracted from posterior regions of the venom duct of *C. magus* produced a contracture of isolated skeletal and smooth musculature of the rat in the absence of electrical stimulation (Endean and Izatt, 1965). The effect was sustained and the musculature became paralyzed in the contracted state. When electrically stimulated or strongly curarized rat phrenic nerve–diaphragm preparations were used, the contracture elicited by a certain concentration of venom was of similar magnitude and time course to that produced in unstimulated or noncurarized preparations. For a period, the length of which was inversely related to venom concentration, the diaphragm responded with a twitch to electrical stimulation. Then the heights of successive muscle twitches declined progressively and the musculature became refractory to electrical stimulation. Tubocurarine or neostigmine did not affect the sustained contracture of musculature elicited by the venom. Venom from the posterior duct of *C. magus* increased the strength of contraction of the ventricle of the toad, *Bufo marinus*, but decreased its rate of contraction. The effects of the venom on skeletal, smooth, and cardiac muscle could be reversed by washing out the venom.

Venom from the posterior half of the venom duct of *C. striatus* showed a direct action on diaphragm musculature (Endean *et al.*, 1967), causing a progressive decline in the heights of successive muscle twitches made in response to direct electrical stimulation until the musculature became refractory to stimulation. In this respect, the action of the venom resembled that of *C. geographus*. However, as well as eliciting the above effect, venoms from some specimens of *C. striatus* also elicited a sustained contracture of the musculature. In this respect, they resembled the venom of *C. magus*. It is possible that venom from these specimens of *C. striatus* and venom from *C. magus* contain two fractions active against the musculature of the diaphragm, one fraction eliciting a sustained contracture of the musculature and the other fraction abolishing the excitability of muscle fiber membranes.

It is of interest that the venoms of *C. geographus*, *C. striatus*, and *C. magus* all have a direct action on muscle. It remains to be seen whether they also affect nerve conduction and transmission at the neuromuscular junction. The results of bioassay work suggest that the venom of *C. tulipa* and that of some specimens of *C. striatus* affect the central nervous system of mammals.

Kohn *et al.* (1960) found protein, carbohydrate, *N*-methylpyridinium, homarine, γ-butyrobetaine, and unidentified indole derivatives in the venom of *C. striatus*. The quaternary ammonium compounds listed above were also found in *C. textile*. Since the venom of the molluscivorous *C. textile* elicits no toxic signs in mice or fish (Endean and Rudkin, 1963), it would seem unlikely that these quaternary ammonium compounds form the principal toxic components of the venom of *Conus*. It was found by Kohn *et al.* (1960) that considerable toxicity remained after the venom of *C. striatus* was heated for 10 minutes at 90–100°C or after it was incubated with trypsin. These findings militate against protein being the principal toxic component of the venom. On the other hand, the results obtained by Whysner and Saunders (1966) for *C. cali-*

fornicus venom indicate that the lethal compound is either protein or bound to protein.

Endean and Duchemin (1967) using histochemical techniques found that the cores of ellipsoidal bodies occurring in abundance in the venom of *C. magus* contained protein, carbohydrate, 3-indolyl derivatives and protein-bound amino groups. Protein and 3-indolyl derivatives were detected histochemically in the cores of similar bodies which occur in the venom of *C. striatus* (Endean *et al.*, 1967).

3. *Cymatiidae*

Asano and Itoh (1960) found that extracts of the salivary glands of *Argobuccinum oregonense* were toxic to mice. They detected tetramine in the glands at a concentration of 3.0–4.0 mg. per gram of gland. The secretion of the proboscis gland of *A. argus* is toxic to echinoids, mollusks, and polychetes (Day, 1969).

Laxton (1968) found that the salivary glands of *Charonia rubicunda* are transversely divided. Aqueous extracts of the anterior and posterior regions of the glands were tested on the starfish *Patiriella regularis.* Extracts from anterior regions caused the starfish to extend its tube feet and evert its stomach. Extracts from posterior regions of the glands caused instant paralysis of the starfish. The paralysis lasted about 4 hours, after which time the starfish began to recover. When the starfish *Acanthaster planci* is seized by the giant triton *Charonia tritonis* (Fig. 4), the starfish is quickly inactivated, and it is suspected that the giant triton injects a salivary toxin (Endean, 1969).

4. *Muricidae*

A colorless or yellowish fluid is produced by the hypobranchial glands of members of the family Muricidae. In the presence of light and air, a chromogen in the fluid is oxidized to a purple dye well known to the ancients as Tyrian purple. Friedländer (1909) showed that the dye produced by *Murex brandaris* is 6,6′-dibromoindigo. However, the hypobranchial glands of Muricidae also produce toxic material. Dubois (1903, 1907, 1909) noted that alcoholic extracts of the hypobranchial glands of *M. brandaris* and *M. trunculus* produced inactivity, muscular paralysis, convulsions and, death in frogs and fish. Vincent and Jullien (1938a,b) showed that extracts from the hypobranchial glands of *M. trunculus* contained a substance which produced contractions in the eserinized dorsal muscle of the leech. They considered the substance was an ester or mixture of esters of choline.

That the substance was indeed a choline derivative was shown by Erspamer and his colleagues who isolated murexine (urocanylcholine)

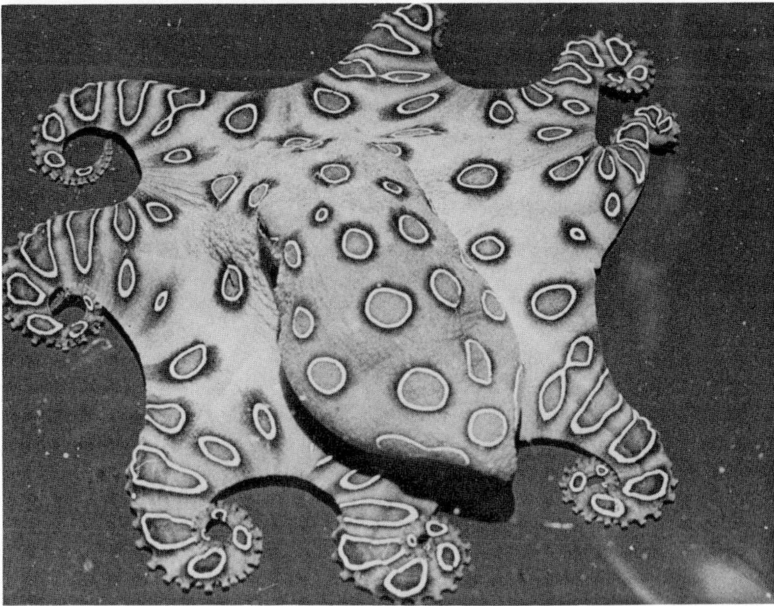

FIG. 4. (*Top*) The giant triton (*Charonia tritonis*) eating a crown-of-thorns starfish (*Acanthaster planci*) which it has inactivated with its salivary toxin. (*Bottom*) The venomous blue-ringed octopus (*Hapalochlaena lunulata*) found in tropical Australian waters.

with the formula below from the hypobranchial glands of *Murex trun-*

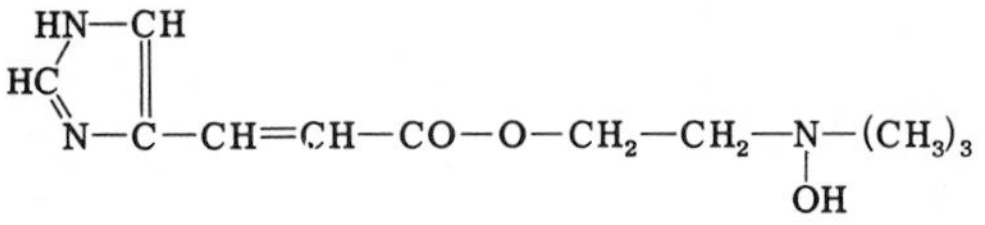

$$\begin{array}{l} \mathrm{HN{-}CH} \\ \mathrm{HC} \quad \| \\ \quad \mathrm{N{-}C{-}CH{=}CH{-}CO{-}O{-}CH_2{-}CH_2{-}\underset{\displaystyle OH}{\underset{|}{N}}{-}(CH_3)_3} \end{array}$$

culus, M. erinaceus, and *M. brandaris* (Erspamer and Dordoni, 1947; Erspamer, 1948a; Erspamer and Benati, 1953). Essentially murexine exhibits nicotinic and curariform activity. It is capable of paralyzing the skeletal musculature of both vertebrates and invertebrates. Its neuromuscular blocking action is apparently due to its ability to depolarize the motor end plate region (Erspamer and Glässer, 1967; Quilliam, 1957; Keyl and Whittaker, 1958).

Other compounds with pharmacological activity are present in the hypobranchial glands of Muricidae. Erspamer and Ghiretti (1951) and Erspamer (1956) have noted the presence of 5-hydroxytryptamine. Mansour-Bek (1934) reported the presence of proteolytic enzymes in the saliva of *Murex anguliferus.*

5. *Thaisidae*

Dubois (1907) showed that toxic material which he termed purpurine was produced in the hypobranchial gland of *Thais lapillus.* Roaf and Nierenstein (1907a,b) showed that aqueous extracts of the hypobranchial gland of this species produced vasoconstriction and hypotension in rabbits. They believed that the active material in the extracts was similar to adrenaline. Whittaker (1960) found that urocanylcholine was present in extracts of the hypobranchial glands of *T. lapillus,* but another choline ester senecioylcholine with the formula below was present in

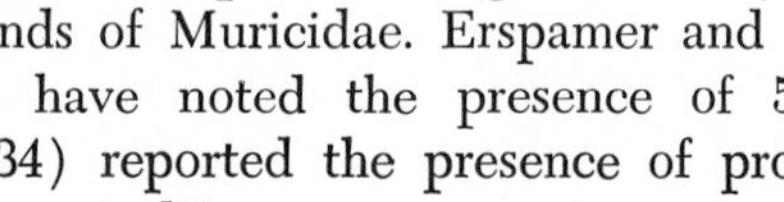

$$\begin{array}{l} \mathrm{H_3C} \\ \quad \mathrm{C{=}CH{-}CO_2{-}CH_2{-}CH_2{-}\overset{+}{N}{-}(CH_3)_3} \\ \mathrm{H_3C} \end{array}$$

extracts of the hypobranchial glands of *Thais floridana.* Senecioylcholine resembles urocanylcholine but is not as potent as a neuromuscular blocking agent.

Precursors of Tyrian purple have been found in extracts of the hypobranchial gland of *Purpura aperta, T. lapillus* (Friedländer, 1922), and *Dicathais orbita* (Baker and Sutherland, 1968). Beers (1967) found homarine and glycine betaine in the hypobranchial gland of *Thais haemastoma.*

6. *Cassidae*

Some of the large gastropods belonging to the genus *Cassis* are known to feed on echinoids (Moore, 1956). *Cassis tuberosa* feeds on the black sea urchin *Diadema antillarum* in Caribbean waters (Schroeder, 1962), while *Cassis cornuta* feeds on *Diadema setosum* in waters of the Great Barrier Reef (Endean, 1969). Cornman (1963) has shown that toxic material in the saliva of *C. tuberosa* incapacitates the surface organs of the black sea urchin, thereby decreasing its mobility and the effectiveness of its spines. The active principle in the saliva would appear to be a neurotoxin, since it first inactivates the sensory receptors or their afferent nerves. The chemistry of the toxic material is not known. The material is produced in large paired salivary glands.

7. *Aplysiidae*

Crude aqueous and acetone extracts of the digestive glands of the sea hares *Aplysia californica* and *Aplysia vaccaria* were found to be toxic to mice (Winkler, 1961). A purified acetone extract, termed aplysin, produced a junctional block, antagonized by neostigmine in the rat phrenic nerve–diaphragm preparation (Winkler *et al.*, 1962). Aplysin also produced a spasm in intestinal smooth musculature of the rabbit and lowered blood pressure in the dog. The active constituent in aplysin awaits identification.

B. Bivalvia

Although marine invertebrates occasionally possess benign growths, the freedom of mollusks in particular from spontaneous cancers led to investigations of such animals as potential sources of anticancer drugs during the early 1960's. A growth inhibiting substance was found to be highly concentrated in bivalves of the genus *Mercenaria* and was called mercenene (Schmeer, 1963, 1964a,b, 1965, 1966, 1967a; Schmeer and Beery, 1965; Schmeer and Cassidy, 1966; Schmeer and Huala, 1965; Hegyeli, 1964; Li *et al.*, 1965). Systems against which the inhibitory substance is active include HeLa cells, Krebs 2 carcinoma, and sarcoma 180 in mice (Schmeer, 1963, 1964a, 1964b; Schmeer and Beery, 1965), murine leukemia induced by Moloney virus (Judge, 1966), and adenovirus 12 tumors in hamsters (Li *et al.*, 1968). Mercenene has no inhibitory activity or toxic effects on normal human amnion cells (Schmeer and Beery, 1965), and although it did not extend the longevity of mice infected with Friend leukemia virus, it inhibited the splenomegalic response of such animals (Judge, 1966).

Studies on the chemical nature of mercenene have shown that the compound is slowly dialyzable, heat stable (Schmeer *et al.*, 1966), and possesses a molecular weight <1000 (Schmeer, 1966). It has been suggested that at least part of the active principle of mercenene extracts is a substance of molecular weight 1000–2000 which may have a glycopeptide type of structure (Schmeer *et al.*, 1966). The proposal that mercenene is a volatile derivative of methylglyoxal (Szent-Györgyi, 1965; Együd, 1965) is considered improbable (Schmeer, 1966).

Attempts to identify mercenene's site of action resulting in its antineoplastic effectiveness have so far failed (Schmeer, 1967b, 1968, 1969). However, it has been established (Schmeer, 1968) that mercenene prevents tumor cells from entering the DNA synthetic period of the mitotic cell cycle. It remains to be shown whether mercenene possesses activity against spontaneous cancers comparable with its effects on cancers transplanted into mice. Nonetheless, since mercenene kills only neoplastic cells both *in vitro* and *in vivo,* it is of potentially great significance as an anticancer drug and as a tool for cytological research.

The adaptive significance of the presence of mercenene in the tissues of marine organisms is worthy of investigation. Is it a metabolic accident, or has the compound been evolved as a defense against neoplasia? Of what biological significance is the seasonal variation in mercenene concentration in clam tissues? As Schmeer and Beery (1965) have pointed out, mercenene occurs in more than one species, and its presence is therefore unlikely to be the result of a selective feeding device, concentrating mercenene produced by other organisms.

C. Cephalopoda

Crustaceans, mollusks, and fish form the prey of cephalopods. The presence of potent toxic material in the saliva of octopuses which prey on crabs has been reported by several workers and the earlier literature on the subject has been reviewed by Ghiretti (1960) and Halstead (1965). Violent excitation of the extremities followed by a quiet phase, paralysis, and death occur in crabs injected with saliva from at least some species of octopus. Fujita (1916) observed that specimens of *Octopus vulgaris* bored holes in the shells of pearl oysters and suggested that venom which weakened the adductor muscles of the bivalve was injected through the holes. Pilson and Taylor (1961) found that *Octopus bimaculoides* and *Octopus bimaculatus* drilled holes through the shells of several species of mollusk and appeared to inject a paralyzing venom via the holes. Recently, the behavior of *O. vulgaris* when feeding upon shelled mollusks has been described by Arnold and Arnold (1969) and by Wodinsky (1969). They found that the octopus does indeed pass a toxic secretion through a hole drilled in the shell of a victim, but that the secretion is not injected into any specific part of the occupant's body. Although the secretion does not usually kill the occupant the latter is weakened or immobilized.

The evidence for the use by octopods of salivary venom to inactivate fish is inconclusive. Baglioni (1909) reported that small fish—species of *Uranoscopus, Scorpaena,* and *Conger*—bitten by *O. vulgaris* showed signs of respiratory distress, loss of position reflex, jerked violently, and with the exception of specimens of *Conger,* died within 3 hours. On the other hand, Taylor and Chat Chen (1969) found no evidence that venom was used to paralyze *Scorpaena guttata* captured and eaten by *O. bimaculatus.*

Interest in the toxicity of octopus saliva has not been confined to predator–prey relationships, for a few cases of human envenomations have occurred. The symptoms displayed included initial pain, swelling of the region in the vicinity of the bite, numbness, giddiness, nausea, vomiting, ataxia, paralysis of differing degrees of severity, and in two instances, death. The two fatalities occurred in Australian waters and were attributed respectively to envenomations from *Hapalochlaena lunulata* (Fig. 4) (Flecker and Cotton, 1955; Cleland and Southcott, 1965) and *Hapalochlaena maculosa* (Lane and Sutherland, 1967; Sutherland and Lane, 1969). Pain or swelling of the region near the bite do not appear to be prominent among the symptoms displayed by victims of envenomations from *Hapalochlaena* species, but flaccid paralysis, which may be partial or complete, of skeletal muscles has been displayed by all victims.

Although extracts from the posterior salivary glands of the cuttlefish *Sepia officinalis* are toxic to fish and crabs (Romijn, 1935), it is doubtful whether the majority of decapod cephalopods normally use venom to inactivate their prey.

The interesting biological activity exhibited by cephalopod saliva has long attracted the attention of pharmacologists and chemists. A number of biologically active amines has been detected in the material obtained from the salivary glands of octopods. Henze (1913) isolated tyramine from the posterior salivary gland of *O. vulgaris* and considered this substance to be the agent present in the glands and saliva which had the ability to paralyze crabs. During studies of the pharmacology of octopus saliva, Bottazzi (1916, 1918, 1919, 1921) and Bottazi and Valentini (1924) noted that aqueous and alcoholic extracts prepared from the posterior salivary glands of *Octopus macropus* produced marked hypotension in dogs. This hypotension was associated with a reduction in strength and an increase in frequency of ventricular systoles. Strong contractures of smooth muscle were also produced by the extracts, and it was suggested that *p*-oxyphenylethylamine, β-iminazolylethylamine, or choline was responsible for these contractures. The vasomotor activity caused by the extracts was attributed to the presence of histamine in the extracts. It was suggested that histamine or histamine associated with tyramine might be responsible for the paralyzing action against crustaceans shown by octopus saliva. The presence of histamine in this saliva was subsequently confirmed by Ungar *et al.* (1937) and by Erspamer and Boretti in 1951.

Octopamine was identified in the posterior salivary glands of octopods by Erspamer (1948b). In the same year, Erspamer (1948c) noted the presence in the glands of a substance with properties resembling those of 5-hydroxytryptamine. Subsequently, Erspamer and Asero (1953) showed that the substance was identical with 5-hydroxytryptamine.

The endogenous amines and the amino acid decarboxylases apparently involved in their synthesis occurring in the posterior salivary glands of *Octopus apollyon* and *O. bimaculatus* have been described by Hartman *et al.* (1960). Among the endogenous amines found were octopamine, 5-hydroxytryptamine, histamine, and dopamine. Material from the glands decarboxylated L-3,4-dihydroxyphenylalanine, DL-5-hydroxytryptophan, DL-*erythro*-3,4-dihydroxyphenylserine, DL-*erythro*-*p*-hydroxyphenylserine, DL-*m*-tyrosine, DL-*erythro*-*m*-hydroxyphenylserine, L-histidine, and DL-*erythro*-phenylserine.

Tyrosine, tryptophane, and histidine were found in the glands by Erspamer (1952). Acetylcholine was detected in the glands by Bacq (1935, 1947), and noradrenaline by von Euler (1952). The occurrence of the guanidine bases octopine and agmatine was reported in the glands by Akasi (1937) and Baldwin (1948), respectively.

Attempts to relate the effects elicited in crabs by octopus saliva to the presence of pharmacologically active amines have met with difficulties. As pointed out by Ghiretti (1949, 1960), no amine present in the glands when injected into a crab in an amount equal to that present in a lethal quantity of saliva is able to kill the crab. Also, there is marked variation in the contents of phenolic and indolic amines present in the saliva of different species of octopus even though each is equally toxic (Erspamer, 1948b,c).

However, as noted by Ghiretti (1960) the amines could be responsible for the initial overexcitability observed when crabs are injected with octopus saliva. Acetone extracts of octopus saliva contained all the active amines of the crude saliva (Ghiretti, 1953), and such extracts cause overexcitability in crabs but do not produce a subsequent quiet phase and paralysis. On the other hand, if a saliva that has been dialyzed against water is injected into crabs no excitatory phase is observed.

The crabs become inactive and paralyzed. A substance or substances other than the amines already identified in octopus saliva might be responsible for the inactivity and paralysis. Livon and Briot in 1906 noted that octopus saliva which had been heated for 10 minutes at 100°C was no longer capable of paralyzing and killing crabs. They suspected that an albuminoid, as proteins were then known, was the really toxic factor in octopus saliva.

In 1949 Erspamer noted that the posterior salivary glands of *Eledone moschata* and *Eledone aldrovandri* contained a substance which when injected into mammals caused marked vasodilation and hypotension and which was capable of stimulating some extravascular smooth muscles. This substance was subsequently termed eledoisin and was shown by Erspamer and Anastasi (1962) to be an endecapeptide having the following amino acid sequence: Pry–Pro–Ser–Lys–Asp(OH)–Ala–Phen–Ileu–Gly–Leu–Met–NH_2. Although eledoisin elicits a marked hypotension in the dog (Erspamer and Glaesser, 1963) and in man (Sicuteri *et al.*, 1963), the pressor response varies considerably from one animal species to another. In the dog, eledoisin is much more effective than histamine, acetylcholine, or bradykinin in its ability to produce hypotension. The hypotension is not appreciably affected by previous treatment with atropine, sympatholytic drugs, ganglion blocking agents, or reserpine. The pressor effects elicited by catecholamines and angiotensin are potently antagonized by eledoisin. The action of eledoisin appears to be chiefly peripheral, affecting vascular smooth muscle or postganglionic pathways to blood vessels or both (Olmsted and Page, 1962).

Despite its potent pharmacological activities, the significance of eledoisin in the salivary glands of octopods is obscure. There is no evidence that the polypeptide is a vital constituent of the salivary toxins of octopods. Indeed, it is possible that eledoisin occurs only in the salivary glands of species of *Eledone*. Anastasi and Erspamer (1962) could not detect eledoisin in the salivary glands of *O. vulgaris* or *O. macropus*. Moreover, they showed that different size groups of *Eledone* specimens possessed conspicuous differences in their eledoisin content. These authors suggested that eledoisin may be concerned with protein synthesis or breakdown in *Eledone* but did not exclude the possibility that the polypeptide had some other biological significance for *Eledone*. Also, they pointed out that the function of eledoisin in *Eledone* may be taken over in other cephalopods by related polypeptides. It is of interest that an eledoisin-like substance has been isolated from the skin of the South American frog *Physalaemus fuscumaculatus* (Erspamer *et al.*, 1962).

Ghiretti (1959, 1960) isolated a protein which he called cephalotoxin from the posterior salivary glands of *S. officinalis* and *O. vulgaris.* He showed that the protein had the same paralyzing action on crabs as dialyzed saliva from *O. vulgaris.* The length of time required for paralysis to occur varied with the amount of cephalotoxin injected and the species of crustacean involved. However, attempts to identify the precise site of action of cephalotoxin were unsuccessful. Neuromuscular preparations of the chelate appendages and legs of crustaceans were unaffected by the protein. Although a temporary increase in tone followed by a decrease in amplitude and arrest in diastole are shown by the perfused isolated hearts of *Maja* and *Eriphia* exposed to cephalotoxin at concentrations of 5×10^{-4} and 2×10^{-3} gm. per liter, these effects on isolated crab hearts could be elicited only with concentrations of cephalotoxin greatly in excess of those which are toxic to the intact crabs. An inhibiting effect upon crustacean respiration was exhibited by cephalotoxin, but no modification of the respiration of isolated crab gills, muscles, or tissue homogenates exposed to cephalotoxin could be detected.

It was suggested by Ghiretti (1960) that cephalotoxin may modify the electrical activity of the central nervous system of crabs. However, it remains to be seen whether cephalotoxin is a common constituent of the saliva of cephalopods generally. Also, the range of activity elicited by cephalotoxin in mollusks, fish, and higher vertebrates awaits elucidation. It is of interest that the protein causes an increase in tone and amplitude of spontaneous contractions of the isolated rabbit duodenum.

A variety of enzymes has been found in the salivary glands of octopods, including proteolytic enzymes (Ghiretti, 1950); a hyaluronidase (Romanini, 1952); tyramine oxidase, tryptamine oxidase, and 5-hydroxytryptamine oxidase (Blaschko and Hawkins, 1952; Blaschko and Philpot, 1953); histamine oxidase, succinic dehydrogenase, phosphatase, adenosinetriphosphatase, butyrylthiocholinesterase, acetylthiocholinesterase, and acetylnaphtholesterase (Arvy, 1960).

Thus, a great variety of compounds has been found in the salivary glands of cephalopods. However, it is likely that different species of octopus possess different compounds in their saliva. Such differences have already been noted for *O. macropus*, *O. vulgaris*, and *E. moschata* by Erspamer and Boretti (1951).

The toxicity to humans of the salivary venoms of *H. maculosa* and *H. lunulata* apparently exceeds that of the venoms of other cephalopods. Simon *et al.* (1964) obtained evidence that extracts of the posterior salivary gland of *H. maculosa* contained a neurotoxin and either one or two neuromuscular blocking agents. These authors attributed deaths of animals injected with the extracts to respiratory failure stemming from blockade of the phrenic nerve by the neurotoxin, blockade of the neuromuscular junction at the diaphragm, or both. The toxic principles were stated to be dialyzable, heat stable, and resistant to mild acid hydrolysis. The presence in the venom of phenolic amines was demonstrated. Trethewie (1965) also found that the venom of *H. maculosa* causes respiratory paralysis in vertebrates. He noted that neuromuscular transmission in phrenic nerve–diaphragm preparations of the rat was inhibited by posterior salivary gland extracts and that there was a lesser, more easily reversible, direct effect of the extracts on the diaphragm.

Aquatic vertebrates do not appear to be used for food by *H. maculosa,* yet the toxicity to vertebrates of the venom of this species is surprisingly high. For example, Sutherland and Lane (1969) found that a young adult specimen of *H. maculosa,* 25 gm. in weight, possessed enough venom in its salivary glands to cause flaccid paralysis in at least 750 kg. of rabbits. Crabs seem to be the principal prey of *H. maculosa,* and its venom will inactivate crabs if squirted into the water in their immediate vicinity (Sutherland and Lane, 1969). It will be of interest to ascertain whether one or more of the principles in the venom which are capable of paralyzing vertebrates are also capable of inactivating crabs.

VII. General Discussion of Molluscan Biotoxins

Studies of some of the biotoxins elaborated by mollusks indicate that choline esters are prominent among the active materials present in the hypobranchial glands of some species of Buccinidae, Muricidae, and Thaisidae. Tetramine occurs in the salivary glands of some species of Buccinidae and Cymatiidae and a mixture of active compounds, some of them as yet unidentified, occurs in the salivary glands of species of Octopodidae. Undoubtedly, a systematic search for pharmacologically active compounds in biotoxins produced by other molluscan families would be rewarding.

In so far as toxicity is an indicator of pharmacological activity, it is to be expected that the extremely potent venoms of Conidae will yield some interesting new compounds with marked pharmacological activity. In this connection, it might be noted that not only do venoms of piscivorous Conidae possess pharmacological activity different from that shown by venoms of molluscivorous Conidae, but also venoms from different species of piscivorous Conidae show differences in the type of activity possessed. Likewise, there are obvious differences in the pharmacological activities of venoms from different species of

Octopodidae. It seems that it will be necessary to examine the venom of each species of venomous mollusk if the full spectrum of pharmacological activity shown by molluscan venoms is to be determined.

Of course, venoms provide obvious sources of new compounds with interesting pharmacological activity. Other materials produçed by nonvenomous mollusks should be screened for activity which might be expressed in more subtle ways. The roles of the poisonous choline esters found in hypobranchial glands of certain mollusks are not known. Neither is the role of a compound found by Atkinson and Granholm (1968) in the egg mass jelly and secretions of the nidamental gland of the squid *Loligo pealei.* This compound inhibits the normal functioning of cilia. The demonstration that oncolytic activity was possessed by a compound present in the tissues of *Mercenaria* was scarcely to be predicted. In this case, the material is apparently toxic to certain types of tumor cell, but apparently nontoxic to normal cells. Again, the normal role of this material is not known, but its discovery has triggered a search of marine organisms generally for compounds with similar activity.

REFERENCES

Agarwal, R. A., and Greenberg, M. J. (1969). *Comp. Biochem. Physiol.* **31**, 835.
Akasi, S. (1937). *Jap. J. Biochem.* **25**, 261.
Alexandrowicz, J. S. (1964). *J. Mar. Biol. Ass. U.K.* **44**, 111.
Alexandrowicz, J. S. (1965). *J. Mar. Biol. Ass. U.K.* **45**, 209.
Alpers, F. (1931). *Z. Naturwiss.* **65**, 587.
Anastasi, A., and Erspamer, V. (1962). *Brit. J. Pharmacol. Chemother.* **19**, 326.
Arnold, J. M., and Arnold, K. O. (1969). *Amer. Zool.* **9**, 991.
Arvy, L. (1960). *Ann. N.Y. Acad. Sci.* **90**, 929.
Asano, M. (1952). *Bull. Jap. Soc. Sci. Fish.* **17**, 283.
Asano, M., and Itoh, M. (1959). *Tohoku J. Agr. Res.* **10**, 209.
Asano, M., and Itoh, M. (1960). *Ann. N.Y. Acad. Sci.* **90**, 674.
Ascher, P. (1968). *J. Physiol.* (*London*) **196**, 51.
Ascher, P., Kehoe, J. S., and Tauc, L. (1967). *J. Physiol. Pathol. Gen.* **59**, 331.
Ascher, P., Glowinski, J., Tauc, L., and Taci, J. (1968). *Advan. Pharmacol.* **6**, 365.
Atkinson, B. G., and Granholm, N. A. (1968). *Biol. Bull.* **135**, 413.
Bacq, Z. M. (1935). *Arch. Int. Physiol.* **42**, 24.
Bacq, Z. M. (1947). *Biol. Rev. Cambridge Phil. Soc.* **22**, 73.
Baglioni, S. (1909). *Arch. Ital. Biol.* **51**, 349.
Baker, J. T., and Sutherland, M. D. (1968). *Tetrahedron Lett.* No. 1, p. 43.
Baldwin, E. (1948). "Dynamic Aspects of Biochemistry." Macmillan, New York.
Beers, J. R. (1967). *Comp. Biochem. Physiol.* **21**, 11.
Bergh, R. (1895). *Nova Acta Leopold.* **65**, 67.
Blaschko, H., and Hawkins, J. (1952). *J. Physiol.* (*London*) **118**, 88.
Blaschko, H., and Philpot, F. J. (1953). *J. Physiol.* (*London*) **122**, 403.

Bottazzi, F. (1916). *Pubbl. Staz. Zool. Napoli* **1**, 59.
Bottazzi, F. (1918). *Atti Accad. Naz. Lincei, Cl. Sci. Fis., Mat. Natur., Rend.* [5] **27**, 190.
Bottazzi, F. (1919). *Pubbl. Staz. Zool. Napoli* **1**, 69.
Bottazzi, F. (1921). *Arch. Int. Physiol.* **18**, 313.
Bottazzi, F., and Valentini, V. (1924). *Arch. Sci. Biol. (Bologna)* **6**, 153.
Bouvier, E. L. (1887). *Ann. Sci. Nat. Bot. Biol. Veg.* [7] **3**, 1.
Cambridge, G. G., Holgate, J. A., and Sharp, J. A. (1959). *J. Physiol. (London)* **148**, 451.
Carroll, P. R., and Cobbin, L. B. (1968). *Comp. Biochem. Physiol.* **27**, 843.
Carroll, P. R., and Cobbin, L. B. (1969a). *Comp. Biochem. Physiol.* **28**, 1075.
Carroll, P. R., and Cobbin, L. B. (1969b). *Comp. Biochem. Physiol.* **29**, 1175.
Carroll, P. R., Chesher, G. B., and Dougan, D. F. H. (1968). *Comp. Biochem. Physiol.* **25**, 913.
Chase, T. N., Breese, G. R., Carpenter, D. O., Schanberg, S. M., and Kopin, I. J. (1968). *Advan. Pharmacol.* **6**, 351.
Chong, G. C., and Phillis, J. W. (1965). *Brit. J. Pharmacol. Chemother.* **25**, 481.
Cleland, J. B., and Southcott, R. V. (1965). *N.H.M.R.C. Spec. Rep. Ser. (Canberra)* No. 12, p. 209.
Clench, W. J., and Condo, Y. (1943). *Amer. J. Trop. Med. Hyg.* **23**, 105.
Cloney, R. A., and Florey, E. (1968). *Z. Zellforsch. Mikrosk. Anat.* **89**, 250.
Cornman, I. (1963). *Nature (London)* **200**, 88.
Cottrell, G. A. (1966). *Comp. Biochem. Physiol.* **17**, 891.
Cottrell, G. A. (1967). *Brit. J. Pharmacol. Chemother.* **29**, 63.
Cottrell, G. A. (1968). *In* "Neurobiology of Invertebrates" (J. Salánki, ed.), pp. 353–362. Plenum Press, New York.
Cottrell, G. A., and Laverack, M. S. (1968). *Annu. Rev. Pharmacol.* **8**, 273.
Cottrell, G. A., and Maser, M. (1967). *Comp. Biochem. Physiol.* **20**, 901.
Cottrell, G. A., and Osborne, N. (1969). *Comp. Biochem. Physiol.* **28**, 1455.
Cottrell, G. A., Pentreath, V. W., and Powell, B. (1968). *Comp. Biochem. Physiol.* **27**, 787.
Curtis, D. R. (1961). *In* "Nervous Inhibition" (E. Florey, ed.), pp. 342–349. Pergamon Press, Oxford.
Dahl, E., Falck, B., Lundquist, M., and von Mecklenburg, C. (1962). *Kgl. Fysiogr. Saellsk. Lund, Foerh.* **32**, 89.
Dahl, E., Falck, B., von Mecklenburg, C., and Myhrberg, H. (1963). *Gen. Comp. Endocrinol.* **3**, 693.
Dahl, E., Falck, B., von Mecklenburg, C., Myhrberg, H., and Rosengren, E. (1966). *Z. Zellforsch. Mikrosk. Anat.* **71**, 489.
Day, J. A. (1969). *Amer. Zool.* **9**, 909.
Dubois, R. (1903). *C. R. Soc. Biol.* **55**, 81.
Dubois, R. (1907). *C. R. Soc. Biol.* **63**, 636.
Dubois, R. (1909). *Arch. Zool. Exp. Gen.* **5**, 471.
Együd, L. G. (1965). *Proc. Nat. Acad. Sci. U.S.* **54**, 200.
Endean, R. (1969). Unpublished observations.
Endean, R. (1966). *Sci. J.* **2**, 57.
Endean, R., and Duchemin, C. (1967). *Toxicon* **4**, 275.
Endean, R., and Izatt, J. (1965). *Toxicon* **3**, 81.
Endean, R., and Rudkin, C. (1963). *Toxicon* **1**, 49.

Endean, R., and Rudkin, C. (1965). *Toxicon* **2**, 225.
Endean, R., Izatt, J., and McColm, D. (1967). *In* "Animal Toxins" (F. E. Russell and P. R. Saunders, eds.), pp. 137–144. Pergamon Press, Oxford.
Erspamer, V. (1948a). *Experientia* **4**, 226.
Erspamer, V. (1948b). *Acta Pharmacol. Toxicol.* **4**, 213.
Erspamer, V. (1948c). *Acta Pharmacol. Toxicol.* **4**, 224.
Erspamer, V. (1949). *Experientia* **5**, 79.
Erspamer, V. (1952). *Arzneim.-Forsch.* **2**, 253.
Erspamer, V. (1956). *Triangle* **2**, 129.
Erspamer, V., and Anastasi, A. (1962). *Experientia* **18**, 53.
Erspamer, V., and Asero, B. (1953). *J. Biol. Chem.* **200**, 311.
Erspamer, V., and Benati, O. (1953). *Biochem. Z.* **324**, 66.
Erspamer, V., and Boretti, G. (1951). *Experientia* **6**, 348.
Erspamer, V., and Dordoni, F. (1947). *Arch. Int. Pharmacodyn. Ther.* **74**, 263.
Erspamer, V., and Ghiretti, F. (1951). *J. Physiol.* (*London*) **115**, 470.
Erspamer, V., and Glässer, A. (1957). *Brit. J. Pharmacol. Chemother.* **12**, 176.
Erspamer, V., and Glässer, A. (1963). *Brit. J. Pharmacol. Chemother.* **20**, 516.
Erspamer, V., Bertaccini, G., and Cei, J. M. (1962). *Experientia* **18**, 562.
Fänge, R. (1960). *Ann. N.Y. Acad. Sci.* **90**, 689.
Fänge, R. (1962). *Pharmacol. Rev.* **14**, 281.
Fänge, R., and Mattisson, A. (1958). *Acta Zool.* **39**, 53.
Flecker, H., and Cotton, B. C. (1955). *Med. J. Aust.* **2**, 239.
Florey, E. (1965). *Annu. Rev. Pharmacol.* **5**, 357.
Florey, E. (1966). *Comp. Biochem. Physiol.* **18**, 305.
Florey, E. (1967). *Comp. Biochem. Physiol.* **20**, 365.
Florey, E., and Florey, E. (1953). *Z. Naturwiss.* **40**, 413.
Florey, E., and Winesdorfer, J. (1967). *Fed. Proc., Fed. Amer. Soc. Exp. Biol.* **26**, 1.
Friedländer, P. (1909). *Ber. Deut. Chem. Ges.* **24**, 765.
Friedländer, P. (1922). *Ber. Deut. Chem. Ges.* **55**, 1655.
Frontali, N., Williams, L., and Welsh, J. H. (1967). *Comp. Biochem. Physiol.* **22**, 833.
Fujita, S. (1916). *Dolytsugaku Zasshi* **28**, 250.
Gerschenfeld, H. M. (1964). *Nature* (*London*) **203**, 415.
Gerschenfeld, H. M. (1966). *Symp. Soc. Exp. Biol.* **20**, 1.
Gerschenfeld, H. M., and Lasansky, A. (1964). *Int. J. Neurol.* **3**, 301.
Gerschenfeld, H. M., and Stefani, E. (1965). *Nature* (*London*) **205**, 1216.
Gerschenfeld, H. M., and Stefani, E. (1966). *J. Physiol.* (*London*) **185**, 684.
Gerschenfeld, H. M., and Stefani, E. (1968). *Advan. Pharmacol.* **6**, 369.
Gerschenfeld, H. M., and Tauc, L. (1961). *Nature* (*London*) **189**, 924.
Gerschenfeld, H. M., and Tauc, L. (1964). *J. Physiol. Pathol. Gen.* **56**, 360.
Gersh, M., and Deuse, R. (1960). *Zool. Jahrb., Abt. Allgem. Zool. Physiol. Tiere* **68**, 519.
Ghiretti, F. (1949). *Boll. Soc. Ital. Biol. Sper.* **25**, 1304.
Ghiretti, F. (1950). *Boll. Soc. Ital. Biol. Sper.* **26**, 559.
Ghiretti, F. (1953). *Arch. Sci. Biol.* (*Bologna*) **37**, 435.
Ghiretti, F. (1959). *Nature* (*London*) **183**, 1192.
Ghiretti, F. (1960). *Ann. N.Y. Acad. Sci.* **90**, 726.
Glaizner, B. (1968). *In* "Neurobiology of Invertebrates" (J. Salánki, ed.), pp. 267–283. Plenum Press, New York.

Graziadei, P. (1966). *J. Ultrastruct. Res.* **15**, 1.
Greenberg, M. J. (1962). *Amer. Zool.* **2**, 526.
Greenberg, M. J. (1965). *Comp. Biochem. Physiol.* **14**, 513.
Greenberg, M. J., and Windsor, D. A. (1962). *Science* **137**, 534.
Halstead, B. W. (1965). "Poisonous and Venomous Marine Animals of the World," Vol. 1. U.S. Govt. Printing Office, Washington, D.C.
Hartman, W. J., Clark, W. G., Cyr, S. D., Jordon, A. L., and Seibhold, R. A. (1960). *Ann. N.Y. Acad. Sci.* **90**, 637.
Hegyeli, A. F. (1964). *Science* **146**, 76.
Henze, M. (1913). *Hoppe-Seyler's Z. Physiol. Chem.* **87**, 253.
Hermitte, L. C. D. (1946). *Trans. Roy. Soc. Trop. Med. Hyg.* **39**, 485.
Hidaka, T., Osa, T., and Twarog, B. M. (1967). *J. Physiol. (London)* **192**, 869.
Hill, R. B. (1958). *Biol. Bull.* **115**, 471.
Hill, R. B., and Thibault, W. N. (1968). *Comp. Biochem. Physiol.* **24**, 19.
Hill, R. B., and Welsh, J. H. (1966). *In* "Physiology of Mollusca" (K. M. Wilbur and C. M. Yonge, eds.), Vol. 2, pp. 125–174. Academic Press, New York.
Hinegardner, R. T. (1958). *Hawaii Med. J. Inter-Isl. Nurses' Bull.* **17**, 533.
Hoyle, G. (1964). *In* "Physiology of Mollusca" (K. M. Wilbur and C. M. Yonge, eds.), Vol. 1, pp. 313–351. Academic Press, New York.
Hoyle, G. (1969). *Annu. Rev. Physiol.* **31**, 43.
Irisawa, H., Shigeto, N., and Otani, M. (1967). *Comp. Biochem. Physiol.* **23**, 199.
Jaeger, C. P. (1962). *Comp. Biochem. Physiol.* **7**, 63.
Jaeger, C. P. (1966). *Comp. Biochem. Physiol.* **17**, 409.
Judge, J. R. (1966). *Proc. Soc. Exp. Biol. Med.* **123**, 299.
Jullien, A., Ripplinger, J., and Joly, M. (1956). *Ann. Sci. Univ. Besancon, Bot.* [2] **5**, 67.
Jullien, A., Ripplinger, J., Joly, M., and Candot, J. (1961). *C. R. Acad. Sci.* **252**, 1512.
Kandel, E. R., Frazier, W. T., and Coggeshall, R. E. (1967). *Science* **155**, 342.
Kerkut, G. A., and Cottrell, G. A. (1962). *Comp. Biochem. Physiol.* **5**, 227.
Kerkut, G. A., and Cottrell, G. A. (1963). *Comp. Biochem. Physiol.* **8**, 53.
Kerkut, G. A., and Laverack, M. S. (1960). *Comp. Biochem. Physiol.* **1**, 62.
Kerkut, G. A., and Leake, L. D. (1966). *Comp. Biochem. Physiol.* **17**, 623.
Kerkut, G. A., and Meech, R. W. (1966). *Comp. Biochem. Physiol.* **19**, 819.
Kerkut, G. A., and Thomas, R. C. (1964). *Comp. Biochem. Physiol.* **11**, 199.
Kerkut, G. A., and Walker, R. J. (1961). *Comp. Biochem. Physiol.* **3**, 143.
Kerkut, G. A., and Walker, R. J. (1962). *Comp. Biochem. Physiol.* **7**, 277.
Kerkut, G. A., Leake, L. D., Shapira, A., Cowan, S., and Walker, R. J. (1965). *Comp. Biochem. Physiol.* **15**, 485.
Kerkut, G. A., Sedden, C. B., and Walker, R. J. (1966a). *Comp. Biochem. Physiol.* **18**, 921.
Kerkut, G. A., Woodhouse, M., and Newman, G. R. (1966b). *Comp. Biochem. Physiol.* **19**, 309.
Kerkut, G. A., Sedden, C. B., and Walker, R. J. (1967a). *Comp. Biochem. Physiol.* **23**, 159.
Kerkut, G. A., Shapira, A., and Walker, R. J. (1967b). *Comp. Biochem. Physiol.* **23**, 729.
Kerkut, G. A., Horn, N., and Walker, R. J. (1969a). *Comp. Biochem. Physiol.* **30**, 1061.
Kerkut, G. A., Brown, L. C., and Walker, R. J. (1969b). *Life Sci.* **8**, 297.

Keyl, M. J., and Whittaker, V. P. (1958). *Brit. J. Pharmacol. Chemother.* **13**, 103.
Kohn, A. J. (1958). *Hawaii Med. J. Inter-Isl. Nurses' Bull.* **17**, 528.
Kohn, A. J. (1959). *Ecol. Monogr.* **29**, 47.
Kohn, A. J. (1963). *In* "Venomous and Poisonous Animals and Noxious Plants of the Pacific Region" (H. L. Keegan and W. V. MacFarlane, eds.), pp. 83–96. Pergamon Press, Oxford.
Kohn, A. J., Saunders, P. R., and Weiner, S. (1960). *Ann. N.Y. Acad. Sci.* **90**, 706.
Korn, E. (1969). *Comp. Biochem. Physiol.* **28**, 923.
Lane, W. R., and Sutherland, S. K. (1967). *Med. J. Aust.* **2**, 475.
Laxton, J. H. (1968). M.Sc. Thesis, University of Auckland.
Li, C. P., Eddy, B., Prescott, B., Caldes, G., Green, W. R., Martino, E. C., and Young, A. M. (1965). *Ann. N.Y. Acad. Sci.* **130**, 374.
Li, C. P., Prescott, B., Eddy, B. E., Chu, E. W., and Martino, E. C. (1968). *J. Nat. Cancer Inst.* **41**, 1249.
Livon, C., and Briot, A. (1906). *J. Physiol. Pathol. Gen.* **8**, 1.
Loe, P. R., and Florey, E. (1966). *Comp. Biochem. Physiol.* **17**, 509.
McClennan, H. (1963). "Synaptic Transmission." Saunders, Philadelphia, Pennsylvania.
Mansour-Bek, J. J. (1934). *Z. Vergl. Physiol.* **20**, 343.
Martin, R. (1968). *Brain Res.* **8**, 201.
Martoja, M. (1960). *Ann. Sci. Nat. Bot. Biol. Veg.* [12] **2**, 513.
Mattisson, A. G. M., and Arvidsson, J. A. (1966). *Z. Zellforsch. Mikrosk. Anat.* **73**, 37.
Meng, K. (1960). *Zool. Jahrb., Abt. Allgem Zool. Physiol. Tiere* **68**, 539.
Miledi, R. (1966). *Nature (London)* **212**, 1240.
Miledi, R. (1969). *Nature (London)* **223**, 1284.
Mirolli, M. (1964). Ph.D. Thesis, Harvard University.
Moore, D. R. (1956). *Nautilus* **69**, 73.
Newman, G., Kerkut, G. A., and Walker, R. J. (1968). *Symp. Zool. Soc. London* No. 22, 1.
Nistratova, S. N. (1968). *In* "Neurobiology of Invertebrates" (J. Salánki, ed.), pp. 315–325. Plenum Press, New York.
Oestlund, E. (1954). *Acta Physiol. Scand.* Suppl. 112, p. 1.
Olmstead, F., and Page, I. H. (1962). *Amer. J. Physiol.* **203**, 951.
Oomura, Y., Ooyama, H., and Suwada, M. (1965). *Symp. Comp. Neurophysiol., Tokyo,* Abst., p. 33.
Paton, W. D. M. (1958). *Annu. Rev. Physiol.* **20**, 431.
Phillis, J. W. (1966). *Comp. Biochem. Physiol.* **17**, 719.
Pilson, M. E. Q., and Taylor, P. B. (1961). *Science* **134**, 1366.
Puppi, A. (1964). *Experientia* **20**, 1.
Quilliam, J. P. (1957). *Brit. J. Pharmacol. Chemother.* **12**, 388.
Rice, R. D., and Halstead, B. W. (1968). *Toxicon* **5**, 223.
Ripplinger, J. (1957). *Ann. Sci. Univ. Besancon, Bot.* [2] **8**, 3.
Roaf, H. E., and Nierenstein, M. (1907a). *J. Physiol. (London)* **36**, 5.
Roaf, H. E., and Nierenstein, M. (1907b). *C. R. Soc. Biol.* **63**, 773.
Romanini, M. G. (1952). *Pubbl. Staz. Zool. Napoli* **23**, 251.
Romijn, C. (1935). *Arch. Neer. Zool.* **1**, 373.
Sakharov, D. A., and Turpaev, T. M. (1968). *In* "Neurobiology of Invertebrates" (J. Salánki, ed.), pp. 305–314. Plenum Press, New York.

Saunders, P. R., and Wolfson, F. (1961). *Veliger* 3, 73.
Schmeer, M. R. (1963). *Biol. Bull.* **125**, 390.
Schmeer, M. R. (1964a). *Biol. Bull.* **127**, 388.
Schmeer, M. R. (1964b). *Science* **144**, 413.
Schmeer, M. R. (1965). *Fed. Proc., Fed. Amer. Soc. Exp. Biol.* **24**, 1504.
Schmeer, M. R. (1966). *Ann. N.Y. Acad. Sci.* **136**, 213.
Schmeer, M. R. (1967a). *J. Cell Biol.* **35**, 121A.
Schmeer, M. R. (1967b). *Biol. Bull.* **133**, 483.
Schmeer, M. R. (1968). *Biol. Bull.* **135**, 434.
Schmeer, M. R. (1969). *Biol. Bull.* **137**, 385
Schmeer, M. R., and Beery, G. (1965). *Life Sci.* **4**, 2157.
Schmeer, M. R., and Cassidy, J. D. (1966). *Biol. Bull.* **131**, 405.
Schmeer, M. R., and Huala, C. V. (1965). *Ann. N.Y. Acad. Sci.* **118**, 603.
Schmeer, M. R., Horton, D., and Tanimura, A. (1966). *Life Sci.* **5**, 1169.
Schroeder, R. E. (1962). *Sea Front.* **8**, 156.
Sedden, C. B., Walker, R. J., and Kerkut, G. A. (1968). *Symp. Zool. Soc. London* No. 22, 19.
Shaw, H. O. N. (1914). *Quart. J. Microsc. Sci.* **60**, 1.
Sicuteri, F., Fanciullacci, M., Franchi, G., and Michelacci, S. (1963). *Experientia* **19**, 44.
Silvey, G. E. (1968). *Comp. Biochem. Physiol.* **25**, 257.
Simon, S. E., Cairncross, K. D., Satchell, D. G., Gay, W. S., and Edwards, E. (1964). *Arch. Int. Pharmacodyn. Ther.* **149**, 318.
S.-Rózsa, K. (1968). *In* "Neurobiology of Invertebrates" (J. Salánki, ed.), pp. 327–334. Plenum Press, New York.
S.-Rózsa, K., and Perenyi, L. (1966). *Comp. Biochem. Physiol.* **19**, 105.
S.-Rózsa, K., and Zs.-Nagy, I. (1967). *Comp. Biochem. Physiol.* **23**, 373.
Sutherland, S. K., and Lane, W. R. (1969). *Med. J. Aust.* **1**, 893.
Sweeney, D. (1963). *Science* **139**, 1051.
Szent-Györgyi, A. (1965). *Science* **149**, 34.
Tauc, L. (1958). *Arch. Ital. Biol.* **96**, 78.
Tauc, L. (1959). *C. R. Acad. Sci.* **249**, 318.
Tauc, L. (1966). *In* "Physiology of Mollusca" (K. M. Wilbur and C. M. Yonge, eds.), Vol. 2, pp. 387–454. Academic Press, New York.
Tauc, L., and Gerschenfeld, H. M. (1961). *Nature (London)* **192**, 366.
Tauc, L., and Gerschenfeld, H. M. (1962). *J. Neurophysiol.* **25**, 236.
Taylor, P. B., and Chat Chen, L. O. (1969). *Pac. Sci.* **23**, 311.
Trethewie, E. R. (1965). *Toxicon* **3**, 55.
Trueman, E. R. (1967). *Nature (London)* **214**, 832.
Twarog, B. M. (1960). *J. Physiol. (London)* **152**, 236.
Twarog, B. M. (1966). *Life Sci.* **5**, 1201.
Twarog, B. M. (1967a). *J. Physiol. (London)* **192**, 847.
Twarog, B. M. (1967b). *J. Physiol. (London)* **192**, 857.
Ungar, G., Ungar, A., and Parrot, J. L. (1937). *C. R. Soc. Biol.* **126**, 1156.
Vincent, D., and Jullien, A. (1938a). *C. R. Soc. Biol.* **127**, 334.
Vincent, D., and Jullien, A. (1938b). *C. R. Soc. Biol.* **127**, 1506.
von Euler, U. S. (1952). *Acta Physiol. Scand.* **28**, 297.
von Euler, U. S. (1961). *Nature (London)* **190**, 170.
Walker, R. J., Hedges, A., and Woodruff, G. N. (1968a). *Symp. Zool. Soc. London* No. 22, 33.

Walker, R. J., Woodruff, G. N., Glaizner, G., Sedden, C. B., and Kerkut, G. A. (1968b). *Comp. Biochem. Physiol.* **24**, 455.
Webb, G. D., Dettbarn, W. D., and Brain, M. (1966). *Biochem. Pharmacol.* **15**, 1813.
Welsh, J. H. (1957). *Ann. N.Y. Acad. Sci.* **66**, 618.
Welsh, J. H., and Frontali, N. (1966). *Proc. Int. Pharmacol. Congr., 3rd, 1964* p. 137.
Welsh, J. H., and Moorehead, M. (1960). *J. Neurochem.* **6**, 146.
Whittaker, V. P. (1960). *Ann. N.Y. Acad. Sci.* **90**, 695.
Whysner, J. A., and Saunders, P. R. (1963). *Toxicon* **1**, 113.
Whysner, J. A., and Saunders, P. R. (1966). *Toxicon* **4**, 177.
Whyte, J. M. (1962). *Aust. J. Sci.* **25**, 99.
Whyte, J. M., and Endean, R. (1962). *Toxicon* **1**, 25.
Winkler, L. R. (1961). *Pac. Sci.* **15**, 211.
Winkler, L. R., Tilton, B. E., and Hardinge, M. G. (1962). *Arch. Int. Pharmacodyn. Ther.* **137**, 76.
Wodinsky, J. (1969). *Amer. Zool.* **9**, 997.
Woodruff, G. N., and Walker, R. J. (1969). *Int. J. Neuropharmacol.* **8**, 279.
Woodruff, G. N., Oniwinde, A. B., and Kerkut, G. A. (1969). *Comp. Biochem. Physiol.* **31**, 599.
Zeimal, E. V.., and Vulfius, E. A. (1968). *In* "Neurobiology of Invertebrates" (J. Salánki, ed.), pp. 255–265. Plenum Press, New York.
Zs.-Nagy, I. (1964). *Ann. Biol. Tihany* **31**, 147.
Zs.-Nagy, I. (1968). *In* "Neurobiology of Invertebrates" (J. Salánki, ed.), pp. 69–84. Plenum Press, New York.
Zs.-Nagy, I., Rózsa, K. S., Salánki, J., Foldes, I., Perenyi, L., and Demeter, M. (1965). *J. Neurochem.* **12**, 245.

CHAPTER 14

Biochemical Ecology of Mollusca

R. Gilles

I. Introduction

The phylum Mollusca has radiated successfully into a variety of habitats. Mollusks have indeed conquered all types of aquatic media and many species are terrestrial. Such a variety of living environments supposes important adaptations at the morphological, physiological, and biochemical level. An example of such adaptation to the various biotopes invaded by mollusks can be found in the various types of respiratory systems developed by molluscan species (Ghiretti, 1966). As we shall see later on, such adaptations are often of ecological value. On the other hand, and from an ecological viewpoint, it is also useful to consider the notion of the biochemical continuum. The biochemical continuum is the result of a slow evolution starting from an organic abiogenic continuum (Florkin, 1966, 1969). If privileged regions of this early abiogenic continuum can be considered as having given birth to a primitive life (see Florkin and Schoffeniels, 1969), one must also consider that the evolution or the changes occurring in the biochemical continuum may influence the ecology of the species to which it has given birth. The oxygen of the atmosphere has been produced by the living plants, and the carbon dioxide it contains has a clear relationship to the metabolism of organisms. There are now many examples of such interactions (Collier, 1953; Johannes and Webb, 1965; Adler, 1966; Richardson *et al.*, 1967) and it is widely recognized that organisms are linked to one another and to the biotope through complex webs of biochemical exchanges. Thus, in an ecosystem, besides the contribution of the trophic chains in supplying molecules endowed with nutritive or regulatory functions, one may describe molecules as being active in the constitution and the maintenance of the biotic community. We shall record them under the general denomination of ecomones (Florkin, 1966). Some ecomones may act in an unspecific way. For instance, the concentration of dissolved carbohydrates in seawater greatly varies from place to place and may have an influence not only on the nature and growth of the phytoplankton but also on the pumping rate of oysters which mainly feed themselves on this plankton (Collier, 1953). Other ecomones are recognized as being specifically active in the process of the coaction of organisms upon each other. Such specific substances, termed coactones (Florkin, 1966), are determinant in the relationship of the coactor (directing organism) with the coactee (receiving organism). A number of coactones are liberated in the medium by the coactor. They have been termed exocoactones in contrast with endocoactones which act without being liberated in the medium (Florkin, 1966). For instance, glutathione, when acting as a phagostimulant for *Hydra* (Loomis, 1959) may be considered to be an endocoactone. On the other hand, the common sea hare *Aplysia juliana* is strongly attracted to its food plant, *Ulva lactuca,* by a compound secreted in the medium by the plant. The chemical nature of this exocoactone is not yet known (Frings and Frings, 1965).

Thus, when dealing with the biochemical aspects of ecology, one must consider on one hand the relationship between organism and medium and the biochemical changes, often of an adaptive nature, which allow a given species to live or to survive in a given biotope; and on the other hand, the relationship between organism and organism and the circulation, through the biochemical continuum, of molecules endowed with a certain amount of information. An ecological study should therefore proceed down to the molecular level where one can expect to find the underlying causes of the integration of a species in an ecosystem. Moreover, because of the complexity of the relationships between the organisms and the environment, such a study should be very detailed, as is the case for instance in the work done by Florkin and co-workers dealing with the osmotic relationship that an organism established with its surroundings (Florkin and Schoffeniels, 1969; Schoffeniels and Gilles, 1971a; Gilles, 1969). Unfortunately, the concept of biochemical ecology is relatively new, and the studies in this field, although numerous, often present some disparity. We shall therefore try in this chapter to group in various examples the data we have gathered, although we have ruled out some studies as being too isolated. We shall divide this chapter, using the main physical and chemical characteristics of the environment which influence the distribution of mollusks. We shall thus consider successively the aquatic, the intertidal, and finally the terrestrial biotopes.

II. Aquatic Mollusks

Many species of aquatic mollusks are sessile or able to make only slow movements; they are thus entirely dependent on their immediate environment. From the ecological viewpoint, one of the main chemical properties of the aquatic medium is the saline concentration. Representatives of the phylum Mollusca can be found in fresh, brackish, and seawater. Many species, mainly estuarine, are moreover able to withstand the sometimes tremendous salinity changes occurring in this biotope. The ways by which aquatic mollusks can adapt to media of different salinities have been the subject of many investigations. These studies have been recently reviewed (Robertson, 1964; Schoffeniels and Gilles, 1971b). We shall therefore not deal any longer with this subject.

The aquatic medium is a well suited environment for the transportation of "chemical information." With the exception of cephalopods, which are fast moving animals and where it seems that a great part of their behavior can be associated with the considerable elaboration of the central nervous system (H. J. Wells, 1966), the behavior of aquatic mollusks appears to be directed in many cases by active ecomones which can rapidly become incorporated into the organic community of this medium. This influence of environmental factors through all the life of aquatic mollusks can be shown in a few examples.

A. Ecotopic Characteristics

1. Sex Determinism

In the great majority of molluscan species, individuals are gonochoristic. Throughout the various phyla, however, some cases of hermaphroditism are known, and in a few species sex transformation normally occurs. It is usually held that hermaphroditism is secondary, derived from a gonochoristic condition, although some authors consider the hermaphroditic state as being the primitive one (see Fretter and Graham, 1964). Some classes of the phylum are exclusively dioecious (scaphopods, cephalopods). In other classes, one species may be dioecious while a related species is hermaphroditic. It is probable that in some cases there is an ecological reason for this. It sometimes appears, indeed, that a hermaphroditic species is normally living in an ecosystem in which it is confronted with some difficulties in reproductive activity. As a matter of fact, hermaphroditism is frequent in terrestrial forms which are restricted by environmental factors to limited activity and is frequent in sessile species. In many hermaphroditic species, mainly in bivalves, opistobranchs, and pulmonates, the change of sex is a general rule. Some species change sex once in their lives, such as *Kellia suborbicularis,* other species such as *Ostrea* may change sex annually or at closer intervals, as first described by Orton (1927). To our knowledge, there is little information dealing with ecological factors implicated in the occurrence of hemaphroditic species or with the circumstances in which sex transformation occurs. Rosenwald (1926) and also Richter (1935) show that variations in environmental conditions, particularly moisture and food, affected the sexual development towards masculinity or feminity in *Limax* species.

The influence of ecological factors in the sexual development of mollusks is also suggested by the fact that in *Crassostrea cucullata,* the proportion of males is raised from 41% to 82% when parasitized by the crab *Pinnotheres* (Awati and Rai, 1931) and by the relationship which Bloomer (1939) demonstrated between the proportion of males, females, and hermaphrodites in *Anodonta cygnea* found in different geographical situations. Various studies show that in the prosobranch *Crepidula,* the sex changes are the result of an interaction between the various members of the colony. The slipper limpets are found in chains up to twelve individuals, each clinging to the shell of the next so that the right lip of the shells are aligned. The basal members of the group are females, the apical members are males. As early as 1919, Gould suggested that the lower members of the colony (females) secreted some substance into the surrounding water with the effect of maintaining the masculinity of the upper members. Although Wilczynski (1959) is unable to show evidence that female sex hormones have any effect in changing the sex of the male in *Crepidula fornicata,* the existence of this exocoactone is now generally recognized (Gould, 1952; Coe, 1953; Fretter and Graham, 1964). Unfortunately, data are lacking as far as the chemical nature of this exocoactone is concerned.

The hermaphroditic state observed in mollusks appears thus to be directly influenced by environmental factors, although the prime cause of hermaphroditism is probably genetic. According to Montalenti and Bacci (1951), hermaphroditic species fall into two main groups called balanced and unbalanced, the difference between them being primarily genetic. They are all supposed to have an uniform genotype and exceptions to the norm of behavior are mainly due to environmental influences.

2. *Reproduction and Development*

These important steps in the maintenance of species appear also to be directly influenced by environmental factors. The maturation of the gametes appears to be set off mainly by annual temperature fluctuations, although the availability of food may also have an effect on the gamete maturation. Temperature as well as starvation have an effect on the metabolic activity of mollusks (Spärck, 1936; Berg *et al.*, 1958; Berg and Ockelmann, 1959; Read, 1962). These changes in the metabolic rate will interfere with the seasonal variation observed in the metabolic activity and appear to be intimately associated with the physiological activity of the animal. In *Acroloxus lacustris* (Berg, 1952), as in *Ancylus fluviatilis* (Berg and Ockelman, 1959), the oxygen consumption, measured at a same temperature, is about 1.3 times to nearly twice as great in spring and summer as in autumn or winter.

This increased oxygen consumption is not due to an increased growth rate during the same period. Indeed, results obtained by Hunter (1953) on the same species (*A. fluviatilis*) show that the growth rate is very low during winter, then increases in early spring but decreases again during late spring and summer, just at the time when the animals show the highest oxygen consumption and are sexually mature. Berg *et al.* (1958) thus assume that the seasonal variations in oxygen consumption are an expression of the sexual activity. In the same way, seasonal biochemical changes in tissue composition have been related to sexual maturation (Krüger, 1960). Gonadal maturation may thus be influenced by environmental conditions (mainly temperature and food availability). However, a combination of ecological factors also trigger the other sequences of the reproductive cycle in mollusks.

Among aquatic gastropods, the role of chemoreception in mating seems to have first been observed in *Chromodoris*. A positive response indicated by successful reciprocal fertilization seems to be elicited by some secretion in this species (Crozier, 1918; Crozier and Arey, 1919). In the freshwater prosobranch *Viviparus*, locomotor activity is increased by a factor of two when an individual is introduced into an aquarium which has contained a member of the opposite sex. Water from a vessel containing a member of the same sex had no effect (Wölper, 1950). These experiments indicate that an exocoactone may facilitate a successful fertilization by attracting toward each other mature members of opposite sexes.

In other aquatic species, fertilization occurs in the medium. After the gonads have reached maturation, the gametes are emitted into the surrounding water, where fertilization can occur. Emission of gametes in mollusks appears to be narrowly controlled by activators of spawning and largely by other factors. Aquatic species which broadcast their gametes produce substances which activate spawning by other members of the same species. These have been shown to occur in oysters (Galtsoff, 1930) and mussels (Young, 1942). In oysters, the spawning agent is heat stable and dialyzable. Male oysters can be activated by other males and also by egg water (Galtsoff, 1938, 1940). It appears, therefore, that a pheromone broadcasted in the medium by individuals can initiate the process of spawning which then echo and re-echo throughout the population. The receptors of the coactee animals are on the gills, and their stimulation induces relaxation of a sphincter on the male duct, resulting in spawning (Nelson and Allison, 1940).

In 1936, Nelson discovered another pheromone in oyster sperm, which he called diantlin. This hormonelike protein increases the ventilation of the mantle cavity by increasing the size of the branchial pores, relaxing the adductors, and accelerating the rate of ciliary beat. Its secretion occurs prior to egg spawning and allows the egg to pass through the gill passages into the inhalant chamber more readily. Similar substances occur and act comparably in mussels (Young, 1946) and in *Tridacna* (Wada, 1954). Egg laying can also be stimulated by a neuroendocrine secretion. Recently, Kupfermann (1967) described experiments in which a seawater extract of the bag cells of *Aplysia* abdominal ganglion induce egg laying behavior in animals of the same species shortly after injection of the extract into the aquarium. These results have been confirmed by Strumwasser *et al.* (1968).

These last authors have recently shown (1969) that the extract stays without any effect when taken from animals from February to mid June, although the *Aplysia*, which are in the aquarium where the extract is injected, were controlled for sexual maturity. These experiments, however, do not allow distinguishing between a seasonal fluctuation in the amount of the active principle in donors or in some seasonal refractory state in the receiving animal. These findings are comparable to a similar phenomenon recently described in the starfish *Asterias pectinifera* (Chaet, 1967). In this species, a seasonal refractoriness has been shown to be due to the presence of a spawning inhibitor identified as glutamic acid (Ikegami *et al.*, 1967).

The existence of endocrinelike substances facilitating the union of the gametes was also demonstrated some time ago. These pheromones can be described following the determination done on Echinoderms (Rothschild, 1956). In this phylum, at least four substances occur: (1) gynogamone I, which activates spermatozoa and which is antagonized by androgamone I; (2) gynogamone II, which agglutinates sperm and its antagonizer androgamone II; and (3) glynogamone II and (4) androgamone II, which appear to be present in various aquatic mollusks and to be of protein nature (Glaser, 1921; Terao, 1926; Tyler, 1939, 1940; Runnström and Monné, 1945; Hultin, 1947; Dan and Wada, 1955).

It appears, therefore, that spawning is more or less controlled by some pheromones. However, one can consider that other factors play some role in the discharge of the gametes. These factors can be described as physical properties of the environment. As a matter of fact, some intertidal species appear to be stimulated in spawning by wave action or mechanical shock (Field, 1922; Orton, 1924; Young, 1946; Orton *et* al., 1956). Temperature is often regarded as the most important physical factor of the environment influencing breeding. The work of Loosanoff and co-workers (Loosanoff and Engle, 1942; Loosanoff *et al.*, 1951; Loosanoff, 1959, 1960) and others (Strauber, 1947, 1950; Scheltema, 1967) has made clear the effect of temperature on spawning. It appears, as suggested by Orton (1920), that mollusks spawn when the temperature exceeds a critical level characteristic of the species, while others do so at a particular change in environmental temperature. Thus, *Crassostrea virginica* has races with spawning temperature to 17°, 20°, and 25°C, the first occurring at the northern and the last at the southern extremities of geographical range of the species.

At 12°C, Long Island oysters were 67% ripe after 68 days, while New Jersey oysters failed to ripen after 78 days, and those from Carolina and Florida could not even be sexed (Loosanoff, 1960). The same changes in breeding behavior according to latitude can be described for *Nassarius obsoletus* (Scheltema, 1967). Sometimes, at the same latitude, one race may live in deeper, cooler water, another in shallower and warmer situations; then the critical temperature for spawning will be reached at different times at different depths as is the case for *Nassarius reticulatus* or *Akera bullata* (Thorson, 1946).

Ecologically, it is doubtlessly advantageous for the species to spawn as soon as the seawater becomes warm enough for larval development. It is also interesting to consider that the gametogenesis is achieved in some cases a long time before spawning, thus allowing for a great flexibility in breeding. For instance, gametogenesis is completed almost 6 months before the time of spawning of *Nassarius* populations north of Cape Cod and at least 2 months before spawning in populations south of Cape Hatteras in North Carolina (Scheltema, 1967). Thus, temperature directly influences gametogenesis and spawning in mollusks.

It appears, however, that this physical property of the environment also acts on the following sequences of the molluscan life. As a matter of fact, as early as 1926, Seno *et al.* determined the effect of temperature on the early development of *Ostrea gigas* from the time of fertilization to the early shelled larva. Such experiments have been repeated on various species, such as *C. virginica* (A. E. Clark, 1935; Stickney, 1964), *Venus mercenaria* (Loosanoff, 1959), *Ostrea edulis* (Davis and Calabrese, 1964), *Mytilus edulis* (Bayne, 1965), and *N. obsoletus* (Scheltema, 1967). In this last species, a relationship between temperature, spawning, and emergence of veliger from their egg capsules can be demonstrated. As shown in Fig. 1, the time required for emergence increases about 0.25 day per degree centigrade between 20° and 28°C and 2 days per degree between 16.5° and 20°C.

In *Nassarius* (Scheltema, 1967) as in *Venus* (Loosanoff, 1959) and *Mytilus* (Bayne, 1965), temperature also affects the growing rate of the larvae. It is not always clear, however, in these experiments whether this is an intrinsic effect on the veliger larvae themselves or whether growth is indirectly influenced by the effect of temperature on the algal food. From what we know about the effect of temperature on metabolic rate and on enzymatic activity (see below) it seems doubtful that at least part of the effect of temperature on growing rate of larvae can be attributed to an effect of temperature on the larvae themselves.

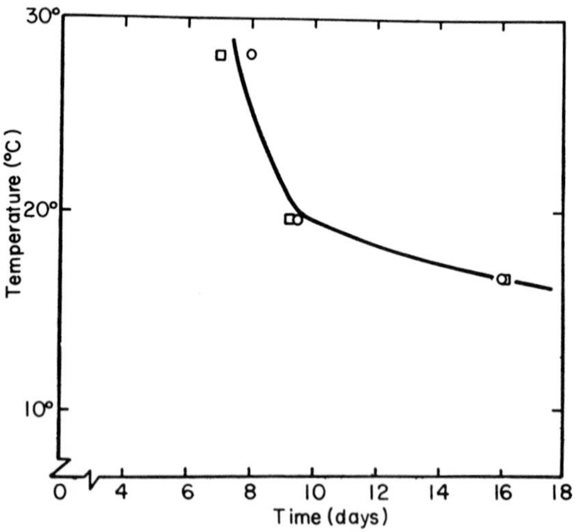

Fig. 1. Time in days required between spawning and emergence of *Nassarius obsoletus* larvae as a function of temperature; (○) *N. obsoletus* from Beaufort, North Carolina; (□) *N. obsoletus* from Cape Cod, Massachusetts. (After Scheltema, 1967.)

The larval life of aquatic mollusks can be divided in most cases, into two periods. The first period is one of rapid growth and morphological development (developmental period). The growth rate during the first period is largely affected by environmental factors as temperature, salinity, or food availability (Davis and Calabresse, 1964; Stickney, 1964). The second period is one of gradual decrease in growth (delay period). The duration of this delay period is largely determined by the availability of a bottom sediment favorable for settlement. Bayne (1965) has demonstrated that in *M. edulis,* the delay period can be very long in the absence of an adequate substratum for settlement and that this period is accompanied by a gradual decrease in growth rate down to zero. It thus appears that metamorphosis can occur as a response to a stimulus received from the desirable environment.

The earliest observations of this kind were those of Harrington (1921) who showed that shipworm larvae (*Teredo norvegica*) were positively chemotactic to a substance which he extracted from wood and which he believed was malic acid. A chemical stimulation to settlement has now been demonstrated in many species (see Wilson, 1968). A study of Scheltema (1961) shows that the presence of a favorable substratum has a marked effect on the settlement behavior of *Nassarius* larvae. This metamorphosis-inducing factor is probably a water-soluble substance. Veliger larvae of the nudibranch *Adalaria proxima* settle and metamorphose on living colonies of the ectoproct bryozoan *Electra pilosa* (Thompson, 1958). A diffusible exocoactone produced by *E. pilosa* is suggested by the author as the inducer of the settlement behavior of the mollusk.

This settlement factor is species specific (Thompson, 1958). According to Bayne (1967) *Ostrea* larvae settle preferentially on surfaces previously colonized by individuals of the same species. This author demonstrated that oyster larvae will settle in larger numbers (about 3.5 times more) on surfaces treated with a crude extract of oyster than on similar surfaces treated with seawater. The settlement factor is probably a protein since it is completely destroyed by pronase. The existence of a settlement factor in species that are either slow moving or sessile is of ecological significance. Indeed, the propagation of these species as well as the renewal of a population by young individuals coming from other population is insured only by the production of a great number of pelagic larvae. The success of benthic organisms will therefore be dependent upon the metamorphosis of their larvae onto a substratum which meets their requirements. The settlement factor, by favoring the settlement of larvae in a biotope already tested by members of the same species, certainly contributes to the maintenance of benthic organisms.

3. *Behavioral Activities*

Thus, already during the larval stages, chemical information is able to induce in aquatic mollusks a behavior that is of ecological significance. Such chemical information appears also to be of importance during the adult life of mollusks. For instance, the violent escape response of *Nassarius* when touched by a sea star is considered to be mediated through chemoreception (Hoffmann, 1930; Weber, 1924). A same response to predators has been demonstrated in many mollusk species (Degner, 1921; Herter, 1929; Bullock, 1953; W. C. Clark, 1958).

Bullock (1953) found that the escape response was only given when the gastropod studied was in the presence of predatory species. Non-predatory species did not elicit the response. A glass rod rubbed on a predatory sea star induces the escape response of *Nassarius* (Bullock, 1953), thus indicating the chemical nature of the coactone. Studies on the nature of this exocoactone have been undertaken recently on *Pisaster ochraceus, Asterias rubens,* and *Marthasterias glacialis* (Feder and Lasker, 1964; Feder and Arvidsson, 1967). The results suggest that the coactone is of low molecular weight and is absorbed or masked by protein. In contrast, Hashimoto (1967) has implicated a high molecular weight lipoprotein as the active principle in *Asterias amurensis.* Mackie *et al.* (1968) show that the withdrawal response of *Buccinum undatum* to *A. rubens* and *M. glacialis* is due to a saponin-like molecule which exists as an equilibrium mixture of free saponin and its protein complex.

Another type of escape response has also been demonstrated in *Helisoma nigricans,* which burrows rapidly in the substrate when specimens of the same species are crushed in the water. The same response was obtained when the turtle *Pseudemys* introduced into the aquarium ate one of the snails. The escape response of *Helisoma* does not occur until the turtle ate one individual, suggesting that the effective stimulus is inherent in the snail. This alarm substance is also species-specific, since crushed individuals of related species do not evoke the response (Kempendorff, 1942).

Aquatic mollusks not only can avoid contact with a predatory species owing to their sensitivity to chemical information, they also use chemoreception in the search for food. Members of the genus *Nassarius* have long been known to congregate about dead fishes and crustaceans. MacGinitie and MacGinitie (1949) observed *Nassarius fossatus* moving upstream in a current toward a dead fish from a distance of more than 30 meters. The same behavior has been observed in *Cyclope neritea* (Morton, 1960). As early as 1918, Copeland showed that the feeding response of *Nassarius obsoletus* is elicited in the presence of filtered extract of *Fundulus* muscle. Chemoreception is probably the means by which predatory aquatic mollusks become aware of the presence of prey. A typical feeding response is elicited in *Conus striatus* by introduction into the aquarium of water in which fishes have been swimming (Kohn, 1956). The same feeding response has been shown in the oyster drill *Urosalpinx cinerea* toward *Crassostrea virginica* (Blake, 1960).

Many predatory species seem to be able to distinguish between chemical stimuli coming from various potential prey species. Choice chambers, in which water flows over various prey species and then enters different regions of an aquarium containing snails have been employed to test the response of several predator mollusks. *Aeolidia papillosa* is attracted to chemical stimuli from *Metridium senile* and *Actinia equina* but not to three other anemones nor to five hydroids species tested (Stehouwer, 1952). The same kind of choice between various potential prey has been demonstrated in *Cratena aurantia* (Braams and Geelen, 1953), *Conus sponsalis* (Kohn, 1959), and *Urosalpinx cinerea* (Blake, 1960).

In this last species, it seems that excreted metabolic end products produced in direct proportion to oxygen consumption are the chemical attractants. Carr (1967) identifies compounds in shrimp extracts which induce a feeding behavior in *Nassarius obsoletus.* These compounds are mainly amino acids (glycine, alanine, aspartic and glutamic acids, proline, serine, and threonine), some amines directly related to the amino acid metabolism (asparagine, glutamine, and taurine), and lactate and betaïne. A combination of all these compounds is more active than each of them taken individually. Glycine and lactate are the only compounds identified in the shrimp extract which possess marked stimulatory capacities when the compounds are tested individually.

The role of chemoreception in the feeding of herbivorous mollusks has also been demonstrated. A feeding behavior is induced in *Littorina obtusata* when it is placed in a water stream which has passed over pieces of Fucaceae. Moreover, a difference in attractiveness of different species of *Fucus* and *Ascophyllum* can be shown in *Littorina* (Van Dongen, 1956). *Aplysia juliana* is strongly attracted to its plant food *Ulva lactuca* by an exocoactone, which apparently comes from the leaflike portion of the plant (Frings and Frings, 1965). A feeding behavior can be induced in *Aplysia* by adding a few drops of *Ulva* water to an aquarium containing the animal. The *Ulva* water remains effective even after a 150,000-fold dilution. The *Ulva* water appears also to contain a phagostimulant. Indeed, if *Ulva* water is dropped onto the mouth, it elicits radular action as in feeding (Frings and Frings, 1965). Behavioral response of mollusks to many types of chemical information has been studied for a long time. Mollusks respond to a variety of inorganic and organic compounds, including lithium chloride, sodium nitrate, potassium chloride, ammonium nitrate, quinine sulfate, coumarin, maltose, fructose, sodium salycilate, cedar wood oil, and orange flavoring. The effect of such compounds on the behavioral response of various species has been recently reviewed (Kohn, 1961). Moreover, the ecological significance of many of these behavioral-response-inducing compounds must be questioned. We shall thus not deal any longer with this problem.

Sites of Chemoreception. Chemoreception has been mainly studied in the past by following the typical behavior which can be induced by various types of coactones. However, several experiments have been undertaken in order to identify the sites of chemoreception. Many of these experiments involve the study of the behavioral response to chemical stimuli before and after extirpation of supposed chemoreceptors. These studies point out a possible role of rhinophores and oral tentacles (Agersborg, 1922, 1925), tentacles, foot, and siphon (Copeland, 1918; Hoffmann, 1930; Wölper, 1950), and also the osphradium.

As early as 1881, Spengel concluded from the position, innervation, and histology of the osphradium that its main function was the testing of chemical and physical properties of the water, including perception of properties of food. This concept has been shared by many investigators (Brock, 1936; Brown and Noble, 1960; Michelson, 1960) who reached this conclusion by means of experiments in which behavioral responses to a variety of stimuli were observed before and after the removal of the osphradium. A more direct approach has been attempted by Bailey and Laverack (1963, 1966) on *Buccinum undatum* by following at the level of single cells of the central ganglia the excitatory and inhibitory responses after placing the osphradium of the animal in contact with muscle extracts or with solution of different amino acids. These studies show that the organ is used by the carnivorous gastropod *Buccinum* in the detection of its prey. More recently, Stinnakre and Tauc (1969) pointed out an osmoreceptor role of the osphradium in *Aplysia.* By using an isolated preparation including this organ, the branchial nerve, and the visceral ganglion, they show that a dilution of the seawater bathing the external face of the osphradium provoked a marked inhibition in an identifiable neuron (R 15) of the visceral ganglion. The dilution of the seawater is the only stimulus found to have a constant effect on the membrane resistance of the neuron. The authors thus concluded that the inhibition observed in the R 15 neuron is the result of excitation of osmoreceptors located in the osphradium.

4. *Physiological Activities*

It thus appears that some specific organs, such as the osphradium or rhinophores, are involved in the reception of chemical informations by mollusks. These organs, through the central nervous system, induce a behavioral response which is in many cases of ecological significance. We have indeed seen that chemoreception in mollusks is implicated in behavior which is important in the maintenance of a species in a given area; e.g., mating, homing, escape from predators, or feeding. Moreover, mollusks are also sensitive to physical parameters of the medium. We have already pointed out the role of the osphradium in the response of *Aplysia* to an osmotic stress as well as the effect of temperature on the gamete maturation and spawning. Temperature appears to be an important factor in the selection of a biotope by mollusks. The fact that spawning is more or less controlled by temperature (see Section A,2) doubtlessly plays a part in the ecology of Mollusca. Moreover, temperature also affects the whole physiological activity of aquatic mollusks.

Since the energy metabolism of most organisms is geared to reactions involving oxygen as an electron acceptor, most of the studies dealing with the effect of temperature on organisms take into consideration the oxygen consumption as being truly representative of the physiological state of the organism. Metabolic rate is generally related to temperature in Mollusca. Krogh (1914) has first described the relationship between temperature and oxygen consumption in terms of a curve known as Krogh's normal curve. As shown in Fig. 2, *Australorbis glabratus* follows the curve and its extension to 37°C. Normally, respiration rate rose with increasing temperature, passed to a maximum, and then decreased to the thermal death point. Deviations from Krogh's curve have been reported for a number of species (Berg and Ockelmann, 1959; Read, 1962), and the maximal respiratory rates differ between species (Newcombe *et al.*, 1936; Read, 1962). This maximum in respiratory rate appears to be related to the temperature ranges that mollusks experience in their natural environment. As a matter of fact, Spärck (1936) shows that for three species of *Pecten*, the rate is highest for the arctic *Pecten groenlandicus*, intermediate for the boreal *Pecten varius*, and lowest for the mediterranean *Pecten flexuosus*.

The effect of temperature has also been examined on growth rate in mollusks. Field studies of temperature effects face the complexity of multiple factors such as food quantity and quality or age. The quantitative relationship between temperature and growth rate are thus of a considerably complex nature in many cases. However, for a few species

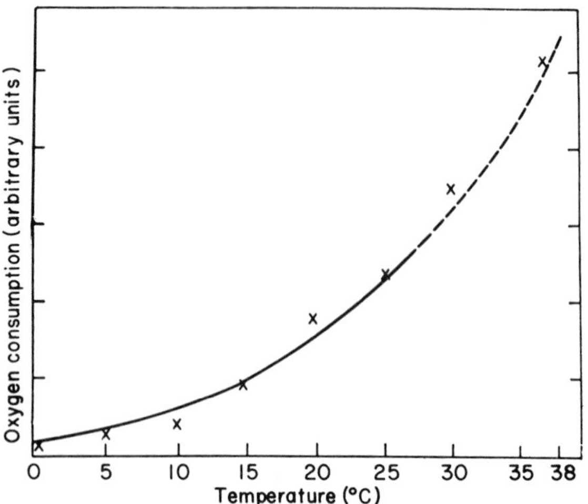

Fig. 2. Relationship between temperature and oxygen consumption of *Australorbis glabratus* in the range 0.3–37°C. (After von Brand *et al.*, 1948.)

(and by choosing a suitable parameter of growth), equations relating growth rate to environmental temperature can be given. Watabe (1952) found that the weight of the nacreous layer of the growing pearl increases directly as the sum of the average daily temperature rises above 13°C (see also Taylor, 1959).

It is thus clear that, except in a few particular cases, it is difficult to draw a simple generalized conclusion about the metabolic pattern an organism establishes under various conditions of environmental temperature. While the relationship between the environmental temperature and the physiological stage may be related to the ecology of some species, in other cases this does not seem to be true. For instance, Evans (1948) determined the thermal death point of a series of British prosobranchs. This author, contrary to some other workers, finds no correlation between the thermal resistance and the zonal sequence, nor any variation in tolerance between individuals of a same species taken from different temperature habitat.

On the other hand, it is worth noting, as pointed out by Newell (1966), that in many poikilotherms the upper limit of environmental temperature which may be related to the temperatures that the animals normally experience in nature can be directly related to the melting point of their molecular collagen. Since collagen plays an important role in controlling the properties of skin structure, the above considerations may be considered to indicate an interesting aspect of biochemical adaptation. Unfortunately, information on the melting point of molluscan collagen is scarce. Collagen has only been isolated from the land snail *Helix aspersa*. In this case the melting point of 27°C (Rigby, 1968) corresponds well to the upper limit of environmental temperature of 30°C (Newell, 1966). From just this data, it is hazardous to draw general statements. Moreover, the definition of the upper limit of environmental temperature is arbitrary, and values quoted for a given animal may thus vary. The available information on thermal death points of marine mollusks has been reviewed in tabular form by Gunter (1957). We therefore refer the reader particularly interested in this field to his paper.

Another physical property of the environment known to interfere with the ecology of mollusks is the water oxygen tension. One may describe two types of changes in oxygen consumption in response to a change in oxygen supply. In the first one, the oxygen consumption is reduced when the oxygen content of the water is decreased. In the second one, the organism maintains its usual oxygen consumption until the oxygen supply falls to a certain low critical value. This last type of response change in oxygen supply is certainly of ecological value since it permits a species to survive in media where the conditions of oxygen availability are poor or may vary. As pointed out by Berg and Ockelmann (1959), *Bithynia*, which occur at fairly great depths in eutrophic lakes and in other places with bad respiratory conditions, is able to maintain its normal oxygen consumption under various conditions of oxygen supply. This is at variance with what happens in *Limnea*, which survive near the surface of ponds or on the shore of lakes and rivers and where the oxygen consumption is reduced very considerably when the oxygen content of the water is low.

Most mollusks have hemocyanin as a respiratory pigment. Many also have hemoglobin in certain muscles which may act as an oxygen store (Ball and Meyerhof, 1940; Manwell, 1958). Fox (1955) reports that in young individuals of the gastropod *Planorbis corneus* reduction of environmental oxygen tension stimulates synthesis of vascular hemoglobin. However, the myoglobin content of the pharyngeal muscle of *Physa fontinalis* appears not to be a function of the oxygen supply. In some cases, a reversal Bohr effect has been reported (Redfield *et al.*, 1926; Manwell, 1958; Redmond, 1962); i.e., oxygenated blood will combine with more carbon dioxide than will reduced blood under comparable conditions, and the presence of carbon dioxide increases the affinity of the hemocyanin for oxygen. Scheer (1948) concludes that this property of the blood is advantageous, since oxygen deficiency will not seriously reduce the oxygen supply to the tissues.

In this section, we have attempted to show that the whole life of an aquatic mollusk, from the development of the egg to adult life and reproduction, is influenced by environmental factors. These factors, which are of chemical or physical nature, induce behavior or physiological adaptations which are of ecological value. This is particularly true in the forms of mollusk we have dealt with and which are sessile or only able to make slow movements and thus are entirely dependent upon their immediate environment. Some aquatic mollusks, however, are able to make fast movements or are buoyant. In these species, new types of relationships between animals and medium and animals and animals may sometimes be added to the relations described above. For example, cephalopods have the ability to perceive linearly polarized light and to determine its plane of polarization (Waterman, 1966). This polarotaxis appears to be of ecological significance. As a matter of fact, it has been demonstrated by Lythgoe and Hemmings (1967) that polarized light sensitivity enables distant objects to be more clearly seen and causes contrasts in intensity to be much sharper. Moreover, as pointed out by Umminger (1968), polarotaxis appears mainly in carnivorous or omnivorous fast moving species, suggesting that the ability to perceive polarized light might be an advantage to a species having predatory habits. The buoyancy of aquatic mollusks also appears to be of ecological importance. We shall therefore give here a brief account of the various means by which mollusks have reached a relative independence from the environmental medium.

B. Buoyancy

Some mollusks which are denser than seawater have to work continuously if they are not to sink. This is the case of the common squid *Loligo*. However, various other mollusks have evolved mechanisms which allow them to bring their specific gravity very close to that of the environmental water, thus enabling them to float. A first type of buoyancy mechanism used by some mollusks consists of building themselves a float by forming mucous-coated bubbles of air. Wilson and Wilson (1956) have given a clear account of how *Janthina janthina* builds its float. Some other mollusks, such as *Pterotrachea coronata* or *Cymbalia peroni*, are gelatinous animals. The fluids obtained by centrifuging these species, although they are isosmotic with seawater, have a lower specific gravity. This low specific gravity is mainly obtained by excluding sulfate ions from the body fluids. As shown by Denton and Shaw (1961), *Pterotrachea* has a sulfate concentration of about 70% that of seawater. This can give a lift of about 0.7 mg. per milliliter of body fluid, which is enough to ensure the floatability of this species, since only a very small proportion of the body components is denser than seawater.

In some branchid squids, the buoyancy is obtained by maintaining in a large coelomic cavity a liquid isosmotic with seawater but much lighter. This cavity amounts to about two-thirds of the squid total volume. The coelomic fluids of *Verrillitheuthis hyperborea* contain about 480 m*M* ammonium. On the other hand, sodium concentration is only 90 m*M*, while chloride accounts for almost all the anion (Denton *et al.*, 1958).

Other species have insured their floatability by the development of a chambered gas-filled shell. This is the case of *Nautilus*, *Spirula*, and *Sepia*. In this last species, changes in buoyancy are produced by changes in density of the cuttlebone. The cuttlebone of *Sepia* contains both gas and liquid. The gas is mainly nitrogen at an average pressure of about four-fifths of an atmosphere. When the cuttlefish has to become less dense, liquid is pumped out of the cuttlebone. Gas then slowly diffuses into this free space until its partial pressure equals that of the surrounding tissues (Adam, 1940; Denton and Gilpin-Brown, 1961). The liquid is only pumped through the posterior ends of the chambers. Moreover, the liquid inside the cuttlebone of an animal kept in shallow water is almost isosmotic with seawater, while it is markedly hypoosmotic when coming from an animal just hauled up from the sea bottom. This suggests that the liquid movements in the cuttlebone are dependent on an active transport of salts. However no attempt has been made so far to explain how the liquid pump works.

III. Intertidal Mollusks

An important characteristic of this biotope is certainly the tidal oscillations submitting the animals to rapid and important changes mainly in the physical factors of the environment. Besides these factors, the ecology of intertidal mollusks is influenced by a complex set of circumstances, many of which are dependent on the aquatic medium. This is, for instance, the case for the conditions of reproduction (mating, spraying) or the chances of settlement. The influence of environmental factors on these activities has been considered in the preceding section which deals with the aquatic medium. In this section, we shall give an account of results dealing mainly with the influence of the exposure immersion time on the ecology of intertidal species.

Descriptions of the zonation of living species in the intertidal zone appeared in the literature at the beginning of the last century. Since then, the shores of the world have been studied from this point of view. Zonation can of course be related not only to tidal levels but also to other factors, such as climate or degree of exposure to wave action. However, in many cases, a good correlation has been obtained for a variety of organisms between the zonation and the exposure immersion time (Evans, 1957; Doty, 1957). We can therefore consider that the exposure–immersion time at least plays an important part in determining zonation. The most important effect of the tidal oscillations is to place the species alternatively in an aquatic or in an aerial medium. Respiration and excretion are thus two main physiological parameters which can be considered to be playing an important part in the ecology of intertidal mollusks.

A. Respiration

Most shore-dwelling prosobranchs are adapted to aerial breathing at varying degrees. High level littorinids and shore pulmonates, although they have become readapted for aquatic breathing by the development of a secondary gill, possess a thin, vascularized mantle cavity which acts as a respiratory surface. When the tide withdraws, shore snails close down after taking in a bubble of air which is released on reimmersion. Limpets can clamp down on the rocks but usually not very firmly so that air can enter the mantle cavity. This air-gaping system is also found in a bivalve mollusk, the ribbed mussel *Modiolus demissus.* Pelecypods of the intertidal zone usually close their valves when exposed to air. *Modiolus* however, does not follow this pattern. Rather, it leaves its valves ajar when exposed to the atmosphere during low tide (Kuenzler, 1961). As shown by Lent (1968), air-gaping is a mechanism used by *Modiolus* for gaseous exchange which allows this species to penetrate the high intertidal habitat. In fact, since this species spends more time exposed than inundated, access to oxygen allows more efficient respiration. The other intertidal bivalves are limited normally to the low intertidal zone. Spending more time in water than in aerial medium, these species usually close their cavities by contraction of the adductor muscles.

Obviously, facultative anaerobiosis has great survival value to these intertidal mollusks. As early as 1931, Moore reported that *Syndosmya alta* can survive for more than 3 days in deoxygenated water, *Mya arenaria* can survive for 8 days without oxygen (Ricketts and Calvin, 1948), and *Littorina* can survive for several weeks in pure nitrogen (Patané, 1955). The anaerobiosis metabolism of intertidal mollusks seems to be different from that classically described for mammalian tissues and which involves the formation of lactic acid, leading to an oxygen debt to be repaid when the tissue comes back to aerobic conditions. As shown by Morton *et al.* (1957), *Lasaea rubra,* after staying 3 hours without taking up oxygen, resumes respiration at the normal rate when reimmersed in water. In fact, the regeneration of NAD^+ during anaerobic glycolysis in many mollusks is not brought about by the reduction of pyruvate to lactate as is the case in mammalian tissues.

Various mollusks, when placed in anaerobic conditions, either produce lactate in amounts much smaller than those dictated by the stoichiometry of glycolysis or produce no lactate at all (Humphrey, 1944; Awapara and Simpson, 1967; Simpson and Awapara, 1965). These organisms seem to produce substantial amounts of succinate from glucose during its anaerobic breakdown. In *Rangia cuneata,* for instance succinate accounts for about 40% of the glucose degraded. During anaerobic incubation with $NaH^{14}CO_3$, the oyster *Crassostrea virginica* accumulates labeled succinate (Hammen and Wilbur, 1959) and little or no lactate (Hammen and Osborne, 1959). Then, when the mantle tissue of the same oyster is incubated anaerobically with labeled fumarate, 55% of the radioactivity is found in succinate, while about 10% is found in malate or oxaloacetate, and only 2% in α-ketoglutarate or citrate (Wegener *et al.*, 1969). It is worth noticing that the same system is at play in species belonging to other phyla but which can all be considered "specialists" of anaerobic metabolism. As a matter of fact, many experiments on parasitic Nematoda, Trematoda, Cestoda, or Acanthocephala emphasize the importance of succinate in the energy metabolism during anaerobiosis (von Brand, 1933; Bueding, 1963; Bryant and Williams, 1962; Agosin and Repetto, 1965; Bryant and Nicholas, 1966; Bryant and Morseth, 1968).

The production of succinate has been postulated to begin with the carboxylation of pyruvate yielding oxaloacetate. Conversion of oxaloacetate to fumarate and then succinate being then achieved through the familiar reactions of the Krebs cycle, but taken in the reverse way. There is now considerable evidence showing that in these species (mollusks and parasitic worms), the pathway going from glucose to succinate follows glycolysis down to phosphoenol-pyruvate (PEP), which cannot give rise directly to appreciable amounts of pyruvate, but would first fix carbon dioxide to form oxaloacetate. As a matter of fact, in most mollusks studied by Awapara and co-workers, activity of the enzyme catalyzing the carboxylation of pyruvate is undetectable or is much too low to be considered as significant (see Chen and Awapara, 1969). Moreover, as shown in Table I, the activity of the enzyme catalyzing the formation of pyruvate from PEP is very low in various tissues of intertidal mollusks when compared to that measured in vertebrate tissues.

Arguments favoring the formation of succinate from glucose following the pathway shown in Fig. 3 have been recently reviewed by Gilles (1970). In this pathway, conversion of oxaloacetate coming from the carboxylation of PEP to malate yields a reducible substrate. Thus, in

TABLE I
PYRUVATE KINASE ACTIVITY IN DIFFERENT TISSUES OF VERTEBRATES AND INTERTIDAL MOLLUSKS[a]

Species and tissue	Pyruvate kinase activity[b]
Rattus sp.	
brain	391
muscle	1763.5
Xenopus laevis	
brain	700
muscle	2290
Mytilus edulis	
mantle	8.6
muscle	21.4
gill	0
Ostrea edulis	
mantle	4.7
muscle	23.6
gill	4.9

[a] After Gilles (1970).
[b] The enzymatic activity is given as the initial velocity per milligram of protein.

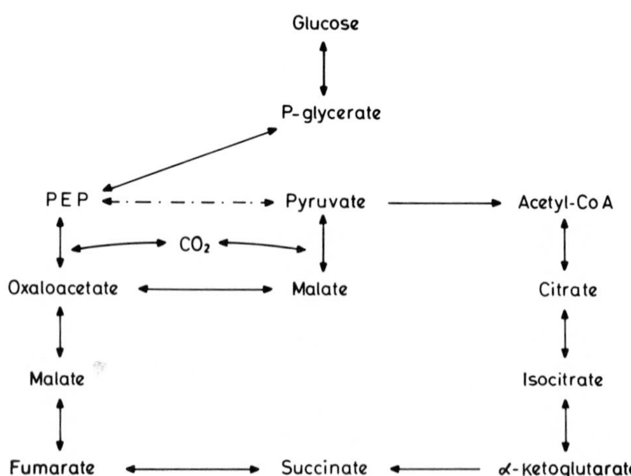

FIG. 3. Main pathway of glucose metabolism in some invertebrates. Broken line indicates a low enzymatic activity. (After Gilles, 1970).

this sequence of reaction, NADH could conceivably be oxidized to maintain glycolysis in a continuous fashion. This does not, however, answer the question of why it is succinate and not, for instance, malate or fumarate which is accumulated. As a matter of fact, the oxidation of NADH is achieved during the transformation of oxaloacetate to malate. Therefore, if this pathway only serves in maintaining a good balance between NAD and NADH so that glycolysis can proceed, there is no reason for malate to be converted to succinate. There is no available answer to this question as far as mollusks are concerned.

On the other hand, one may suggest from the experiments done with parasitic worms that "these animals possess a modified electron transport system in which fumaric acid, acting as a terminal electron acceptor, is converted to succinic acid" (Bryant and Morseth, 1968; Gilles, 1970). There is indeed considerable evidence demonstrating that the terminal oxidation mechanisms in parasitic worms as in mollusks are different from the classically described ones. As far as helminths are concerned, Bueding and Charms (1952) and later Kmetec and Bueding (1961) failed to demonstrate the presence of any cytochromes in homogenates of muscle or in muscle mitochondria of *Ascaris*. As far as mollusks are concerned, the results of Tappel (1960) show a cytochrome concentration pattern which is quite different from that found in vertebrate tissue. Particularly the concentration of the cytochromes of the a type, which are the last electron acceptors before oxygen, is extremely low (see Table II) and the physiological significance of a so small a cytochrome concentration may be questioned.

There is little information in the literature suggesting a mechanism which can account for these considerations. On the basis of the important work of Slater and co-workers showing that a high-energy intermediate used for ATP synthesis may be used to promote the reduction of NAD by succinate, we have recently proposed a mechanism in which fumarate plays the role of electron acceptor to promote the formation of an energy-rich intermediate which can be used for the synthesis of ATP from ADP and P_i. In this system, succinate accumulates. Such a mechanism supposes the possibility of coupling a phosphorylation with the oxidation of NADH by fumarate. That this is the case has been demonstrated by Seidman and Entner (1961) on *Ascaris* sarcosome preparations incubated in the presence of malate or fumarate, NAD, ADP, and P_i. These authors also show that this reaction is in fact a dismutation system in which the phosphorylation of ATP implicates the formation of equimolar amounts of succinate and pyruvate. Pyruvate may then be transaminated to alanine. This formation of alanine together with the formation of succinate during the anaerobic utilization of glucose

TABLE II

Concentration of the Respiratory Chain Components in Muscle of Various Mollusks and Vertebrates[a]

Species and tissue	Respiratory chain component concentration (μM/kg. wet weight)				
	a	a_3	b	$c + c_1$	fp
Mollusks					
Pecten irradians adductor muscle	0.3	0.3	1.2	1.8	5.5
Crassostrea virginica adductor muscle	0.08	0.5	1.0	1.9	5.3
Mercenaria mercenaria adductor muscle	0.08	0.8	5.3	3.7	3.8
Vertebrates					
Bufo marinus sartorius	12.0	10.44	4.32	13.2	18.0
Rana pipiens sartorius	3.2	2.84	2.08	4.8	4.16
Pseudemys elegans coraco-hyoideus	6.5	—	2.01	5.91	4.87
Rattus sp. heart sarcosomes	27	30	27	36	30

[a] After Gilles (1970).

by mollusks would account for the fact that there is an accumulation of both succinate and alanine in the mantle tissue of *R. cuneata* incubated anaerobically (Stokes and Awapara, 1968).

The overall metabolic sequence proposed is shown in Fig. 4. Although this scheme explains satisfactorily the observations mentioned above, it is considered tentative and is by no means exclusive of other mechanisms. More information is of course needed before a complete picture can be produced. It is however worth noting that many species belonging to different phyla (mollusks and helminths) seem to have developed the same system of energy production in anaerobiosis, which is different from that classically described. One may therefore consider this biochemical adaptation as being of ecological value, since it has permitted the survival of various species in media where they are faced with a problem of oxygen availability. Another problem with which intertidal mollusks are faced deals with the nature of their end products of nitrogen metabolism. We shall now briefly consider this particular point.

B. Excretion

It has long been known that in man about 85% of the total nitrogen of urine is in the form of urea, whereas in birds and snakes, uric acid of the excreta accounts for approximately the same proportion. On the other hand, in the urine of the aquatic *Sepia officinalis*, the ammonia nitrogen represents more than 65% of the total nitrogen excreted. Such

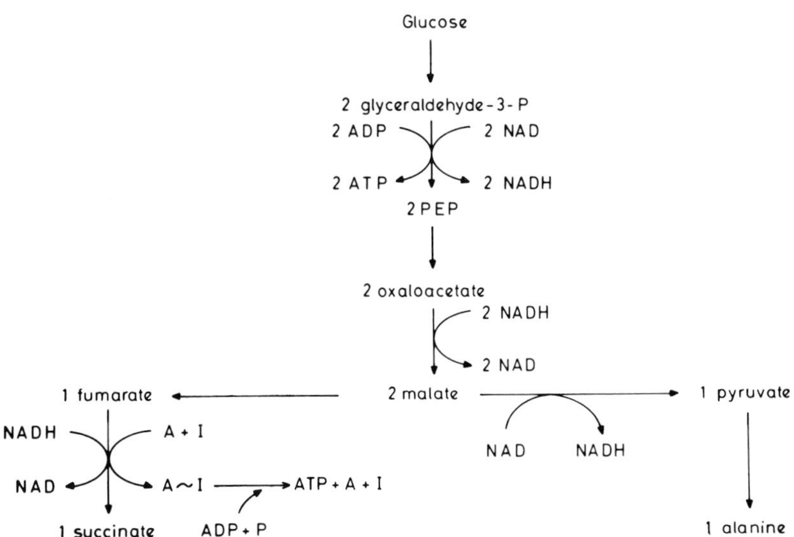

FIG. 4. NAD/NADH balance during anaerobic glucose degradation in mollusks.

observations lead many authors to consider that the nature of the end products of nitrogen metabolism of any species is related to the relative toxicity of ammonia, urea, and uric acid, and to the availability of water (Delaunay, 1931; Heidermanns, 1937; Needham, 1935, 1938, 1950). In this view, ammonia is excreted only by aquatic animals. Urea, which is very soluble is also eliminated along with water, while uric acid, being almost insoluble, can be eliminated in the solid state with practically no water loss. Mollusks are no exception to this general rule, as shown by the important work of Needham (1935) on the correlation between nitrogen excretion and habitat in the phylum Mollusca. Needham (1935) even shows that high level species of intertidal littorinids excrete more uric acid than do species living lower down on the shore (see Table III). This correlation does not seem to be true for all the intertidal species (Duerr, 1968). Anyway, such a biochemical adaptation, together with that observed at the level of the respiratory metabolism, is certainly of ecological value, since it has allowed many intertidal species to conquer the high shore level and the terrestrial medium. We shall now consider some ecological aspects of this last biotope.

IV. Land Mollusks

In the terrestrial biotope, mollusks are mainly faced with the problem of availability of water in an environment of variable humidity and temperature. This problem of economy of water has been solved by various types of adaptive processes, mainly involved with permeability to water, temperature regulation, and excretion.

A. Water Relationships

Land mollusks have a skin that is readily permeable to water. As early as 1916, Künkel noted important weight loss in pulmonate snails and slugs deprived of access to water. When kept in dry air, *Limax variegatus* lost about 2.5% of its initial weight per hour, and if stimulated to move it lost 16% per hour. Under these conditions, death results within a few hours.

Water is mainly lost through general evaporation from the moist skin; this may explain why snails are more successful than slugs in relatively dry places. Although some water is lost through the shell, the rate of evaporative loss may be considered minimal (Machin, 1967). In fact, snails expose a minimum of flesh to air, mainly the so-called mantle collar. Although the evaporation of water from active specimens of *Helix aspersa* is nearly identical to that from a free water surface (Machin, 1964), it appears that this species can considerably reduce the rate of evaporation when inactive. This regulation is well shown in Fig. 5. The underlying causes of this regulation have been attributed by Machin (1964, 1966) to the cessation of mucus extrusion, which can be observed on inactive animals. Indeed, during regulation, the mucous layer becomes much thinner due to continued evaporation in the absence of further

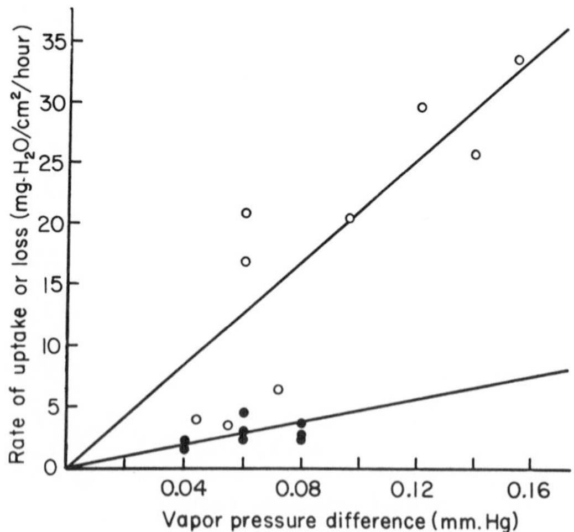

Fig. 5. Regulation of the water flux under different vapor pressure gradients in intacts (○) and isolated dorsal integument (●) of *Helix aspersa*. (After Machin, 1966.)

replacement of mucus. Since the permeability to water of regulating tissue is low (0.39 mg./cm.^{2}/hour), the evaporation of mucus results in an increase in solute concentration, which causes a lowering of the vapor pressure at the mucus surface (Machin, 1966). This study demonstrates that the general statements according to which the mucus covering the skin of land mollusks slows down water loss are fallacious. In fact, it appears that the water loss is considerably decreased when the mucus secretion stops. Such a mechanism helps the snail to survive when the dryness of air increases. Of course, in such conditions most pulmonate snails withdraw into the shell and enter a resting phase by closing the mouth of the shell by secreting an epiphragm. The permeability of this structure is lower than that of the mantle collar (Machin, 1968).

B. Estivation and Hibernation

These resting phases, which allow the terrestrial gastropods to survive while waiting better environmental conditions, have been well studied from the physiological and biochemical view points (Duval, 1930; Wells, 1944; Saxena, 1955; Vorwohl, 1961; Burton, 1964). Meenakshi (1956) investigated the rate of respiration of *Pila* and found no detectable uptake of oxygen during estivation. Lactic acid accumulates in the tissues and the glycogen reserves became depleted. In contrast *Helix pomatia* appears to respire aerobically during estivation (Fischer, 1931). Estivating species normally accumulate uric acid (Needham, 1935; Saxena, 1955), and the hemolymph osmotic pressure may rise to twice its normal value, although an osmoregulatory mechanism takes place during this period (Klekowski, 1963; Burton, 1964; Little, 1968).

C. Temperature

More than other species of mollusks, terrestrial slugs and snails have to withstand important variations in temperature. This environmental factor must thus influence their distribution. As a matter of fact, terrestrial slugs such as *Limax maximus* or *Phylomycus carolinianus* tend to locate themselves in the part of their environment where there is least stress (Rising and Armitage, 1969). However, these mollusks can adapt to changes in environmental temperature. As mentioned above, the continued evaporation from the moist skin effects a degree of temperature regulation. At 33.7°C, the body temperature of *Arion ater* is found to be 21°C, i.e., well below the thermal death point (Hogben and Kirk, 1944). Many species also show metabolic compensations that enable them to maintain subsistence levels of activity (Roy, 1963; Segal, 1959, 1961; Rising and Armitage, 1969). At the same temperature, cold acclimated animals show generally a higher metabolic rate than those acclimated at high temperature. For instance, *L. maximus* acclimated at 5°C consumed more oxygen at 25°C than animals acclimated to 15°C or 25°C (Rising and Armitage, 1969).

This is also the case for *Arion circumscriptus* (Roy, 1963) and *Limax flavus* (Segal, 1961). However, *Ph. carolinianus* adapted at 15°C use, on the average, more oxygen than animals of this species adapted at either 5° or 25°C (Rising and Armitage, 1969). This suggests that temperature must impose different problems for different species. It is worth noting that *A. circumscriptus, L. maximus,* and *L. flavus* are adapted to live in urban areas and are found in the same habitats, while *Ph. carolinianus* is a species of moist, hardwood forests.

D. Excretion

It is generally admitted that the nature of the end products of nitrogen metabolism is dependant on the availability of water. When dealing with the adaptive processes linked to the survival of mollusks in the intertidal habitat, we have already emphasized this point (see Section III,B). In all the biotopes that mollusks have conquered, land is certainly that where the problem of water availability is the most important. It may thus be expected that land snails excrete end products of nitrogen metabolism which can be eliminated with practically no water loss. As a matter of fact, uric acid forms the greater part of the nonprotein nitrogen excreted by terrestrial mollusks.

In *Helix pomatia,* the first excreta after hibernation contain 65% of nonprotein nitrogen in the form of uric acid and 30–40% during feeding. Practically all the rest is purines (Jezewska *et al.,* 1963). This seems to be a general rule in land pulmonate snails (Speeg and Campbell, 1968; Tramell and Campbell, 1970). This high degree of uricotelism is no doubt an adaptation to terrestrial life, although the distribution of uric acid in other gastropods raises some problems. As shown in Table III, uric acid is comparatively concentrated in the kidney of the aquatic pulmonates *Limnaea* and *Planorbis.* Needham (1935, 1938) attributed the occurrence of uric acid in these forms to the retention of uricotelism after a secondary return to an aquatic existence.

On the other hand, Morton (1954, 1955) suggests that the pulmonate lung originated as an adaptation to life in ponds. He believes that Limnidae and Planorbidae are descendants from more terrestrial ancestors but which have never been completely emancipated from water. Potts (1967) considers that the survival value of uric acid in the basommatophorans "is possibly an example of preadaptation to terrestrial life rather than a relic of it." This nonprotein nitrogen excretion is of

TABLE III
URIC ACID CONTENT OF THE KIDNEYS OF CERTAIN MOLLUSKS AS RELATED TO HABITAT[a]

Habitat and species	Uric acid content[b]
Seawater	
Prosobranchs	
Buccinum undatum	4.1
Nucella lapillus	4.4
Osilinus lineatus	1.6
Gibbula umbilicalis	2.3
Turritella communis	6.0
Tectibranchs	
Scaphander lignarius	4.6
Lamellibranchs	
Pecten maximus	0.5
Pecten opercularis	3.9
Intertidal (from seawater to terrestrial)	
Prosobranchs	
Littorina littorea	1.4
Littorina rudis	5.1
Littorina saxatilis	5
Littorina neritoides	25
Fresh water	
Lamellibranchs	
Anodonta cygnea	0
Prosobranchs	
Viviparus fasciatus	35
Bithynia tentaculata	150
Pulmonates	
Limnea stagnalis	114
Planorbis corneus	40
Terrestrial	
Pulmonates	
Limax flavus	31
Milax sowerbyi	66
Arion ater	36
Bulimulus sporadicus	183
Helix pomatia	743
Helix aspersa	128
Prosobranchs	
Cyclostoma elegans	1000

[a] After Needham (1935, 1938).

[b] Uric acid content is given in milligrams per gram of dry weight of kidney.

course the reflection of an evolution at the level of the metabolic sequences leading to the formation of ammonia, urea, and uric acid. In terrestrial mollusks, as in insects (Schoffeniels and Gilles, 1971c), the products of protein degradation are directed toward the catabolism pathway and puric and pyrimidic bases. It is not clear, however, whether uric acid found in aquatic species together with ammonia and urea is derived from purine or from amino acids. These metabolic aspects of the nonprotein nitrogen excretion are studied in Chapter 10 of this volume. We shall therefore not deal any longer with this subject.

V. Concluding Remarks

In this chapter, we have accumulated evidence showing that environmental factors either of a chemical or physical nature influence the whole life of mollusks. In many cases, the underlying causes of the behavioral or physiological responses to stimuli from the environment can be found at the molecular level. The various types of end products of nitrogen catabolism, the metabolic basis of anaerobiosis in intertidal mollusks, or the basic properties of chemoreceptor cells can be considered examples of this adaptation at the molecular level. Such a biochemical adaptation is evidently only one among many features that explains the relationship a given species establishes with its surroundings.

It is at the level of the organism that the impact of natural selection takes place. However, we have seen that in the case of an evolutionary adaptation allowing a species to live in the intertidal or terrestrial medium, the advantage conferred by the adaptation at the molecular scale is obvious, since it permits colonization of new biotopes. In many cases reported here, for instance the temperature adaptability of terrestrial slugs, or the sex determinism in some aquatic mollusks, the biochemical responses to the solicitation of the surrounding are phenotypic variations. However, in other cases, genetic differences must be invoked. Staiger (1957) showed that the very variable species *Purpura lapillus* has two main forms which are fully interfertile, one having a haploid chromosome number of 13 and the other of 18. The form with 13 chromosomes occupies exposed situations and that with 18 chromosomes sheltered places. Chromosomally heterogeneous populations colonize intermediate habitats. We can say with Prosser (1955), "It is probable that if populations were examined for physiological variations as carefully as they have been for minute structural variations, many more races and subspecies would be found than are now recognized."

Organisms have thus sometimes responded to the solicitation of the surroundings either by the diversification of the genotype or by an adjustment of the translation of the code to the new conditions. Through the unitary canvas of cellular structure and metabolism, the cells have embroidered the diversity of their adaptive differentiation. As another example, we will recall the anaerobic metabolism of intertidal bivalves (see Section III,A). In these species, one can find all the enzymes implicated in the classically described intermediary metabolism. These enzymes, although being the homologs of those found in the other species, control an analogous physiological function (i.e., formation of ATP in the absence of oxygen) but in another way than that described for most species. Indeed, in intertidal bivalves, the concentration or the kinetic properties of several of these enzymes have been modified so that the overall metabolic sequence of anaerobiosis has been changed. We believe that this biochemical adaptation is of ecological value since it is also found in species of other phyla which can withstand long periods without oxygen (i.e., the parasitic helminths).

Within the complex picture of all the adaptive features, it is thus possible to individualize biochemical traits which can explain at least part of the adaptation to a given biotope and which are sometimes the reflection of genotypical adjustment. It appears, therefore, that different aspects of molecular evolution can be investigated in taking advantage of environmental fitness. Almost everything remains to be accomplished in this attractive and fruitful field.

REFERENCES

Adam, W. (1940). *Mem. Mus. Hist. Nat. Belg., Ser. 2* No. 21, p. 82.

Adler, J. (1966). *Science* **153**, 708.

Agersborg, H. P. K. (1922). *J. Exp. Zool.* **36**, 423.

Agersborg, H. P. K. (1925). *Acta Zool.* **6**, 167.

Agosin, M., and Repetto, Y. (1965). *Comp. Biochem. Physiol.* **14**, 299.

Awapara, J., and Simpson, J. W. (1967). *Annu. Rev. Physiol.* **29**, 87.

Awati, P. R., and Rai, H. S. (1931). *Indian Zool. Mem.* **3**, 1.

Bailey, D. F., and Laverack, M. S. (1963). *Nature (London)* **200**, 1122.

Bailey, D. F., and Laverack, M. S. (1966). *J. Exp. Biol.* **44**, 131.

Ball, E. G., and Meyerhof, O. (1940). *J. Biol. Chem.* **134**, 483.

Bayne, B. L. (1965). *Ophelia (Copenhagen)* **2**, 1.

Bayne, B. L. (1967). *In* "Discussion Meeting on the Biochemistry of Aquatic Organisms," pp. 71–77. Crombie Hall, University of Aberdeen.

Berg, K. (1952). *Hydrobiologica* **4**, 225.

Berg, K., and Ockelmann, K. W. (1959). *J. Exp. Biol.* **36**, 690.

Berg, K., Lumbye, J., and Ockelmann, K. W. (1958). *J. Exp. Biol.* **35**, 43.

Blake, J. W. (1960). *Linnol. Oceanogr.* **5**, 273.

Bloomer, H. H. (1939). *Proc. Malacol. Soc. London* **23**, 285.

Braams, W. G., and Geelen, H. F. M. (1953). *Arch. Neer. Zool.* **10**, 241.

Brock, F. (1936). *Zoologica, Stuttgart* **34**, No. 92, 1.
Brown, A. C., and Noble, R. G. (1960). *Nature (London)* **188**, 1045.
Bryant, C., and Morseth, D. J. (1968). *Comp. Biochem. Physiol.* **25**, 541.
Bryant, C., and Nicholas, W. L. (1966). *Comp. Biochem. Physiol.* **17**, 825.
Bryant, C., and Williams, J. P. C. (1962). *Exp. Parasitol.* **12**, 372.
Bueding, E. (1963). *In* "Control Mechanisms in Respiration and Fermentation" (B. Wright, ed.) pp. 167–177. Ronald Press. New York.
Bueding, E., and Charms, B. (1952). *J. Biol. Chem.* **196**, 615.
Bullock, T. H. (1953). *Behaviour* **5**, 130.
Burton, R. F. (1964). *Can. J. Zool.* **42**, 1085.
Carr, W. E. S. (1967). *Biol. Bull.* **133**, 106.
Chaet, A. B. (1967). *Symp. Zool. Soc. London* **20**, 13.
Chen, C., and Awapara, J. (1969). *Comp. Biochem. Physiol.* **30**, 727.
Clark, A. E. (1935). *Progr. Rep. Atl. Biol. Sta.* **16**, 10.
Clark, W. C. (1958). *Nature (London)* **181**, 137.
Coe, W. R. (1953). *J. Exp. Zool.* **122**, 5.
Collier, A. (1953). *Trans. Amer. Wildl. Conf., 18th, 1953, p.* 463.
Copeland, M. (1918). *J. Exp. Zool.* **27**, 247.
Crozier, W. J. (1918). *J. Exp. Zool.* **27**, 247.
Crozier, W. J., and Arey, L. B. (1919). *J. Exp. Zool.* **29**, 261.
Dan, J. C., and Wada, S. K. (1955). *Biol. Bull.* **109**, 40.
Davis, H. C., and Calabrese, A. (1964). *U.S. Fish. Wildl. Serv., Fishery Bull.* **63**, 643.
Degner, E. (1921). *Arch. Molluskenk.* **53**, 117.
Delaunay, H. (1931). *Biol. Rev. Cambridge Phil. Soc.* **6**, 265.
Denton, E. J., and Gilpin-Brown, J. B. (1961). *J. Mar. Biol. Ass. U.K.* **41**, 365.
Denton, E. J., and Shaw, T. I. P. (1961). *J. Physiol. (London)* **161**, 14.
Denton, E. J., Shaw, T. I. P., and Gilpin-Brown, J. B. (1958). *Nature (London)* **182**, 1810.
Doty, M. S. (1957). *Geol. Soc. Amer., Mem.* **67**, No. 1, 535.
Duerr, F. D. (1968). *Comp. Biochem. Physiol.* **26**, 1051.
Duval, M. (1930). *Ann. Physiol. Physicochim. Biol.* **6**, 346.
Evans, R. G. (1948). *J. Anim. Ecol.* **45**, 165.
Evans, R. G. (1957). *J. Ecol.* **45**, 245.
Feder, H. M., and Arvidsson, J. (1967). *Ark. Zool.* [2] **19**, 369.
Feder, H. M., and Lasker, R. (1964). *Life Sci.* **3**, 1047.
Field, I. A. (1922). *Bull. U.S. Fish. Bur.* **38**, 127.
Fischer, P. H. (1931). *J. Conchyliol.* **75**, 5.
Florkin, M. (1966). "Aspects moléculaires de l'adaptation et de la phylogénie." Masson, Paris.
Florkin, M. (1969). *Bull. Cl. Sci., Acad. Roy. Belg.* [5] **55**, 257.
Florkin, M., and Schoffeniels, E. (1969). "Molecular Approaches to Ecology." Academic Press, New York.
Fox, H. M. (1955). *Proc. Roy. Soc., Ser. B* **143**, 203.
Fretter, V., and Graham, A. (1964). *In* "Physiology of Mollusca" (K. M. Wilbur and C. M. Yonge, eds.), Vol. 1, pp. 127–164. Academic Press, New York.
Frings, H., and Frings, C. (1965). *Biol. Bull.* **128**, 211.
Galtsoff, P. S. (1930). *Proc. Nat. Acad. Sci. U.S.* **16**, 555.
Galtsoff, P. S. (1938). *Biol. Bull.* **75**, 286.
Galtsoff, P. S. (1940). *Biol. Bull.* **78**, 117.

Ghiretti, F. (1966). *In* "Physiology of Mollusca" (K. M. Wilbur and C. M. Yonge, eds.), Vol. 2, pp. 175–208. Academic Press, New York.
Gilles, R. (1969). *Arch. Int. Physiol. Biochim.* **77**, 441.
Gilles, R. (1970). *Arch. Int. Physiol. Biochem.* **78**, 313.
Glaser, O. C. (1921). *Biol. Bull.* **41**, 63.
Gould, H. N. (1919). *J. Exp. Zool.* **29**, 113.
Gould, H. N. (1952). *J. Exp. Zool.* **119**, 93.
Gunter, G. (1957). *Geol. Soc. Amer., Mem.* **67**, No. 1, 159.
Hammen, C. S., and Osborne, P. J. (1959). *Science* **130**, 1409.
Hammen, C. S., and Wilbur, K. M. (1959). *J. Biol. Chem.* **234**, 1268.
Harrington, C. R. (1921). *Biochem. J.* **15**, 736.
Hashimoto, Y. (1967). *Bull. Jap. Soc. Sci. Fish.* **33**, 243.
Heidermanns, C. (1937). *Tabulae Biol.* **14**, 209.
Herter, K. (1929). *Z. Vergl. Physiol.* **9**, 145.
Hoffmann, H. (1930). *Z. Vergl. Physiol.* **11**, 662.
Hogben, L., and Kirk, R. L. (1944). *Proc. Roy. Soc., Ser. B* **132**, 239.
Hultin, T. (1947). *Pubbl. Staz. Zool. Napoli* **21**, 153.
Humphrey, G. F. (1944). *Aust. J. Exp. Biol. Med. Sci.* **22**, 135.
Hunter, W. R. (1953). *Proc. Zool. Soc. London* **123**, 623.
Ikegami, S., Tamura, S., and Kanatani, H. (1967). *Science* **158**, 1052.
Jezewska, M. M., Gorzkowski, B., and Heller, J. (1963). *Acta Biochim. Pol.* **10**, 309.
Johannes, R. E., and Webb, K. L. (1965). *Science* **150**, 76.
Kempendorff, W. (1942). *Arch. Molluskenk.* **74**, 1.
Klekowski, R. Z. (1963). *Pol. Arch. Hydrobiol.* **11**, 219.
Kmetec, E., and Bueding, E. (1961). *J. Biol. Chem.* **236**, 584.
Kohn, A. J. (1956). *Proc. Nat. Acad. Sci. U.S.* **42**, 168.
Kohn, A. J. (1959). *Ecol. Monogr.* **29**, 47.
Kohn, A. J. (1961). *Amer. Zool.* **1**, 291.
Krogh, A. (1914). *Int. Z. Phys.-Chem. Biol.* **1**, 491.
Krüger, F. (1960). *Helgolaender Wiss. Meeresunters.* **7**, 125.
Kuenzler, E. J. (1961). *Linnol. Oceanog.* **6**, 191.
Künkel, K. (1916). "Zur Biologie der Lungenschnecken." Carl Winter, Heidelberg.
Kupfermann, I. (1967). *Nature (London)* **216**, 814.
Lent, C. M. (1968). *Biol. Bull.* **134**, 60.
Little, C. (1968). *J. Exp. Biol.* **48**, 569.
Loomis, W. F. (1959). *Ann. N.Y. Acad. Sci.* **77**, 73.
Loosanoff, V. L. (1959). *Biol. Bull.* **117**, 308.
Loosanoff, V. L. (1960). *Perspect. Mar. Biol., Symp., 1960,* pp. 483–495.
Loosanoff, V. L., and Engle, J. B. (1942). *Biol. Bull.* **82**, 413.
Loosanoff, V. L., and Miller, W. S., and Smith, P. B. (1951). *J. Mar. Res.* **10**, 59.
Lythgoe, J. N., and Hemmings, C. C. (1967). *Nature (London)* **213**, 893.
MacGinitie, G. E., and MacGinitie, N. (1949). "Natural History of Marine Animals." McGraw-Hill, New York.
Machin, J. (1964). *J. Exp. Biol.* **41**, 759.
Machin, J. (1966). *J. Exp. Biol.* **45**, 269.
Machin, J. (1967). *J. Zool.* **152**, 55.
Machin, J. (1968). *Biol. Bull.* **134**, 87.
MacKie, A. M., Lasker, R., and Grant, P. T. (1968). *Comp. Biochem. Physiol.* **26**, 415.

Manwell, C. (1958). *J. Cell. Comp. Physiol.* **52**, 341.
Meenakshi, V. R. (1956). *Curr. Sci.* **25**, 321.
Michelson, E. H. (1960). *Amer. J. Trop. Med. Hyg.* **9**, 480.
Montalenti, G., and Bacci, G. (1951). *Sci. Genet.* **4**, 5.
Moore, H. B. (1931). *J. Mar. Biol. Ass. U.K.* **17**, 325.
Morton, J. E. (1954). *Proc. Zool. Soc. London* **125**, 127.
Morton, J. E. (1955). *Phil. Trans. Roy. Soc. London, Ser. B* **239**, 39.
Morton, J. E. (1960). *Proc. Malacol. Soc. London* **34**, 96.
Morton, J. E., Boney, A. D., and Corner, E. D. S. (1957). *J. Mar. Biol. Ass. U.K.* **36**, 383.
Needham, J. (1935). *Biochem. J.* **29**, 238.
Needham, J. (1938). *Biol. Rev. Cambridge Phil. Soc.* **13**, 225.
Needham, J. (1950). "Biochemistry and Morphogenesis." Cambridge Univ. Press, London and New York.
Nelson, T. C. (1936). *Proc. Soc. Exp. Biol. Med.* **34**, 189.
Nelson, T. C., and Allison, J. B. (1940). *J. Exp. Zool.* **85**, 299.
Newcombe, C. L., Miller, C. E., and Chappel, D. W. (1936). *Nature* (*London*) **137**, 33.
Newell, R. C. (1966). *Nature* (*London*) **212**, 426.
Orton, J. H. (1920). *J. Mar. Biol. Ass. U.K.* **12**, 339.
Orton, J. H. (1924). *Nature* (*London*) **114**, 191.
Orton, J. H. (1927). *J. Mar. Biol. Ass. U.K.* **14**, 967.
Orton, J. H., Southward, A. J., and Dodd, J. M. (1956). *J. Mar. Biol. Ass. U.K.* **35**, 149.
Patané, L. (1955). *Boll. Accad. Sci. Nat. Gioenia, Sper. IV* **3**, 65.
Potts, W. T. W. (1967). *Biol. Rev. Cambridge Phil. Soc.* **42**, 1.
Prosser, C. L. (1955). *Biol. Rev. Cambridge Phil. Soc.* **30**, 229.
Read, K. R. H. (1962). *Comp. Biochem. Physiol.* **7**, 89.
Redfield, A. C., Loulidge, T., and Hurd, A. L. (1926). *J. Biol. Chem.* **69**, 475.
Redmond, J. R. (1962). *Physiol. Zool.* **35**, 34.
Richardson, D. H. S., Smith, D. C., and Lewis, D. H. (1967). *Nature* (*London*) **214**, 879.
Richter, E. (1935). *Z. Naturwiss.* **69**, 507.
Ricketts, E. F., and Calvin, J. (1948). "Between Pacific Tides." Stanford Univ. Press, Stanford, California.
Rigby, B. J. (1968). *Biol. Bull.* **135**, 223.
Rising, T. L., and Armitage, K. B. (1969). *Comp. Biochem. Physiol.* **30**, 1091.
Robertson, J. D. (1964). *In* "Physiology of Mollusca" (K. M. Wilbur and C. M. Yonge, eds.), Vol. 1, pp. 283–311. Academic Press, New York.
Rosenwald, K. (1926). *Z. Indukt. Abstamm. Vererbungsl.* **43**, 238.
Rothschild, Lord. (1956). "Fertilization." Methuen, London.
Roy, A. (1963). *Can. J. Zool.* **41**, 671.
Runnström, J., and Monné, L. (1945). *Ark. Zool.* [1] **36A**, No. 18, 1.
Saxena, B. B. (1955). *J. Anim. Morphol. Physiol.* **2**, 87.
Scheer, B. T. (1948). "Comparative Physiology." Wiley, New York.
Scheltema, R. S. (1961). *Biol. Bull.* **120**, 92.
Scheltema, R. S. (1967). *Biol. Bull.* **132**, 253.
Schoffeniels, E., and Gilles, R. (1970a). *Chem. Zool.* **5**, 255–286.
Schoffeniels, E., and Gilles, R. (1970b). *Chem. Zool.* **5**, 199–227.
Schoffenels, E., and Gilles, R. (1971). *Chem. Zool.* **6**, 393–420.

Segal, E. (1959). *Anat. Rec.* **134**, 636.
Segal, E. (1961). *Amer. Zool.* **1**, 235.
Seidman, I., and Entner, N. (1961). *J. Biol. Chem.* **236**, 915.
Seno, H., Hori, J., and Kusakabe, D. (1926). *J. Fish. Inst. Tokyo* **22**, 41.
Simpson, J. W., and Awapara, J. (1965). *Comp. Biochem. Physiol.* **15**, 1.
Spärck, R. (1936). *Kgl. Dan. Vidensk. Selsk., Biol. Medd.* 13, No. 5, 1.
Speeg, K. V., Jr., and Campbell, J. W. (1968). *Comp. Biochem. Physiol.* **26**, 579.
Spengel, J. W. (1881). *Z. Wiss. Zool.* **35**, 333.
Staiger, H. (1957). *Année Biol.* 33, Nos. 5–6, 252.
Stehouwer, H. (1952). *Arch. Neer. Zool.* **10**, 161.
Stickney, A. P. (1964). *Ecology* **45**, 283.
Stinnakre, J., and Tauc, L. (1969). *J. Exp. Biol.* **51**, 347.
Stokes, T. H., and Awapara, J. (1968). *Comp. Biochem. Physiol.* **25**, 883.
Strauber, L. A. (1947). *Anat. Rec.* **99**, 614.
Strauber, L. A. (1950). *Ecology* **31**, 109.
Strumwasser, F., Jacklet, J. W., and Alvarez, R. (1968). *Int. Congr. Physiol. Sci.* [*Proc.*] *24th, 1968,* p. 420.
Strumwasser, F., Jacklet, J. W., and Alvarez, R. (1969). *Comp. Biochem. Physiol.* **29**, 197.
Tappel, A. L. (1960). *J. Cell. Comp. Physiol.* **55**, 111.
Taylor, C. C. (1959). *J. Cons., Cons. Perm. Int. Explor. Mer.* **25**, 93.
Terao, A. (1926). *Sci. Rep. Tohoku. Imp. Univ., Ser 4* **2**, 127.
Thompson, T. E. (1958). *Phil. Trans. Roy. Soc. London, Ser. B* **242**, 1.
Thorson, G. (1946). *Medd. Komm. Havundersog., Kbh., Ser. D* **4**, 1.
Tramell, P. R., and Campbell, J. W. (1970). *Comp. Biochem. Physiol.* **32**, 569.
Tyler, A. (1939). *Proc. Nat. Acad. Sci. U.S.* **25**, 317.
Tyler, A. (1940). *Biol. Bull.* **78**, 159.
Umminger, B. L. (1968). *Biol. Bull.* **135**, 239.
Van Dongen, A. (1956). *Arch. Neerl. Zool.* **11**, 373.
von Brand, T. (1933). *Z. Vergl. Physiol.* **18**, 562.
von Brand, T., Nolan, M. O., and Mann, E. R. (1948). *Biol. Bull.* **95**, 199.
Vorwohl, G. (1961). *Z. Vergl. Physiol.* **45**, 12.
Wada, S. K. (1954). *Jap. J. Zool.* **11**, 273.
Watabe, N. (1952). *J. Fuji Pearl Inst.* **2**, 21.
Waterman, T. H. (1966). *In* "Environmental Biology" (P. L. Altman and D. S. Dittmer, eds.), pp. 155–163. Fed. Am. Soc. Exptl. Biol., Bethesda, Maryland.
Weber, H. (1924). *Zool. Anz.* **60**, 261.
Wegener, B. A., Barnitt, A. E., and Hammen, C. S. (1969). *Life Sci.* **8**, Part II, 335.
Wells, G. P. (1944). *J. Exp. Biol.* **20**, 79.
Wells, H. J. (1966). *In* "Physiology of Mollusca" (K. M. Wilbur and C. M. Yonge, eds.), Vol. 2, pp. 547–590. Academic Press, New York.
Wilczynski, J. Z. (1959). *J. Exp. Biol.* **36**, 34.
Wilson, D. P. (1968). *J. Mar. Biol. Ass. U.K.* **48**, 387.
Wilson, D. P., and Wilson, M. A. (1956). *J. Mar. Biol. Assoc. U.K.* **35**, 291.
Wölper, C. (1950). *Z. Vergl. Physiol.* **32**, 272.
Young, R. T. (1942). *Ecology* **23**, 490.
Young, R. T. (1946). *Ecology* **26**, 58.

Author Index

Numbers in italics refer to pages on which the complete references are listed.

G

H

K

N

Q

R

S

T

U

V

W

Y

Subject Index

B

C

G

H

N

O

U

V